Saas-Fee Advanced Course 27
Lecture Notes 1997

Springer
Berlin
Heidelberg
New York
Barcelona
Budapest
Hong Kong
London
Milan
Paris
Singapore
Tokyo

R. J. LeVeque D. Mihalas
E. A. Dorfi E. Müller

Computational Methods for Astrophysical Fluid Flow

Saas-Fee Advanced Course 27
Lecture Notes 1997
Swiss Society for Astrophysics and Astronomy

Edited by O. Steiner and A. Gautschy

With 124 Figures

Springer

Professor R. J. LeVeque
Department of Applied Mathematics
University of Washington
Box 352420, Seattle, WA 98195-2420, USA
rjl@amath.washington.edu,
http://www.amath.washington.edu/~rjl/

Professor E. A. Dorfi
Institut für Astronomie der Universität Wien
Türkenschanzstrasse 17,
A-1180 Wien, Austria
ead@astro.ast.univie.ac.at,
http://amok.ast.univie.ac.at/~ead/ead_home.html

Professor D. Mihalas
Department of Astronomy
University of Illinois
1002 West Green Street
Urbana, IL 61801, USA
dmihalas@astro.uiuc.edu

Professor E. Müller
Max-Planck-Institut für Astrophysik
Karl-Schwarzschild-Strasse 1
D-85740 Garching, Germany
ewald@MPA-Garching.MPG.de,
http://www.MPA-Garching.MPG.de

Volume Editors:

Dr. O. Steiner
High Altitude Observatory,
National Center for Atmospheric Research
Box 3000, Boulder, CO 80301, USA, and
Kiepenheuer-Institut für Sonnenphysik
Schöneckstrasse 6, D-79104 Freiburg, Germany
steiner@kis.uni-freiburg.de

Dr. A. Gautschy
Astronomisches Institut der Universität Basel
Venusstrasse 7
CH-4102 Biningen, Switzerland
gautschy@astro.unibas.ch

This series is edited on behalf of the Swiss Society for Astrophysics and Astronomy:

Société Suisse d'Astrophysique et d'Astronomie
Observatoire de Genève, ch. des Maillettes 51, CH-1290 Sauverny, Switzerland

---

Cover photograph: Two-dimensional simulation of the convection inside a nascent neutron star. The panels show the time evolution of the electron number fraction Y_e in a radial region between 15 km and 155 km. The panels are arranged in counter-clockwise order, starting from the right upper side showing snapshots at 12 ms, 14 ms, 18 ms, and 21 ms after the start of the simulation. (From Müller and Janka, 1997). Background: Finite difference formulas of the multidimensional version of the wave-propagation algorithm. (From LeVeque, 1997).

ISBN 978-3-642-08412-6 e-ISBN 978-3-540-31632-9

Cataloging-in-Publication Data applied for
Die Deutsche Bibliothek – CIP-Einheitsaufnahme
Computational methods for astrophysical fluid flow: lecture notes 1997/Saas Fee Advanced Course 27. R. J. LeVeque... Swiss Society for Astrophysics and Astronomy. Ed. by O. Steiner and A. Gautschy. – Berlin; Heidelberg; New York; Barcelona; Budapest; Hong Kong; London; Milan; Paris; Santa Clara; Singapore; Tokyo: Springer, 1998
(Saas Fee Advanced Courses; 27)

Cover design: *design & production* GmbH, Heidelberg

Preface

Many standard numerical algorithms for fluid dynamics have their roots in astrophysics; the reasons are probably twofold. First, since astrophysical objects are normally not amenable to experimental studies, astrophysicists seek some understanding by simulating them by methods of computational physics. With the advancement of computer technology and numerical algorithms, complex astrophysical phenomena such as supernova explosions, accretion of material onto a star, stellar pulsations, or the granular pattern of solar convection are now accessible to simulation almost as if they were approachable via experiments in the laboratory. Second, the extreme conditions that prevail in astrophysical flows such as extremely high shock strengths, or very high compressibility of the fluid lead to the development of novel numerical methods that would normally not have been required for terrestrial engineering applications.

Computational astrophysics has become an important branch of astrophysical research and many students and researchers of astrophysics are at some point in their career confronted with computer simulation results, or the prospect of executing simulation calculations, or even the writing of a simulation code. Considering the wealth of numerical schemes and computer codes available for astrophysical fluid flow, for novices it is not easy to assess and evaluate simulation results or to choose the correct scheme and to avoid at least "commonly known" pitfalls when carrying out their own simulations. The present book should provide some help in such circumstances. Moreover, it might also prove to be a valuable reference for the more experienced computational astrophysicist.

The first part of this book leads directly to the most modern numerical techniques for compressible fluid flow, with special consideration given to astrophysical applications. Emphasis is put on high-resolution shock-capturing finite-volume schemes based on Riemann solvers. Examples of unphysical solutions resulting from lackluster methods and incorrect applications are discussed, as well as more advanced topics such as MHD-Riemann solver or computational methods for (general) relativistic fluid flow. An extensive literature list leads the reader to specific topics.

The applications of such schemes, in particular the PPM method, is portrayed in the last part of the book (the colorful finale). Examples of large-scale simulations include supernova explosions by core collapse and thermonuclear burning and astrophysical jets. They demonstrate the interplay of observations with simulations and exemplify the deeper physical understanding of astrophysical objects that can be gained through computer simulations.

For most astrophysical fluid flow, radiation transfer needs to be (should be) taken into account, hence part two and three, which treat the daring subject of radiation hydrodynamics. Dimitri Mihalas presents – in a light-hearted style – the basic equations of "radiation hydro" that were otherwise only accessible in his voluminous "Foundations of Radiation Hydrodynamics" (now out of print). Mihalas' contribution is largely identical to his handwritten notes that he uses in teaching radiation hydrodynamics at the University of Illinois at Urbana Champaign. The numerical implementations of these equations and simulation examples are the subject of part three. The power of adaptive (moving) grids, which are capable locally refining the resolution by several orders of magnitude, is demonstrated with a number of stellar-physical simulations showing very crispy shock-front structures. Interestingly, these powerful radiation hydrodynamic codes are based on more traditional finite volume techniques using artificial viscosity and it seems that Riemann solver methods have not yet been applied in this difficult field of computational astrophysics.

This book is a written version of the the lectures delivered by Ernst Dorfi, Randall LeVeque, and Ewald Müller during the 27th "Saas-Fee Advanced Course". Dimitri Mihalas, who was scheduled as a lecturer, had to withdraw shortly before the course began because of an accident which forced him to remain in hospital for several weeks. The organizers of the course were very glad when Ernst Dorfi accepted to step in as a lecturer on numerical radiation hydrodynamics and when Dimitri Mihalas offered to contribute his (not delivered) lectures to the present book. Ernst Dorfi wrote a more comprehensive text on the subject of his ad hoc lecture.

The 27th "Saas-Fee Advanced Course" of the Swiss Society for Astrophysics and Astronomy took place in Les Diablerets, a small village in the Swiss Alps, during March 3–8, 1997. Ninety participants from 14 countries attended the course. The Eurotel provided a much appreciated hospitality. As can be deduced from the photographs in the book, sun and snow were also abundant and appropriately enjoyed.

This course would not have been possible without the financial contributions from the Swiss Society for Astrophysics and Astronomy (through the Swiss Academy of Sciences) which are gratefully acknowledged. O.S. would like to acknowledge the generous support by the High Altitude Observatory of the National Center for Atmospheric Research. A.G. was financially supported by the Swiss National Science Foundation through a PROFIL2 fellowship. Last but not least we are grateful to Wolfgang Löffler for professionally maintaining the Saas-Fee Web site.

Freiburg i.Br., Basel
May, 1998

Oskar Steiner
Alfred Gautschy

Table of Contents

List of Previous Saas-Fee Advanced Courses

**	1998	Star Clusters *B.W. Carney, C. Pryor, W.E. Harris*
!!	1996	Galaxies Interactions and Induced Star Formation *J. Barnes, R. Kennicutt, F. Schweizer*
!!	1995	Stellar Remnants *S.D. Kawaler, I. Novikov, G. Srinivasan*
*	1994	Plasma Astrophysics *J.G. Kirk, D.B. Melrose, E.R. Priest*
*	1993	The Deep Universe *A.R. Sandage, R.G. Kron, M.S. Longair*
*	1992	Interacting Binaries *S.N. Shore, M. Livio, E.J.P. van den Heuvel*
*	1991	The Galactic Interstellar Medium *W.B. Burton, B.G. Elmegreen, R. Gentzel*
*	1990	Active Galactic Nuclei *R. Blandford, H. Netzer, L. Woltjer*
!	1989	The Milky Way as a Galaxy *G. Gilmore, I. King, P. van der Kruit*
!	1988	Radiation in Moving Gaseous Media *H. Frisch, R.P. Kudritzki, H.W. Yorke*
!	1987	Large Scale Structures in the Universe *A.C. Fabian, M. Geller, A. Szalay*
!	1986	Nucleosynthesis and Chemical Evolution *J. Audouze, C. Chiosi, S.E. Woosley*
!	1985	High Resolution in Astronomy *R.S. Booth, J.W. Brault, A. Labeyrie*
!	1984	Planets, Their Origin, Interior and Atmosphere *D. Gautier, W.B. Hubbard, H. Reeves*
!	1983	Astrophysical Processes in Upper Main Sequence Stars *A.N. Cox, S. Vauclair, J.P. Zahn*
*	1982	Morphology and Dynamics of Galaxies *J. Binney, J. Kormendy, S.D.M. White*
!	1981	Activity and Outer Atmospheres of the Sun and Stars *F. Praderie, D.S. Spicer, G.L. Withbroe*
*	1980	Star Formation *J. Appenzeller, J. Lequeux, J. Silk*
*	1979	Extragalactic High Energy Physics *F. Pacini, C. Ryter, P.A. Strittmatter*
*	1978	Observational Cosmology *J.E. Gunn, M.S. Longair, M.J. Rees*
*	1977	Advanced Stages in Stellar Evolution *I. Iben Jr., A. Renzini, D.N. Schramm*
*	1976	Galaxies *K. Freeman, R.C. Larson, B. Tinsley*

- * 1975 Atomic and Molecular Processes in Astrophysics
  A. Dalgarno, F. Masnou-Seeuws, R.V.P. McWhirter
- * 1974 Magnetohydrodynamics
  L. Mestel, N.O. Weiss
- * 1973 Dynamical Structure and Evolution of Stellar Systems
  G. Contopoulos, M. Hénon, D. Lynden-Bell
- * 1972 Interstellar Matter
  N.C. Wickramasinghe, F.D. Kahn, P.G. Metzger
- * 1971 Theory of the Stellar Atmospheres
  D. Mihalas, B. Pagel, P. Souffrin

!! May be ordered from Springer Verlag   * Out of print
! May be ordered from Geneva Observatory   ** In preparation

Saas-Fee Courses
Geneva Observatory
CH-1290 Sauverny, Switzerland

Ernst Dorfi and Randall LeVeque (from left to right) on top of the Glacier des Diablerets, ready to hit the ski slopes

Nonlinear Conservation Laws and Finite Volume Methods

Randall J. LeVeque

Department of Applied Mathematics and Department of Mathematics, University of Washington, Box 352420, Seattle, WA 98195-2420, rjl@amath.washington.edu

1. Introduction

These notes are primarily concerned with the numerical solution to hyperbolic systems of conservation laws that arise in astrophysical calculations, primarily in "gas dynamic" simulations. The **Euler equations** are one fundamental system that will be discussed in detail, along with generalizations to magnetohydrodynamics (MHD) and relativistic flows, and to include source terms modeling effects such as combustion, radiation, or gravity.

The Euler equations are a nonlinear hyperbolic system of equations that arise from the compressible Navier–Stokes equations by neglecting viscosity and heat conduction. Mathematically the most interesting feature of such systems is that they admit **shock waves** in their solution, discontinuities in the solution that can form even from smooth initial data. These are mathematical idealizations of the steep gradients that can be present in smooth solutions to the full Navier–Stokes equations, where rapid changes occur over very thin zones (relative to the spatial dimensions that must be modeled). Often the width of these "viscous shock layers" is much smaller than the distance between grid points on a reasonable compuational grid, in which case we cannot possibly model the structure in these layers and are satisfied to use a model that produces a good approximation to the macroscopic behavior. There is a rich mathematical structure to the hyperbolic equations that can be exploited to develop numerical methods that are very effective in spite of the fact that the computed solution should be nearly discontinuous. Computationally, solving the hyperbolic approximation also has the advantage that explicit methods rather than implicit methods can typically be used, resulting in a substantial reduction in the computing time required.

Naive discretizations of the differential equations typically either result in very smeared-out representations of shocks, or else produce oscillations near discontinuities. This is because a finite-difference approximation to a derivative simply isn't appropriate when the function is discontinuous. Figure 1.1 shows some sample calculations for a "shock tube" problem discussed in Sect. 2.6. Figure 1.1(a) shows the smeared result from the first-order Godunov method introduced in Sect. 4.5. Figure 1.1(b) shows the results if we instead use a Lax–Wendroff style second-order method, which should be a superior method if the solution being computed were smooth, but here

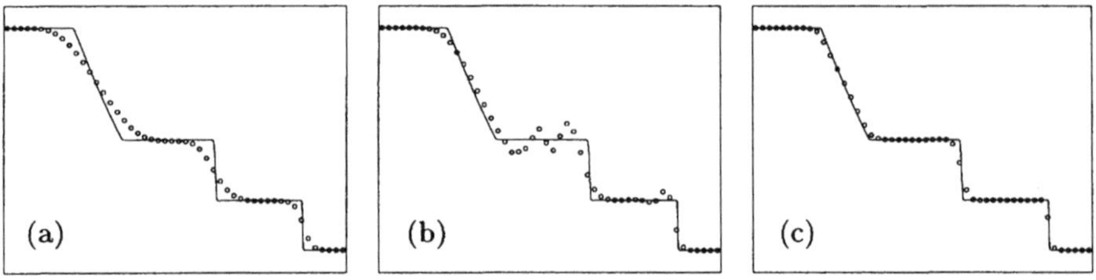

Fig. 1.1. Solution to the Sod problem for the Euler equations, as computed with three numerical methods. (**a**) Godunov's method, (**b**) A Lax–Wendroff style second-order method, (**c**) A high-resolution method with the superbee limiter

produces a very oscillatory approximation to this discontinuous solution. A variety of **high-resolution** numerical methods have been developed over the past several decades that resolve discontinuities sharply and yet produce at least second-order accuracy in smooth regions of the flow. Figure 1.1(c) shows the same calculation with a high-resolution method.

The main goal of these notes is to introduce the ideas used in this type of method. Understanding these numerical methods requires a good foundation in the mathematical theory of hyperbolic conservation laws, which is also reviewed here, together with a brief survey of some fundamental concepts from numerical analysis. We will then take a look at some complications that arise in the more complex equations important in astrophysical applications.

In such a short course it is not possible to fully develop all the ideas discussed. Instead my goal is to present an overview of main ideas and approaches, including computational difficulties as well as successes, in a framework that I hope will serve as a roadmap to students or researchers new to this area. Most of the topics covered here are discussed extensively in the literature, often in textbooks as well as research papers. I will try to give enough pointers that the reader can at least get started tracking down the details on any topic of interest. I have not, however, attempted to provide a complete bibliography to any of the subjects discussed, and there are many approaches and interesting works which are not cited or surveyed here.

I am not an astrophysicist, but rather an applied mathematician and numerical analyst with an interest in several applications where very similar mathematical equations and computational problems arise from the study of seemingly diverse phenomena. The equations discussed here are used directly in the study of aerodynamics, gas turbines, internal combustion engines, fusion reactors, and a host of other computational fluid dynamics applications. Moreover, the mathematical structure of hyperbolic conservation laws arises in many other fields including elasticity, flow in porous media, oceanography, and atmospheric sciences. Studying the literature in these other fields and in the applied mathematics community can be valuable in understanding many of the mathematical subtleties of these equations. Numerical methods have been developed in all of these areas and often good algorithms can be adapted to problems in other fields.

I have little direct experience in computational astrophysics, and am very grateful to the organizers of this course for giving me the opportunity to learn more about this fascinating application and some of the techniques developed in this field. In this learning process I have benefitted enormously from the other lectures in this course, conversations with many of the participants, and also from the help of countless other people who have generously shared their knowledge with me and in some cases commented on earlier drafts of these notes. In particular I would like to mention Derek Bale, Bruce Balick, Dinshaw Balsara, Jerry Brackbill, Moysey Brio, Phil Colella, Ewald Müller, Philippos Papadopoulos, Ken Powell, Ed Seidel, Michael Shearer, Phil Roe, Dongsu Ryu, Oskar Steiner, Bram van Leer, and Rolf Walder. Of course any errors that remain in these notes are my own responsibility.

My own work on numerical methods for conservation laws discussed here has been supported primarily by the National Science Foundation, the Department of Energy, and the National Center for Atmospheric Research.

1.1 Software

Developing computer models of complex phenomena generally requires developing your own computer programs. Rarely will you find software that easily solves the full problem you are studying. You may be able to obtain a research code from another group studying a similar problem, but such codes should be viewed with suspicion until you have convinced yourself that the methods implemented are appropriate for your problem. It may also turn out to be more trouble converting someone else's code to your needs than it would be to write one from scratch, particularly if it is poorly documented and not written with such extensions in mind (as is true of most research codes).

On the other hand, there exists by now a large quantity of excellent software available for solving many standard mathematical problems of the sort that often arise as part of a larger simulation. Much of this software is freely available on the Web.[1] When it is possible to make use of such software, it is often much more efficient and reliable than anything you can write yourself. This is particularly true in areas such as linear algebra or ordinary differential equations, where a very general framework covers a huge number of practical problems. Linear systems of the form $Ax = b$ arise everywhere, and enormous effort has been put into developing optimal software for solving such systems in a variety of contexts: direct methods such as Gaussian elimination for dense systems, faster methods for symmetric or banded systems, and a wide variety of iterative methods for sparse systems. For the general initial value problem of ODEs, $y'(t) = f(q,t)$ with $y(t_0)$ given, numerous packages

[1] A few links to software sources can be found from my homepage.

are available that automatically select methods and adaptively choose time steps to efficiently solve the equation to a desired tolerance. Other packages are available for two-point boundary-value problems.

Partial differential equations are more difficult to handle with general software. There are many different types of equations, exhibiting diverse behavior and requiring quite different methods for their solution. Most practical problems are also posed in multi-dimensional regions with complicated boundaries and boundary conditions, further complicating efforts to develop general software. Nonetheless, more and more software is appearing for particular classes of problems, with elliptic equations probably receiving the best coverage to date.

For hyperbolic equations and conservation laws, a number of packages have appeared in the past few years. Much of my own effort recently has gone into the CLAWPACK (Conservation LAWs PACKage) software [138] and the adaptive mesh refinement version AMRCLAW [29]. This software is aimed at providing a unified framework for solving a wide variety of conservation laws, and hyperbolic systems more generally, using high-resolution methods of the type discussed in these notes.

The CLAWPACK software was initially developed as a teaching tool, and studying the fortran code in conjunction with reading the brief descriptions of methods presented here may be valuable in understanding how these methods are implemented. In addition, many sample programs using this software are available with the package, or available elsewhere, and experimenting with these can help develop intuition both about the physics and the behavior of numerical methods. In particular, almost all of the sample computational results presented in these notes were produced with this package and can be easily reproduced.

1.2 Notation

In general I will use $q(x,t)$ to denote a generic function we are attempting to compute with some numerical method. This will typically be a vector-valued function. For example, in the Euler equations we have conservation of mass, momentum, and energy and these are the components of q. Often u is used instead of q in the PDE and numerical analysis literature, but it is very convenient to reserve u for the x-component of velocity.

Typically vectors will not be denoted by boldface or arrows, but can be recognized as such from the context. In the sections on MHD and relativistic flow, boldface is used.

Components of vectors will be denoted either by subscripts or superscripts, and often the notation changes from one section to another, depending on which is more convenient. In particular, when numerical methods are discussed a subscript is used for the index indicating spatial position, and then superscripts are often used for vector components. Only in the sections on relativity does position follow tensor notation.

1.3 Classification of Differential Equations

First we review the classification of differential equations into elliptic, parabolic, and hyperbolic equations. Not all PDE's fall into one of these classes, by any means, but many important equations that arise in practice do. These classes of equations model different sorts of phenomena, display different behavior, and require different numerical techniques for their solution. Standard texts on partial differential equations such as Kevorkian [126] give further discussion.

1.3.1 Second-Order Equations. In most elementary texts the classification is given for a linear second-order differential equation in two independent variables of the form

$$aq_{xx} + bq_{xy} + cq_{yy} + dq_x + eq_y + fq = g \ .$$

The classification depends on the sign of the discriminant,

$$b^2 - 4ac \begin{cases} < 0 & \Longrightarrow & \text{elliptic} \\ = 0 & \Longrightarrow & \text{parabolic} \\ > 0 & \Longrightarrow & \text{hyperbolic} \end{cases} ,$$

and the names arise by anology with conic sections. The canonical example are the Poisson problem $q_{xx} + q_{yy} = g$ for an elliptic problem, the heat equation $q_t = q_{xx}$ for a parabolic problem, and the wave equation $q_{tt} = q_{xx}$ for a hyperbolic problem. In the parabolic and hyperbolic case t is used instead of y since these are typically time-dependent problems. These can all be extended to more space dimensions. These equations describe different types of phenomena and require different techniques for their solution (both analytically and numerically), and so it is convenient to have names for classes of equations exhibiting the same general features. There are other equations that have some of the same features and the classification scheme can be extended beyond the second-order linear form given above. Some hint of this is given in the next few sections.

In particular, these notes are primarily concerned with hyperbolic equations that are first-order differential equations (involving only first derivatives with respect to space and time), and which generally are nonlinear. This form is described in Sect. 1.3.4.

1.3.2 Elliptic Equations. The classic example of an elliptic equation is the **Poisson problem**

$$\nabla^2 q = f \ , \tag{1.1}$$

where ∇^2 is the Laplacian operator and f is a given function of $\boldsymbol{x}$ in some spatial domain Ω. We seek a function $q(\boldsymbol{x})$ in Ω satisfying (1.1) together with some **boundary conditions** all along the boundary of Ω. Elliptic equations typically model steady-state or equilibrium phenomena, and so there is no temporal dependence. Elliptic equations may also arise in solving time-dependent problems if we are modeling some phenomena that are always in local equilibrium and equilibrate on time scales that are much faster than the time scale being modeled. For example, in "incompressible" flow the fast acoustic waves are not modeled and instead the pressure is computed by solving a Poisson problem at each time step which models the global effect of these waves. This is described in Sect. 8.6.1. Similar elliptic equations are solved in MHD to keep the magnetic field divergence-free at each time (see Sect. 8.6). In a self-gravitating flow the gravitational potential at each time is typically determined by solving a Poisson problem where the right-hand side depends on the density distribution at each time.

Elliptic equations give boundary-value problems (BVP's) where the solution at all points must be simultaneously determined based on the boundary conditions all around the domain. This typically leads to a very large sparse sytem of linear equations to be solved for the values of Q at each grid point. If an elliptic equation must be solved in every time step of a time-dependent calculation, as in the examples above, then it is crucial that these systems be solved as efficiently as possible. Very sophisticated iterative methods such as preconditioned conjugate gradient methods and multigrid methods have been developed for solving sparse linear systems arising from elliptic equations. There is a huge literature on these methods; see [41], [101], [244], [255] for general discussions of such methods and [171], [172] for some astrophysical applications.

More generally, a linear elliptic equation has the form

$$Lq = f \,, \tag{1.2}$$

where L is some **elliptic operator**. This notion will not be discussed further here, but the idea is that mathematical conditions are required on the differential operator L which insure that the boundary value problem has a unique solution.

1.3.3 Parabolic Equations. If L is an elliptic operator then the time-dependent equation

$$q_t = Lq - f \tag{1.3}$$

is called **parabolic**. If $L = \nabla^2$ is the Laplacian, then (1.3) is known as the **heat equation** or **diffusion equation** and models the diffusion of heat in a material, for example.

Now $q(\boldsymbol{x}, t)$ varies with time and we require **initial data** $q(\boldsymbol{x}, 0)$ for every $\boldsymbol{x} \in \Omega$ as well as boundary conditions around the boundary at each time $t > 0$. If the boundary conditions are independent of time, then we might expect the heat distribution to reach a steady state in which q is independent of t. We could then solve for the steady state directly by setting $q_t = 0$ in (1.3), which results in the elliptic equation (1.2).

Marching to steady state by solving the time-dependent equation (1.3) numerically would be one approach to solving the elliptic equation (1.2), but this is typically not the fastest method if all we require is the steady state.

1.3.4 Hyperbolic Equations. My lectures will primarily concern first-order hyperbolic systems of m equations. The linear problem in one space dimension has the form

$$q_t + Aq_x = 0 \ , \tag{1.4}$$

where $q(x,t) \in \mathbb{R}^m$ and A is an $m \times m$ matrix. The problem is called **hyperbolic** if A has *real* eigenvalues and is *diagonalizable*, *i.e.*, has a complete set of linearly independent eigenvectors. As we will see in Sect. 3.2, these conditions allow us to view the solution in terms of propagating waves, and indeed hyperbolic systems typically arise from physical processes that give wave motion or advective transport.

The simplest example of a hyperbolic equation is the constant-coefficient **advection equation**

$$q_t + uq_x = 0 \ , \tag{1.5}$$

where u is the advection velocity. The solution is simply $q(x,t) = q(x-ut,0)$, so any q profile simply advects with the flow at velocity u.

As a simple example of a linear system, the equations of linear acoustics in a γ-law gas can be written as a first-order system of two equations in one space dimension as

$$\begin{bmatrix} p \\ u \end{bmatrix}_t + \begin{bmatrix} 0 & \gamma p_0 \\ 1/\varrho_0 & 0 \end{bmatrix} \begin{bmatrix} p \\ u \end{bmatrix}_x = 0 \tag{1.6}$$

in terms of pressure and velocity perturbations, where p_0 and ϱ_0 are the background pressure and density. Note that if we differentiate the first equation with respect to t, the second with respect to x, and then eliminate $u_{xt} = u_{tx}$ we obtain the second-order wave equation for the pressure:

$$p_{tt} = c^2 p_{xx} \ ,$$

where

$$c = \sqrt{\gamma p / \varrho}$$

is the speed of sound in the gas.

We will also study **nonlinear** hyperbolic systems in depth, where shock waves arise in the solution. In one space dimension we will see (Sect. 2.) that physical conservation laws typically lead to systems of the form

$$q_t + f(q)_x = 0 \ , \tag{1.7}$$

where $f(q)_x = \frac{\partial}{\partial x}(f(q))$ and $f(q)$ is the **flux function**, as discussed in Sect. 2. This nonlinear system is called hyperbolic if the Jacobian matrix $f'(q)$ with (i,j) element $\partial f_i/\partial q_j$ is diagonalizable with real eigenvalues for each value q of physical interest. Linearizing a nonlinear hyperbolic system about some state q_0 then gives a linear hyperbolic equation $\tilde{q}_t + f'(q_0)\tilde{q}_x = 0$ where $\tilde{q}$ is the perturbation from q_0. Linear acoustics, for example, arises from linearizing the Euler equations of compressible flow given in Sect. 2.

2. Derivation of Conservation Laws

We now turn to the derivation and mathematical analysis of hyperbolic systems of conservation laws. This theory is needed to understand both the behavior of solutions of nonlinear conservation laws and the development of modern numerical methods for their solution.

This section of these notes is largely a condensed version of my ETH Notes [141], and is intended to be an overview of the important points for readers who are already familiar with this theory. More details can be found in [141] for those who are not. Numerous other books are also available on this theory, including [60], [97], [136], [221], [256].

Conservation laws are time-dependent systems of partial differential equations (usually nonlinear) with a particularly simple structure. In one space dimension the equations take the form

$$\frac{\partial}{\partial t}q(x,t) + \frac{\partial}{\partial x}f(q(x,t)) = 0 \,. \tag{2.1}$$

Here $q(x,t) \in \mathbb{R}^m$ is an m-dimensional vector of conserved quantities, or state variables, such as mass, momentum, and energy in a fluid dynamics problem. More properly, q_j is the density function for the jth state variable, with the interpretation that $\int_{x_1}^{x_2} q_j(x,t)\,\mathrm{d}x$ is the total quantity of this state variable in the interval $[x_1, x_2]$ at time t.

The fact that these state variables are conserved means that $\int_{-\infty}^{\infty} q_j(x,t)\mathrm{d}x$ should be constant with respect to t. The functions q_j themselves, representing the spatial distribution of the state variables at time t, will generally change as time evolves. The main assumption underlying (2.1) is that knowing the value of $q(x,t)$ at a given point and time allows us to determine the rate of flow, or **flux**, of each state variable at (x,t). The flux of the jth component is given by some function $f_j(q(x,t))$. The vector-valued function $f(q)$ with jth component $f_j(q)$ is called the **flux function** for the system of conservation laws.

To see how conservation laws arise from physical principles, we will begin by deriving the equation for conservation of mass in a one-dimensional gas dynamics problem, for example flow in a tube where properties of the gas such as density and velocity are assumed to be constant across each cross section of the tube. Let x represent the distance along the tube and let $\varrho(x,t)$ be the density of the gas at point x and time t. This density is defined in such a way that the total mass of gas in any given section from x_1 to x_2, say, is given by the integral of the density:

$$\text{mass in } [x_1, x_2] \text{ at time } t = \int_{x_1}^{x_2} \varrho(x,t)\,\mathrm{d}x \,. \tag{2.2}$$

If we assume that the walls of the tube are impermeable and that mass is neither created nor destroyed, then the mass in this one section can change only because of gas flowing across the endpoints x_1 or x_2.

Now let $u(x,t)$ be the velocity of the gas at the point x at time t. Then the rate of flow, or **flux** of gas past this point is given by

$$\text{mass flux at } (x,t) = \varrho(x,t)u(x,t)\ . \tag{2.3}$$

By our comments above, the rate of change of mass in $[x_1, x_2]$ is given by the difference in fluxes at x_1 and x_2:

$$\frac{\mathrm{d}}{\mathrm{d}t}\int_{x_1}^{x_2} \varrho(x,t)\,\mathrm{d}x = \varrho(x_1,t)u(x_1,t) - \varrho(x_2,t)u(x_2,t)\ . \tag{2.4}$$

This is one **integral form** of the conservation law. Another form is obtained by integrating this in time from t_1 to t_2, giving an expression for the mass in $[x_1, x_2]$ at time $t_2 > t_1$ in terms of the mass at time t_1 and the total (integrated) flux at each boundary during this time period:

$$\begin{aligned}\int_{x_1}^{x_2} \varrho(x,t_2)\,\mathrm{d}x &= \int_{x_1}^{x_2} \varrho(x,t_1)\,\mathrm{d}x \\ &\quad + \int_{t_1}^{t_2} \varrho(x_1,t)u(x_1,t)\,\mathrm{d}t - \int_{t_1}^{t_2} \varrho(x_2,t)u(x_2,t)\,\mathrm{d}t\ .\end{aligned} \tag{2.5}$$

To derive the differential form of the conservation law, we must now assume that $\varrho(x,t)$ and $u(x,t)$ are differentiable functions. Then using

$$\varrho(x,t_2) - \varrho(x,t_1) = \int_{t_1}^{t_2} \frac{\partial}{\partial t}\varrho(x,t)\,\mathrm{d}t \tag{2.6}$$

and

$$\varrho(x_2,t)u(x_2,t) - \varrho(x_1,t)u(x_1,t) = \int_{x_1}^{x_2} \frac{\partial}{\partial x}(\varrho(x,t)u(x,t))\,\mathrm{d}x \tag{2.7}$$

in (2.5) gives

$$\int_{t_1}^{t_2}\int_{x_1}^{x_2} \left\{\frac{\partial}{\partial t}\varrho(x,t) + \frac{\partial}{\partial x}(\varrho(x,t)u(x,t))\right\}\,\mathrm{d}x\,\mathrm{d}t = 0\ . \tag{2.8}$$

Since this must hold for any section $[x_1, x_2]$ and over any time interval $[t_1, t_2]$, we conclude that in fact the integrand in (2.8) must be identically zero, *i.e.*,

$$\varrho_t + (\varrho u)_x = 0 \qquad \text{conservation of mass}\ . \tag{2.9}$$

This is the desired **differential form** of the conservation law for the conservation of mass, which is often called the **continuity equation**.

Exactly the same derivation applies more generally if we replace ϱ by the density of any conserved quantity, and let $f(\varrho)$ be the corresponding flux function describing the flow of this quantity.

The simplest example of a conservation law consists of the equation (2.9) alone in the case where u is constant. This models flow of an incompressible fluid in a one-dimensional pipe at constant velocity. In this case ϱ would model not the density of the fluid but the concentration of something being transported with the fluid (a chemical, or heat perhaps). If we relabel this concentration as q, then we have the simple linear scalar **advection equation**,

$$q_t + uq_x = 0\,. \tag{2.10}$$

2.1 The Euler Equations of Gas Dynamics

In gas dynamics, the equation (2.9) must be solved in conjunction with equations for the conservation of momentum and energy:

$$(\varrho u)_t + (\varrho u^2 + p)_x = 0 \qquad \text{conservation of momentum}\,, \tag{2.11}$$

$$E_t + (u(E+p))_x = 0 \qquad \text{conservation of energy}\,. \tag{2.12}$$

The resulting system of three conservation laws gives the **Euler equations** of gas dynamics. Note that these equations involve another quantity, the pressure p, which must be specified as a given function of ϱ, ϱu, and E in order that the fluxes be well defined functions of the conserved quantities alone. This additional equation is called the **equation of state** and depends on physical properties of the gas under study. We will consider the simplest case of a polytropic gas:

$$E = \frac{p}{\gamma - 1} + \frac{1}{2}\varrho u^2\,, \tag{2.13}$$

where γ is the ratio of specific heats. From statistical mechanics it can be shown that $\gamma = (\alpha + 2)/\alpha$, where α is the number of internal degrees of freedom of the gas molecules. For a monatomic gas $\alpha = 3$ and $\gamma = 5/3$, which is frequently the proper value for astrophysical calculations. In aerodynamics $\gamma = 1.4 = 7/5$ is generally used since air consists primarily of diatomic molecules.

If we introduce the vector $q \in \mathbb{R}^3$ as

$$q(x,t) = \begin{bmatrix} \varrho(x,t) \\ \varrho(x,t)u(x,t) \\ E(x,t) \end{bmatrix} \equiv \begin{bmatrix} q_1 \\ q_2 \\ q_3 \end{bmatrix}, \tag{2.14}$$

then the system of equations (2.9), (2.11), (2.12) can be written simply as

$$q_t + f(q)_x = 0\,, \tag{2.15}$$

where

$$f(q) = \begin{bmatrix} \varrho u \\ \varrho u^2 + p \\ u(E+p) \end{bmatrix} = \begin{bmatrix} q_2 \\ q_2^2/q_1 + p(q) \\ q_2(q_3 + p(q))/q_1 \end{bmatrix}. \tag{2.16}$$

Again, the form (2.15) is the differential form of the conservation laws, which holds in the usual sense only where u is smooth. More generally, the integral form for a system of m equations says that

$$\frac{\mathrm{d}}{\mathrm{d}t}\int_{x_1}^{x_2} q(x,t)\,\mathrm{d}x = f(q(x_1,t)) - f(q(x_2,t)) \tag{2.17}$$

for all $x_1,\ x_2,\ t$. Equivalently, integrating from t_1 to t_2 gives

$$\begin{aligned}\int_{x_1}^{x_2} q(x,t_2)\,\mathrm{d}x &= \int_{x_1}^{x_2} q(x,t_1)\,\mathrm{d}x \\ &\quad + \int_{t_1}^{t_2} f(q(x_1,t))\,\mathrm{d}t - \int_{t_1}^{t_2} f(q(x_2,t))\,\mathrm{d}t\end{aligned} \tag{2.18}$$

for all $x_1,\ x_2,\ t_1$, and t_2. These integral forms of the conservation law are the fundamental physical conservation laws. It is important to note that these equations make sense even if the function $q(x,t)$ is discontinuous, while transforming to the differential form (2.15) is valid only when q is smooth.

2.2 Dissipative Fluxes

So far we have ignored dissipative effects such as molecular diffusion, heat conduction or fluid viscosity. These effects generally lead to fluxes that depend on the *spatial gradient* of the state variable q rather than on the value of q.

As a simple example, consider a fluid that is flowing through a pipe at constant speed u and let q represent the concentration of some chemical, or heat, in the fluid. According to Fourier's law of heat conduction, q diffuses at a rate $-\mu q_x$ which is proportional to the derivative of q, as well as advecting with the flow. Then the flux function is

$$f(q,q_x) = uq - \mu q_x$$

and we obtain the **advection-diffusion equation**

$$q_t + uq_x = \mu q_{xx}\ . \tag{2.19}$$

If $u = 0$ then this is simply the heat equation.

Similarly, adding fluid viscosity and heat conduction to the Euler equations gives the full **Navier–Stokes equations**, which include terms that depend on the second derivative of the velocity (in the case of viscosity) and the second derivative of the temperature (for heat conduction).

Adding these dissipative fluxes causes problems numerically because the equations are now *parabolic* rather than hyperbolic. This typically causes severe time step restrictions as discussed in Sect. 1.3.3 and requires the use of implicit methods for efficient solution.

2.3 Source Terms

In deriving the conservation law (2.15) we assumed that the "total mass" of q over an arbitrary interval $[x_1, x_2]$, given by the integral (2.2), should change only due to fluxes through the endpoints of this interval. In many cases this is not true. There may be some creation or destruction (source or sink) of the “conserved quantity” within this interval. This leads to the introduction of **source terms** in the equation. Here are a few examples of the sorts of effects that lead to source terms:

- Reacting flow or ionization, in which case we have a separate continuity equation for the mass of each species. The equations model the advection with the flow, as in (2.9), but also include source terms for the chemical kinetics, corresponding to the fact that each species is not individually conserved.
- External forces such as gravity, which lead to a source term in the momentum and energy equations. (In astrophysics one must often consider self-gravitation as well, in which case the gravitational force must be computed from the density function and changes with time.)
- Radiative heat transfer, which transfers energy on much faster time scales than the fluid dynamics being modeled and results in source terms in the energy equation.
- Geometric source terms, which arise when a problem in several space dimensions is reduced to a lower-dimensional problem using spherical or cylindrical symmetry, for example.

To see how source terms come into the conservation law, let's reconsider the expression (2.17) in the situation where there is another source of q affecting $\int_{x_1}^{x_2} q(x,t)\,\mathrm{d}x$ other than fluxes at x_1 and x_2. This source will be denoted in general by $\psi(q)$, with the idea that $\psi(q(x,t))$ represents the increase in q (or decrease, if $\psi < 0$) at the point (x,t) per unit time, due to effects of the sort described above. Then in place of (2.4) we have

$$\frac{\mathrm{d}}{\mathrm{d}t}\int_{x_1}^{x_2} q(x,t)\,\mathrm{d}x = F_1(t) - F_2(t) + \int_{x_1}^{x_2} \psi(q(x,t))\,\mathrm{d}x \;. \tag{2.20}$$

Note that the source term must be integrated over the interval since there is an effect at each point in the interval.

Proceeding as before, assuming q, f, and ψ are smooth, we will obtain the differential equation

$$q_t + f(q)_x = \psi(q) \;. \tag{2.21}$$

This is a nonhomogeneous version of the conservation laws considered thus far. Note that if there are no spatial variations, $q(x,t) \equiv q(t)$, then the equation (2.21) reduces to the ordinary differential equation

$$q_t = \psi(q) \;. \tag{2.22}$$

2.4 Radiative Transfer and Isothermal Equations

If we wish to include radiative heat transfer in a calculation then we must include source terms corresponding to the energy radiated or absorbed at each point. (See [166], [217], and the lectures of Dorfi and Mihalas in this volume for a complete discussion of radiative transfer.) Typically this is very complicated because we must model the fact that radiation from all points in the computational domain may be absorbed at each point $\boldsymbol{x}$, and so the source term at $\boldsymbol{x}$ involves an integral over the entire domain. Such problems can presumably be handled by the techniques discussed in Sect. 5., but this may be computationally expensive due to the need to approximate these integrals in each time step.

A simple special case arises if we assume that radiation is not reabsorbed within the computational domain (the optically thin limit). This is a reasonable model in studying the dynamics of an interstellar molecular cloud, for example. Since molecular motion radiates heat very effectively, such a cloud may have a temperature of only a few degrees Kelvin, and any heat generated by a shock wave passing through the cloud will be rapidly radiated away. In this case it may be appropriate to use a source term in the energy equation which simply drives the temperature exponentially quickly towards some constant value T_0 on time scale τ (Newton's law of cooling). Then the one-dimensional Euler equations become

$$\begin{aligned} \varrho_t + (\varrho u)_x &= 0\,, \\ (\varrho u)_t + (\varrho u^2 + p)_x &= 0\,, \\ E_t + u(E+p)_x &= -\varrho(T-T_0)/\tau\,. \end{aligned} \tag{2.23}$$

If the radiation time scale τ is very small relative to the dynamic time scales that we hope to model numerically, then these source terms will be very "stiff", as discussed in Sect. 5.4. An alternative in this case would be to use the isothermal Euler equations, which is obtained by replacing the ideal gas equation of state (2.13) by the simple equation of state that results from assuming that the temperature is in fact constant everywhere. Since $T = p/\mathcal{R}\varrho$, this results in the equation of state

$$p = \mathcal{R}\varrho T_0 \equiv a^2 \varrho\,, \tag{2.24}$$

where $\mathcal{R}$ is the gas constant. Then conservation of mass and momentum gives the system of two *isothermal equations*

$$\varrho_t + (\varrho u)_x = 0\,, \tag{2.25}$$

$$(\varrho u)_t + \left(\frac{1}{2}\varrho u^2 + a^2\varrho\right)_x = 0\,. \tag{2.26}$$

In this case energy is not conserved due to radiation, the equation of state does not involve E, and we do not need an energy equation since the isothermal assumption allows us to compute the energy directly from ϱ and ϱu if desired.

The full Euler equations are sometimes used to simulate isothermal situations by setting γ to a value slightly larger than 1, say $\gamma = 1.001$. This corresponds to taking the number of internal degrees of freedom α in the gas to be very large (see Sect. 2.1). Then if the gas is heated by a shock, for example, this heat will be spread around among so many degrees of freedom that it has little impact on the pressure or temperature, so that isothermal behavior results provided the initial data has constant temperature.

2.5 Multi-dimensional Conservation Laws

Conservation laws generalize naturally to more than one space dimension. In three dimensions, the equation takes the form

$$q_t + f(q)_x + g(q)_y + h(q)_z = 0 \,, \tag{2.27}$$

where $g(q)$ and $h(q)$ give the fluxes in the y- and z-directions respectively for a given state q. This can again be derived from the more fundamental integral form. Let $R = [x_1, x_2] \times [y_1, y_2] \times [z_1, z_2]$ be a cube in 3-space aligned with the coordinate directions. Then

$$\begin{aligned} \iiint_R q(\boldsymbol{x}, t_2)\,\mathrm{d}\boldsymbol{x} - \iiint_R q(\boldsymbol{x}, t_1)\,\mathrm{d}\boldsymbol{x} = \\ \int_{t_1}^{t_2}\int_{z_1}^{z_2}\int_{y_1}^{y_2} f(q(x_1, y, z, t)\,\mathrm{d}y\,\mathrm{d}z\,\mathrm{d}t \\ - \int_{t_1}^{t_2}\int_{z_1}^{z_2}\int_{y_1}^{y_2} f(q(x_2, y, z, t)\,\mathrm{d}y\,\mathrm{d}z\,\mathrm{d}t \\ + \int_{t_1}^{t_2}\int_{z_1}^{z_2}\int_{x_1}^{x_2} f(q(x, y_1, z, t)\,\mathrm{d}x\,\mathrm{d}z\,\mathrm{d}t \\ - \int_{t_1}^{t_2}\int_{z_1}^{z_2}\int_{x_1}^{x_2} f(q(x, y_2, z, t)\,\mathrm{d}x\,\mathrm{d}z\,\mathrm{d}t \\ + \int_{t_1}^{t_2}\int_{y_1}^{y_2}\int_{x_1}^{x_2} f(q(x, y, z_1, t)\,\mathrm{d}x\,\mathrm{d}y\,\mathrm{d}t \\ - \int_{t_1}^{t_2}\int_{y_1}^{y_2}\int_{x_1}^{x_2} f(q(x, y, z_2, t)\,\mathrm{d}x\,\mathrm{d}y\,\mathrm{d}t \end{aligned} \tag{2.28}$$

and manipulations similar to those performed above lead to the differential equation (2.27) when q is sufficiently smooth.

We do not need to use a cube in this construction. More generally let Ω represent any volume in 3-space. Then the (outward) flux through the surface of Ω at any point $\boldsymbol{x} = (x, y, z)$ is given by $\mathcal{F}(q(\boldsymbol{x}, t)) \cdot \boldsymbol{n}$ where $\mathcal{F} = (f, g, h)$ and $\boldsymbol{n}$ is the (outward) normal to the surface of Ω at $\boldsymbol{x}$. The total net outward flux can then be computed by integrating this flux over the surface $\partial\Omega$, so

$$\iiint_\Omega q(\boldsymbol{x}, t_2)\,\mathrm{d}\boldsymbol{x} - \iiint_\Omega q(\boldsymbol{x}, t_1)\,\mathrm{d}\boldsymbol{x} = -\iint_{\partial\Omega} \mathcal{F}(q) \cdot \boldsymbol{n}\,\mathrm{d}A \,. \tag{2.29}$$

By Gauss's theorem, when q is smooth we can rewrite this as

$$\iiint_\Omega q(\boldsymbol{x}, t_2)\,\mathrm{d}\boldsymbol{x} - \iiint_\Omega q(\boldsymbol{x}, t_1)\,\mathrm{d}\boldsymbol{x} = -\iiint_\Omega \nabla \cdot \mathcal{F}(q)\,\mathrm{d}\boldsymbol{x} \,, \tag{2.30}$$

where

$$\nabla \cdot \mathcal{F}(q) = f(q)_x + g(q)_y + h(q)_z \, . \tag{2.31}$$

This leads to

$$\int_{t_1}^{t_2} \iiint_{\Omega} q_t + \nabla \cdot \mathcal{F}(q) \, dx \, dt = 0 \tag{2.32}$$

and hence again to (2.27) since Ω, t_1, and t_2 are arbitrary.

The Euler equations in three space dimensions take the form

$$\begin{aligned} \varrho_t + (\varrho u)_x + (\varrho v)_y + (\varrho w)_z &= 0 \, , \\ (\varrho u)_t + (\varrho u^2 + p)_x + (\varrho uv)_y + (\varrho uw)_z &= 0 \, , \\ (\varrho v)_t + (\varrho uv)_x + (\varrho v^2 + p)_y + (\varrho vw)_z &= 0 \, , \\ (\varrho w)_t + (\varrho uw)_x + (\varrho vw)_y + (\varrho w^2 + p)_z &= 0 \, , \\ E_t + (u(E+p))_x + (v(E+p))_y + (w(E+p))_z &= 0 \, . \end{aligned} \tag{2.33}$$

Note that the forces that affect the momentum come from the *stress tensor*

$$T = \begin{bmatrix} \varrho u^2 + p & \varrho uv & \varrho uw \\ \varrho uv & \varrho v^2 + p & \varrho vw \\ \varrho uw & \varrho vw & \varrho w^2 + p \end{bmatrix} = \varrho \boldsymbol{u}\boldsymbol{u} + pI \, , \tag{2.34}$$

where $\boldsymbol{u}\boldsymbol{u} = \boldsymbol{u} \otimes \boldsymbol{u}$ represents the tensor product of these vectors.

2.6 The Shock Tube Problem

A simple example that illustrates the interesting behavior of solutions to conservation laws is the "shock tube problem" of gas dynamics. The physical set-up is a tube filled with gas, initially divided by a membrane into two sections. The gas has a higher density and pressure in one half of the tube than in the other half, with zero velocity everywhere. At time $t = 0$, the membrane is suddenly removed or broken, and the gas allowed to flow. We expect a net motion in the direction of lower pressure. Assuming the flow is uniform across the tube, there is variation in only one direction and the one-dimensional Euler equations apply.

The structure of this flow turns out to be very interesting, involving three distinct waves separating regions in which the state variables are constant. Across two of these waves there are discontinuities in some of the state variables. A **shock wave** propagates into the region of lower pressure, across which the density and pressure jump to higher values and all of the state variables are discontinuous. This is followed by a **contact discontinuity**, across which the density is again discontinuous but the velocity and pressure are constant. The third wave moves in the opposite direction and has a very different structure: all of the state variables are continuous and there is a smooth transition. This wave is called a **rarefaction wave** since the density of the gas decreases (the gas is rarefied) as this wave passes through.

The interface between the two initial gases moves along the contact discontinuity, hence the name. Figure 2.1 shows particle paths in the x-t plane for one example with $\varrho_l = p_l = 3$ and $\varrho_r = p_r = 1$. This particular problem is called the **Sod problem** since Sod [225] used it as a test in an influential early comparison of different numerical methods. With this data the temperature T, which is proportional to p/ϱ, is initially the same everywhere but the gas to the left has a higher density and pressure.

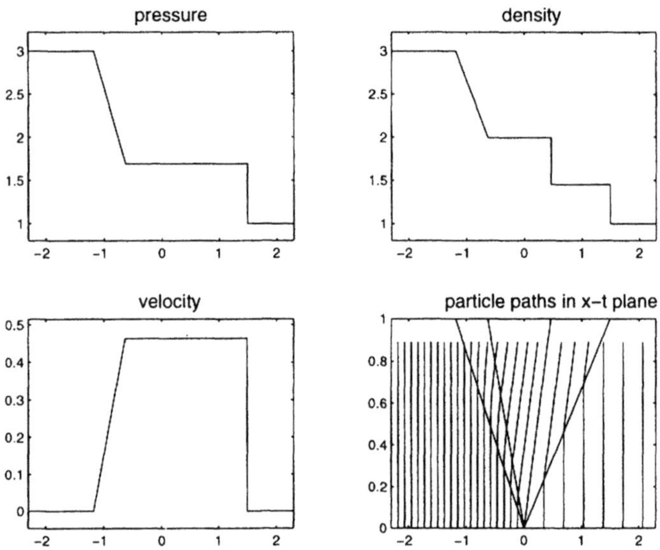

Fig. 2.1. Solution to Sod's shock tube problem for the Euler equations

One plot in Fig. 2.1 shows some particle paths in the x-t plane. Notice that the gas is accelerated abruptly across the shock and smoothly through the rarefaction. At any fixed t the particles shown are spaced proportional to $1/\varrho$ (the specific volume), so wider spacing indicates lower density. Note that there is a jump in density across the contact discontinuity while the fluid velocity is the same on both sides, and equal to the speed at which this discontinuity propagates, so particles never cross this contact surface. Note also the decrease in density as particles go through the rarefaction wave but the compression across the shock.

If we put the initial discontinuity at $x = 0$, then the resulting solution $q(x,t)$ is a "similarity solution" in the variable x/t, meaning that $q(x,t)$ can be expressed as a function of x/t alone, say $q(x,t) = w(x/t)$. It follows that $q(x,t) = q(\alpha x, \alpha t)$ for any $\alpha > 0$, so the solution at two different times t and αt look the same if we rescale the x-axis. This also means that the waves move at constant speed and the solution $q(x,t)$ is constant along any ray $x/t =$ constant in the x-t plane. Figure 2.1 shows the solution as a function of x at time $t = 1$, but also shows the solution at any other time $t > 0$ by simply rescaling the x-coordinate by multiplying by t.

In a real experimental shock tube, the state variables would not be discontinuous across the shock wave or contact discontinuity because of effects such as viscosity and heat conduction. These are ignored in the Euler equations. If we include these effects, using the full Navier–Stokes equations, then the solution of the partial differential equations would also be smooth. However, these smooth solutions would be nearly discontinuous, in the sense that the rise in density would occur over a distance that is microscopic compared to the natural length scale of the shock tube. If we plotted the smooth solutions they would look indistinguishable from the discontinuous plots shown in Fig. 2.1. For this reason we would like to ignore these viscous terms altogether and work with the simpler Euler equations.

The shock tube problem is a special case of what is known mathematically as the **Riemann problem**. In general, the Riemann problem consists of the conservation law $q_t + f(q)_x = 0$ together with the special initial data

$$q(x,0) = \begin{cases} q_l & \text{if } x < 0 \\ q_r & \text{if } x > 0\,. \end{cases} \tag{2.35}$$

This data consists of two constant states separated by a discontinuity. It turns out that this problem is relatively easy to solve in general, and that the resulting wave structure reveals a lot about the structure of solutions more generally. Many numerical methods are based on solving Riemann problems, including the methods discussed below.

Figures 2.2 and 2.3 show two more examples of solutions to Riemann problems for the Euler equations. The initial data chosen in these figures gives an indication of structures typically seen in certain calculations of stellar atmosphere dynamics. See Sect. 3.4 for details on how the solutions to these Riemann problems are computed.

Figure 2.2 shows a shock tube problem with data $u \equiv 0$, $\varrho_l = \varrho_r = 1$ and a pressure jump from $p_l = 10$ to $p_r = 1$. In this case the gas to the left has a higher temperature $T = p/\mathcal{R}\varrho$, as might occur if this gas has just been burned in the initiation of an explosion, for example. The higher pressure on the left leads to an expansion of this gas so that the rarefaction wave reduces the density to a lower value than $\varrho_l = \varrho_r$. This expansion requires the gas to move to the right, creating a shock wave that moves into the initially cooler gas, compressing it and raising its density to a value substantially higher than $\varrho_l = \varrho_r$. Across the contact discontinuity there is a large jump in density, and still a jump in temperature between the hot and cool gases, but the pressures are now equal.

The high-density region between the shock and contact consists of initially-cool gas that has been "swept up" by the expanding hot gas. Structures of this form arise in supernova explosions, for example. An explosion might be modeled at the simplest level by a "spherical Riemann problem" with piecewise constant data in three dimensions having one constant state outside a sphere

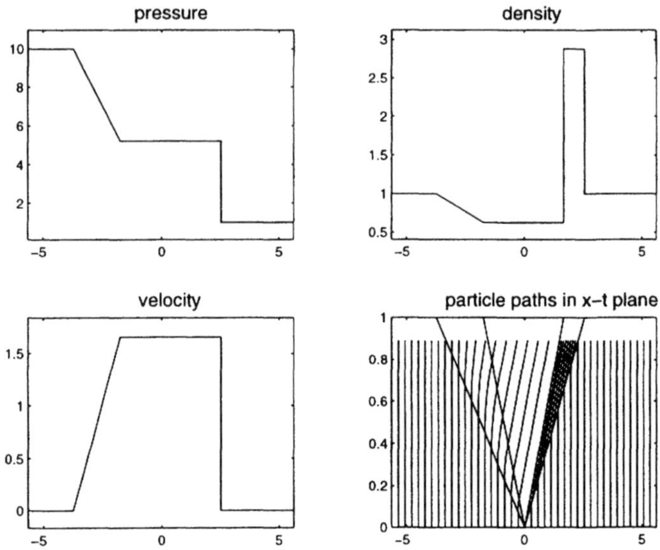

Fig. 2.2. Solution to a Riemann problem for the Euler equations where the gas to the left is at a higher temperature

of some radius R_0 and a different constant state, with much higher temperature, inside this sphere. For short times the solution along any ray from the origin will look roughly like what is seen in Fig. 2.2, with a spherical shell of high-density gas moving outward. (See Fig. 6.1 for a multi-dimensional calculation.) Three dimensional effects will lead to an attenuation of this structure and more complicated phenomena as the rarefaction wave reflects at the origin. In modeling a supernova explosion one must also include gravitational forces so that the density should initially be decaying with r rather than constant, and of course a variety of other effects such as radiative transfer are crucial. Nonetheless, the same basic structure seen in Fig. 2.2 can be identified in realistic simulations of supernovae, as seen for example in Fig. 3.10 of Müller's lectures in this volume.

Note that when gravitational forces are included, directed towards $r = 0$, this dense shell will be separated from lighter gas below by a contact surface. In a three-dimensional model this surface would initially be spherical but would be quickly broken up by Rayleigh–Taylor instabilities. This is discussed further in Section 3.4 of Müller's lectures. See also Fig. 3.15 of his notes, where the stable shock and unstable contact surface are clearly seen (though here the instability is a results of energy deposition by neutrinos).

Figure 2.3 shows the solution to a different Riemann problem where the initial data is

$$\begin{aligned} \varrho_l &= 1, \qquad u_l = 3, \quad p_l = 1\,, \\ \varrho_r &= 2, \qquad u_r = 1, \quad p_r = 1\,. \end{aligned}$$

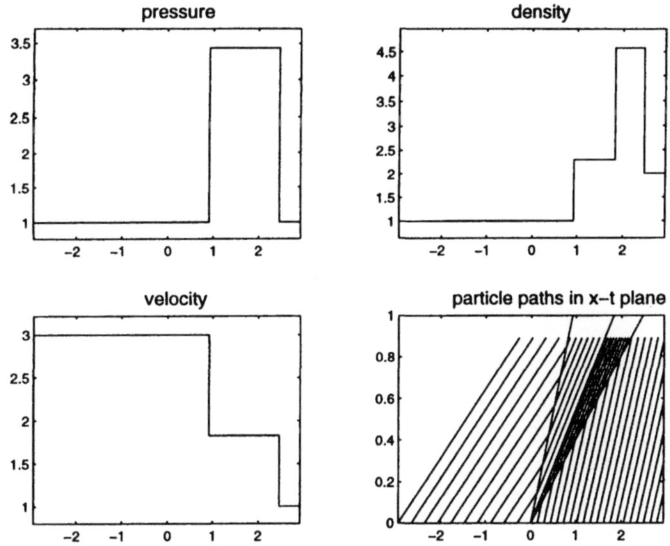

Fig. 2.3. Solution to a Riemann problem for the Euler equations where the gas to the left has a higher velocity

The gas on the left is moving faster than the gas on the right, and the collision produces two shock waves, one moving into each gas. As in the previous example, the shock moving to the right sweeps up material into a high-density layer. Again the spherical shell analog of this Riemann problem arises in astrophysics, in the formation of planetary nebulae, for example. An old star, in which the hydrogen and helium fuel has been exhausted, reaches a period of instability and ejects up to 80% of its mass in a so-called "slow wind", characterized by velocities of about 10 km/sec. This is followed by a "fast wind" phase with velocities more on the order of 1000 km/sec. The fast wind forms a hot (10^8 K) invisible bubble which expands supersonically into the slow wind from inside, forming a thin dense shell of swept-up gas ahead of it. The emerging hot core of the central star ionizes the swept up shell and the undisturbed slow wind upstream of it. The ionized gas emits fluorescent and recombination-line radiation making the "planetary nebula" visible, see for example Fig. 2.4. The photo on the right shows NGC 6826, a relatively smooth and almost spherical nebula. The wind-compressed swept-up shell behind the outward moving shock is seen as the bright leading edge of the hot expanding bubble.

In many cases the ejected slow wind is not at all spherically symmetric, leading to structures such as the Hourglass Nebula MyCn 18 shown on the left in Fig. 2.4. One possible mechanism for this is an asymmetric density distribution in the slow wind, allowing the wind-swept gas and the shock that precedes it to propagate faster in some directions than others. This is

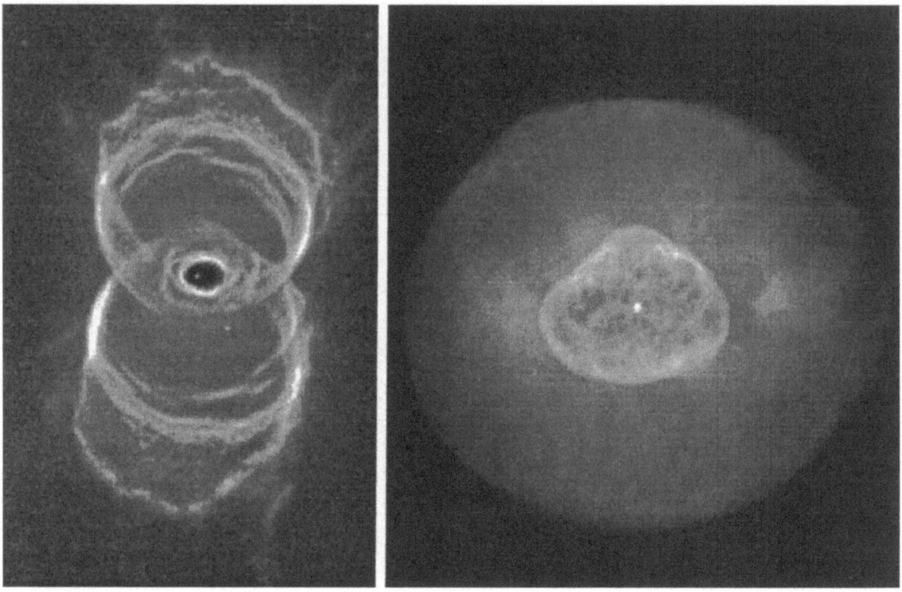

Fig. 2.4. Photographs of planetary nebulae, visible evidence of dense swept-up shells generated by the stars which appear as bright dots in the center of each image. Observations by various teams using the WFPC2 camera on NASA's Hubble Space Telescope, provided by B. Balick

an active area of computational research. Some references include [76], [90], [113], [164].

A jet of gas intruding into an ambient medium also gives a structure of the form seen in Fig. 2.3 at the leading tip. Extragalactic jets are discussed by Müller, and their structure is investigated in Section 5.3 of his notes. See in particular Figs. 5.3 and 5.5.

Similar shock structures arise from the head-on collision of stellar winds between the two stars of a binary system. In this case, the shocks approach a steady state location between the stars, at least in the coordinate frame co-rotating with the orbiting system. The orbital motion creates spiral-shaped shocks. Figure 2.5 shows the mass density for such a system over a cross-section through the orbital plane. This is a computational result from a three-dimensional calculation where adaptive mesh refinement has been used, as described in Sect. 6.3.2, in order to concentrate grid points where they are most needed near the shocks waves between the stars. The rectagular regions on the plot show regions of grid refinement. In this calculation two shocks are clearly visible. In general there is also a contact discontinuity, but in this particular calculation this is very weak.

Mass is continually flowing out from the stars, through the shocks, and into the region between the shocks from which it must flow outwards in

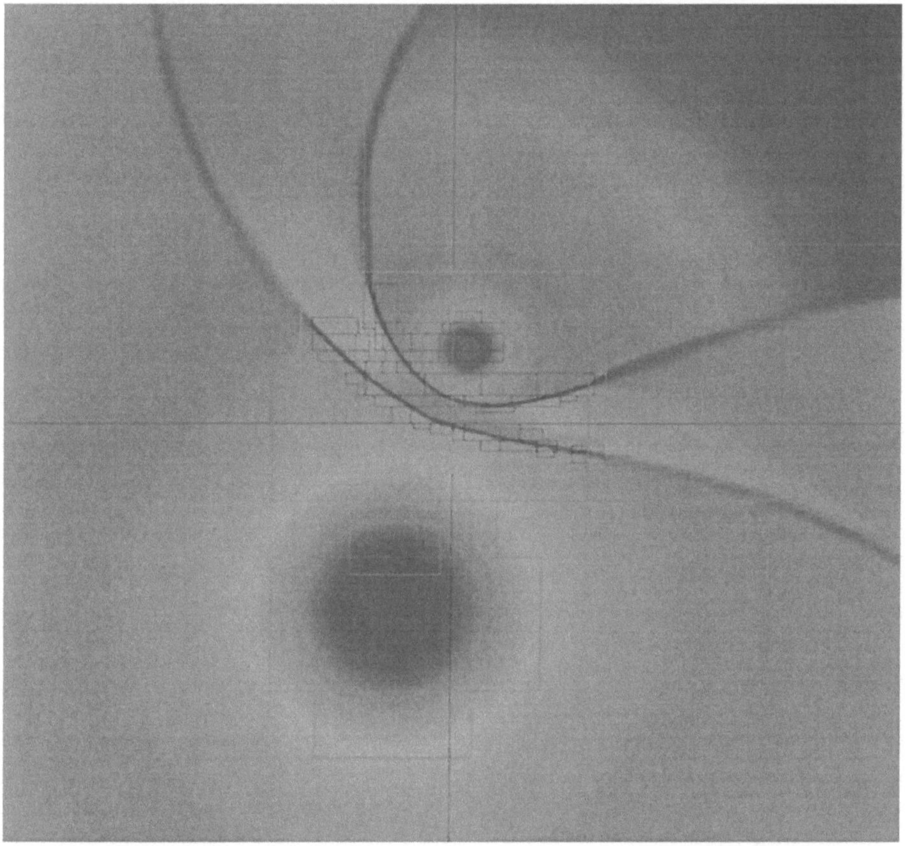

Fig. 2.5. Adaptive mesh refinement calculation of colliding winds between binary stars. Density in the orbital plane from a full three-dimensional calculation. The rectangles show regions of refinement (see Sect. 6.3.2). Calculation by R. Walder

directions roughly parallel to the shocks. This stationary structure can only exist in multidimensional problems and is coupled with strong transverse flow through the shocks. Any contact discontinuity would also be a slip line across which the two flows move parallel, typically at different speeds. This can lead to Kelvin–Helmholtz instabilities and roll-up of the slip lines. The thin region between the two shocks is also subject to the "thin shell instability" described for example in [230], [70], [251]. (See also Sect. 7.7.) See [177], [253] for more discussion of binary systems.

3. Mathematical Theory of Hyperbolic Systems

3.1 Scalar Equations

We begin our study of conservation laws by considering the scalar case. Many of the difficulties encountered with systems of equations are already encountered here, and a good understanding of the scalar equation is required before proceeding.

3.1.1 The Linear Advection Equation. We first consider the linear advection equation,

$$q_t + uq_x = 0 . \tag{3.1}$$

The Cauchy problem is defined by this equation on the domain $-\infty < x < \infty,\ t \geq 0$ together with initial conditions

$$q(x,0) = q_0(x) . \tag{3.2}$$

The solution to this problem is simply

$$q(x,t) = q_0(x - ut) \tag{3.3}$$

for $t \geq 0$, as can be easily verified. As time evolves, the initial data simply propagates unchanged to the right (if $u > 0$) or left (if $u < 0$) with velocity u. The solution $q(x,t)$ is constant along each of the rays $x - ut = x_0$, which are known as the **characteristics** of the equation.

Note that the characteristics are curves in the x-t plane satisfying the ordinary differential equations $x'(t) = u,\ x(0) = x_0$. If we differentiate $q(x,t)$ along one of these curves to find the rate of change of q along the characteristic, we find that

$$\begin{aligned} \frac{\mathrm{d}}{\mathrm{d}t} q(x(t),t) &= \frac{\partial}{\partial t} q(x(t),t) + \frac{\partial}{\partial x} q(x(t),t)\, x'(t) \\ &= q_t + uq_x \\ &= 0 , \end{aligned} \tag{3.4}$$

confirming that q is constant along these characteristics.

3.1.2 Burgers' Equation. Now consider the nonlinear scalar equation

$$q_t + f(q)_x = 0 , \tag{3.5}$$

where $f(q)$ is a nonlinear function of q. We will assume for the most part that $f(q)$ is a convex function, $f''(q) > 0$ for all q (or, equally well, f is concave with $f''(q) < 0\ \forall q$). The convexity assumption corresponds to a "genuine nonlinearity" assumption for systems of equations that holds in many important cases, such as the Euler equations. The nonconvex case is also important in some applications (e.g., in MHD, see Sect. 8.) but is more complicated mathematically.

By far the most famous model problem in this field is **Burgers' equation**, in which $f(q) = \frac{1}{2}q^2$, so (3.5) becomes

$$q_t + qq_x = 0 . \tag{3.6}$$

Actually this should be called the "inviscid Burgers' equation", since the equation originally studied by Burgers also includes a viscous term:

$$q_t + qq_x = \varepsilon q_{xx} . \tag{3.7}$$

This is about the simplest model that includes the nonlinear and viscous effects of fluid dynamics.

Consider the inviscid equation (3.6) with smooth initial data. For small time, a solution can be constructed by following characteristics. Note that (3.6) looks like an advection equation, but with the advection velocity q equal to the value of the advected quantity. The characteristics satisfy

$$x'(t) = q(x(t), t) \tag{3.8}$$

and along each characteristic q is constant, since

$$\begin{aligned} \frac{\mathrm{d}}{\mathrm{d}t}q(x(t),t) &= \frac{\partial}{\partial t}q(x(t),t) + \frac{\partial}{\partial x}q(x(t),t)\,x'(t) \\ &= q_t + qq_x \\ &= 0 . \end{aligned} \tag{3.9}$$

Moreover, since q is constant on each characteristic, the slope $x'(t)$ is constant by (3.8) and so the characteristics are straight lines, determined by the initial data.

If the initial data is smooth, then this can be used to determine the solution $q(x,t)$ for small enough t that characteristics do not cross: for each (x,t) we can solve the equation

$$x = \xi + q(\xi, 0)t \tag{3.10}$$

for ξ and then

$$q(x,t) = q(\xi, 0) . \tag{3.11}$$

3.1.3 Shock Formation. For larger t the equation (3.10) may not have a unique solution. This happens when the characteristics cross, as will eventually happen if $q_x(x,0)$ is negative at any point. At the time T_b where the characteristics first cross, the function $q(x,t)$ has an infinite slope — the wave "breaks" and a shock forms. Beyond this point there is no classical solution of the PDE, and the weak solution we wish to determine becomes discontinuous.

For times beyond the breaking time some of the characteristics have crossed and so there are points x where there are three characteristics leading back to $t = 0$. One can view the "solution" q at such a time as a triple-valued function. However, the density of a gas cannot possibly be triple valued at a point. We can determine the correct physical behavior by adopting the **vanishing viscosity** approach. The equation (3.6) is a model of (3.7) valid only for small ε and smooth q. When it breaks down, we must return to (3.7). If the initial data is smooth and ε very small, then before the wave begins to break the εq_{xx} term is negligible compared to the other terms and the solutions to the two PDEs look nearly identical. However, as the wave begins to break, the second derivative term q_{xx} grows much faster than q_x, and at some point the εq_{xx} term is comparable to the other terms and begins to play a role. This term keeps the solution smooth for all time, preventing the breakdown of solutions that occurs for the hyperbolic problem. This smooth solution has a steep transition zone where the viscous term is important. As $\varepsilon \to 0$ this zone becomes sharper and approaches the discontinuous solution known as a shock. It is this vanishing-viscosity solution that we hope to capture by solving the inviscid equation.

3.1.4 Weak Solutions. A natural way to define a generalized solution of the inviscid equation that does not require differentiability is to go back to the integral form of the conservation law, and say that $q(x,t)$ is a generalized solution if (2.5) is satisfied for all $x_1,\ x_2,\ t_1,\ t_2$.

There is another approach that results in a different integral formulation that is often more convenient to work with. This is a mathematical technique that can be applied more generally to rewrite a differential equation in a form where less smoothness is required to define a "solution". The basic idea is to take the PDE, multiply by a smooth "test function", integrate one or more times over some domain, and then use integration by parts to move derivatives off the function q and onto the smooth test function. The result is an equation involving fewer derivatives on q, and hence requiring less smoothness.

In our case we will use test functions $\phi \in C_0^1(\mathbb{R} \times \mathbb{R})$. Here C_0^1 is the space of functions that are continuously differentiable with compact support. If we multiply $q_t + f_x = 0$ by $\phi(x,t)$ and then integrate over space and time, we obtain

$$\int_0^{\infty}\int_{-\infty}^{+\infty} [\phi q_t + \phi f(q)_x]\, \mathrm{d}x\, \mathrm{d}t = 0 \,. \tag{3.12}$$

Now integrate by parts, yielding

$$\int_0^{\infty}\int_{-\infty}^{+\infty} [\phi_t q + \phi_x f(q)]\, \mathrm{d}x\, \mathrm{d}t = -\int_{-\infty}^{\infty} \phi(x,0)q(x,0)\, \mathrm{d}x \,. \tag{3.13}$$

Note that nearly all the boundary terms which normally arise through integration by parts drop out due to the requirement that ϕ have compact support, and hence vanishes at infinity. The remaining boundary term brings in the initial conditions of the PDE, which must still play a role in our weak formulation.

Definition 3.1. *The function $q(x,t)$ is called a weak solution of the conservation law if (3.13) holds for all functions $\phi \in C_0^1(\mathbb{R} \times \mathbb{R})$.*

The vanishing-viscosity generalized solution we defined above is a weak solution in the sense of (3.13), and so this definition includes the solution we are looking for. Unfortunately, weak solutions are often not unique, and so an additional problem is to identify *which* weak solution is the physically correct vanishing-viscosity solution. Again, one would like to avoid working with the viscous equation directly, but it turns out that there are other conditions one can impose on weak solutions that are easier to check and will also pick out the correct solution. These are usually called **entropy conditions** by analogy with the gas dynamics case, where a discontinuity is physically realistic only if the entropy of the gas *increases* as it crosses the shock.

3.1.5 The Riemann Problem. The conservation law together with piecewise constant data having a single discontinuity is known as the Riemann problem. As an example, consider Burgers' equation $q_t + qq_x = 0$ with piecewise constant initial data

$$q(x,0) = \begin{cases} q_l & \text{if } x < 0 \\ q_r & \text{if } x > 0\,. \end{cases} \tag{3.14}$$

The form of the solution depends on the relation between q_l and q_r.

Case I. $q_l > q_r$.

In this case there is a unique weak solution,

$$q(x,t) = \begin{cases} q_l & \text{if } x < st \\ q_r & \text{if } x > st\,, \end{cases} \tag{3.15}$$

where

$$s = (q_l + q_r)/2 \tag{3.16}$$

is the **shock speed**, the speed at which the discontinuity travels. A general expression for the shock speed will be derived below. Note that characteristics in each of the regions where q is constant go *into* the shock (see Fig. 3.1) as time advances.

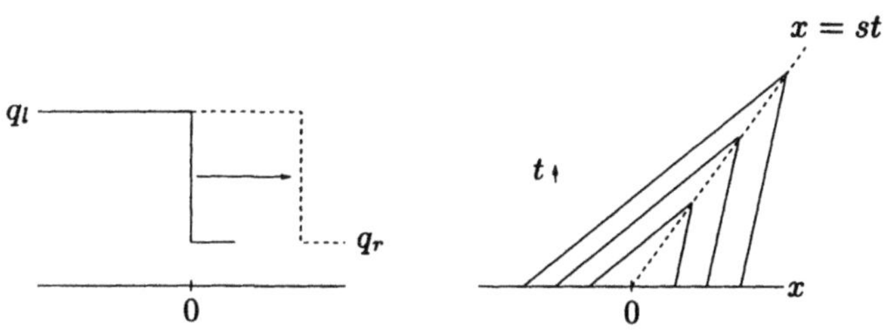

Fig. 3.1. Shock wave

Case II. $q_l < q_r$.

In this case there are infinitely many weak solutions. One of these is again (3.15), (3.16) in which the discontinuity propagates with speed s. Note that characteristics now go *out* of the shock (Fig. 3.2(a)) and that this solution is

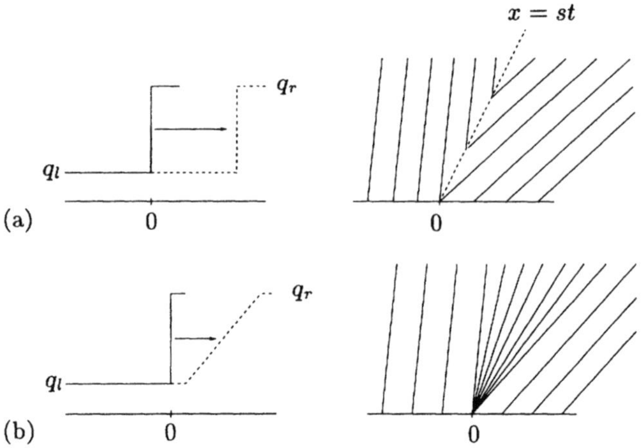

Fig. 3.2. (**a**) Entropy-violating shock. (**b**) Rarefaction wave

not stable to perturbations. If the data is smeared out slightly, or if a small amount of viscosity is added to the equation, the solution changes completely.

Another weak solution is the **rarefaction wave**

$$q(x,t) = \begin{cases} q_l & \text{if } x < q_l t \\ x/t & \text{if } q_l t \le x \le q_r t \\ q_r & \text{if } x > q_r t\ . \end{cases} \tag{3.17}$$

This solution is stable to perturbations and is in fact the vanishing-viscosity generalized solution (Fig. 3.2(b)).

3.1.6 Shock Speed. The propagating shock solution (3.15) is a weak solution to Burgers' equation only if the speed of propagation is given by (3.16). The same discontinuity propagating at a different speed would not be a weak solution.

The speed of propagation can be determined by conservation. The relation between the shock speed s and the states q_l and q_r is called the **Rankine–Hugoniot jump condition**:

$$f(q_l) - f(q_r) = s(q_l - q_r)\ . \tag{3.18}$$

For scalar problems this gives simply

$$s = \frac{f(q_l) - f(q_r)}{q_l - q_r} = \frac{[f]}{[q]}\ , \tag{3.19}$$

where $[\cdot]$ indicates the jump in some quantity across the discontinuity. Note that any jump is allowed, provided the speed is related via (3.19).

For systems of equations, $q_l - q_r$ and $f(q_r) - f(q_l)$ are both vectors while s is still a scalar. Now we cannot always solve for s to make (3.18) hold. Instead, only certain jumps $q_l - q_r$ are allowed, namely those for which the vectors $f(q_l) - f(q_r)$ and $q_l - q_r$ are linearly dependent.

For a linear system with $f(q) = Aq$, (3.18) gives

$$A(q_l - q_r) = s(q_l - q_r) , \tag{3.20}$$

i.e., $q_l - q_r$ must be an eigenvector of the matrix A and s is the associated eigenvalue. For a linear system, these eigenvalues are the characteristic speeds of the system. Thus discontinuities can propagate only along characteristics, just as for the scalar advection equation.

3.2 Linear Hyperbolic Systems

We now begin to investigate systems of equations. We start with constant coefficient linear systems. Here we can solve the equations explicitly by transforming to characteristic variables. We will also obtain explicit solutions of the Riemann problem and introduce a "phase space" interpretation that will be very useful in our study of nonlinear systems.

Consider the linear system

$$\begin{aligned} q_t + Aq_x &= 0 , \\ q(x,0) &= q_0(x) , \end{aligned} \tag{3.21}$$

where $q : \mathbb{R} \times \mathbb{R} \to \mathbb{R}^m$ and $A \in \mathbb{R}^{m\times m}$ is a constant matrix. This is a system of conservation laws with the flux function $f(q) = Aq$. This system is called **hyperbolic** if A is diagonalizable with real eigenvalues, so that we can decompose

$$A = R\Lambda R^{-1} , \tag{3.22}$$

where $\Lambda = \operatorname{diag}(\lambda_1, \lambda_2, \ldots, \lambda_m)$ is a diagonal matrix of eigenvalues and $R = [r_1|r_2|\cdots|r_m]$ is the matrix of right eigenvectors. Note that $AR = R\Lambda$, i.e.,

$$Ar_p = \lambda_p r_p \quad \text{for } p = 1,\ 2,\ \ldots,\ m . \tag{3.23}$$

The system is called **strictly hyperbolic** if the eigenvalues are distinct. We will always make this assumption as well.

3.2.1 Characteristic Variables. We can solve (3.21) by changing to the "characteristic variables"

$$v = R^{-1}q . \tag{3.24}$$

Multiplying (3.21) by R^{-1} and using (3.22) gives

$$v_t + \Lambda v_x = 0 . \tag{3.25}$$

Since Λ is diagonal, this decouples into m independent scalar equations. Each of these is a constant coefficient linear advection equation, with solution

$$v_p(x,t) = v_p(x - \lambda_p t, 0) . \tag{3.26}$$

Since $v = R^{-1}q$, the initial data for v_p is simply the pth component of the vector

$$v(x,0) = R^{-1}q_0(x) . \tag{3.27}$$

The solution to the original system is finally recovered via (3.24):

$$q(x,t) = Rv(x,t) . \tag{3.28}$$

Note that the value $v_p(x,t)$ is the coefficient of r_p in an eigenvector expansion of the vector $q(x,t)$, i.e., (3.28) can be written out as

$$q(x,t) = \sum_{i=1}^{m} v_p(x,t) r_p . \tag{3.29}$$

Combining this with the solutions (3.26) of the decoupled scalar equations gives

$$q(x,t) = \sum_{p=1}^{m} v_p(x - \lambda_p t, 0) r_p . \tag{3.30}$$

Note that $q(x,t)$ depends only on the initial data at the m points $x - \lambda_p t$, so the domain of dependence is given by $\mathcal{D}(\bar{x},\bar{t}) = \{x = \bar{x} - \lambda_p \bar{t},\ p = 1,\ 2,\ \ldots,\ m\}$.

The curves $x = x_0 + \lambda_p t$ satisfying $x'(t) = \lambda_p$ are the "characteristics of the pth family", or simply "p-characteristics". These are straight lines in the case of a constant coefficient system. Note that for a strictly hyperbolic system, m distinct characteristic curves pass through each point in the x-t plane. The coefficient $v_p(x,t)$ of the eigenvector r_p in the eigenvector expansion (3.29) of $q(x,t)$ is constant along any p-characteristic.

3.2.2 The Riemann Problem. For the constant coefficient linear system, the Riemann problem can be explicitly solved. We will see shortly that the solution to a nonlinear Riemann problem has a simple structure which is quite similar to the structure of this linear solution, and so it is worthwhile studying the linear case in some detail.

The Riemann problem consists of the equation $q_t + Aq_x = 0$ together with piecewise constant initial data of the form

$$q(x,0) = \begin{cases} q_l & x < 0 \\ q_r & x > 0 . \end{cases} \tag{3.31}$$

Recall that the general solution to the linear problem is given by (3.30). For the Riemann problem we can simplify the notation if we decompose q_l and q_r as

$$q_l = \sum_{p=1}^{m} v_p^l r_p , \qquad q_r = \sum_{p=1}^{m} v_p^r r_p . \tag{3.32}$$

Then

$$v_p(x,0) = \begin{cases} v_p^l & x < 0 \\ v_p^r & x > 0 \end{cases} \tag{3.33}$$

and so

$$v_p(x,t) = \begin{cases} v_p^l & \text{if } x - \lambda_p t < 0 \\ v_p^r & \text{if } x - \lambda_p t > 0 \,. \end{cases} \tag{3.34}$$

If we let $P(x,t)$ be the maximum value of p for which $x - \lambda_p t > 0$, then

$$q(x,t) = \sum_{p=1}^{P(x,t)} v_p^r r_p + \sum_{p=P(x,t)+1}^{m} v_p^l r_p \,. \tag{3.35}$$

The determination of $q(x,t)$ at a given point is illustrated in Fig. 3.3. In the case shown, $v_1 = v_1^r$ while $v_2 = v_2^l$ and $v_3 = v_3^l$. The solution at the point illustrated is thus

$$q(x,t) = v_1^r r_1 + v_2^l r_2 + v_3^l r_3 \,. \tag{3.36}$$

Note that the solution is the same at any point in the wedge between the $x' = \lambda_1$ and $x' = \lambda_2$ characteristics. As we cross the pth characteristic, the value of $x - \lambda_p t$ passes through 0 and the corresponding v_p jumps from v_p^l to v_p^r. The other coefficients v_i $(i \neq j)$ remain constant.

The solution is constant in each of the wedges as shown in Fig. 3.4. Across the pth characteristic the solution jumps with the jump given by

$$[q] = (v_p^r - v_p^l) r_p \,. \tag{3.37}$$

Note that these jumps satisfy the Rankine–Hugoniot conditions (3.18), since $f(q) = Aq$ leads to

$$[f] = A[q] = (v_p^r - v_p^l) A r_p = \lambda_p [q]$$

and λ_p is precisely the speed of propagation of this jump. The solution $q(x,t)$ in (3.35) can alternatively be written in terms of these jumps as

$$q(x,t) = q_l + \sum_{\lambda_p < x/t} (v_p^r - v_p^l) r_p \tag{3.38}$$

$$= q_r - \sum_{\lambda_p > x/t} (v_p^r - v_p^l) r_p \,. \tag{3.39}$$

It might happen that the initial jump $q_r - q_l$ is already an eigenvector of A, if $q_r - q_l = (v_i^r - v_i^l) r_i$ for some i. In this case $v_p^l = v_p^r$ for $p \neq i$. Then this discontinuity simply propagates with speed λ_i, and the other characteristics carry jumps of zero strength.

In general this is not the case, however, and the jump $q_r - q_l$ cannot propagate as a single discontinuity with any speed without violating the Rankine–Hugoniot condition. We can view "solving the Riemann problem" as finding a way to split up the jump $q_r - q_l$ into a sum of jumps,

$$q_r - q_l = (v_1^r - v_1^l) r_1 + \cdots + (v_m^r - v_m^l) r_m \equiv \alpha_1 r_1 + \cdots + \alpha_m r_m \,, \tag{3.40}$$

each of which *can* propagate at an appropriate speed λ_i with the Rankine–Hugoniot condition satisfied.

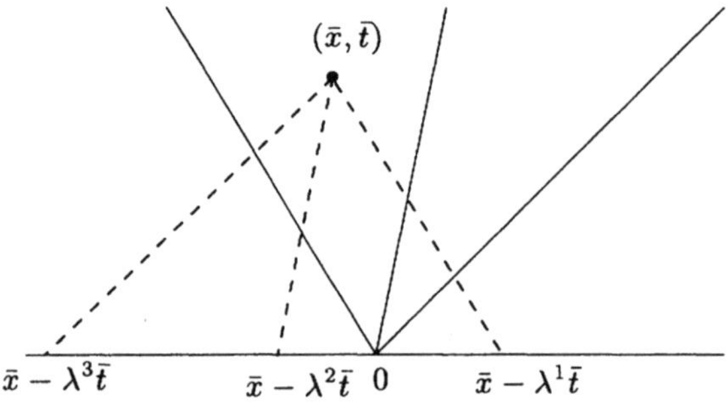

Fig. 3.3. Construction of the solution to the Riemann problem at $(\bar{x}, \bar{t})$. We trace back along the p'th characteristic to determine the coefficient of r_p depending on whether this lies in the left state or right state initially

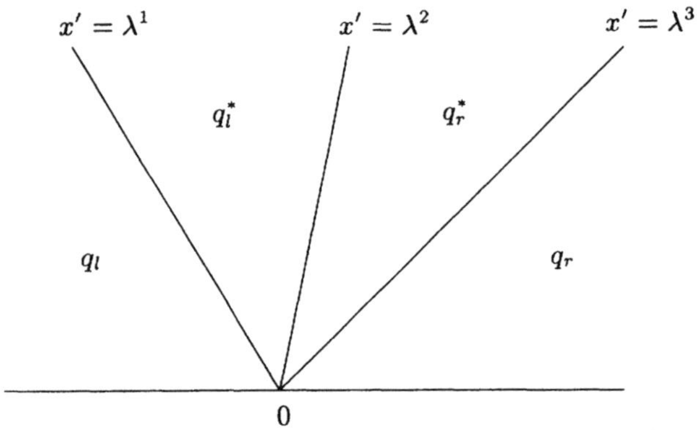

Fig. 3.4. Values of solution q in each wedge of x–t plane:

$$q_l = v_1^l r_1 + v_2^l r_2 + v_3^l r_3, \qquad q_r^* = v_1^r r_1 + v_2^r r_2 + v_3^l r_3,$$
$$q_l^* = v_1^r r_1 + v_2^l r_2 + v_3^l r_3, \qquad q_r = v_1^r r_1 + v_2^r r_2 + v_3^r r_3.$$

Note that the jump across each discontinuity in the solution is an eigenvector of A

For nonlinear systems we solve the Riemann problem in much the same way: The jump $q_r - q_l$ will usually not have the property that $[f]$ is a scalar multiple of $[q]$, but we can attempt to find a way to split this jump up as a sum of jumps, across each of which this property does hold. (Although life is complicated by the fact that we may need to introduce rarefaction waves as well as shocks.) In studying the solution of the Riemann problem, the jump in the pth family, traveling at constant speed λ_p, is often called the p-wave.

3.2.3 The Phase Plane. For systems of two equations, it is illuminating to view this splitting in the phase plane. This is simply the q_1–q_2 plane, where $q = (q_1, q_2)$. Each vector $q(x, t)$ is represented by a point in this plane. In particular, q_l and q_r are points in this plane and a discontinuity with left and right states q_l and q_r can propagate as a single discontinuity only if $q_r - q_l$

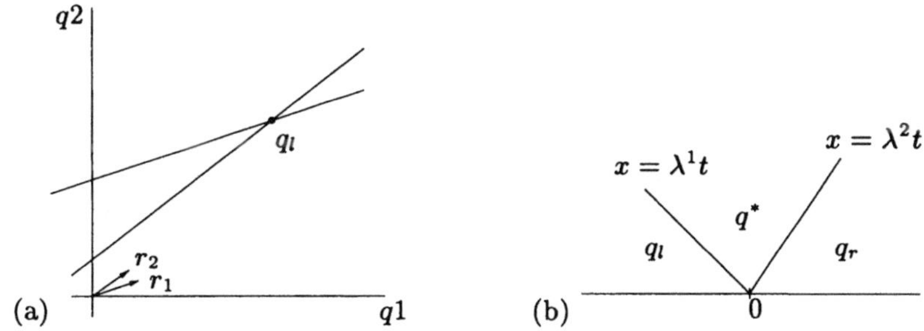

Fig. 3.5. (**a**) The Hugoniot locus of the state q_l consists of all states that differ from q_l by a scalar multiple of r_1 or r_2. (**b**) Solution to the Riemann problem in the x–t plane

is an eigenvector of A, which means that the line segment from q_l to q_r must be parallel to the eigenvector r_1 or r_2. Figure 3.5 shows an example. For the state q_l illustrated there, the jump from q_l to q_r can propagate as a single discontinuity if and only if q_r lies on one of the two lines drawn through q_l in the drection r_1 and r_2. These lines give the locus of all points that can be connected to q_l by a 1-wave or a 2-wave. This set of states is called the **Hugoniot locus**. We will see that there is a direct generalization of this to nonlinear systems in the next chapter.

Similarly, there is a Hugoniot locus through any point q_r that gives the set of all points q_l that can be connected to q_r by an elementary p-wave. These curves are again in the directions r_1 and r_2.

For a general Riemann problem with arbitrary q_l and q_r, the solution consists of two discontinuities travelling with speeds λ_1 and λ_2, with a new constant state in between that we will call q^*. By the discussion above,

$$q^* = v_1^r r_1 + v_2^l r_2 \,, \tag{3.41}$$

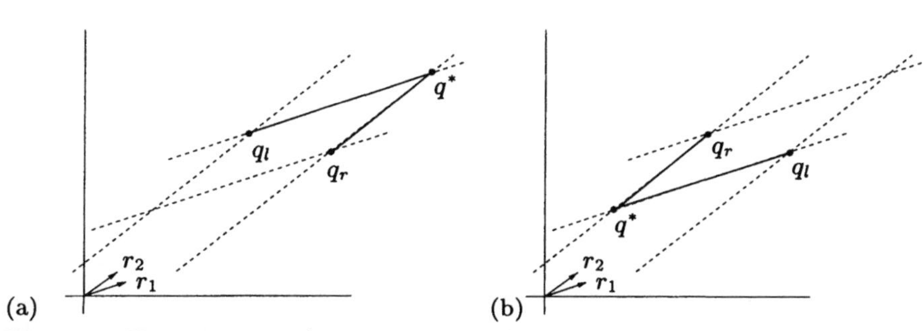

Fig. 3.6. The new state q^* arising in the solution to the Riemann problem for two different choices of q_l and q_r. In each case the jump from q_l to q^* lies in the direction of the eigenvalue r_1 corresponding to the slower speed, while the jump from q^* to q_r lies in the direction of the eigenvalue r_2

so that $q^* - q_l = \alpha_1 r_1$ and $q_r - q^* = \alpha_2 r_2$. The location of q^* in the phase plane must be where the 1-wave locus through q_l intersects the 2-wave locus through q_r. This is illustrated in Fig. 3.6a.

Note that if we interchange q_r and q_l in this picture, the location of q^* changes as illustrated in Fig. 3.6(b). In each case we travel from q_l to q_r by first going in the direction r_1 and then in the direction r_2. This is required by the fact that $\lambda_1 < \lambda_2$ since clearly the jump between q_l and q^* must travel slower than the jump between q^* and q_r if we are to obtain a single-valued solution.

For systems with more than two equations, the same interpretation is possible but becomes harder to draw since the phase space is now m dimensional. Since the m eigenvectors r_p are linearly independent, we can decompose any jump $q_r - q_l$ into the sum of jumps in these directions, obtaining a piecewise linear path from q_l to q_r in m-dimensional space.

3.3 Nonlinear Systems

Now consider a nonlinear system of conservation laws

$$q_t + f(q)_x = 0 , \tag{3.42}$$

where $q : \mathbb{R} \times \mathbb{R} \to \mathbb{R}^m$ and $f : \mathbb{R}^m \to \mathbb{R}^m$. This can be written in the quasilinear form

$$q_t + A(q)q_x = 0 , \tag{3.43}$$

where $A(q) = f'(q)$ is the $m \times m$ Jacobian matrix. Again the system is **hyperbolic** if $A(q)$ is diagonalizable with real eigenvalues for all values of q, at least in some range where the solution is known to lie, and **strictly hyperbolic** if the eigenvalues are distinct for all q. We will assume that the system is strictly hyperbolic, and order the eigenvalues of A in increasing order,

$$\lambda_1 < \lambda_2 < \cdots < \lambda_m . \tag{3.44}$$

Since the eigenvalues are distinct, the eigenvectors are linearly dependent. We choose a particular basis for these eigenvectors, $\{r_p(q)\}_{p=1}^m$, usually chosen to be normalized in some manner, e.g., $||r_p(q)|| \equiv 1$.

In the previous section we constructed the solution to the general Riemann problem for a linear hyperbolic system of conservation laws. Our next goal is to perform a similar construction for the nonlinear Riemann problem. In the linear case the solution consists of m waves, which are simply discontinuities traveling at the characteristic velocities of the linear system. In the nonlinear case our experience with the scalar equation leads us to expect more possibilities. In particular, the physically relevant vanishing-viscosity solution may contain rarefaction waves as well as discontinuities. We will first ignore the entropy condition and ask a simpler question: is it possible to construct a weak solution of the Riemann problem consisting only of m

discontinuities propagating with constant speeds $s_1 < s_2 < \cdots < s_m$? As we will see, the answer is yes for $\|q_l - q_r\|$ sufficiently small.

3.3.1 The Hugoniot Locus. Recall that if a discontinuity propagating with speed s has constant values $\hat{q}$ and w on either side of the discontinuity, then the Rankine–Hugoniot jump condition must hold,

$$f(w) - f(\hat{q}) = s(w - \hat{q}) . \tag{3.45}$$

Now suppose we fix the point $\hat{q} \in \mathbb{R}^m$ and attempt to determine the set of all points w which can be connected to $\hat{q}$ by a discontinuity satisfying (3.45) for some s. This gives a system of m equations in $m + 1$ unknowns: the m components of w, and s. This leads us to expect one-parameter families of solutions.

We know that in the *linear* case there are indeed m such families for any $\hat{q}$. In the pth family the jump $w - \hat{q}$ is some scalar multiple of r_p, the pth eigenvector of A. We can parameterize these families of solutions using this scalar multiple, say ξ, and we obtain the following solution curves:

$$\begin{aligned} w_p(\xi;\hat{q}) &= \hat{q} + \xi r_p , \\ s_p(\xi;\hat{q}) &= \lambda_p \end{aligned}$$

for $p = 1,\ 2,\ \ldots,\ m$. Note that $w_p(0;\hat{q}) = \hat{q}$ for each p and so through the point $\hat{q}$ in phase space there are m curves (straight lines in fact) of possible solutions. This is illustrated in Fig. 3.6 for the case $m = 2$. The two lines through each point are the states that can be connected by a discontinuity with jump proportional to r_1 or r_2.

In the nonlinear case we also obtain m curves through any point $\hat{q}$, one for each characteristic family. We again parameterize these curves by $w_p(\xi;\hat{q})$ with $w_p(0;\hat{q}) = \hat{q}$ and let $s_p(\xi;\hat{q})$ denote the corresponding speed. To simplify notation, we will frequently write these as simply $w_p(\xi),\ s_p(\xi)$ when the point $\hat{q}$ is clearly understood.

The Rankine–Hugoniot condition gives

$$f(w_p(\xi)) - f(\hat{q}) = s_p(\xi)(w_p(\xi) - \hat{q}) . \tag{3.46}$$

Differentiating this expression with respect to ξ and setting $\xi = 0$ gives

$$f'(\hat{q})w_p'(0) = s_p(0)w_p'(0) , \tag{3.47}$$

so that $w_p'(0)$ must be a scalar multiple of the eigenvector $r_p(\hat{q})$ of $f'(\hat{q})$, while $s_p(0) = \lambda_p(\hat{q})$. The curve $w_p(\xi)$ is thus tangent to $r_p(\hat{q})$ at the point $\hat{q}$. For a system of $m = 2$ equations, this is easily illustrated in the phase plane. An example for the isothermal equations of gas dynamics is shown in Fig. 3.7. See [141] for details on the construction of these curves.

For smooth f, it can be shown using the implicit function theorem that these solution curves exist locally in a neighborhood of $\hat{q}$, and that the functions w_p and s_p are smooth. See Lax [136] or Smoller [221] for details. These curves are called Hugoniot curves. The set of all points on these curves is

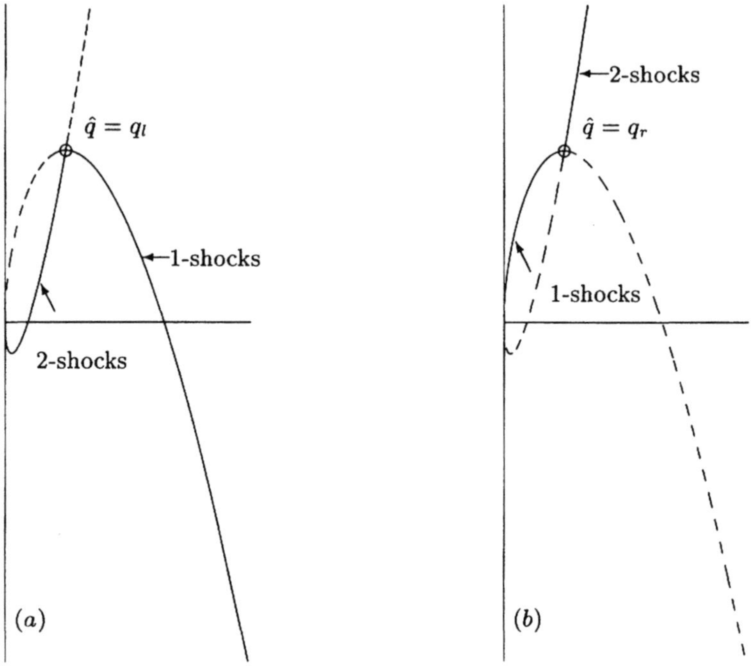

Fig. 3.7. (**a**) States q_r that can be connected to $q_l = \hat{q}$ by an entropy-satisfying shock in the density-momentum phase plane for the isothermal equations. (**b**) States q_l that can be connected to $q_r = \hat{q}$ by an entropy-satisfying shock. In each case the entropy-violating branches of the Hugoniot locus are shown as dashed lines

often collectively called the **Hugoniot locus** for the point $\hat{q}$. If $\tilde{q}_p$ lies on the pth Hugoniot curve through $\hat{q}$, then we say that $\hat{q}$ and $\tilde{q}_p$ are connected by a p-shock.

3.3.2 All-Shock Solution of the Riemann Problem. Now suppose that we wish to solve the Riemann problem with left and right states q_l and q_r (and recall that we are ignoring the entropy condition at this point). Just as in the linear case, we can accomplish this by finding an intermediate state q^* such that q_l and q^* are connected by a discontinuity satisfying the Rankine–Hugoniot condition, and so are q^* and q_r. Graphically we accomplish this by drawing the Hugoniot locus for each of the states q_l and q_r and looking for intersections. See Fig. 3.8 for an example with the isothermal equations.

More generally, for a system of m equations we can attempt to solve the Riemann problem by finding a sequence of states $q_1, q_2, \ldots, q_{m-1}$ such that q_l is connected to q_1 by a 1-shock, q_1 is connected to q_2 by a 2-shock, and so on, with q_{m-1} connected to q_r by an m-shock. If q_l and q_r are sufficiently close together then this can always be achieved. Lax proved a stronger version of this in his fundamental paper [135]. (By considering rarefaction waves also,

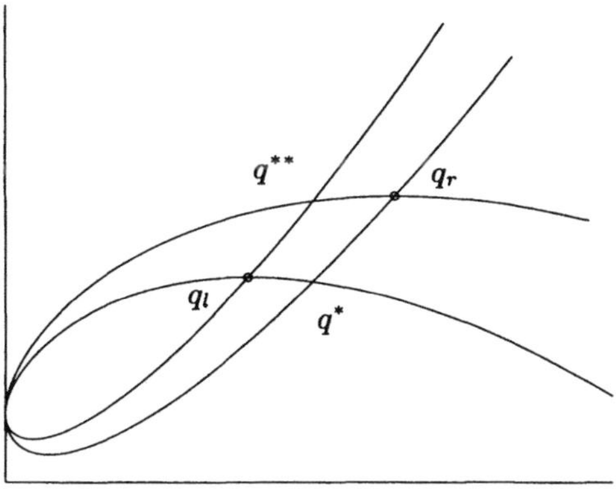

Fig. 3.8. Construction of a 2-shock solution to the Riemann problem for the isothermal equations with left and right states q_l and q_r. The intermediate state q^* is obtained between the two shocks. If the states q_l and q_r were switched in the Riemann problem, then the intermediate state q_{**} would be obtained instead

the entropy satisfying solution can be constructed in a similar manner, as we will do in Sect. 3.3.6)

3.3.3 Genuine Nonlinearity. The pth characteristic field is said to be **genuinely nonlinear** (a natural generalization of the *convex* scalar equation) if

$$\nabla\lambda_p(q) \cdot r_p(q) \neq 0 \qquad \text{for all } q \ , \tag{3.48}$$

where $\nabla\lambda_p(q) = (\partial\lambda_p/\partial q_1, \ \ldots, \ \partial\lambda_p/\partial q_m)$ is the gradient of $\lambda_p(q)$. Note that in the scalar case, $m = 1$ and $\lambda_1(q) = f'(q)$ while $r_1(q) = 1$ for all q. The condition (3.48) reduces to the convexity requirement $f''(q) \neq 0 \ \forall q$. This implies that the characteristic speed $f'(q)$ is monotonically increasing or decreasing as q varies, and leads to a relatively simple solution of the Riemann problem.

For a system of equations, (3.48) implies that $\lambda_p(q)$ is monotonically increasing or decreasing as q varies along an integral curve of the vector field $r_p(q)$. An integral curve of $r_p(q)$ is a curve with the property that it is everywhere tangent to $r_p(q)$. This will be discussed in detail shortly, where we will see that through a rarefaction wave q varies along an integral curve. Since monotonicity of the propagation speed λ_p is clearly required through a rarefaction wave, genuine nonlinearity is a natural assumption.

For the Euler equations the two acoustic waves are genuinely nonlinear.

3.3.4 The Lax Entropy Condition. In defining the Hugoniot locus above, we ignored the question of whether a given discontinuity is physically relevant. The state $\tilde{q}$ is in the Hugoniot locus of $\hat{q}$ if the jump satisfies the Rankine–Hugoniot condition, regardless of whether this jump could exist in a vanishing viscosity solution. We would now like to define an entropy condition that can be applied directly to a discontinuous weak solution to determine whether the jumps should be allowed. Lax [135] proposed a simple entropy condition for systems of equations that are genuinely nonlinear. Lax's entropy condition says that a jump in the pth field (from q_l to q_r, say) is admissible only if

$$\lambda_p(q_l) > s > \lambda_p(q_r) , \tag{3.49}$$

where s is again the shock speed. Characteristics in the pth family disappear into the shock as time advances, just as in the scalar case. Note that for a system of m equations we thus expect to have a total of $m+1$ characteristics impinging on the shock, those from families 1 through p from the left and from families p through m from the right.

3.3.5 Linear Degeneracy. The assumption of genuine nonlinearity is obviously violated for a constant coefficient linear system, in which $\lambda_p(q)$ is constant and hence $\nabla\lambda_p \equiv 0$. More generally, for a nonlinear system it might happen that in one of the characteristic fields the eigenvalue $\lambda_p(q)$ is constant along integral curves of this field, and hence

$$\nabla\lambda_p(q) \cdot r_p(q) \equiv 0 \qquad \forall q . \tag{3.50}$$

(Of course the value of $\lambda_p(q)$ might vary from one integral curve to the next.) In this case we say that the pth field is **linearly degenerate**.

In the Euler equations, the contact discontinuity is linearly degenerate. The characteristic speed is u, the fluid velocity, which is the same on each side of an arbitrary contact discontinuity.

3.3.6 Rarefaction Waves and Integral Curves. We now turn our attention to rarefaction waves. All of the Riemann solutions considered so far have the following property: the solution is constant along all rays of the form $x = \xi t$. Consequently, the solution is a function of x/t alone, and is said to be a "similarity solution" of the PDE. A rarefaction wave solution to a system of equations also has this property and takes the form

$$q(x,t) = \begin{cases} q_l & x \le \xi_1 t \\ w(x/t) & \xi_1 t < x < \xi_2 t \\ q_r & x \ge \xi_2 t , \end{cases} \tag{3.51}$$

where w is a smooth function with $w(\xi_1) = q_l$ and $w(\xi_2) = q_r$.

When does a system of equations have a solution of this form? As in the case of shocks, for arbitrary states q_l and q_r there might not be a solution of this form. But in general, starting at each point q_l there are m curves consisting of points q_r which can be connected to q_l by a rarefaction wave. These turn out to be subsets of the integral curves of the vector fields $r_p(q)$.

An integral curve for $r_p(q)$ is a curve which has the property that the tangent to the curve at any point q lies in the direction $r_p(q)$. The existence of smooth curves of this form follows from smoothness of f and strict hyperbolicity, since $r_p(q)$ is then a smooth function of q. If $q_p(\xi)$ is a parameterization (for $\xi \in \mathbb{R}$) of an integral curve in the pth family, then the tangent vector is proportional to $r_p(q_p(\xi))$ at each point, *i.e.*,

$$q_p'(\xi) = \alpha(\xi) r_p(q_p(\xi)) \,, \tag{3.52}$$

where $\alpha(\xi)$ is some scalar factor.

A function of q that is constant along any integral curve is called a **Riemann invariant** of the system. Such functions can always be found for systems of two equations, and the integral curves are level sets of the invariants. Riemann invariants can be used to solve Riemann problems involving rarefaction waves.

To see that rarefaction curves lie along integral curves, and to explicitly determine the function $w(x/t)$ in (3.51), we differentiate $q(x,t) = w(x/t)$ to obtain

$$q_t(x,t) = -\frac{x}{t^2}\, w'(x/t) \,, \tag{3.53}$$

$$q_x(x,t) = \frac{1}{t}\, w'(x/t) \,, \tag{3.54}$$

so that $q_t + f'(q) q_x = 0$ yields

$$-\frac{x}{t^2} w'(x/t) + \frac{1}{t} f'(w(x/t)) w'(x/t) = 0 \,. \tag{3.55}$$

Multiplying by t and rearranging gives

$$f'(w(\xi)) w'(\xi) = \xi w'(\xi) \,, \tag{3.56}$$

where $\xi = x/t$. One possible solution of (3.56) is $w'(\xi) \equiv 0$, *i.e.*, w constant. Any constant function is a similarity solution of the conservation law, and indeed the rarefaction wave (3.51) takes this form for $\xi < \xi_1$ and $\xi > \xi_2$. In between, w is presumably smoothly varying and $w' \neq 0$. Then (3.56) says that $w'(\xi)$ must be proportional to some eigenvector $r_p(w(\xi))$ of $f'(w(\xi))$,

$$w'(\xi) = \alpha(\xi) r_p(w(\xi)) \tag{3.57}$$

and hence the values $w(\xi)$ all lie along some integral curve of r_p. In particular, the states $q_l = w(\xi_1)$ and $q_r = w(\xi_2)$ both lie on the same integral curve. This is a necessary condition for the existence of a rarefaction wave connecting q_l and q_r, but note that it is not sufficient. We need $\xi = x/t$ to be monotonically increasing as $w(\xi)$ moves from q_l to q_r along the integral curve; otherwise the function (3.51) is not single-valued. If this characteristic field is genuinely nonlinear, then the eigenvalue will vary monotonically along the integral curve.

Figure 3.9 shows the integral curves for the isothermal equations through the same state as shown in Fig. 3.7. See [141] for more details on the construction of these curves. Which points can be connected to this state by a physically realizable rarefaction wave depends on whether the fixed state is to be on the left or right, since the speed must increase from left to right. The realizable rarefaction waves are indicated by solid lines in the figure.

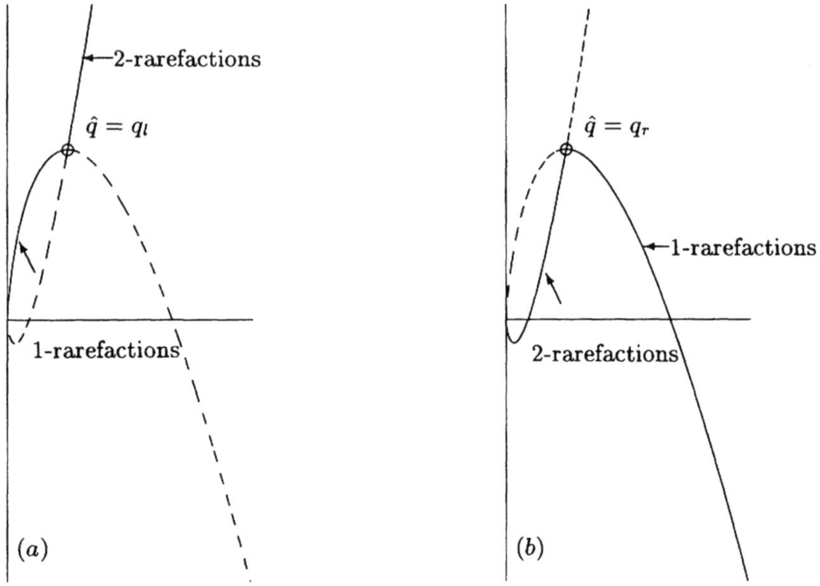

Fig. 3.9. (**a**) States q_r that can be connected to $q_l = \hat{q}$ by a rarefaction wave for the isothermal equations. (**b**) States q_l that can be connected to $q_r = \hat{q}$ by a rarefaction wave. In each case the full integral curves are shown as dashed lines

These integral curves look quite similar to the Hugoniot locus shown in Fig. 3.7. Locally, near the point $\hat{q}$, they must in fact be very similar. We know already that in the pth family each of these curves is tangent to $r_p(\hat{q})$ at $\hat{q}$. Moreover, it can be shown that the curvature of both curves is the same (See Lax [136]). The curves are not identical, however. Figure 3.10 shows a comparison of the two sets of curves through one particular state.

3.3.7 General Solution of the Riemann Problem. We can combine the appropriate curves from Fig. 3.7 and Fig. 3.9 to obtain a plot showing all states that can be connected to a given $\hat{q}$ by entropy-satisfying waves, either shocks or rarefactions. Again, the nature of this plot depends on whether $\hat{q}$ is to be the left state or right state, so we obtain two plots as shown in Fig. 3.11. Here $\mathcal{S}_p$ is used to denote the locus of states that can be connected by a p-shock and $\mathcal{R}_p$ is the locus of states that can be connected by a p-rarefaction. Notice that the shock and rarefaction curves match up smoothly (with the same slope and curvature) at the point $\hat{q}$.

To solve the general Riemann problem between two different states q_l and q_r, we simply superimpose the appropriate plots and look for the intersection q_m of a 1-wave curve from q_l and a 2-wave curve from q_r. An example for the isothermal equations is shown in Fig. 3.12. This is the same example considered in Fig. 3.8. We now see that the entropy-satisfying weak solution consists of a 1-shock from q_l to q_m followed by a 2-rarefaction from q_m to q_r.

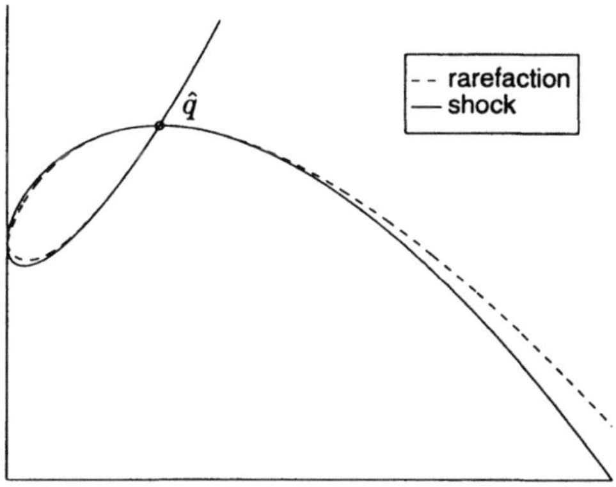

Fig. 3.10. Comparison of the states that can be connected to $\hat{q}$ through a rarefaction or a shock. The solid lines are the Hugoniot locus while the dashed lines are the integral curves

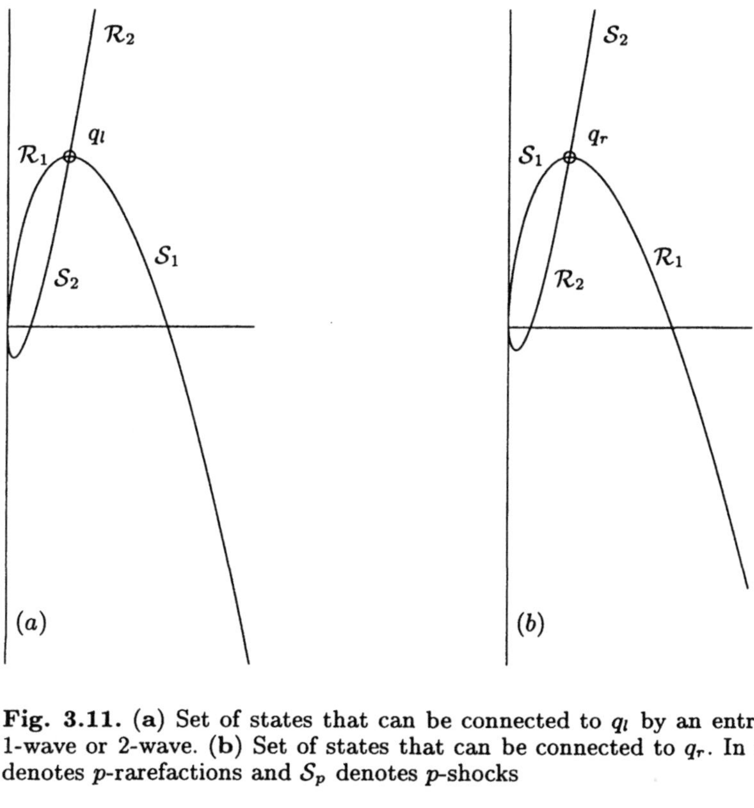

Fig. 3.11. (**a**) Set of states that can be connected to q_l by an entropy-satisfying 1-wave or 2-wave. (**b**) Set of states that can be connected to q_r. In each case, $\mathcal{R}_p$ denotes p-rarefactions and $\mathcal{S}_p$ denotes p-shocks

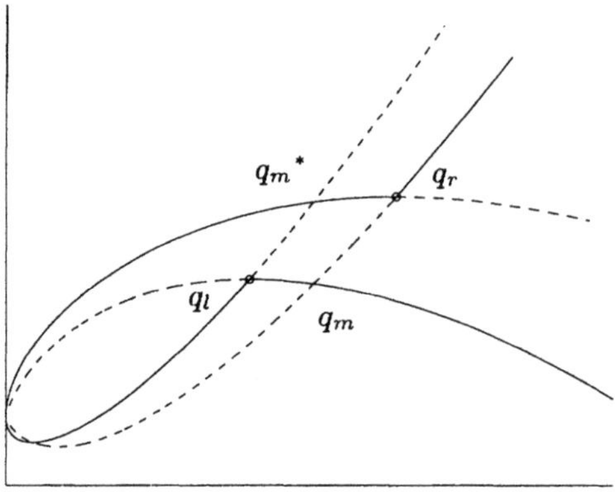

Fig. 3.12. Construction of a weak solution to the Riemann problem with left and right states q_l and q_r. Solid lines show the Hugoniot locus (shocks) and dashed lines show the integral curves (rarefactions)

3.4 The Riemann Problem for the Euler Equations

If we compute the Jacobian matrix $f'(q)$ from (2.16), with the polytropic equation of state (2.13), we obtain

$$f'(q) = \begin{bmatrix} 0 & 1 & 0 \\ -\frac{1}{2}(\gamma+1)u^2 & (3-\gamma)u & (\gamma-1) \\ -u(E+p)/\varrho + \frac{1}{2}(\gamma-1)u^3 & (E+p)/\varrho - (\gamma-1)u^2 & \gamma u \end{bmatrix}$$

The eigenvalues are

$$\lambda_1(q) = u - c\,, \quad \lambda_2(q) = u\,, \quad \lambda_3(q) = u + c\,, \tag{3.58}$$

where c is the sound speed,

$$c = \sqrt{\frac{\gamma p}{\varrho}}\,. \tag{3.59}$$

Of particular note in these equations is the fact that the second characteristic field is linearly degenerate. It is easy to check that

$$r_2(q) = \begin{bmatrix} 1 \\ u \\ \frac{1}{2}u^2 \end{bmatrix} \tag{3.60}$$

is an eigenvector of $f'(q)$ with eigenvalue $\lambda_2(q) = u = (\varrho u)/\varrho$. Since

$$\nabla\lambda_2(q) = \begin{bmatrix} -u/\varrho \\ 1/\varrho \\ 0 \end{bmatrix} \tag{3.61}$$

we find that $\nabla\lambda_2 \cdot r_2 \equiv 0$.

Since the second field is linearly degenerate, we can have neither rarefaction waves nor shocks in this field. Instead we have contact discontinuities, which are linear discontinuities that propagate with speed equal to the characteristic speed λ_2, which is simply the fluid velocity u. Across a contact discontinuity there is a jump in the density of the gas but the pressure and velocity are continuous.

The first and third characteristic fields are genuinely nonlinear and have behavior similar to the two characteristic fields in the isothermal equations. The solution to a Riemann problem typically has a contact discontinuity and two nonlinear waves, each of which might be either a shock or a rarefaction wave depending on q_l and q_r. A typical solution was shown in Fig. 2.1.

Because u and p are constant across the contact discontinuity, it is often easier to work in the variables (ϱ, u, p) rather than $(\varrho, \varrho u, E)$, although of course the jump conditions must be determined using the conserved variables. The resulting Hugoniot locus and integral curves can be transformed into (ϱ, u, p) space.

If the Riemann data is (ϱ_l, u_l, p_l) and (ϱ_r, u_r, p_r), then the two new constant states that appear in the Riemann solution will be denoted by $q_l^* = (\varrho_l^*, u^*, p^*)$ and $q_r^* = (\varrho_r^*, u^*, p^*)$ (see Fig. 2.1.). Note that across the 2-wave we know there is a jump only in density.

Solution of the Riemann problem proceeds in principle just as before. Given the states q_l and q_r in the phase space, we need to determine the two intermediate states in such a way that q_l and q_l^* are connected by a 1-wave, q_l^* and q_r^* are connected by a 2-wave, and finally q_r^* and q_r are connected by a 3-wave.

This procedure is greatly simplified by using the known structure of the solution. The 1-wave Hugoniot locus and integral curves can be parameterized by the pressure. Let $q_l = (\varrho_l, u_l, p_l)$ be the left state. Then for any choice of p_l^* we can determine values ϱ_l^* and u_l^* so that (ϱ_l, u_l, p_l) and $(\varrho_l^*, u_l^*, p_l^*)$ are connected by an admissible shock wave (if $p_l^* > p_l$) or by a rarefaction wave (if $p_l^* < p_l$). This defines in particular a function U^1 for the velocity u_l^* in terms of the pressure p_l^*, which depends of course on the left state q_l:

$$u_l^* = U^1(p_l^*; q_l) .$$

This function is defined by integrating the Riemann invariant through the rarefaction wave or by using the jump conditions across a shock wave.

Similarly, the 3-wave Hugoniot locus and integral curves can be parameterized by pressure, defining an analogous function U^3 so that

$$u_r^* = U^3(p_r^*; q_r) .$$

This gives the velocity in the state connected to q_r by a 3-wave when p_r^* is the pressure in this state.

Since the 2-wave is always a contact discontinuity, we know that u and p must be constant across this wave. We can thus reduce the solution process for the general Riemann problem to finding the single value $p^* = p_l^* = p_r^*$

(the intermediate pressure) so that the resulting velocities u_l^* and u_r^* are also equal:

$$U^1(p^*; q_l) = U^3(p^*; q_r) .$$

This is a single nonlinear equation for p^* which can be solved by Newton's method, for example. Then the intermediate velocity u^* can be determined by this common value of $U^1 = U^3$. The densities ϱ_l^* and ϱ_r^* are also determined from the parameterization of states based on $p_l^* = p_r^*$. In general $\varrho_l^* \neq \varrho_r^*$, but since any jump in density is allowed by the Hugoniot condition across the contact discontinuity, we have solved the problem.

Figure 3.13 shows the solution to a typical Riemann problem in the velocity-pressure plane, for a shock tube problem where the initial velocity is zero everywhere, initial density is 1 everywhere, and there is a jump in pressure from 100 to 20. The resulting intermediate state q^* is connected to q_l by a left-going rarefaction wave and connected to q_r by a right-going shock, similar to what was seen in Fig. 2.1.

For the adiabatic Euler equations, these curves are often called the **adiabats**. In Sect. 9.5 we will see that these adiabats change if the gas is moving at relativistic speeds.

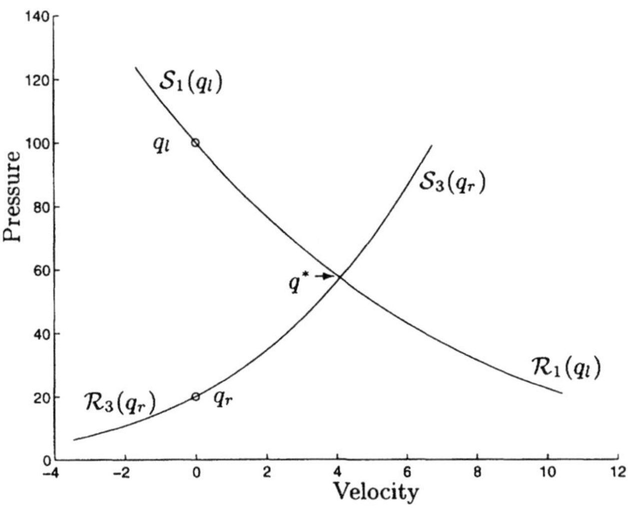

Fig. 3.13. Intersection of loci connecting q_l to q_r through a 1–rarefaction and a 3–shock for $p_l = 100, \varrho_l = 1, u_l = 0$ and $p_r = 20, \varrho_r = 1, u_r = 0$

Godunov first proposed a numerical method based on the solution of Riemann problems and presented one iterative method for computing p^* in his paper, [98] (also described in §12.15 of [192]). Chorin [52] describes an improvement of this method. More details on the solution of the Riemann problem can also be found in §81 of [60]. Solving the Riemann problem for non-ideal gases can be more difficult. See [57], for example.

4. Numerical Methods in One Dimension

4.1 Finite Difference Theory

For students not familiar with numerical methods, it may be useful to review some of the fundamental ideas and issues that arise in discretizing differential equations, before going into more detail on hyperbolic problems. This section contains a very brief overview of these issues in the context of *finite difference* discretization of ODE's and linear PDE's. More details in the same style can be found in [142]. Many elementary texts are available that cover this material in depth, *e.g.*, [115], [169], [192], [232]. The basic ideas in the context of hyperbolic systems are also discussed in [141].

After introducing some concepts in the context of finite difference methods, we will then turn to the *finite-volume* viewpoint that is often more valuable in developing methods for conservation laws.

We will generally consider finite difference grids with mesh spacing $h = \Delta x = \Delta y = \Delta z$ in all spatial directions for simplicity, and for time-dependent equations the time step will be denoted by $k = \Delta t$. Subscript indices will refer to spatial grid points while superscript indices refer to the time step, so for example

$$Q_{ij}^n \approx q(x_i, y_j, t_n) .$$

With a **finite difference method**, we discretize the derivatives appearing in the differential equation so that we obtain a system of algebraic relations between (approximate) values at all the grid points. We can then solve this finite set of equations for the discrete grid-point values on the computer.

As examples of finite difference methods, consider the advection equation (2.10). There are many different ways to discretize the derivatives, leading to different finite difference methods. As one example, using one-sided approximations to both q_t and q_x gives

$$\frac{Q_i^{n+1} - Q_i^n}{k} + u\left(\frac{Q_i^n - Q_{i-1}^n}{h}\right) = 0 \tag{4.1}$$

or

$$\frac{Q_i^{n+1} - Q_i^n}{k} + u\left(\frac{Q_{i+1}^n - Q_i^n}{h}\right) = 0 . \tag{4.2}$$

Which of these is better to use depends on the velocity u. If $u > 0$ then the fluid flows from left to right and we should use (4.1), which updates Q_i^n based on the value Q_{i-1}^n to the left (in the upstream or upwind direction). If $u < 0$ then (4.2) is better. These methods are called the first-order **upwind**

methods. They can be rewritten to give an explicit formula for Q_i^{n+1} in terms of data at the previous time step, *e.g.*, (4.1) gives

$$Q_i^{n+1} = Q_i^n - \frac{k}{h}u(Q_i^n - Q_{i-1}^n) . \tag{4.3}$$

These methods are easily extended to systems of equations of the form (1.4), e.g.,

$$Q_i^{n+1} = Q_i^n - \frac{k}{h}A(Q_i^n - Q_{i-1}^n) . \tag{4.4}$$

This one-sided form would be appropriate only if all eigenvalues of A are positive.

To achieve second-order accuracy we could use centered approximations to the derivatives, e.g.,

$$\frac{Q_i^{n+1} - Q_i^{n-1}}{2k} + A\left(\frac{Q_{i+1}^n - Q_{i-1}^n}{2h}\right) = 0 , \tag{4.5}$$

which gives the **leapfrog method**

$$Q_i^{n+1} = Q_i^{n-1} - \frac{k}{h}A(Q_{i+1}^n - Q_{i-1}^n) . \tag{4.6}$$

Another approach is based on the Taylor series expansion

$$q(x,t+k) = q(x,t) + kq_t(x,t) + \frac{1}{2}k^2 q_{tt}(x,t) + \cdots \tag{4.7}$$

and the observation that from $q_t = -Aq_x$ we can compute

$$q_{tt} = -Aq_{xt} = -Aq_{tx} = -A(-Aq_x)_x = A^2 q_{xx} \tag{4.8}$$

so that (4.7) becomes

$$q(x,t+k) = q(x,t) - kAq_x(x,t) + \frac{1}{2}k^2A^2q_{xx}(x,t) + \cdots . \tag{4.9}$$

The **Lax–Wendroff method** then results from retaining only the first three terms of (4.9) and using centered difference approximations for the derivatives appearing there:

$$Q_i^{n+1} = Q_i^n - \frac{k}{2h}A(Q_{i+1}^n - Q_{i-1}^n) + \frac{k^2}{2h^2}A^2(Q_{i+1}^n - 2Q_i^n + Q_{i-1}^n) . \tag{4.10}$$

For conservation laws arising from physical principles, it is often important to insure that the numerical methods also conserve the appropriate quantities, such as mass, momentum, and energy. This is particularly true when the solution involves shock waves, as discussed in Sect. 4.3. For conservation laws it is thus often preferable to use a **finite volume** method rather than a finite difference method. In a finite volume method the discrete value Q_i^n is viewed as an approximation to the average value of q over a grid cell rather than as an approximation to a pointwise value of q. The cell average is simply the integral of q over the cell divided by its area, so conservation can be maintained by updating this value based on **fluxes** through the cell edges. This is described in detail starting in Chapter 4.2. Although the derivation of such methods may be quite different from that of finite difference methods, the resulting formulas may be identical. We will see, for example, that the upwind and Lax–Wendroff methods are easily interpreted as finite volume methods.

The main issues that should be understood in order to effectively develop and use finite-difference methods are briefly outlined below.

4.1.1 Convergence and Global Order of Accuracy. As we refine the grid we hope to get better approximations to the true solution. The difference between the true and computed solution is called the **global error** and the method is **convergent** if this goes to zero as the mesh is refined. The method has (global) order p if the global error is $O(h^p + k^p)$ as $h,\ k \to 0$. In analyzing a method it is generally impossible to get an expression for the global error directly, however, since the system of equations to be solved grows without bound in size as we refine the grid.

4.1.2 Consistency and the Local Truncation Error. It is much easier to compute the **local truncation error** (LTE), which is a measure of how well the difference equations model the differential equation locally. This is defined by inserting the true solution into the difference equations at a single arbitrary point in spacetime. We don't generally know the true solution, but if it is a smooth function then we can expand in Taylor series and use the differential equation to determine an expansion for the LTE, which shows the **local order** of the method. If the LTE vanishes like $O(h^p + k^p)$ then the method is (locally) **p'th order accurate**. If the method is at all reasonable then we expect the LTE to at least go to zero as the mesh is refined, in which case the method is called **consistent**.

4.1.3 Stability. We hope that a method which is locally p'th order accurate will also give a global error that is p'th order, and in particular will be convergent. But this is far from obvious and may not hold. The problem is that as we refine the grid we must compute an increasing number of values to solve the discrete system, and unless the discretization is **stable** in some way then this process may not converge even if the method is consistent. For convergence we generally need both consistency and stability, and stability is by far the harder concept to deal with.

For partial differential equations there is typically some restriction on the relation between the time step k and the spatial grid size h. Both the Lax–Wendroff method (4.10) and the leapfrog method (4.5), for example, are stable only if $|\lambda k/h| \le 1$ for all eigenvalues λ of the matrix A. This is motivated by the discussion of the *CFL condition* in Sect. 4.1.10 below. The upwind method (4.4) is stable only if $0 \le \lambda k/h \le 1$ and in particular can *only* be used if $\lambda > 0$ for all eigenvalues of A (which also follows from the CFL condition). By contrast the method with one-sided differences in the other direction is stable only if $-1 \le \lambda k/h \le 0$ for all λ. *Godunov's method* gives the proper generalization of upwind methods to problems where information flows in both directions. See [142] or other references for a more complete discussion of stability.

4.1.4 One-Step and Multistep Methods. For time-dependent problems we typically use time-marching procedures, in which the solution at all points $x \in \Omega$ at one time t_n are used to compute the solution at the next time $t_{n+1} = t_n + k$. Such a methods is called a *one-step* or *two-level* method. One can also use multistep methods that base the solution at time t_{n+1} on several past levels, but in more than one space dimension the storage requirements of multistep methods are often prohibitive. For nonlinear conservation laws the standard high-resolution methods are all one-step methods and we will concentrate on these.

4.1.5 Implicit and Explicit Methods. An explicit method is one which yields an explicit expression for each value Q_i^{n+1} at time t_{n+1} in terms of nearby values at time t_n. All of the methods introduced above for hyperbolic problems are explicit.

An implicit method couples together values at different grid points at time t_{n+1} and hence an algebraic system of equations must be solved in each time step in order to advance the solution. For example, the one-dimensional heat equation $q_t = q_{xx}$ might be solved by the explicit method

$$Q_i^{n+1} = Q_i^n + \frac{k}{h^2}(Q_{i-1}^n - 2Q_i^n + Q_{i+1}^n) \tag{4.11}$$

or by the implicit method

$$\begin{aligned} Q_i^{n+1} = Q_i^n &+ \tfrac{1}{2}\tfrac{k}{h^2}\left[(Q_{i-1}^n - 2Q_i^n + Q_{i+1}^n)\right. \\ &+ \left.(Q_{i-1}^{n+1} - 2Q_i^{n+1} + Q_{i+1}^{n+1})\right], \end{aligned} \tag{4.12}$$

which is known as the **Crank–Nicolson** method. To determine the values Q_i^{n+1} we must use these equations, together with the boundary conditions at each end of the interval, to set up and solve a tridiagonal linear system of equations in each time step. In one space dimension such a system can be solved very efficiently by Gaussian elimination and this is not much more work than an explicit method. In more than one dimension one obtains a large sparse linear system that has structure similar to what arises from discretization of an elliptic equation, although since the solution changes by only $O(k)$ in each time step this can be more easily solved than an elliptic equation.

Explicit methods are generally preferable if they can be used with the time step that seems reasonable for the problem, which depends on how fine the spatial grid is and how smooth the solution is. For many problems, however, an explicit method turns out to be **unstable** unless the time step is orders of magnitude smaller than what seems reasonable based on accuracy considerations. This is the case with the heat equation, where the method (4.11) turns out to be stable only if $k < \frac{1}{2}h^2$ whereas (4.12) is stable for any time step. If the solution is varying smoothly then we expect variations in time to be on roughly the same scale as variations in space and so we would typically like to take a time step $k \approx h$. This is possible with the implicit Crank–Nicolson method whereas the explicit method (4.11) would require taking k smaller than this by a factor of h.

For a nonlinear differential equation an implicit method will give rise to a nonlinear algebraic system to solve in each time step which may be much more expensive than an explicit method with the same time step. However, in situations where an explicit method would force us to take much smaller time steps than desired, the use of an implicit method may be much more efficient. Such problems are called ***stiff***, as discussed next.

4.1.6 Stiffness. For time-dependent problems the idea of stiffness is fundamental as it often affects what sort of method must be used. Roughly speaking, a problem is said to be stiff if there are a variety of different time scales in the solution and we are attempting to model some phenomenon on a relatively slow time scale, in a situation where the faster processes simply maintain local equilibrium but all visible variations are on the slower scale. We would then like to take a time step that is reasonable relative to the active time scales, and not have to use the much smaller time step that would be necessary if we were modeling activity on the faster time scales. Since these time scales may easily vary by a factor of a million or more for practical problems, this is a major consideration in the efficiency of the method. The problem is that a random perturbation of the solution may activate these fast processes, and numerical methods are constantly introducing perturbations in the solution. Unless we use a method designed for stiff problems, we will have to take small time steps appropriate for the fastest time scales in order to maintain stability. Methods for stiff problems are necessarily **implicit** in time.

4.1.7 Stiff ODE's. The classic example is the case of a stiff ordinary differential equation. See, for example, [131], [142], or any text on numerical methods for ordinary differential equations.

As an example, consider the ODE

$$u'(t) = \lambda(\cos t - u) - \sin t \,. \tag{4.13}$$

One particular solution is the function $u(t) = \cos t$, and this is the solution with the particular choice of initial data $u(0) = 1$. This smooth function is a solution for any value of λ. If we consider initial data of the form $u(t_0) = \eta$ that does not lie on this curve, then the solution through this point is a different function, of course. It is easy to verify that the solution is

$$u(t) = \mathrm{e}^{\lambda(t-t_0)}(\eta - \cos(t_0)) + \cos t \,. \tag{4.14}$$

However, if $\lambda < 0$ (or $\mathrm{Re}(\lambda) < 0$ more generally), this function approaches $\cos t$ exponentially quickly, with decay rate λ. Figure 4.1 shows a number of different solution curves for this equation with different choices of t_0 and η,

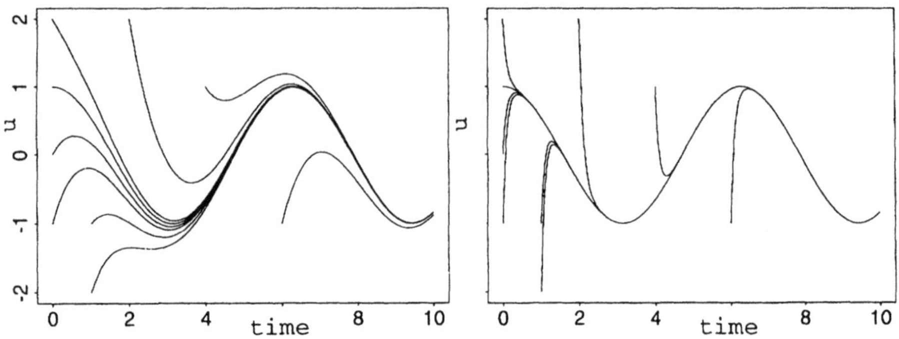

Fig. 4.1. Solution curves for the ODE (4.13) for various initial values. Left: With $\lambda = -1$. Right: With $\lambda = -10$ and the same set of initial values

with the fairly modest value $\lambda = -1$. Figure 4.1b shows the corresponding solution curves when $\lambda = -10$.

Stiffness causes numerical difficulties because any finite difference method is constantly introducing errors. The local truncation error acts as a perturbation to the system that moves us away from the smooth solution we are trying to compute. Why does this cause more difficulty in a stiff system than in other systems? At first glance it seems that the stiffness should work to our advantage. If we are trying to compute the solution $u(t) = \cos t$ to the ODE (4.13) with initial data $u(0) = 1$, for example, then the fact that any errors introduced decay exponentially should help us. The true solution is very robust and the solution is almost completely insensitive to errors made in the past. In fact this *stability* of the true solution does help us, as long as the numerical method is also stable.

The difficulty arises from the fact that many numerical methods, including all explicit methods, are unstable (in the sense of absolute stability) *unless the time step is small relative to the time scale of the rapid transient*, which in a stiff problem is much smaller than the time scale of the smooth solution we are trying to compute. Although the true solution is smooth and it seems that a reasonably large time step would be appropriate, the numerical method must always deal with the rapid transients introduced in every time step and may need a very small time step to do so stably.

For *systems* of ODE's, stiffness is sometimes defined in terms of the "stiffness ratio" of the system, which is the ratio

$$\frac{\max |\lambda_p|}{\min |\lambda_p|}$$

over all eigenvalues of the Jacobian matrix $f'(u)$. If this is large then there is a large range of time scales present in the problem, a necessary component for stiffness to arise. While this is often a useful quantity, one should not rely entirely on this measure to determine whether a problem is stiff. For one thing, it is possible even for a scalar problem to be stiff (as we have seen in the example above), even though for a scalar problem the stiffness ratio is always 1 since there is only one eigenvalue. There can still be more than one time scale present. In (4.13) the fast time scale is determined by λ, the eigenvalue, and the slow time scale is determined by the inhomogeneous term $\sin(t)$. For systems of equations there may also be additional time scales arising from inhomogeneous forcing terms or other time-dependent coefficients that are distinct from the scales imposed by the eigenvalues.

4.1.8 Parabolic Equations. Stiffness can be seen in partial differential equations as well, and in fact the need for an implicit method such as Crank–Nicolson for the heat equation can be interpreted in terms of stiffness. By Fourier analysis we can analyze the effect of the linear heat equation on arbitrary data by decomposing it into Fourier modes. Sinusoidal initial data of the form $\sin(\xi x)$ will decay exponentially at the rate $e^{-\xi^2 t}$, leading to a very rapid damping of high frequencies. In solving the heat equation with smooth data (only low frequencies present) we have relatively slow variation in time leading us to desire a time step $k \approx h$. But because any high frequency perturbation of the solution leads to rapid transient behavior, the equations are stiff and an implicit method must be used.

Note that the time scale for the heat equation can be arbitrarily fast by taking a sufficiently large value of ξ in the perturbation. This might lead one to expect the discrete equations would be "infinitely stiff", so that an explicit method would be unstable for any positive time step. However, when we discretize in space on a particular grid we limit the frequencies that can be represented so that ξ cannot be arbitrarily large. The largest frequency visible on the grid has $\xi \sim 1/h$ and hence decay rate $1/h^2$ which explains the $k \sim h^2$ limit on the time step for an explicit method. If we take a time step that is too large with an explicit method, then an instability will typically appear in the form of an exponentially growing high-frequency oscillation, typically at the highest frequency that can be represented on the grid since this gives the most rapid *growth* with an unstable method.

Parabolic terms arise from dissipative effects such as fluid viscosity, heat conduction, or finite resistivity in MHD, and if such terms are included then implicit methods must typically be used to obtain efficient solutions. This is one reason that we prefer to solve hyperbolic systems of conservation laws in which these dissipative terms are ignored whenever possible.

4.1.9 Hyperbolic Equations. Hyperbolic systems are *usually* not stiff. The time step k for an explicit method is typically limited by something proportional to $s_{\max}h$, where $s_{\max}$ is the maximum wave speed in the problem. (See Sect. 4.1.10 below.) Provided some disturbance of interest is propagating at this maximum speed, then this is generally the right size time step to use in order that the temporal resolution is consistent with the spatial resolution. In this case, there is no need to use an implicit method.

In some problems, however, the fastest waves may be physically unimportant. For example in atmospheric flow acoustic waves are typically unimportant, and since these travel much faster than the phenomena of interest the use of compressible gas dynamics equations would lead to stiffness. Instead other systems are typically used that do not model acoustic waves.

In "incompressible" flow, as is often assumed of liquids, for example, the acoustic waves are filtered out by the introduction of an elliptic problem that is solved for the pressure, as described in Sect. 8.6.1. This equation couples all grid points together, so that solving the elliptic equation is tantamount to using an implicit method to model the net effect of the fast acoustic waves that have been eliminated from the system. This is relatively efficient since the elliptic equation is linear, and the remaining nonlinear hyperbolic equations for the advection of momentum can then be solved with an explicit method.

Nonlinear hyperbolic equations may also exhibit stiffness if the wave speeds differ by orders of magnitude at different points in the domain, as can happen with a nonlinear problem since the wave speeds depend on $q(x,t)$. With an explicit method the time step may be limited by the fastest wave speed seen anywhere in the domain, while the phenomenon of interest may be in a region with much slower wave speeds. This often happens in MHD calculations.

Stiffness also arises when gas dynamics is coupled together with source terms that model radiative transfer or chemical or nuclear reactions that may have much faster time scales than the fluid dynamics. This is discussed in Sect. 5.4.

4.1.10 The CFL Condition. The *CFL condition* is named after Courant, Friedrichs, and Lewy, who wrote a fundamental paper in 1928 that was essentially the first paper on the stability and convergence of finite difference methods for partial differential equations [61].

To understand this general condition, we must discuss the *domain of dependence* of a time-dependent PDE. For the advection equation (1.5), the solution $q(X,T)$ at some fixed point (X,T) depends on the initial data $q^0(x) = q(x,0)$ at only a single point: $q(X,T) = q(X - uT)$. We say that the **domain of dependence** of the point (X,T) is the point $X - uT$:

$$\mathcal{D}(X,T) = \{X - uT\} .$$

If we modify the data q^0 at this point then the solution $q(X,T)$ will change, while modifying the data at any other point will have no effect on the solution at this point.

This is a rather unusual situation for a PDE. More generally we might expect the solution at (X,T) to depend on the data at several points or over a whole interval. For the linear hyperbolic system $q_t + Aq_x = 0$, the propagation speeds are given by the eigenvalues of A, $\lambda_1, \lambda_2, \ldots, \lambda_m$. If these values are distinct then we will see that the solution $q(X,T)$ depends on the data at the m distinct points $X - \lambda_1 T, \ldots, X - \lambda_m T$ and hence

$$\mathcal{D}(X,T) = \{X - \lambda_p T, \text{ for } p = 1, 2, \ldots, m\} .$$

The heat equation $q_t = q_{xx}$ has a much larger domain of dependence. For this equation the solution at any point (X, T) depends on the data *everywhere* and the domain of dependence is the whole real line,

$$\mathcal{D}(X,T) = (-\infty, \infty) .$$

This equation is said to have infinite propagation speed, since data at any point is felt everywhere at any small time in the future (though its effect decays exponentially away from this point).

A finite difference method also has a domain of dependence. On a particular fixed grid we define the domain of dependence of a grid point (x_j, t_n) to be the set of grid points x_i at the initial time $t = 0$ with the property that the data Q_i^0 at x_i has an effect on the solution Q_j^n. For example, with the Lax–Wendroff method (4.10) or any other 3-point method, the value Q_j^n depends on Q_{j-1}^{n-1}, Q_j^{n-1}, and Q_{j+1}^{n-1}. These values depend in turn on Q_{j-2}^{n-2} through Q_{j+2}^{n-2}. Tracing back to the initial time we obtain a triangular array of grid points as seen in Fig. 4.2(a), and we see that Q_j^n depends on the initial data at the points $x_{j-n}, \ldots, x_{j+n}$.

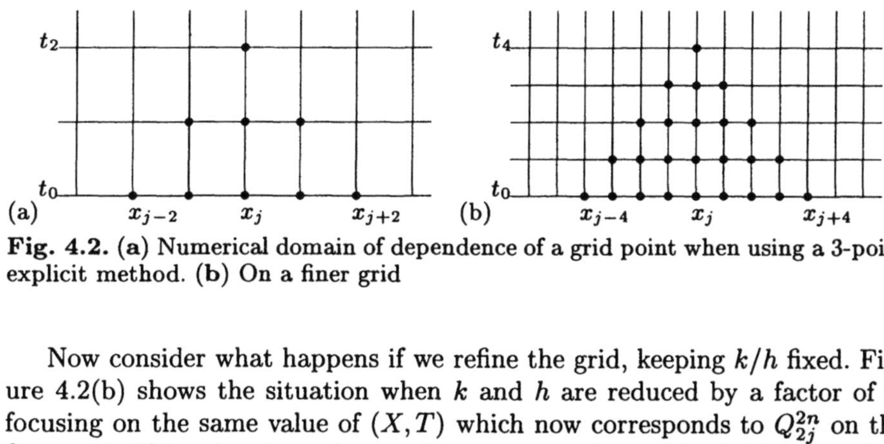

Fig. 4.2. (a) Numerical domain of dependence of a grid point when using a 3-point explicit method. **(b)** On a finer grid

Now consider what happens if we refine the grid, keeping k/h fixed. Figure 4.2(b) shows the situation when k and h are reduced by a factor of 2, focusing on the same value of (X, T) which now corresponds to Q_{2j}^{2n} on the finer grid. This value depends on twice as many values of the initial data, but these values all lie within the same interval and are merely twice as dense.

If the grid is refined further with $k/h \equiv r$ fixed, then clearly the numerical domain of dependence of the point (X, T) will fill in the interval $[X - T/r, \ X + T/r]$. As we refine the grid, we hope that our computed solution at (X, T) will converge to the true solution $q(X, T) = q^0(X - uT)$. Clearly this can only be possible if

$$X - T/r \le X - uT \le X + T/r . \tag{4.15}$$

Otherwise, the true solution will depend only on a value $q^0(X - uT)$ that is never seen by the numerical method, no matter how fine a grid we take. We could change the data at this point and hence change the true solution without having any effect on the numerical solution, so the method cannot be convergent for general initial data.

Note that the condition (4.15) translates into $|u| \leq 1/r$ and hence $|uk/h| \leq 1$. For a hyperbolic system with eigenvalues λ, we need $|\lambda k/h| \leq 1$ as a necessary condition for stability.

The CFL condition generalizes this idea:

The CFL Condition: *A numerical method can be convergent only if its numerical domain of dependence contains the true domain of dependence of the PDE, at least in the limit as k and h go to zero.*

It is important to note that in general the CFL condition is only a *necessary* condition. If it is violated then the method cannot be convergent. If it is satisfied, then the method *might* be convergent, but a proper consistency and stability analysis is required to prove this and determine the proper stability restriction on k and h.

For the heat equation the true domain of dependence is the whole real line. It appears that any 3-point explicit method violates the CFL condition, and indeed it does if we fix k/h as the grid is refined. However, it can be shown that the 3-point explicit method (4.11) is convergent as we refine the grid provided we have $k/h^2 \leq 1/2$. In this case when we make the grid finer by a factor of 2 in space it will become finer by a factor of 4 in time, and hence the numerical domain of dependence will cover a wider interval at time $t = 0$. As $k \to 0$ the numerical domain of dependence will spread to cover the entire real line, and hence the CFL condition is satisfied in this case.

An implicit method such as the Crank–Nicolson method (4.12) satisfies the CFL condition for any time step k. In this case the numerical domain of dependence is the entire real line because the tridiagonal linear system couples together all points in such a manner that the solution at each point depends on the data at all points (*i.e.*, the inverse of a tridiagonal matrix is dense).

4.2 Finite Volume Methods

In the previous sections we reviewed the basic theory of finite difference methods for linear hyperbolic problems. In this section, these methods will be reinterpreted as finite volume methods, a viewpoint that is important in properly extending these methods to nonlinear conservation laws.

Many high-resolution methods for shock capturing are based on solving Riemann problems between states in neighboring grid cells. In this chapter we will develop one particular set of methods of this type. The development of such methods has a long and rich history, and numerous related methods can be found in the literature. Books such as [2], [86], [97], [111], [130], [141], [178], [185], [226], [242] contain descriptions of these methods and pointers to the literature.

Rather than viewing Q_i^n as an approximation to the single value $q(x_i, t_n)$, we will now view it as approximating the average value of q over an interval of length $h = \Delta x = (b-a)/N$. We will split the physical domain $[a, b]$ into N intervals denoted by

$$\mathcal{C}_i = [x_i, x_{i+1}] ,$$

where now $x_i = a + (i-1)h$. The value Q_i^n will approximate the average value over the i'th interval at time t_n:

$$Q_i^n \approx \frac{1}{h}\int_{x_i}^{x_{i+1}} q(x, t_n)\,\mathrm{d}x \equiv \frac{1}{h}\int_{\mathcal{C}_i} q(x, t_n)\,\mathrm{d}x . \tag{4.16}$$

Notationally it might be better to denote the endpoints of the i'th interval by $x_{i-1/2}$ and $x_{i+1/2}$, which would be more symmetric and remind us that Q_i^n is an approximation to the average value between these points. However, the formulas are less cluttered if we stick to integer subscripts.

If $q(x,t)$ is a smooth function, then the integral in (4.16) agrees with the value of q at the midpoint of the interval to $O(h^2)$. By working with cell averages, however, it is easier to use important properties of the conservation law in deriving numerical methods. In particular, we can insure that the numerical method is **conservative** in a way that mimics the true solution, and this is extremely important in accurately calculating shock waves. This is because $h\sum_{i=1}^N Q_i^n$ approximates the integral of q over the entire interval $[a, b]$, and if we use a method that is in *conservation form* (as described below), then this discrete sum will change only due to fluxes at the boundaries $x = a$ and $x = b$. The total mass within the computational domain will be preserved, or at least will vary correctly provided the boundary conditions are properly imposed.

The integral form of the conservation law (2.18), when applied to one grid cell over a single time step, gives

$$\begin{aligned}\int_{\mathcal{C}_i} q(x, t_{n+1})\,\mathrm{d}x - \int_{\mathcal{C}_i} q(x, t_n)\,\mathrm{d}x \quad &= \quad \int_{t_n}^{t_{n+1}} f(q(x_i, t))\,\mathrm{d}t \\ &\quad - \int_{t_n}^{t_{n+1}} f(q(x_{i+1}, t))\,\mathrm{d}t .\end{aligned}$$

Rearranging this and dividing by h gives

$$\begin{aligned}\frac{1}{h}\int_{\mathcal{C}_i} q(x, t_{n+1})\,\mathrm{d}x \quad &= \quad \frac{1}{h}\int_{\mathcal{C}_i} q(x, t_n)\,\mathrm{d}x \\ &\quad - \frac{1}{h}\left[\int_{t_n}^{t_{n+1}} f(q(x_i, t))\,\mathrm{d}t - \int_{t_1}^{t_2} f(q(x_{i+1}, t))\,\mathrm{d}t\right] .\end{aligned} \tag{4.17}$$

This tells us exactly how the cell average of q from (4.16) should be updated in one time step. In general, however, we cannot evaluate the time integrals on the right-hand side of (4.17) exactly since $q(x_i, t)$ varies with time along each edge of the cell, and we don't have the exact solution to work with.

But this does suggest that we should develop numerical methods in the **flux-differencing form**

$$Q_i^{n+1} = Q_i^n - \frac{k}{h}(F_{i+1}^n - F_i^n) , \tag{4.18}$$

where F_i^n is some approximation to the average flux along $x = x_i$:

$$F_i^n \approx \frac{1}{k} \int_{t_n}^{t_{n+1}} f(q(x_i, t))\, dt . \tag{4.19}$$

If we can approximate this average flux based on the values Q^n, then we will have a fully-discrete method.

Since information propagates with finite speed, it is reasonable to first suppose that we can obtain F_i^n based only on the values Q_{i-1}^n and Q_i^n, the cell averages on either side of this interface. Then we might use a formula of the form

$$F_i^n = F(Q_{i-1}^n, Q_i^n) ,$$

where F is some **numerical flux function**. The method (4.18) then becomes

$$Q_i^{n+1} = Q_i^n - \frac{k}{h}(F(Q_i^n, Q_{i+1}^n) - F(Q_{i-1}^n, Q_i^n)) . \tag{4.20}$$

The specific method obtained depends on how we choose the formula F, but in general any method of this type is an explicit method with a 3-point stencil. Moreover, it is said to be in **conservation form**, since it mimics the property (4.17) of the exact solution. Note that if we sum hQ_i^{n+1} from (4.18) over any set of cells we obtain

$$h \sum_{i=I}^{J} Q_i^{n+1} = h \sum_{i=I}^{J} Q_i^n - \frac{k}{h}\left(F_{J+1}^n - F_I^n\right) . \tag{4.21}$$

The sum of the flux differences cancels out except for the fluxes at the extreme edges. Over the full domain we have exact conservation except for fluxes at the boundaries. (Boundary conditions are discussed at the end of this chapter.)

Note that (4.20) can be viewed as a direct finite difference approximation to the conservation law $q_t + f(q)_x = 0$, since rearranging it gives

$$\frac{Q_i^{n+1} - Q_i^n}{k} + \frac{F(Q_{i+1}^n, Q_i^n) - F(Q_i^n, Q_{i-1}^n)}{h} = 0 .$$

Many methods can be equally well viewed as finite difference approximations to this equation or as finite volume methods. In obtaining a method in conservation form, the above discussion suggests that we should always discretize the conservation law in this form, rather than in the quasi-linear form $q_t + f'(q)q_x = 0$, for example.

4.3 Importance of Conservation Form — Incorrect Shock Speeds

Using methods in conservation form is particularly important when solving problems with shocks or other discontinuities in the solution, as a nonconservative method may give a numerical solution that looks reasonable but is entirely wrong. For example, Burgers' equation

$$q_t + \left(\frac{1}{2}q^2\right)_x = 0 \tag{4.22}$$

can be discretized by the upwind conservative method

$$Q_i^{n+1} = Q_i^n - \frac{k}{h}\left(\frac{1}{2}(Q_i^n)^2 - \frac{1}{2}(Q_{i-1}^n)^2\right).$$

Figure 4.3(a) shows the resulting numerical solution for a shock wave resulting between states 1.2 and 0.4 propagating at speed 0.8. The numerical solution is slightly smeared about the correct location.

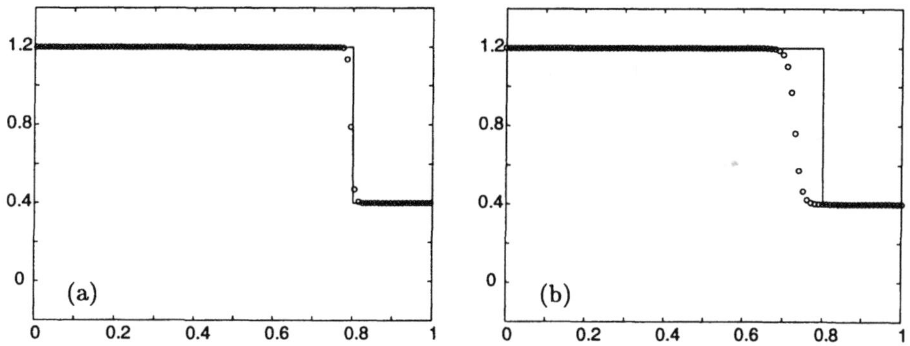

Fig. 4.3. (**a**) True and computed solutions to Burgers' equation using a conservative upwind method. (**b**) True and computed solutions to Burgers' equation using a nonconservative upwind method

On the other hand if we discretize the quasilinear form of Burgers' equation

$$q_t + qq_x = 0 \tag{4.23}$$

using the nonconservative upwind method

$$Q_i^{n+1} = Q_i^n - \frac{k}{h}Q_i^n(Q_i^n - Q_{i-1}^n)$$

we obtain the results seen in Fig. 4.3(b). The shock is moving at the wrong speed! This happens because the equations (4.22) and (4.23) are equivalent for smooth solutions but **not** for problems with shock waves. This example is discussed in more detail in [141].

4.4 Numerical Flux Functions

Given that we want to use a method in conservation form, how should we define $F(q_l, q_r)$, the average flux at a point based on data q_l and q_r to the left and right of this point? A first attempt might be the simple average

$$F(q_l, q_r) = \frac{1}{2}(f(q_l) + f(q_r)) .$$

Using this in (4.20) would give

$$Q_i^{n+1} = Q_i^n - \frac{k}{2h}(f(Q_{i+1}^n) - f(Q_{i-1}^n)) .$$

In general, however, this method turns out to be *unconditionally unstable* for any value of k/h.

If we instead use the modified flux

$$F(q_l, q_r) = \frac{1}{2}(f(q_l) + f(q_r)) - \frac{h}{2k}(q_r - q_l) , \tag{4.24}$$

then we obtain the **Lax–Friedrichs method,**

$$Q_i^{n+1} = \frac{1}{2}(Q_{i-1}^n + Q_{i+1}^n) - \frac{k}{2h}(f(Q_{i+1}^n) - f(Q_{i-1}^n)) . \tag{4.25}$$

Note that the additional term we have added in (4.24) is a *diffusive flux* based on an approximation to $\frac{h^2}{2k}q_x$, and hence this modification amounts to adding some *artificial viscosity* to the centered flux formula.

4.5 Godunov's Method

Many of the methods we will explore in detail are based on solving the *Riemann problem* between the states q_l and q_r in order to define the numerical flux $F(q_l, q_r)$. To see how this comes about, it is useful to view the data Q^n at time t_n as defining a piecewise constant function $\tilde{q}^n(x, t_n)$ which has the value Q_i^n for all x in the interval C_i. Suppose we could solve the conservation law *exactly* over the time interval $[t_n, t_{n+1}]$ with initial data $\tilde{q}^n(x, t_n)$. Call the resulting function $\tilde{q}^n(x, t)$ for $t_n \leq t \leq t_{n+1}$. Then we might consider defining the numerical flux F_i^n in (4.18) by

$$F_i^n = \frac{1}{k}\int_{t_n}^{t_{n+1}} f(\tilde{q}^n(x_i, t))\, dt . \tag{4.26}$$

This integral is trivial to compute compared to the integral (4.19), at least provided the time step k is small enough, because of the fact that with piecewise constant initial data we can find the exact solution easily by simply piecing together the solutions to each Riemann problem defined by the jump at each interface. Figure 4.4 illustrates this for the case of a linear hyperbolic system of 2 equations.

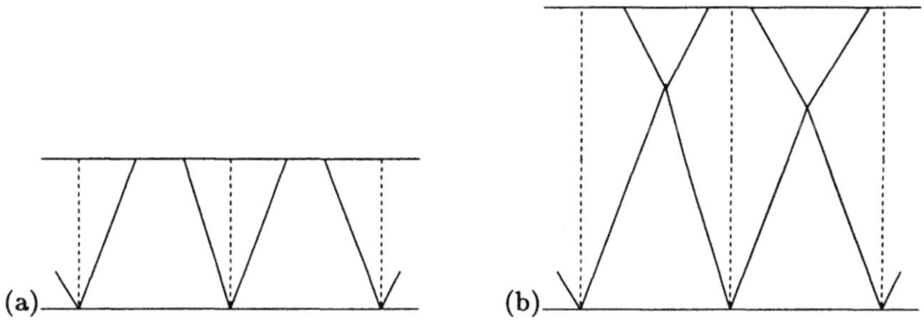

Fig. 4.4. Solving the Riemann problems at each interface for Godunov's method. (**a**) With Courant number less than 1/2 there is no interaction of waves. (**b**) With Courant number less than 1 the interacting waves do not reach the cell interfaces, so the fluxes are still constant in time

The crucial fact now is that the solution to the Riemann problem at x_i is a similarity solution, which is constant along each ray $(x - x_i)/t$ =constant. In general, let $q^*(q_l, q_r)$ denote the exact solution to the Riemann problem along the ray $x/t = 0$, obtained when we use data

$$q(x, 0) = \begin{cases} q_l & \text{if } x < 0 \\ q_r & \text{if } x > 0 \,. \end{cases}$$

Then we have

$$\tilde{q}^n(x_i, t) \equiv q^*(Q_{i-1}^n, Q_i^n) \tag{4.27}$$

for all $t \in [t_n, t_{n+1}]$, provided that the time step is small enough that waves from the Riemann problems do not travel farther than distance h in this time step. If this condition is violated, then the value along $x = x_i$ may change after waves from neighboring Riemann problems pass this point. For a linear system, the maximum wave speed is $\max_p |\lambda^p|$, where λ^p are the eigenvalues of A, so this condition requires that

$$k \max_p |\lambda^p| \leq h \,.$$

We recognize this as being simply the CFL condition for a 3-point method, a condition which we know must be satisfied anyway for stability.

The method obtained by the procedure outlined above is known as **Godunov's method**, and was introduced in [98] as an approach to solving the Euler equations of gas dynamics in the presence of shock waves.

For the simplest case of scalar advection, solving the Riemann problem between states q_l and q_r gives

$$q^*(q_l, q_r) = \begin{cases} q_l & \text{if } u > 0 \\ q_r & \text{if } u < 0 \,. \end{cases}$$

If $u = 0$ then q^* is not well defined, as the discontinuity is stationary along the ray $x/t = 0$. This is no cause for concern, however, since all we really require is the flux value $f(q^*) = uq^*$, and if $u = 0$ then $f(q^*) = 0$ regardless of how we define q^*. So we obtain the numerical flux

$$F(q_l, q_r) = uq^*(q_l, q_r) = \begin{cases} uq_l & \text{if } u \le 0 \\ uq_r & \text{if } u \ge 0 \,. \end{cases}$$

This can also be written in the compact form

$$F(q_l, q_r) = u^+ q_l + u^- q_r \tag{4.28}$$

using the notation

$$u^+ = \max(u, 0) \,, \qquad u^- = \min(u, 0) \,. \tag{4.29}$$

Using this in the conservative method (4.20) gives the **upwind method**. Note that solving the Riemann problem at the interface gives a flux that is defined by the value of Q^n on the upwind side of the interface, so that the method reduces to one-sided differencing in the proper direction. The method takes the form

$$Q_i^{n+1} = \begin{cases} Q_i^n - \frac{k}{h} u(Q_i^n - Q_{i-1}^n) & \text{if } u > 0 \\ Q_i^n - \frac{k}{h} u(Q_{i+1}^n - Q_i^n) & \text{if } u < 0 \end{cases} \tag{4.30}$$

as introduced in (4.1) and (4.2).

This method is easily generalized to nonlinear systems if we can solve the nonlinear Riemann problem at each cell interface, and this method gives the natural generalization of the first-order upwind method to general systems of conservation laws.

Recall that Q_i^n represents an approximation to the cell average of $q(x, t_n)$ over cell C_i,

$$Q_i^n \approx \frac{1}{h} \int_{x_i}^{x_{i+1}} q(x, t_n) \, \mathrm{d}x \,,$$

and the idea is to use the piecewise constant function defined by these cell values as initial data $\tilde{q}^n(x, t_n)$ for the conservation law. Solving over time k with this data gives a function $\tilde{q}^n(x, t_{n+1})$ that is then averaged over each cell to obtain

$$Q_i^n = \frac{1}{h} \int_{x_i}^{x_{i+1}} \tilde{q}^n(x, t_{n+1}) \, \mathrm{d}x \,, \tag{4.31}$$

If the time step k is sufficiently small, then the exact solution $\tilde{q}^n(x, t)$ can be determined by piecing together the solutions to the Riemann problem arising from each cell interface, as indicated in Fig. 4.4(a).

Recall from Sect. 4.5 that we do not need to perform the integration in (4.31) explicitly, which might be difficult since $\tilde{q}^n(x, t_{n+1})$ may be very complicated as a function of x. Instead, we can use the fact that $\tilde{q}^n(x_i, t)$ is constant in time along each cell interface so that the integral (4.26) can be computed exactly. Hence the cell average is updated by (4.20) with

$$F(q_l, q_r) = f(q^*(q_l, q_r)) \,, \tag{4.32}$$

where $q^*(q_l, q_r)$ is the solution to the Riemann problem between q_l and q_r, evaluated along $x/t = 0$.

In Fig. 4.4(a) the time step was taken to be small enough that there was no interaction of waves from neighboring Riemann problems. This would be necessary if we wanted to construct the solution at $\tilde{q}^n(x, t_{n+1})$ in order to explicitly calculate the cell averages (4.31). However, in order to use the flux formula (4.32) it is only necessary that $\tilde{q}^n(x_i, t)$ remain constant in time over the entire time step, which allows a time step roughly twice as large, as indicated in Fig. 4.4(b). If s_{max} represents the largest wave speed that is encountered then on a uniform grid with the cell interfaces distance h apart, we must require

$$\frac{ks_{\text{max}}}{h} \leq 1 \tag{4.33}$$

in order to insure that the formula (4.32) is valid. Note that this is precisely the CFL condition required for stability of this 3-point method, as discussed in Sect. 4.1.10. In general $s_{\text{max}}k/h$ is called the **Courant number**. Figure 4.4(a) shows a case where the Courant number is less than 1/2 while Fig. 4.4(b) shows the Courant number close to 1. Note that for a linear system of equations, $s_{\text{max}} = \max_p |\lambda^p|$ and this agrees with our previous definition of the Courant number.

4.5.1 Godunov's Method on Scalar Equations. On a convex scalar equation with $f'(q)$ an increasing function of q, the solution to the Riemann problem between q_l and q_r is either a shock traveling at speed $s = [f]/[q]$ (if $q_l > q_r$) or a rarefaction wave (if $q_l < q_r$) bounded by $x/t = f'(q_l)$ on the left and $x/t = f'(q_r)$ on the right. Five possible configurations in the x-t plane are shown in Fig. 4.5. In most cases the solution q^* along $x/t = 0$ will be either q_r (if the solution is a shock or rarefaction wave moving entirely to the left, Fig. 4.5(a) or (b)), or q_l (if the solution is a shock or rarefaction wave moving entirely to the right, Fig. 4.5(d) or (e)).

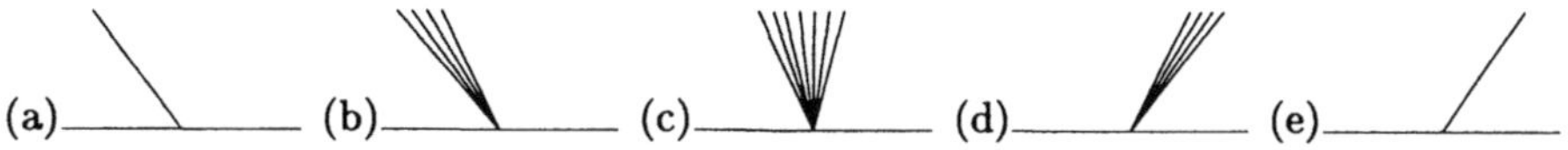

Fig. 4.5. Five possible configurations for the solution to a scalar conservation law, in the x-t plane. **(a)** Left-going shock, $q^* = q_r$. **(b)** Left-going rarefaction, $q^* = q_r$. **(c)** Transonic rarefaction, $q^* = q_0$. **(d)** Right-going rarefaction, $q^* = q_l$. **(e)** Right-going shock, $q^* = q_l$

The only case where q^* has a different value than q_l or q_r is if $q_l < q_0 < q_r$, where q_0 is the value for which $f'(q_0) = 0$. This is called the **stagnation point** since the value q_0 propagates with speed 0. It is also called the **sonic point** since in gas dynamics the eigenvalue $u \pm c$ takes the value 0 only when the fluid speed is equal to the sound speed. The solution to the Riemann problem in this case, shown in Fig. 4.5(c), consists of a rarefaction wave that is partly left-going and partly right-going. This is called a **transonic rarefaction** since in gas dynamics the fluid is accelerated from a subsonic velocity to a supersonic velocity through such a rarefaction. In a transonic rarefaction the value along $x/t = 0$ is simply q_0.

We thus see that the Godunov flux function for a convex scalar conservation law is

$$F(q_l, q_r) = \begin{cases} f(q_l) & \text{if } q_l > q_0 \text{ and } s > 0 \\ f(q_r) & \text{if } q_r < q_0 \text{ and } s < 0 \\ f(q_0) & \text{if } q_l < q_0 < q_r \,. \end{cases} \tag{4.34}$$

4.6 Approximate Riemann Solvers

To apply Godunov's method on a system of equations we need only determine $q^*(q_l, q_r)$, the state along $x/t = 0$ based on the Riemann data q_l and q_r. We do not need the entire structure of the Riemann problem. However, to compute q^* we must typically determine the full wave structure and wave speeds in order to determine where q^* lies in state space. Typically it is one of the intermediate states in the Riemann solution obtained in the process of connecting q_l to q_r by a sequence of shocks or rarefactions, and is one of the intersections of Hugoniot loci and/or integral curves. In the special case of a transonic rarefaction the value q^* will lie along the integral curve somewhere between these intersections, and additional work will be required to find the value q^*.

A wide variety of **approximate Riemann solvers** have been proposed that can be applied much more cheaply than the exact Riemann solver and yet give results that in many cases are equally good when used in the Godunov or high-resolution methods. In this section we will look at several possibilities.

4.6.1 The All-Shock Solver. In solving a nonlinear Riemann problem we must worry about whether the wave in each family should be a shock or rarefaction so that we know whether to use the Hugoniot locus or integral curve correspondingly. One simplification that can be made to the Riemann solver is to ignore the possibility of rarefaction waves and simply find a Riemann solution in which each pair of states is connected along the Hugoniot locus. The solution then consists entirely of discontinuities that satisfies the Rankine–Hugoniot conditions and is a weak solution of the conservation law. This approach is discussed by Colella [54] for gas dynamics. The all-shock solver is particularly valuable for problems where the Riemann problem is harder to solve, such as in problems where a more complicated equation of state than a gamma-law gas must be used. This occurs at high temperatures, or in relativistic flow, for example.

A potential problem with this approach, of course, is that by using a solution to the Riemann problem that does not satisfy the entropy condition, we might obtain a numerical solution which does not approximate the correct weak solution. Actually, however, in most cases (except for transonic rarefactions), Godunov's method with the Riemann solver will typically work well even if the correct solution involves rarefaction waves. This is because of the numerical dissipation that is introduced in every step of Godunov's method through the averaging process. Although the discontinuous solution used in the solution of a particular Riemann problem may not be correct, by averaging this solution over the grid cell at the end of the time step the discontinuity is smeared out and after many time steps a good approximation to the correct rarefaction wave will be computed. The all-shock approximation will typically be inadequate in the case of a transonic rarefaction. Such a discontinuity tends to persist, and results in the computation of an entropy-violating weak solution. An "entropy fix" is often used to eliminate this problem, as mentioned in Sect. 4.6.4

4.6.2 The Osher Solver. Another approach would be to take the opposite viewpoint and use only the integral curves of the eigenvectors to solve the Riemann problem, completely ignoring the Hugoniot locus and the possibility of shocks. This has the advantage over the previous approach that rarefaction waves, including transonic rarefactions, will then be incorporated.

There is, however, an obvious difficulty with this approach. The smooth solution obtained by integrating along the integral curves is only physically realizable if the characteristic speed is increasing as we go from the left state to the right state along this curve. Otherwise the discontinuous data evolves into an overturned, triply valued "compression wave" instead of a rarefaction wave. However, this overturned solution can still be used to define a flux if it is interpreted properly. This is the basis for the **Osher solver**, which is a generalization of the **Engquist–Osher method** for scalar conservation laws, first developed in the context of the small disturbance equation for transonic flow [81]. Details for general conservation laws can be found in [181].

4.6.3 The HLLE Solver. Another approach is to approximate the full Riemann solution by a single intermediate state bounded by two waves moving at speeds s_1 and s_2. The wave speeds should be some approximations to the minimum and maximum wave speeds that would arise from this particular Riemann data, and the intermediate state can then be calculated by the condition of conservation. This approach is developed by Harten, Lax, and van Leer originally [108] and improved by Einfeldt [78].

4.6.4 The Roe Solver. One of the most popular Riemann solvers currently in use is due to Roe [195]. The idea is to determine an approximate Riemann solution $\hat{q}(x,t)$ by solving a constant coefficient linear system of conservation laws instead of the original nonlinear system. Of course the coefficient matrix used to define this linear system must depend on the q_l and q_r in order to have consistency with the nonlinear problem, so we will write the linear system for $\hat{q}$ as

$$\hat{q}_t + \hat{A}(q_l, q_r)\hat{q}_x = 0 \,. \tag{4.35}$$

This linear Riemann problem is relatively easy to solve (see Sect. 3.2).

We still have the problem of determining $\hat{A}(q_l, q_r)$ in a reasonable way. Roe suggested that the following conditions should be imposed on $\hat{A}$:

$$\begin{aligned} &i) \quad \hat{A}(q_l, q_r)(q_r - q_l) = f(q_r) - f(q_l) \\ &ii) \quad \hat{A}(q_l, q_r) \text{ is diagonalizable with real eigenvalues} \\ &iii) \quad \hat{A}(q_l, q_r) \to f'(\bar{q}) \text{ smoothly as } q_l, q_r \to \bar{q}\ . \end{aligned} \tag{4.36}$$

Condition (4.36*i*) has two effects. First, it guarantees that any flux-difference splitting defined by the matrix will be a flux-difference splitting of the true flux f as well. Another effect is that, in the special case where q_l and q_r are connected by a single shock wave or contact discontinuity, the approximate Riemann solution agrees with the exact Riemann solution. This follows from the fact that the Rankine–Hugoniot condition is satisfied for q_l and q_r in this case, so

$$f(q_r) - f(q_l) = s(q_r - q_l)$$

for some s (the speed of the shock or contact). Combined with (4.36*i*), this shows that $q_r - q_l$ must, in this situation, be an eigenvector of $\hat{A}$ with eigenvalue s and so the approximate solution $\hat{q}(x,t)$ also consists of this single jump $q_r - q_l$ propagating with speed s.

Condition (4.36*ii*) is clearly required in order that the problem $\hat{q}_t + \hat{A}\hat{q}_x = 0$ is hyperbolic and solvable.

Condition (4.36*iii*) guarantees that the method behaves reasonably on smooth solutions, since if $\|Q_i - Q_{i-1}\| = O(h)$ then the linearized equation $q_t + f'(Q_i)q_x = 0$ is approximately valid. It is natural to require that the linear system (4.35) agree with the linearization in this case. Since (4.36*i*) guarantees that the method behaves reasonably on an isolated discontinuity, it is only when a Riemann problem has a solution with more than one strong shock or contact discontinuity that the approximate Riemann solution will differ significantly from the true Riemann solution. In practice this happens infrequently — near the point where two shocks collide, for example.

One way to guarantee that both conditions (4.36*ii*) and (4.36*iii*) are satisfied is to take

$$\hat{A}(q_l, q_r) = f'(q_{\text{ave}}) \tag{4.37}$$

for some average value of q, e.g., $q_{\text{ave}} = \frac{1}{2}(q_l + q_r)$. Unfortunately, this simple choice of q_{ave} will not give an $\hat{A}$ that satisfies (4.36*i*) in general. Harten and Lax [107] show (see also Theorem 2.1 in [108]) that for a general system with an entropy function, a more complicated averaging of the Jacobian matrix in state space can be used. This shows that such linearizations exist, but is too complicated to use in practice.

Fortunately, for special systems of equations it is possible to derive suitable $\hat{A}$ matrices that are very efficient to use relative to the exact Riemann solution. The Roe matrix for the simple isothermal equations is given in [141].

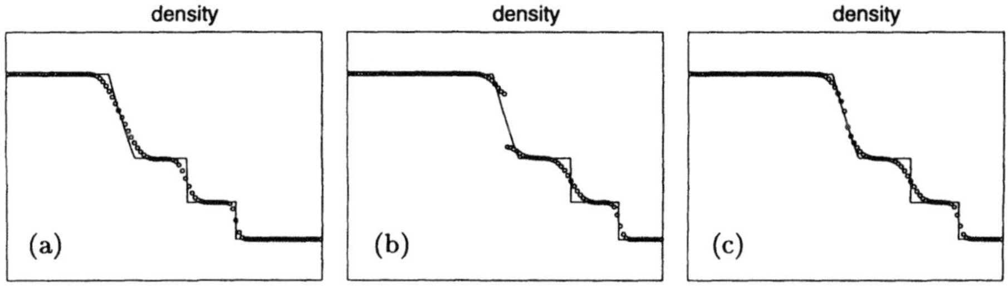

Fig. 4.6. Computed density profiles using Roe's approximate Riemann solver. (**a**) Sod's shock tube problem. (**b**) Sod problem from a moving reference frame so that the rarefaction wave is transonic. An entropy-violating solution is calculated. (**c**) Same problem with an entropy-fix in the Roe solver

Roe [195] presents the analogous formulas for the full Euler equations. Approximate Riemann solvers of this type have been developed for many other applications as well.

4.6.5 A Sonic Entropy Fix. One disadvantage of Roe's linearization or the all-shock solver is that the resulting approximate Riemann solution consists only of discontinuities, with no rarefaction waves. This can lead to a violation of the entropy condition, particularly if the solution involves transonic rarefactions.

Figure 4.6(a) shows the density computed by using the Roe approximate Riemann solver for the Euler equations on the Sod shock tube problem of Fig. 2.1, where

$$\begin{array}{llll} \varrho_l = 3\,, & u_l = 0\,, & p_l = 3\,, & \text{for } x < 0.5\,, \\ \varrho_r = 1\,, & u_r = 0\,, & p_r = 1\,, & \text{for } x > 0.5\,. \end{array} \tag{4.38}$$

The rarefaction wave is captured smoothly in this case even though the Riemann solution in each time step has only discontinuties.

Figure 4.6(b) shows the results for a similar problem using the data

$$\begin{array}{llll} \varrho_l = 3\,, & u_l = 0.9\,, & p_l = 3\,, & \text{for } x < 0.5\,, \\ \varrho_r = 1\,, & u_r = 0.9\,, & p_r = 1\,, & \text{for } x > 0.5\,. \end{array} \tag{4.39}$$

Changing the velocity in this way is equivalent to observing the solution to the Sod problem from a moving reference frame, and the solution is identical to that of the Sod problem except that the velocity is increased by 0.9 everywhere. As a result the rarefaction wave is now transonic. As we can see, the numerical solution now contains an entropy-violating shock at $x = 0.5$, the location of the discontinuity in the initial data.

In the case of a sonic rarefaction wave, it is necessary to modify the approximate Riemann solver in order to obtain entropy satisfying solutions. There are various ways to do this. One approach, discussed by Harten and Hyman [105], is outlined in [141]. Various other approaches have also been used, e.g., [54], [112], [263].

Figure 4.6(c) shows the same test problem as in Fig. 4.6(b) but with the entropy fix applied. Some nonsmoothness is still visible in the rarefaction wave, but no worse than what is seen with the true Riemann solver, it turns out. This results from the fact that Godunov's method has less numerical viscosity near a sonic point than elsewhere, since the rarefaction wave is split up over the two neighboring cells (See [100]).

4.7 High-Resolution Methods

Regardless of what Riemann solver is used, Godunov's method will be at best first-order accurate on smooth solutions and generally gives very smeared approximations to shock waves or other discontinuities. In this section we will see how this method can be extended to a method that gives second-order accuracy on smooth flow, but which avoids nonphysical oscillations near discontinuities. The key is to use a better representation of the solution, say piecewise linear instead of the piecewise constant representation used in Godunov's method, but to form this reconstruction carefully by paying attention to how the data behaves nearby. In smooth regions the finite-difference approximation to the slope can be used to obtain better accuracy, but near a discontinuity the "slope" computed by subtracting two values of Q and dividing by h may be huge and meaningless. Using it blindly in a difference approximation will introduce oscillations into the numerical solution.

One particular method will be developed here in a framework that can be interpreted as a correction phase following the solution of Riemann problems and construction of Godunov fluxes. Many other approaches can be found in the literature and a couple of these are briefly described in Sect. 4.8.

To introduce these ideas we will first consider the scalar advection equation

$$q_t + uq_x = 0$$

with $u > 0$, in which case Godunov's method reduces to the simple first-order upwind method

$$Q_i^{n+1} = Q_i^n - \frac{k}{h}u(Q_i^n - Q_{i-1}^n)\,.$$

After developing a high-resolution version of this method, the ideas will be extended to systems of equations and nonlinear problems starting in Sect. 4.7.10.

4.7.1 Reconstruct–Solve–Average. There is another interpretation of Godunov's method that will be useful in developing higher-order accurate methods of this type. Recall that $\tilde{q}^n(x, t_n)$ denotes the piecewise constant function with value Q_i^n in cell C_i. We defined the numerical flux of Godunov's method by advancing the solution with this data to obtain the interface value $\tilde{q}^n(x, t)$ over the time interval $[t_n, t_{n+1}]$. Another way to describe Godunov's method is to take the advanced solution $\tilde{q}^n(x, t_{n+1})$ at the end of the time step, and average this function over grid cell C_i to obtain

$$Q_i^{n+1} \equiv \frac{1}{h} \int_{C_i} \tilde{q}^n(x, t_{n+1})\, \mathrm{d}x \,.$$

It follows from the integral form of the conservation law that this gives exactly the same value as the flux-differencing method, though implementing it in this form would be more difficult since $\tilde{q}^n(x, t_{n+1})$ is not constant over C_i, and this integral would be difficult to evaluate directly in general. The beauty of the flux-differencing approach is that we do not need to evaluate this integral, but can find it by integrating the flux function, which is constant on the time interval of integration.

But in generalizing Godunov's method to higher-order methods, it is useful to consider what would happen if we took this approach with a different choice of $\tilde{q}^n(x, t_n)$, that better approximates a smooth function. We can think of $\tilde{q}^n(x, t_n)$ as a *reconstruction* of a function from the discrete values Q_i^n, the cell averages of the function. Instead of a piecewise constant function we might reconstruct a piecewise linear function or some other function $\tilde{q}^n(x, t_n)$. We can then generalize Godunov's method to an algorithm that takes the following general form in each time step:

Algorithm RSA (Reconstruct–Solve–Average):

1. **Reconstruct** a function $\tilde{q}^n(x, t_n)$ defined for all x from the cell averages Q_i^n.
2. **Solve** the hyperbolic equation exactly (or approximately) with this initial data to obtain $\tilde{q}^n(x, t_{n+1})$ a time Δt later.
3. **Average** this function over each grid cell to obtain

$$Q_i^{n+1} = \frac{1}{h} \int_{C_i} \tilde{q}^n(x, t_{n+1})\, \mathrm{d}x \,.$$

With a piecewise constant reconstruction we can solve the problem in Step 2 exactly, giving Godunov's method. For linear systems we can solve this problem exactly even with more complicated initial data, such as the piecewise linear function considered in the next section. For nonlinear problems we may not be able to solve the problem in Step 2 exactly, but we will still be able to improve the accuracy by using an approximate solution together with piecewise linear data.

4.7.2 Piecewise Linear Reconstruction. To achieve better than first-order accuracy, we must use a better reconstruction than a piecewise constant function. From the cell averages Q_i^n we can construct a piecewise linear function of the form

$$\tilde{q}^n(x, t_n) = Q_i^n + \sigma_i^n (x - \bar{x}_i) \qquad \text{for } x_i \le x < x_{i+1} \,, \tag{4.40}$$

where

$$\bar{x}_i = \frac{1}{2}(x_i + x_{i+1}) = x_i + \frac{1}{2}h \tag{4.41}$$

is the center of the i'th grid cell and σ_i^n is the slope on the i'th cell. The linear function defined by (4.40) on the i'th cell is defined in such a way that its value at the cell center $\bar{x}_i$ is Q_i^n. More importantly, the average value of $\tilde{q}^n(x, t_n)$ over cell $\mathcal{C}_i$ is Q_i^n (regardless of the slope σ_i^n), so that the reconstructed function has the cell average Q_i^n. This is crucial in developing conservative methods for conservation laws. Note that Steps 2 and 3 are conservative in general, and so Algorithm RSA is conservative provided we use a *conservative reconstruction* in Step 1, as we have in (4.40). Later we will see how to write such methods in the standard conservation form (4.18).

For the scalar advection equation $q_t + uq_x = 0$, we can easily solve the equation with this data, and compute the new cell averages as required in Step 3 of Algorithm RSA. We have

$$\tilde{q}^n(x, t_{n+1}) = \tilde{q}^n(x - uk, t_n) \,.$$

Until further notice we will assume that $u > 0$ and present the formulas for this particular case. The corresponding formulas for $u < 0$ should be easy to derive, and in Sect. 4.7.7 we will see a better way to formulate the methods in the general case.

Suppose also that $|uk/h| \leq 1$. Then it is straightforward to compute (see also Sect. 4.7.7) that

$$\begin{aligned} Q_i^{n+1} &= \frac{uk}{h}\left(Q_{i-1}^n + \frac{1}{2}(h - uk)\sigma_{i-1}^n\right) + \left(1 - \frac{uk}{h}\right)\left(Q_i^n - \frac{1}{2}uk\sigma_i^n\right) \\ &= Q_i^n - \frac{uk}{h}(Q_i^n - Q_{i-1}^n) - \frac{1}{2}\frac{uk}{h}(h - uk)(\sigma_i^n - \sigma_{i-1}^n) \,. \end{aligned} \tag{4.42}$$

4.7.3 Choice of Slopes. Choosing $\sigma_i^n \equiv 0$ gives Godunov's method (the upwind method for the advection equation). To obtain a second-order accurate method we want to choose nonzero slopes in such a way that σ_i^n approximates the derivative q_x over the i'th grid cell. Three obvious possibilities are:

$$\text{Centered slope:} \quad \sigma_i^n = \frac{Q_{i+1}^n - Q_{i-1}^n}{2h} \quad \text{(Fromm)}\,, \tag{4.43}$$

$$\text{Upwind slope:} \quad \sigma_i^n = \frac{Q_i^n - Q_{i-1}^n}{h} \quad \text{(Beam–Warming)}\,, \tag{4.44}$$

$$\text{Downwind slope:} \quad \sigma_i^n = \frac{Q_{i+1}^n - Q_i^n}{h} \quad \text{(Lax–Wendroff)}\,. \tag{4.45}$$

The centered slope might seem like the most natural choice to obtain second order accuracy, but in fact all three choices give the same formal order of accuracy, and it is the other two choices that give methods we have already derived in other ways. Only the downwind slope results in a centered 3-point method, and this choice gives the **Lax–Wendroff** method (4.10). The upwind slope gives a fully-upwinded 3-point method, which is simply **Beam–Warming**.

The centered slope may seem the most symmetric choice at first glance, but due to the fact that the reconstructed function is then advected in the positive direction, the final updating formula turns out to be a non-symmetric 4-point formula, which is known as **Fromm's method**.

4.7.4 Oscillations. Second-order methods such as Lax–Wendroff (or Beam–Warming or Fromm's method) give oscillatory approximations to discontinuous solutions. This can be easily understood using the interpretation of Algorithm RSA.

Consider the Lax–Wendroff method, for example, applied to piecewise constant data with values

$$Q_i^n = \begin{cases} 1 & \text{if } i \le J \\ 0 & \text{if } i > J\,. \end{cases}$$

Choosing slopes in each grid cell based on the Lax–Wendroff prescription (4.45) gives the piecewise linear function shown in Fig. 4.7(a). The slope σ_i^n is nonzero only for $i = J$.

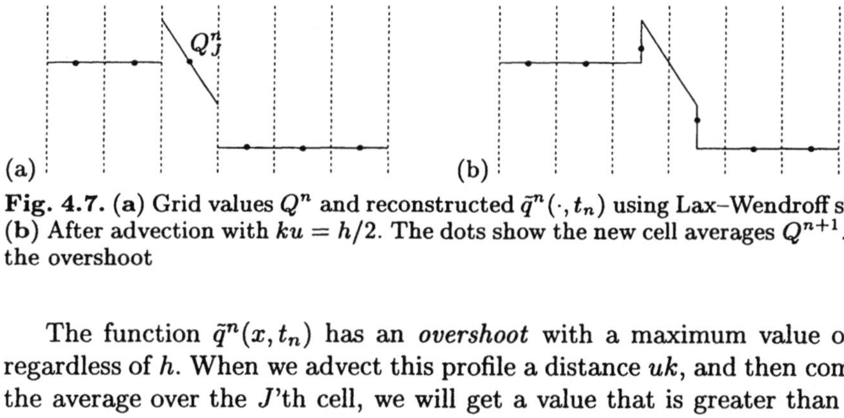

Fig. 4.7. (**a**) Grid values Q^n and reconstructed $\tilde{q}^n(\cdot, t_n)$ using Lax–Wendroff slopes. (**b**) After advection with $ku = h/2$. The dots show the new cell averages Q^{n+1}. Note the overshoot

The function $\tilde{q}^n(x, t_n)$ has an *overshoot* with a maximum value of 3/2 regardless of h. When we advect this profile a distance uk, and then compute the average over the J'th cell, we will get a value that is greater than 1 for any k with $0 < uk < h$. The worst case is when $uk = h/2$, in which case $\tilde{q}^n(x, t_{n+1})$ is shown in Fig. 4.7(b) and $Q_j^{n+1} = 9/8$. In the next time step this overshoot will be accentuated, while in cell $J-1$ we will now have a positive slope, leading to a value Q_{J-1}^{n+1} that is less than 1. This oscillation then grows with time.

The slopes proposed in the previous section were based on the assumption that the solution is smooth. Near a discontinuity there is no reason to believe that introducing this slope will improve the accuracy. On the contrary, if one of our goals is to avoid nonphysical oscillations, then in the above example we must set the slope to zero in the J'th cell. Any $\sigma_J^n < 0$ will lead to $Q_J^{n+1} > 1$, while a positive slope wouldn't make much sense. On the other hand we don't want to set all slopes to zero all the time, or we simply have the first-order upwind method. Where the solution is smooth we want second order accuracy. Moreover, we will see below that even near a discontinuity, once the solution is somewhat smeared out over more than one cell, introducing nonzero slopes can help keep the solution from smearing out too far, and hence will significantly increase the resolution and keep discontinuities fairly sharp, as long as care is taken to avoid oscillations.

This suggests that we must pay attention to *how the solution is behaving* near the i'th cell in choosing our formula for σ_i^n. (And hence the resulting updating formula will be *nonlinear* even for the linear advection equation!). Where the solution is smooth we want to choose something like the Lax–Wendroff slope. Near a discontinuity we may want to "limit" this slope, using a value that is smaller in magnitude in order to avoid oscillations. Methods based on this idea are known as **slope-limiter** methods. This approach was introduced by van Leer in a series of papers [246] through [248], where he developed the **MUSCL** scheme for nonlinear conservation laws (Monotonic Upstream-centered Scheme for Conservation Laws). The same idea in the context of **flux limiting**, reducing the magnitude of the numerical flux to avoid oscillations, was introduced in the **flux-corrected transport (FCT)** algorithms of Boris and Book [37]. We can view this as creating a hybrid algorithm that is second order accurate in smooth regions but which reduces to a more robust first-order algorithm near discontinuities. This idea of hybridization was also used in early work of Harten and Zwas [110]. An enormous variety of methods based on these principles have been developed in the past two decades. One of the algorithms of this type that is best known in the astrophysics community is the **piecewise parabolic method (PPM)** of Woodward and Colella [59], which uses a piecewise quadratic reconstruction, with appropriate limiting.

4.7.5 Total Variation. How much should we limit the slope in a piecewise linear reconstruction? Ideally we would like to have a mathematical prescription that will allow us to use the Lax–Wendroff slope whenever possible, for second-order accuracy, while guaranteeing that no non-physical oscillations will arise. To achieve this we need a way to measure "oscillations" in the solution. This is provided by the notion of the **total variation** of a function. For a grid function Q we define

$$TV(Q) = \sum_{i=-\infty}^{\infty} |Q_i - Q_{i-1}| \, . \tag{4.46}$$

For an arbitrary function $q(x)$ we can define

$$TV(q) = \sup \sum_{j=1}^{N} |q(\xi_j) - q(\xi_{j-1})| \, , \tag{4.47}$$

where the supremum is taken over all subdivisions of the real line $-\infty = \xi_0 < \xi_1 < \cdots < \xi_N = \infty$. Note that for the total variation to be finite Q or q must approach constant values $v^{\pm}$ as $x \to \pm\infty$.

The true solution to the advection equation simply propagates at speed u with unchanged shape, so that the total variation $TV(q(\cdot,t))$ must be constant in time. A numerical solution to the advection equation may not have constant total variation, however. If the method introduces oscillations, then we would expect the total variation of Q^n to *increase* with time. We can thus attempt to avoid oscillations by requiring that the method does not increase the total variation:

Definition 4.1. *A 2-level method is called* **total variation diminishing (TVD)** *if, for any set of data Q^n, the values Q^{n+1} computed by the method satisfy*

$$TV(Q^{n+1}) \le TV(Q^n)\,. \tag{4.48}$$

For a scalar conservation law, the exact solution has nonincreasing variation and so this is a reasonable condition to impose on a numerical method. Harten [103] introduced the use of this criterion in developing and analyzing numerical methods. For a scalar equation, steps 2 and 3 of Algorithm RSA are TVD, and so the overall method is TVD provided that the reconstruction step does not increase the variation, *i.e.*, as long as

$$TV(\tilde{q}^n) \le TV(Q^n)\,. \tag{4.49}$$

4.7.6 Slope-Limiter Methods. Now let's return to the derivation of numerical methods based on piecewise linear reconstruction, and consider how to "limit" the slopes so that (4.48) is satisfied. Note that setting $\sigma_i^n \equiv 0$ works, since the piecewise constant function has the same TV as the discrete data. Hence *the first-order upwind method is TVD* for the advection equation. Hence upwind may smear solutions but cannot introduce oscillations, a familiar result.

One choice of slope that gives second-order accuracy for smooth solutions while still satisfying the TVD property is the **minmod slope**

$$\sigma_i^n = \text{minmod}\left(\frac{Q_i^n - Q_{i-1}^n}{h}, \frac{Q_{i+1}^n - Q_i^n}{h}\right), \tag{4.50}$$

where the minmod function of two arguments is defined by

$$\text{minmod}(a,b) = \begin{cases} a & \text{if } |a| < |b| \text{ and } ab > 0 \\ b & \text{if } |b| < |a| \text{ and } ab > 0 \\ 0 & \text{if } ab \le 0\,. \end{cases} \tag{4.51}$$

Note that if a and b have the same sign then this selects the one which is smaller in modulus, else it returns zero.

Rather than defining the slope on the i'th cell by always using the downwind difference (which would give Lax–Wendroff), or by always using the upwind difference (which would give Beam–Warming), the minmod method compares the two slopes and chooses the one which is smaller in magnitude.

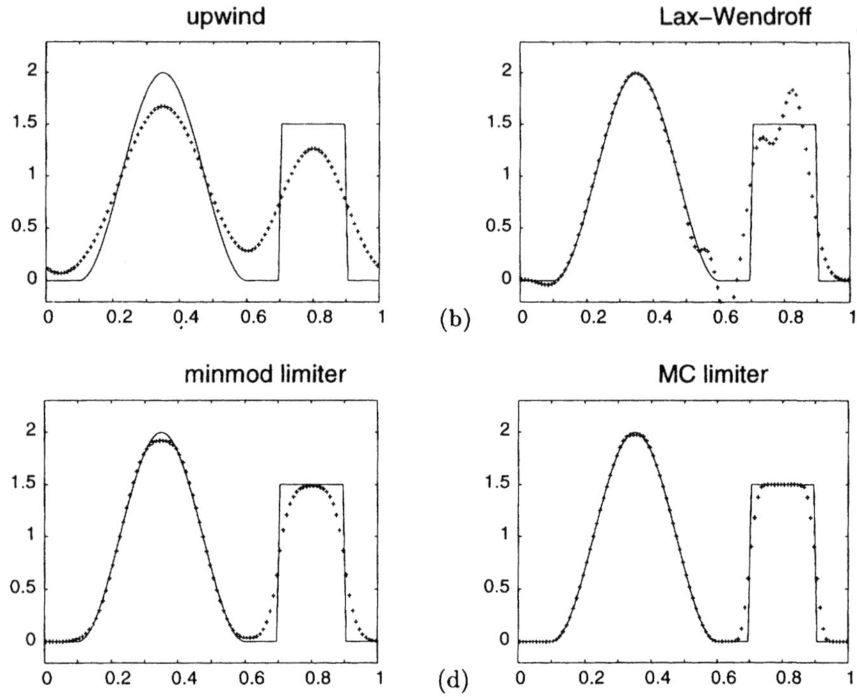

Fig. 4.8. Tests on the advection equation with different limiters. All results are at time $t = 1$, after one revolution with periodic boundary conditions

If the two slopes have different sign, then the value Q_i^n must be a local maximum or minimum, and it is easy to check in this case that we must set $\sigma_i^n = 0$ in order to satisfy (4.49).

Figure 4.8 shows a comparison of the upwind, Lax–Wendroff, and minmod methods for an advection problem with initial data consisting of both a smooth hump and a square wave. The advection velocity is $u = 1$ and periodic boundary conditions are used on $[0, 1]$ so that at integer times $t = 0,\ 1,\ 2,\ \ldots$ the solution should agree with the initial data. The figure shows solutions at $t = 1$ on a grid with $h = 0.01$ and $k = 0.005$ (so the Courant number is 0.5 — better results with all methods would be obtained with a Courant number closer to 1). We see that the minmod method does a fairly good job of maintaining good accuracy in the smooth hump and also sharp discontinuities in the square wave, with no oscillations.

Sharper resolution of discontinuities can be achieved with other limiters that do not reduce the slope as severely as minmod near a discontinuity. One popular choice is the **monotonized central-difference limiter (MC-limiter)**, which was proposed by van Leer [247]:

$$\sigma_i^n = \text{minmod}\left(\left(\frac{Q_{i+1}^n - Q_{i-1}^n}{2h}\right),\ 2\left(\frac{Q_i^n - Q_{i-1}^n}{h}\right),\ 2\left(\frac{Q_{i+1}^n - Q_i^n}{h}\right)\right).$$

This compares the central-difference of Fromm's method with *twice* the one-sided slope to either side. In smooth regions this reduces to the centered slope of Fromm's method but near discontinuities it gives sharper resolution than minmod while remaining TVD.

4.7.7 Flux Formulation with Piecewise Linears. The slope-limiter methods described above can be written as flux-differencing methods of the form (4.18). The updating formulas derived above can be manipulated algebraically to determine what the numerical flux function must be. Alternatively, we can derive the numerical flux by computing the exact flux through the interface x_i using the piecewise linear solution $\tilde{q}^n(x,t)$, by integrating $u\tilde{q}^n(x_i,t)$ in time from t_n to t_{n+1}. For the advection equation this is easy to do and we find that

$$\begin{aligned} F_i^n &= \frac{1}{k}\int_{t_n}^{t_{n+1}} u\tilde{q}^n(x_i,t)\,\mathrm{d}t \\ &= uQ_{i-1}^n + \frac{1}{2}u(h-ku)\sigma_{i-1}^n \,. \end{aligned}$$

Using this in the flux-differencing formula (4.18) gives

$$Q_i^{n+1} = Q_i^n - \frac{ku}{h}(Q_i^n - Q_{i-1}^n) - \frac{1}{2}\frac{ku}{h}(h-ku)(\sigma_i^n - \sigma_{i-1}^n)\,,$$

which agrees with (4.42).

If we also consider the case $u < 0$, then we will find that in general the numerical flux for a slope-limiter method is

$$F_i^n = \begin{cases} uQ_{i-1}^n + \frac{1}{2}u(h-ku)\sigma_{i-1}^n & \text{if } u \ge 0 \\ uQ_i^n - \frac{1}{2}u(h+ku)\sigma_i^n & \text{if } u \le 0\,, \end{cases} \tag{4.52}$$

where σ_i^n is the slope in the i'th cell C_i, chosen by one of the formulas discussed previously.

Rather than associating a slope σ_i^n with the i'th cell, the idea of writing the method in terms of fluxes between cells suggests that we should instead associate our approximation to q_x with the cell interface at x_i where F_i^n is defined. Across the interface x_i we have a jump

$$\Delta Q_i^n = Q_i^n - Q_{i-1}^n \tag{4.53}$$

and this jump divided by h gives an approximation to q_x. This suggests that we write the flux (4.52) as

$$F_i^n = u^- Q_i^n + u^+ Q_{i-1}^n + \frac{1}{2}|u|\left(1 - \left|\frac{ku}{h}\right|\right)\delta_i^n\,, \tag{4.54}$$

where $u^{\pm}$ are defined in (4.29). If δ_i^n is the jump ΔQ_i^n itself then this gives the Lax–Wendroff method. We see that the Lax–Wendroff flux can be interpreted as a modification to the upwind flux (4.28). This observation is crucial in the development of high-resolution methods.

4.7.8 Flux Limiters. From the above discussion it is natural to view Lax–Wendroff as the basic second-order method based on piecewise linear reconstruction. Other second-order methods have fluxes of the form (4.54) with different choices of δ_i^n. The slope-limiter methods can then be reinterpreted as **flux-limiter methods** by choosing δ_i^n to be a limited version of (4.53). In general we will set

$$\delta_i^n = \phi(\theta_i^n)\Delta Q_i^n \,, \tag{4.55}$$

where

$$\theta_i^n = \frac{\Delta Q_I^n}{\Delta Q_i^n} \,. \tag{4.56}$$

The index I here is used to represent the interface on the *upwind* side of x_i:

$$I = \begin{cases} i-1 & \text{if } u > 0 \\ i+1 & \text{if } u < 0 \,. \end{cases} \tag{4.57}$$

The ratio θ_i^n can be thought of as a measure of the smoothness of the data near x_i. Where the data is smooth we expect $\theta_i^n \approx 1$ (except at extrema). Near a discontinuity we expect that θ_i^n may be far from 1.

The function $\phi(\theta)$ is the **flux-limiter function**, whose value depends on the smoothness. Setting $\phi(\theta) \equiv 1$ for all θ gives the Lax–Wendroff method, while setting $\phi(\theta) \equiv 0$ gives upwind. More generally we might want to devise a limiter function ϕ that has values near 1 for θ near 1, but which reduces (or perhaps increases) the slope where the data is not smooth.

There are many other ways one might choose to measure the smoothness of the data besides the variable θ defined in (4.56). However, the framework proposed above results in very simple formulas for the ϕ function corresponding to many standard methods, including all the methods discussed so far.

In particular, note the very nice feature that choosing

$$\phi(\theta) = \theta \tag{4.58}$$

results in (4.55) becoming

$$\delta_i^n = \left(\frac{\Delta Q_I^n}{\Delta Q_i^n}\right) \Delta Q_i^n = \Delta Q_I^n \,.$$

Hence this choice results in the jump at the interface *upwind* from x_i being used to define δ_i^n instead of the jump at this interface. As a result, the method (4.54) with the choice of "limiter" (4.58) reduces to the Beam–Warming method.

Since the centered difference (4.43) is the average of the one-sided slopes (4.44) and (4.45), we also find that Fromm's method can be obtained by choosing

$$\phi(\theta) = \frac{1}{2}(1+\theta) \,. \tag{4.59}$$

Also note that $\phi(\theta) = 2$ corresponds to using $\delta_i^n = 2\Delta Q_i^n$, *i.e.*, twice the jump at this interface, while $\phi(\theta) = 2\theta$ results in using twice the jump at the upwind interface. Recall that these are necessary ingredients in some of the slope limiters discussed in Sect. 4.7.6.

Translating the various slope limiters into flux-limiter functions, we find the following expressions for some standard methods:

$$
\begin{array}{rl}
\textbf{Linear methods:} & \\
\text{upwind}: & \phi(\theta) = 0 \\
\text{Lax–Wendroff}: & \phi(\theta) = 1 \\
\text{Beam–Warming}: & \phi(\theta) = \theta \\
\text{Fromm}: & \phi(\theta) = \frac{1}{2}(1+\theta) \\
 & \\
\textbf{High-resolution limiters:} & \\
\text{minmod}: & \phi(\theta) = \text{minmod}(1,\theta) \\
\text{superbee}: & \phi(\theta) = \max(0,\ \min(1,2\theta),\ \min(2,\theta)) \\
\text{MC}: & \phi(\theta) = \max(0,\ \min((1+\theta)/2,\ 2,\ 2\theta)) \\
\text{van Leer}: & \phi(\theta) = \frac{\theta+|\theta|}{1+|\theta|}\ .
\end{array}
\tag{4.60}
$$

A wide variety of other limiters have also been proposed in the literature.

4.7.9 TVD Limiters. For simple limiters such as minmod, it is clear from the derivation as a slope-limiter (Sect. 4.7.6) that the resulting method is TVD, since it is easy to check that (4.49) is satisfied. For more complicated limiters we would like to have an algebraic proof that the resulting method is TVD. A fundamental tool in this direction is a theorem of Harten [103], which can be used to derive explicit algebraic conditions on the function ϕ required for a TVD method. Sweby [234] derived explicit constraints for limiter functions and shows that we require (see also [141]):

$$0 \le \phi(\theta) \le \text{minmod}(2, 2\theta)\ . \tag{4.61}$$

This defines the **TVD region** in the θ-ϕ plane: the curve $\phi(\theta)$ must lie in this region, which is shown as the shaded region in Fig. 4.9(a). This figure also shows the functions $\phi(\theta)$ from (4.60) for the Lax–Wendroff, Beam–Warming, and Fromm methods. All of these functions lie outside the TVD region for some values of θ, and these methods are not TVD. This graphical analysis of ϕ was first presented by Sweby [234], who analyzed a wide class of flux-limiter methods.

Note that for any second-order accurate method we must have $\phi(1) = 1$. Sweby found, moreover, that it is best to take ϕ to be a convex combination of $\phi = 1$ (Lax–Wendroff) and $\phi = \theta$ (Beam–Warming). Other choices apparently give too much compression, and smooth data such as a sine wave tends to turn into a square wave as time evolves. Imposing this additional restriction gives the "second order TVD" region of Sweby which is shown in Fig. 4.9(b).

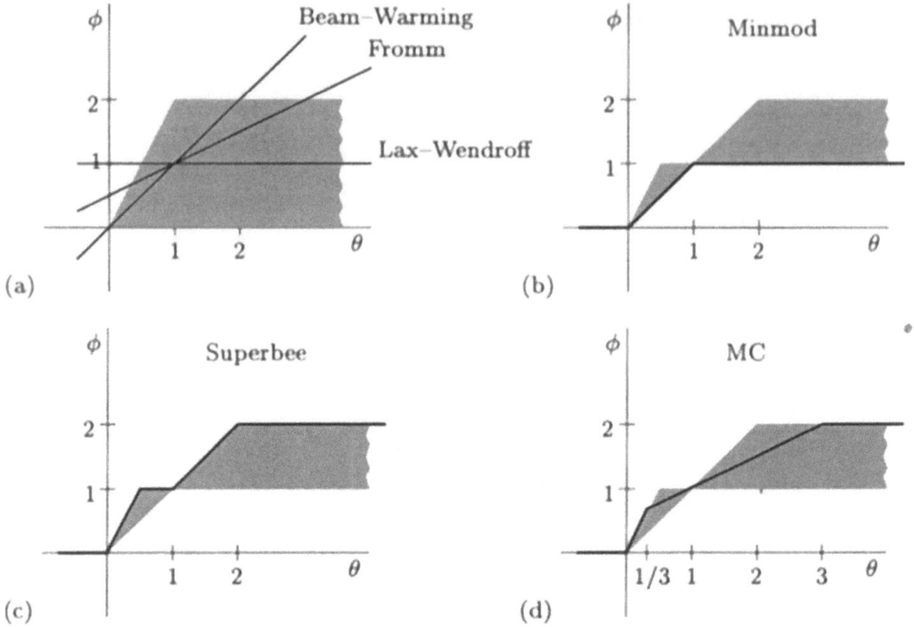

Fig. 4.9. Limiter functions $\phi(\theta)$. **(a)** The shaded region shows where function values must lie for the method to be TVD. The second-order linear methods have functions $\phi(\theta)$ that leave this region. **(b)** The shaded region is the Sweby region of second-order TVD methods. The minmod limiter lies along the lower boundary. **(c)** The superbee limiter lies along the upper boundary. **(d)** The MC limiter is smooth at $\phi = 1$

The "high-resolution" limiter functions from (4.60) are all seen to satisfy the constraints (4.61), and these limiters all give TVD methods. The ϕ functions are graphed in Fig. 4.9. Note that minmod lies along the lower boundary of the Sweby region while superbee lies along the upper boundary. The fact that these functions are not smooth at $\theta = 1$ corresponds to the fact that there is a switch in the choice of one-sided approximation used as θ crosses this point. This can lead to a loss in accuracy near inflection points. For full second order accuracy we would like the function ϕ to be smooth near $\theta = 1$, as for the MC limiter. The van Leer limiter is an even smoother version of this.

4.7.10 Linear Systems. The slope-limiter or flux-limiter methods can also be extended to systems of equations. This is most easily done in the flux-limiter framework, and will be illustrated first for the linear system $q_t + Aq_x = 0$.

We can write $A = A^+ + A^-$, where $A = R\Lambda R^{-1}$ is decomposed based on the sign of each eigenvalue: $A^+ = R\Lambda^+ R^{-1}$ and $A^- = R\Lambda^- R^{-1}$ where Λ^+, for example, has the positive part of each λ on the diagonal, so negative eigenvalues are replaced by zero. Using this, Godunov's method for a linear system (the generalization of the upwind method) can be written in terms of the flux function

$$F(Q_{i-1}, Q_i) = (A^+ Q_{i-1} + A^- Q_i) \,. \tag{4.62}$$

The Lax–Wendroff method (4.10) can also be written in flux-differencing form (4.18) if we define the flux by

$$F(Q_{i-1}, Q_i) = \frac{1}{2} A(Q_{i-1} + Q_i) + \frac{1}{2}\frac{k}{h} A^2 (Q_i - Q_{i-1}) \,. \tag{4.63}$$

We can rewrite this as

$$F(Q_{i-1}, Q_i) = (A^+ Q_{i-1} + A^- Q_i) + \frac{1}{2}|A| \left(1 - \frac{k}{h}|A|\right)(Q_i - Q_{i-1}) \,, \tag{4.64}$$

where $|A| = A^+ - A^-$.

In the form (4.64), we see that the Lax–Wendroff flux can be viewed as being composed of the Godunov flux (4.62) plus a correction term, just as for the scalar advection equation. To define a flux-limiter method we must limit the magnitude of this correction term based on how the data is varying. But for a system of equations, $\Delta Q_i = Q_i - Q_{i-1}$ is a *vector* and it is not so clear how to compare this vector with the neighboring jump ΔQ_{i-1} or ΔQ_{i+1} to generalize (4.55), nor which neighboring jump to consider, since the "upwind" direction is different for each eigen-component. The solution, of course, is that we must decompose the correction term in (4.64) into eigen-components and limit each scalar eigen-coefficient separately based on the algorithm for scalar advection.

We can rewrite the correction term as

$$\frac{1}{2}|A| \left(1 - \frac{k}{h}|A|\right)(Q_i - Q_{i-1}) = \frac{1}{2}|A| \left(1 - \frac{k}{h}|A|\right) \sum_{i=1}^{m} \alpha_i^p r^p \,,$$

where

$$(Q_i - Q_{i-1}) = \sum_{i=1}^{m} \alpha_i^p r^p$$

gives the decomposition of the jump in Q across this interface into eigenvectors of A, *i.e.*, the solution of the Riemann problem.

The flux-limiter method is defined by replacing the scalar coefficient α_i^p by a limited version, based on the scalar formulas of Sect. 4.7.8. We set

$$\tilde{\alpha}_i^p = \alpha_i^p \phi(\theta_i^p) \,, \tag{4.65}$$

where

$$\theta_i^p = \frac{\alpha_I^p}{\alpha_i^p} \qquad \text{with } I = \begin{cases} i-1 & \text{if } \lambda^p > 0 \\ i+1 & \text{if } \lambda^p < 0 \end{cases} \tag{4.66}$$

and ϕ is one of the limiter functions of Sect. 4.7.8. The flux function for the flux-limiter method is then

$$F_i = A^+Q_{i-1} + A^-Q_i + \tilde{F}_i \; , \tag{4.67}$$

where the first term is the upwind flux and the correction flux $\tilde{F}_i$ is defined by

$$\tilde{F}_i = \frac{1}{2}|A|\left(1 - \frac{k}{h}|A|\right)\sum_{i=1}^{m} \tilde{\alpha}_i^p r^p \; . \tag{4.68}$$

Note that in the case of a scalar equation, we can take $r^1 = 1$ as the eigenvector of $A = u$, so that $\Delta Q_i = \alpha_i^1$ which is what we called δ_i in Sect. 4.7.8. The formula (4.67) then reduces to (4.54). For a system of equations, the method just presented can also be obtained by diagonalizing the linear system and applying the scalar flux limiter method to each resulting advection equation. This is what we are doing at each cell interface by solving the Riemann problem.

Also note that the flux F_i depends not only on Q_{i-1} and Q_i, but also on Q_{i-2} and Q_{i+1} in general, because neighboring jumps are used in defining the limited values $\tilde{\alpha}_i^p$ in (4.68). The flux-limiter method thus has a 5-point stencil rather than the 3-point stencil of Lax–Wendroff. This is particularly important in specifying boundary conditions (see Sect. 4.9).

4.7.11 Implementation. For the constant coefficient linear system, we could compute the matrices A^+, A^-, and $|A|$ once and for all and compute the fluxes directly from the formulas given above. However, with limiters we must solve the Riemann problem at each interface to obtain a decomposition of ΔQ_i into waves $\alpha_i^p r^p$ and wave speeds λ^p and these can be used directly in the computation of Q_i^{n+1} without ever forming the matrices. This approach also generalizes directly to nonlinear systems of conservation laws, where we do not have a single matrix A but can still solve a Riemann problem at each interface for waves and wave speeds. This generalization is discussed briefly in the next section.

To accomplish this most easily, note that if we use the flux (4.67) in the flux-differencing formula (4.18) and then rearrange the upwind terms, we can write the formula for Q_i^{n+1} as

$$Q_i^{n+1} = Q_i^n - \frac{k}{h}(A^+\Delta Q_i + A^-\Delta Q_{i+1}) - \frac{k}{h}(\tilde{F}_{i+1} - \tilde{F}_i) \; , \tag{4.69}$$

where $\tilde{F}_i$ is given by (4.68). Here we drop the superscript n from the current time step since we will need to use superscript p below to denote the wave family. Each of the terms in this expression can be written in terms of the waves $\alpha_i^p r^p$ and wave speeds λ^p:

$$A^+\Delta Q_i = \sum_{p=1}^{m} (\lambda^p)^+ \alpha_i^p r^p \; , \tag{4.70}$$

$$A^-\Delta Q_i = \sum_{p=1}^{m} (\lambda^p)^- \alpha_i^p r^p \; , \tag{4.71}$$

and

$$\tilde{F}_i = \frac{1}{2}\sum_{p=1}^{m} |\lambda^p| \left(1 - \frac{k}{h}|\lambda^p|\right) \tilde{\alpha}_i^p r^p .$$

Once we have solved the Riemann problem for the waves $\alpha_i^p r^p$, limited waves $\tilde{\alpha}_i^p r^p$, and speeds λ^p, we can compute everything needed for the high-resolution method (4.69).

4.7.12 Extension to Nonlinear Systems. This approach is now easily extended to nonlinear hyperbolic systems. Given states Q_{i-1} and Q_i, the solution to the Riemann problem will yield a set of waves $\mathcal{W}_i^p \in \mathbb{R}^m$ and speeds $\lambda_i^p \in \mathbb{R}$, analogous to the the linear problem, though now the speeds will vary with i and so will the directions of the vectors $\mathcal{W}_i^p$ in phase space; they will no longer all be scalar multiples of a single set of eigenvectors r^p. The quantities $A^+\Delta Q_i$ and $A^-\Delta Q_i$ are generalized to *flux differences* denoted by $\mathcal{A}^+\Delta q_i$ and $\mathcal{A}^-\Delta q_i$, with the property that

$$\mathcal{A}^-\Delta q_i + \mathcal{A}^+\Delta q_i = f(Q_i) - f(Q_{i-1}) . \tag{4.72}$$

Note that for the linear case $f(Q_i) - f(Q_{i-1}) = A\Delta Q_i$ and this property is satisfied. In general we can think of setting

$$\mathcal{A}^-\Delta q_i = \sum_{p=1}^{m} (\lambda_i^p)^- \mathcal{W}_i^p, \tag{4.73}$$

$$\mathcal{A}^+\Delta q_i = \sum_{p=1}^{m} (\lambda_i^p)^+ \mathcal{W}_i^p , \tag{4.74}$$

a direct extension of (4.70). There are, however, some issues concerned with rarefaction waves and entropy conditions that make the nonlinear case more subtle, see [144] for a more complete discussion of this approach.

Once the waves, speeds, and flux differences have been suitably defined, the algorithm is virtually identical with what has already been defined in the linear case. We set

$$Q_i^{n+1} = Q_i^n - \frac{k}{h}(\mathcal{A}^-\Delta q_i + \mathcal{A}^+\Delta q_{i+1}) - \frac{k}{h}(\tilde{F}_{i+1} - \tilde{F}_i) , \tag{4.75}$$

where the first term corresponds to the Godunov method and the second term gives second-order corrections, with

$$\tilde{F}_i = \frac{1}{2}\sum_{p=1}^{m} |\lambda_i^p| \left(1 - \frac{k}{h}|\lambda_i^p|\right) \tilde{\mathcal{W}}_i^p . \tag{4.76}$$

Here $\tilde{\mathcal{W}}_i^p$ represents a limited version of the wave $\mathcal{W}_i^p$, obtained by comparing this jump with the jump $\mathcal{W}_I^p$ in the same family at the neighboring Riemann problem in the upwind direction, so

$$I = \begin{cases} i-1 & \text{if } \lambda_i^p > 0 \\ i+1 & \text{if } \lambda_i^p < 0 . \end{cases}$$

This limiting procedure is slightly more complicated than in the constant coefficient case since $\mathcal{W}_i^p$ and $\mathcal{W}_I^p$ are in general no longer scalar multiples of the same vector r^p. So we cannot simply apply the scalar limiter function $\phi(\theta)$ to the ratio of these scalar coefficients as in the constant coefficient linear case. Instead we can, for example, project the vector $\mathcal{W}_I^p$ onto $\mathcal{W}_i^p$ and compare the length of this projection with the length of $\mathcal{W}_i^p$.

4.8 Other Approaches

One approach to extending Godunov's method to a high-resolution method has been outlined above. Other approaches are also commonly used. One variant of the above method, often called the MUSCL approach following van Leer [248], consists of computing the flux

$$F_i^n \approx \frac{1}{\Delta t} \int_{t_n}^{t_{n+1}} f(q(x_i,t))\,\mathrm{d}t$$

by solving the Riemann problem using modified data $Q_i^{n+1/2,l}$ and $Q_i^{n+1/2,r}$ and then setting F_i^n to be the flux along $x/t = 0$ in this Riemann solution. The data $Q_i^{n+1/2,l}$ and $Q_i^{n+1/2,r}$ are constructed as approximations to $q(x_i, t_{n+1/2})$ at the midpoint in time, approaching the interface from the left and right, respectively. These are generally based on piecewise linear (or piecewise quadratic, in the case of PPM) reconstructions of the solution in each grid cell, together with characteristic extrapolation forward in time. Limiters are typically used in this reconstruction phase to avoid oscillations. Because this flux approximation is centered in time, and based on at least a linear approximation in space, such a method can be second-order accurate for smooth solutions.

It is hard to achieve better than second-order accuracy with the approaches discussed so far. For problems with discontinuous solutions this may not matter. Accuracy is degraded near the discontinuity anyway, and even smooth components of the flow may lose accuracy after interacting with shock wave in a nonlinear problem (see, *e.g.*, [72]). High-resolution methods with at best second-order accuracy are frequently used in practice. However, for some problems where there is considerable small-scale structure that is smoothly varying, it may be desirable to use methods with a higher formal order of accuracy. The semi-discrete approach described next is then a good approach.

4.8.1 Semi-discrete Methods. Rather than attempting to discretize simultaneously in space and time, another approach is to first discretize the PDE in space alone, obtaining a system of ODE's for the evolution of cell averages based on instantaneous values of the flux through each side. Then this ordinary differential equation is discretized in time by an accurate ODE method, typically a Runge–Kutta method.

Let $Q_i(t)$ be the i'th cell average at time t, for the moment assumed to be defined for all values of t. Then the integral form of the conservation law applied on the i'th cell gives

$$Q_i'(t) = -\frac{1}{h}[F_{i+1}(t) - F_i(t)] , \tag{4.77}$$

where $F_i(t) \approx f(q(x_i, t))$ is the flux through the left cell edge at this instant in time, for example. Suppose we have some formula for this flux in terms of the cell average, say

$$F_i(t) = \mathcal{F}(Q_{i-1}(t), Q_i(t)) .$$

(In practice it may depend on more than just the two neighboring values.) Then we have the **semi-discrete method**

$$Q_i'(t) = -\frac{1}{h}[\mathcal{F}(Q_i^n, Q_{i+1}^n) - \mathcal{F}(Q_{i-1}^n, Q_i^n)] , \tag{4.78}$$

which has been discretized in space, but not yet in time.

If we discretize (4.78) by the forward Euler method for ODE's, we would obtain

$$Q_i^{n+1} = Q_i^n - \frac{k}{h}[\mathcal{F}(Q_i^n, Q_{i+1}^n) - \mathcal{F}(Q_{i-1}^n, Q_i^n)] .$$

This has the same form as (4.20). However, if we now view $\mathcal{F}(Q_{i-1}^n, Q_i^n)$ as an approximation to the flux at time t_n, then this will only be first-order accurate in time, no matter how accurate the spatial discretization. We previously achieved second-order accuracy by determining $\mathcal{F}$ so that $\mathcal{F}(Q_{i-1}^n, Q_i^n)$ approximates the average flux over $[t_n, t_{n+1}]$, or the pointwise value of the flux at $t_{n+1/2}$. That approach is hard to extend to higher order since it requires using more terms in a Taylor series expansion, involving mixed x-t derivatives.

The advantage of the semi-discrete approach is that achieving higher accuracy in space and time are decoupled processes. Instead of forward Euler we could apply a higher-order Runge–Kutta method to (4.78). These are multistage methods that require evaluating $\mathcal{F}$ at one or more intermediate times within each time step, but if $\mathcal{F}$ produces a sufficiently accurate instantaneous flux based on any given set of cell averages, then arbitrarily high-order accuracy can be obtained. Special TVD Runge–Kutta methods have been developed for use in the context of semi-discrete methods for conservation laws [216].

A flux function $\mathcal{F}$ delivering high-order accuracy in space can be obtained for smooth solutions by using higher-order interpolating polynomials. One popular class of methods are the **ENO schemes** (Essentially Non-Oscillatory) which adaptively choose stencils for high-order interpolation based on the behavior of the solution in order to avoid interpolating across a discontinuity. Some references on ENO schemes, and more recent work on WENO (weighted ENO), include [104], [109], [119], [155], [215], [216].

4.8.2 Flux-Vector Splitting. We now return to the question of defining a first-order accurate interface flux and consider an alternative to the Godunov method presented above, but which has many similarities and in particular is identical for linear systems. The Godunov method can be interpreted as a **flux-difference splitting** method, since according to (4.72) the difference $f(Q_i) - f(Q_{i-1})$ is split into a left-going portion $\mathcal{A}^- \Delta q_i$ which modifies Q_{i-1} and a right-going portion $\mathcal{A}^+ \Delta q_i$ which modifies Q_i. This splitting is accomplished by solving the Riemann problem between Q_{i-1} and Q_i.

An alternative is to split each flux $f(Q_i)$, evaluated at the cell average, into a left-going part f_i^- and a right-going part f_i^+, so we have

$$f(Q_i) = f_i^- + f_i^+ \,.$$

We can then define the interface flux

$$F_i = f_{i-1}^+ + f_i^-$$

based on the portion of each cell-centered flux approaching the interface. A method of this form is called a **flux-vector splitting** method, since it is the flux vector $f(Q_i)$ which is split instead of the flux difference.

One possibility for the Euler equations is based on the observation that these equations have the special property that $f(\alpha q) = \alpha f(q)$ for any scalar α and are said to be *homogeneous of degree 1*. From this it follows that

$$f(q) = f'(q)q$$

for any state q, which is not true for nonlinear functions that do not have this homogeneity. Hence if A_i is the Jacobian matrix $f'(Q_i)$ (or some approximation to it), then a natural flux-vector splitting is given by

$$f_i^- = A_i^- Q_i \,, \qquad f_i^+ = A_i^+ Q_i \,.$$

We thus have

$$F_i = A_{i-1}^+ Q_{i-1} + A_i^- Q_i \,.$$

For a constant coefficient linear problem ($A_i \equiv A$), this agrees with (4.62) and this flux-vector splitting is identical to the flux-difference splitting of Godunov's method. For a nonlinear problem they are typically different.

The above splitting for the Euler equations is called **Steger–Warming flux-vector splitting** [227] in the aerodynamics community. An equivalent method, known as the **beam scheme,** was introduced earlier in astrophysics [206] based on a different viewpoint; each state Q_i is decomposed into distinct beams of particles traveling at the different wave speeds.

For transonic flow problems in aerodynamics, the flux-vector splitting given above suffers from the fact that the splitting does not behave smoothly as the Mach number passes through 1 (where the $u - c$ or $u + c$ characteristic speed changes sign). A smoother flux-vector splitting was introduced by van Leer [249], and other variants and improvements have since been introduced, e.g., [152], [153].

The beam scheme was motivated by considering the collisionless Boltzman equation as a model of gas dynamics, as discussed in [108]. More recently the BGK approximation to the collision term, due to Bhatnagar, Gross, and Krook [31], has been used to obtain more accurate Boltzman schemes. BGK methods have the advantage of not requiring Riemann solutions as such, and may be particularly useful for low density gases. Prendergast and Xu [189] give a nice introduction to this approach.

4.8.3 The Marquina Flux. A flux-vector splitting recently introduced by Marquina [73], [157] appears to be very robust on difficult astrophysical computations. Let r_i^p and $\lambda^p(Q_i)$ be the right eigenvectors and eigenvalues of $f'(Q_i)$, respectively, and decompose both Q_i and $f(Q_i)$ into these eigenvectors:

$$f(Q_i) = \sum_{p=1}^{m} \phi_i^p r_i^p , \qquad Q_i = \sum_{p=1}^{m} \omega_i^p r_i^p . \tag{4.79}$$

Let

$$\delta_i^p = \operatorname{sgn}(\lambda^p(Q_i)) , \qquad \Lambda^p = \max_{q=Q_{i-1},Q_i} |\lambda^p(q)| .$$

Then set

$$\phi_i^{p-} = \begin{cases} 0 & \text{if } \lambda^p(Q_i) > 0 \text{ and } \lambda^p(Q_{i-1}) > 0 \\ \phi_i^p & \text{if } \lambda^p(Q_i) < 0 \text{ and } \lambda^p(Q_{i-1}) < 0 \\ \frac{1}{2}(\phi_i^p - \Lambda^p \omega_i^p) & \text{if } \lambda^p(Q_i)\lambda^p(Q_{i-1}) < 0 , \end{cases}$$

$$\phi_{i-1}^{p+} = \begin{cases} 0 & \text{if } \lambda^p(Q_i) < 0 \text{ and } \lambda^p(Q_{i-1}) < 0 \\ \phi_{i-1}^p & \text{if } \lambda^p(Q_i) > 0 \text{ and } \lambda^p(Q_{i-1}) > 0 \\ \frac{1}{2}(\phi_{i-1}^p - \Lambda^p \omega_{i-1}^p) & \text{if } \lambda^p(Q_i)\lambda^p(Q_{i-1}) < 0 . \end{cases}$$

The Marquina flux is then given by

$$F_i^n = \sum_{p=1}^{m} \left[\phi_{i-1}^{p+} r_{i-1}^p + \phi_i^{p-} r_i^p\right] . \tag{4.80}$$

If $\lambda^p(Q_i)$ and $\lambda^p(Q_{i-1})$ have the same sign, then this just reduces to the Steger–Warming splitting. When the eigenvalue changes sign, this introduces some additional dissipation into the flux in a manner motivated by the highly-diffusive Lax–Friedrichs method [157]. In the case $\lambda^p(Q_{i-1}) < 0 < \lambda^p(Q_i)$ this dissipation enforces the entropy condition. When $\lambda^p(Q_{i-1}) > 0 > \lambda^p(Q_i)$ there is a shock rather than a rarefaction wave, but this additional dissipation can still be useful in certain circumstances such as slowly-moving shocks (see Sect. 7.5).

Marquina used this flux function in conjunction with an ENO-based semi-discrete method and Runge–Kutta time stepping (see Sect. 4.8.1) to achieve higher-order accuracy. The flux function can also be used in conjunction with the high-resolution method developed in Sect. 4.7.12 by using F_i to define a flux-difference splitting via

$$\begin{aligned} \mathcal{A}^-\Delta q_i &= F_i - f(Q_{i-1}) , \\ \mathcal{A}^+\Delta q_i &= f(Q_i) - F_i . \end{aligned}$$

Any numerical flux definition can be rewritten as a flux-difference splitting by this device. To apply the high-resolution method we must also define waves $\mathcal{W}_i^p$ and speeds λ_i^p for use in the second-order correction terms of (4.76). From (4.79) we have

$$Q_i - Q_{i-1} = \sum_{p=1}^{m} (\omega_i^p r_i^p - \omega_{i-1}^p r_{i-1}^p) ,$$

which suggests defining the p'th wave $\mathcal{W}_i^p$ as

$$\mathcal{W}_i^p = \omega_i^p r_i^p - \omega_{i-1}^p r_{i-1}^p .$$

The corresponding speed λ_i^p might be defined as

$$\lambda_i^p = \frac{1}{2}(\lambda^p(Q_{i-1}) + \lambda^p(Q_i)) .$$

This solver has been implemented in CLAWPACK by D. Bale and used for some of the relativistic calculations presented in Sect. 9. It was found to work well in cases where other solvers failed.

4.9 Boundary Conditions

So far we have only studied methods for updating the cell average Q_i^n assuming that we have neighboring cell values Q_{i-1}^n and Q_{i+1}^n and perhaps values further away as needed in order to compute the fluxes F_i^n and F_{i+1}^n. In practice we must always compute on some finite set of grid cells covering a bounded domain, and in the first and last cells we will not have the required neighboring information. Instead we have some set of physical boundary conditions that must be used in updating these cell values. One approach is to develop special formulas for use near the boundaries, which will depend both on what type of boundary condition is specified and on what sort of method we are trying to match. However, in general it is much easier to think of extending the computational domain to include a few additional cells on either end, called **ghost cells**, whose values are set at the beginning of each time step in some manner that depends on the boundary conditions and perhaps the interior solution. Then these values provide the neighboring cell values needed in updating the cells near the physical domain. The updating formula is then exactly the same in all cells, and there is no need to develop a special flux-limiter method, say, that works with boundary data instead of initial data. Instead the boundary conditions must be used in deciding how to set the values of the ghost cells, but this can generally be done in a way that depends only on the boundary conditions and is decoupled entirely from the choice of numerical method that is then applied.

Suppose the problem is on the physical domain $[a, b]$, which is subdivided into cells $\mathcal{C}_1, \mathcal{C}_2, \ldots, \mathcal{C}_N$ with $x_1 = a$ and $x_{N+1} = b$, so that $h = (b-a)/N$. If we use a method for which F_i depends only on Q_{i-1} and Q_i, then we need only one ghost cell on either end. The ghost cell $\mathcal{C}_0 = [a-h, a)$ allows us to calculate the flux F_1 at the left boundary while the ghost cell $\mathcal{C}_{N+1} = [b, b+h)$ is used to calculate F_{N+1} at $x = b$. With a flux-limiter method of the type developed above, we will generally need two ghost cells at each boundary since, for example, the jump $Q_0 - Q_{-1}$ will be needed in limiting the flux correction in F_1. For a method with an even wider stencil, additional ghost cells would be needed.

We will refer to the solution in the original domain $[a, b]$ as the **interior solution**, which is computed in each time step by the numerical method. At the start of each time step we have the interior values $Q_1^n, \ldots, Q_N^n$ obtained from the previous time step (or from the initial conditions if $n = 0$), and we apply a **boundary condition procedure** to fill the ghost cells with values Q_0^n, Q_{N+1}^n, *etc.* before applying the method on the next time step. We will look at several examples to see how the ghost cell values might be set in order to implement various physical boundary condtions.

4.9.1 Periodic Boundary Conditions. Periodic boundary conditions of the form $q(a, t) = q(b, t)$ are very easy to impose with any numerical method. In updating Q_1 we need values Q_0 to the left and Q_2 to the right (for a 3-point method). By periodicity the value Q_0 should agree with the value Q_N in the last cell. One could code the formula for updating Q_1 separately to use Q_N in place of the value Q_{i-1} that would normally be used for $i > 1$, but it is simpler to use the ghost-cell approach and simply set $Q_0^n = Q_N^n$ before computing fluxes and updating the cell values, so that the same formula can then be used everywhere. With a 5-point stencil we need to fill two ghost cells at each boundary, and we set

$$Q_{-1}^n = Q_{N-1}^n, \quad Q_0^n = Q_N^n, \quad Q_{N+1}^n = Q_1^n, \quad Q_{N+2}^n = Q_2^n \tag{4.81}$$

at the start of each time step.

4.9.2 Outflow Boundaries. Often we have artificial computational boundaries that arise simply because we can only solve the problem on a bounded domain. At such boundaries we often want to have no incoming signal, while there may be out-going waves that should leave the domain cleanly without generating spurious reflections at the artificial boundary. We thus want **nonreflecting** boundary conditions. At such boundaries we can often set ghost cell values by **extrapolation** from the interior solution. If the ghost cell value Q_{N+1}^n is set based on $Q_N^n, Q_{N-1}^n, \ldots$, then the new value Q_N^{n+1} will effectively be computed on the basis of values to the left alone, even if the formula depends on Q_{N+1}^n, and hence this reduces to some sort of upwind method. The simplest approach is to use a **zero-order extrapolation**, meaning extrapolation by a constant function. We simply set

$$Q_{N+1}^n = Q_N^n, \qquad Q_{N+2}^n = Q_N^n \tag{4.82}$$

at the start of each time step. The idea of extrapolation at outflow boundaries turns out to be extremely powerful in conjunction with methods based on solving Riemann problems. If there is no jump in the values at the boundary, there are no waves in the Riemann solution and in particular no incoming waves.

4.9.3 Solid Walls and Symmetry. Now suppose $x = 0$ is a line of symmetry, or a solid wall. Both can be handled by the same procedure for the Euler equations. If the flow is initially symmetric about $x = 0$, meaning that ϱ and p are even functions while the velocity u is odd, then this will be maintained at later times. If we are solving the equations only for $x > 0$, then we can specify values in the ghost cells according to

$$Q_0^n = \mathcal{R}Q_1^n\,, \qquad Q_{-1}^n = \mathcal{R}Q_2^n\,, \tag{4.83}$$

where $\mathcal{R}$ represents the reflection operator which simply negates the velocity while preserving the mass and pressure. These are the values we would find in these cells if we performed the full calculation with symmetric initial data, and so setting the ghost cell values in this manner gives an identical solution within the domain $x > 0$.

Since velocity is an odd function, the velocity will be zero right at the point $x = 0$. For this reason the numerical boundary conditions (4.83) are also correct for a solid wall, where $u = 0$ is the only physical boundary condition to be imposed. Note that when solving the Riemann problem with this symmetric data, the resulting symmetric similarity solution will have a 1-wave and 3-wave which are symmetric, and a contact discontinuity which must travel at speed zero. Hence the velocity between the wall and the 3-wave moving into the computational domain will be zero, as we expect physically. The in-going wave will be shock if the velocity in the first cell was negative (flow into the wall) or a rarefaction wave if that velocity was positive (flow away from the wall).

5. Source Terms and Fractional Steps

Many "conservation laws" involve source terms, as discussed in Sect. 2.3. There are various ways to handle these terms, which fall into two basic categories:

- Unsplit methods, in which a single finite-difference formula is developed to advance the full equation over one time step.
- Fractional step (splitting) methods, in which the problem is broken down into pieces corresponding to the different processes, and a numerical method appropriate for each separate piece is applied independently. This approach is also often used to split multi-dimensional problems into a sequence of one-dimensional problems (*dimensional splitting*, as discussed in Sect. 6.).

To illustrate, we begin with a simple advection-reaction equation of the form

$$q_t + uq_x = -\lambda q \,, \tag{5.1}$$

with data $q(x,0) = q^0(x)$. This would model, for example, the transport of a radioactive material in a fluid flowing at constant speed u down a pipe. The material decays as it flows along, at rate λ. We can easily compute the exact solution of (5.1), since along the characteristic $\mathrm{d}x/\mathrm{d}t = u$ we have $\mathrm{d}q/\mathrm{d}t = -\lambda q$, and hence

$$q(x,t) = \mathrm{e}^{-\lambda t} q^0(x-ut) \,. \tag{5.2}$$

5.1 Unsplit Methods

It is easy to develop unsplit methods for (5.1). For example, an obvious extension of the upwind method for advection would be (assuming $u > 0$),

$$Q_i^{n+1} = Q_i^n - \frac{ku}{h}(Q_i^n - Q_{i-1}^n) - k\lambda Q_i^n \,. \tag{5.3}$$

This method is first-order accurate and stable for $0 < uk/h < 1$.

A second order Lax–Wendroff style method can be developed by using the Taylor series

$$q(x,t+k) \approx q(x,t) + kq_t(x,t) + \frac{1}{2}k^2 q_{tt}(x,t) \,. \tag{5.4}$$

As in the derivation of Lax–Wendroff, we must compute q_{tt} from the PDE. Differentiating q_t gives

$$q_{tt} = -uq_{xt} - \lambda q_t \,, \qquad q_{tx} = -uq_{xx} - \lambda q_x \,,$$

and combining these, we obtain

$$q_{tt} = u^2 q_{xx} + 2u\lambda q_x + \lambda^2 q \,. \tag{5.5}$$

Using this expression for q_{tt} in (5.4) gives

$$\begin{aligned} q(x,t+k) &\approx q - k(uq_x + \lambda q) + \tfrac{1}{2}k^2(u^2 q_{xx} + 2u\lambda q_x + \lambda^2 q) \\ &= \left(1 - k\lambda + \tfrac{1}{2}k^2\lambda^2\right) q - ku\left(1 - \tfrac{1}{2}k\lambda\right) q_x + \tfrac{1}{2}k^2u^2 q_{xx} \,. \end{aligned} \tag{5.6}$$

We can now approximate x-derivatives by finite differences to obtain the second-order method

$$\begin{aligned} Q_i^{n+1} &= \left(1 - k\lambda + \frac{1}{2}k^2\lambda^2\right) Q_i^n - \frac{ku}{2h}\left(1 - \frac{1}{2}k\lambda\right)(Q_{i+1}^n - Q_{i-1}^n) \\ &\quad + \frac{k^2u^2}{2h^2}(Q_{i-1}^n - 2Q_i^n + Q_{i+1}^n) \,. \end{aligned} \tag{5.7}$$

Note that in order to correctly model the equation (5.1) to second order accuracy, we must properly model the interaction between the uq_x and the λq terms, which brings in the mixed term $-\frac{1}{2}k^2 u\lambda q_x$ in the Taylor series expansion.

5.2 Fractional Step Methods

A fractional step method for (5.1) is applied by first splitting the equation into two *subproblems* which can be solved independently. For the advection-reaction problem (5.1) we might take these to be:

$$\text{Problem A:} \quad q_t + uq_x = 0 \,, \tag{5.8}$$

$$\text{Problem B:} \quad q_t = -\lambda q \,. \tag{5.9}$$

The idea with the fractional step method is to combine these by applying the two methods in an alternating manner. This has great advantage over attempting to derivé an unsplit method, especially for more complicated problems. If we split the general problem $q_t + f(q)_x = \psi(q)$ into the homogeneous conservation law and a simple ODE, then we can use standard methods for each. In particular, the high-resolution shock-capturing methods can be used directly for the homogeneous conservation law, whereas trying to develop an unsplit method based on the same ideas while incorporating the source term directly can be more difficult.

As a simple example of the fractional step procedure, suppose we use the upwind method for the A-step and forward Euler for the ODE in the B-step for the advection-reaction problem. Then the simplest fractional step method over one time step would consist of the following 2 stages:

$$\text{A-step:} \quad Q_i^* = Q_i^n - u\frac{k}{h}(Q_i^n - Q_{i-1}^n) \,, \tag{5.10}$$

$$\text{B-step:} \quad Q_i^{n+1} = Q_i^* - k\lambda Q_i^* \,. \tag{5.11}$$

Note that we first take a time step of length k with upwind, starting with initial data Q_i^n to obtain the intermediate value Q_i^*. Then we take a time step of length k using forward Euler, starting with the data Q^* obtained from the first stage.

It may seem that we have advanced the solution by time $2k$ after taking these two steps of length k. However, in each stage we used only some of the terms in the original PDE, and the two stages combined give a consistent approximation to solving the original PDE (5.1) over a single time step of length k.

To check this consistency, we can combine the two stages by eliminating Q^* to obtain a method in a more familiar form:

$$\begin{aligned} Q_i^{n+1} &= (1-k\lambda)Q_i^* \\ &= (1-k\lambda)\left[Q_i^n - u\tfrac{k}{h}(Q_i^n - Q_{i-1}^n)\right] \\ &= Q_i^n - u\tfrac{k}{h}(Q_i^n - Q_{i-1}^n) - k\lambda Q_i^n + \tfrac{uk^2\lambda}{h}(Q_i^n - Q_{i-1}^n) \,. \end{aligned} \tag{5.12}$$

The first three terms on the right-hand side agree with the unsplit method (5.3). The final term is $O(k^2)$ (since $(Q_i^n - Q_{i-1}^n)/h \approx q_x = O(1)$) and so a local truncation error analysis will show that this method, though slightly different from (5.3), is also consistent and first-order accurate on the original equation (5.1).

A natural question is whether we could improve the accuracy by using a more accurate method in each step. For example, suppose we use Lax–Wendroff in the A-step and the trapezoidal method, or the 2-stage Runge–Kutta method, in the B-step. Would we then obtain a second order accurate method for the original equation? For this particular equation, the answer is yes. In fact if we use p'th order accurate methods for each step, the result will be a p'th order accurate method for the full original equation. But this equation is very special in this regard. In general one would think that splitting the equation into pieces in this manner would introduce some error that depends on the size of the time step k and is indepenendent of how well we then approximate the subproblem in each step. In general this is true — there is a "splitting error" that in general would be $O(k)$ for the type of splitting used above, and so the resulting fractional step method will be only first-order accurate, no matter how well we then approximate each step. This will be analyzed in more detail below.

For the case of equation (5.1) there is no splitting error. This follows from the observation that we can solve (5.1) over any time period k by first solving equation (5.8) over time k, and then using the result as data to solve the equation (5.9) over time k. Physically we can interpret this as follows. Think of the original equation as modeling a radioactive tracer that is advecting with constant speed u (carried along in a fluid, say) and also decaying with rate λ. Since the decay properties are independent of the position x, we can think of first advecting the tracer over time k without allowing any decay, and then holding the fluid and tracer stationary while we allow it to decay for time k. We will get the same result, and this is what we have done in the fractional step method.

On the other hand, suppose we modify the equation slightly so that the decay rate λ depends on x,

$$q_t + uq_x = -\lambda(x)q \,. \tag{5.13}$$

Then our previous argument for the lack of a splitting error breaks down — advecting the tracer a distance uk and then allowing it to decay, with rates given by the values of λ at the final positions, will not in general give the same result as if the decays occurs continuously as it advects, using the instantaneous rate given by $\lambda(x)$ at each point passed. In this case there would be a nonzero splitting error.

5.3 General Formulation of Fractional Step Methods

Consider a more general PDE of the form

$$q_t = (\mathcal{A} + \mathcal{B})q \,, \tag{5.14}$$

where $\mathcal{A}$ and $\mathcal{B}$ may be differential operators, e.g., $\mathcal{A} = -u\partial_x$ and $\mathcal{B} = \lambda(x)$ in the previous example. For simplicity suppose that $\mathcal{A}$ and $\mathcal{B}$ do not depend explicitly on t, e.g., $\lambda(x)$ is a function of x but not of t. Then we can compute that

$$q_{tt} = (\mathcal{A}+\mathcal{B})q_t = (\mathcal{A}+\mathcal{B})^2 q ,$$

and in general

$$\partial_t^j q = (\mathcal{A}+\mathcal{B})^j q .$$

We can then write the solution at time k using Taylor series as

$$\begin{aligned} q(x,k) &= q(x,0) + k(\mathcal{A}+\mathcal{B})q(x,0) + \tfrac{1}{2}k^2(\mathcal{A}+\mathcal{B})^2 q(x,0) + \cdots \\ &= \left(I + k(\mathcal{A}+\mathcal{B}) + \tfrac{1}{2}k^2(\mathcal{A}+\mathcal{B})^2 + \cdots\right) q(x,0) \\ &= \textstyle\sum_{j=0}^{\infty} \frac{k^j}{j!}(\mathcal{A}+\mathcal{B})^j q(x,0) , \end{aligned} \tag{5.15}$$

which formally could be written as

$$q(x,k) = \mathrm{e}^{k(\mathcal{A}+\mathcal{B})} q(x,0) .$$

With the fractional step method, we instead compute

$$q^*(x,k) = \mathrm{e}^{k\mathcal{A}} q(x,0) ,$$

and then

$$q^{**}(x,k) = \mathrm{e}^{k\mathcal{B}}\mathrm{e}^{k\mathcal{A}} q(x,0) ,$$

and so the *splitting error* is

$$q(x,k) - q^{**}(x,k) = \left(\mathrm{e}^{k(\mathcal{A}+\mathcal{B})} - \mathrm{e}^{k\mathcal{B}}\mathrm{e}^{k\mathcal{A}}\right) q(x,0) . \tag{5.16}$$

This should be calculated using the Taylor series expansions. We have (5.15) already, while

$$\begin{aligned} q^{**}(x,k) &= \left(I + k\mathcal{B} + \frac{1}{2}k^2\mathcal{B}^2 + \cdots\right)\left(I + k\mathcal{A} + \frac{1}{2}k^2\mathcal{A}^2 + \cdots\right) q(x,0) \\ &= \left(I + k(\mathcal{A}+\mathcal{B}) + \frac{1}{2}k^2(\mathcal{A}^2 + 2\mathcal{B}\mathcal{A} + \mathcal{B}^2) + \cdots\right) q(x,0) . \end{aligned} \tag{5.17}$$

The $I + k(\mathcal{A}+\mathcal{B})$ terms agree with (5.15). In the k^2 term, however, the term from (5.15) is

$$\begin{aligned} (\mathcal{A}+\mathcal{B})^2 &= (\mathcal{A}+\mathcal{B})(\mathcal{A}+\mathcal{B}) \\ &= \mathcal{A}^2 + \mathcal{A}\mathcal{B} + \mathcal{B}\mathcal{A} + \mathcal{B}^2 . \end{aligned} \tag{5.18}$$

In general this is *not* the same as

$$\mathcal{A}^2 + 2\mathcal{B}\mathcal{A} + \mathcal{B}^2 ,$$

and so the splitting error is

$$q(x,k) - q^{**}(x,k) = \frac{1}{2}k^2(\mathcal{A}\mathcal{B} - \mathcal{B}\mathcal{A})q(x,0) + O(k^3) . \tag{5.19}$$

The splitting error is zero only in the special case when the differential operators $\mathcal{A}$ and $\mathcal{B}$ commute (in which case it turns out that all the higher-order terms in the splitting error also vanish).

If we now design a fractional step method based on this splitting, we will see that the splitting error alone will introduce an $O(k^2)$ error in each time step, which can be expected to accumulate to an $O(k)$ error after the T/k time steps needed to reach some fixed time T (in the best case, assuming the method is stable). Hence even if we solve each subproblem *exactly* within the fractional step method, the resulting method will be only first-order accurate. If the subproblems are actually solved with numerical methods that are p'th order accurate, the solution will still only be first-order accurate no matter how large p is.

The above form of fractional step method is sometimes called the **Godunov splitting**, which in general is only first-order accurate. It turns out that a slight modification of the splitting idea will yield second order accuracy quite generally (assuming each subproblem is solved with a method of at least this accuracy). The idea is to solve the first subproblem $q_t = \mathcal{A}u$ over only a half time step of length $k/2$. Then we use the result as data for a full time step on the second subproblem $q_t = \mathcal{B}q$, and finally take another half time step on $q_t = \mathcal{A}u$. We can equally well reverse the roles of $\mathcal{A}$ and $\mathcal{B}$ here. This approach is often called **Strang splitting** as it was popularized in a paper by Strang [231] on solving multi-dimensional problems.

To analyze the Strang splitting, note that we are now approximating the solution operator $e^{k(\mathcal{A}+\mathcal{B})}$ by $e^{\frac{1}{2}k\mathcal{A}}e^{k\mathcal{B}}e^{\frac{1}{2}k\mathcal{A}}$. Taylor series expansion of this product shows that

$$\begin{aligned} e^{\frac{1}{2}k\mathcal{A}}e^{k\mathcal{B}}e^{\frac{1}{2}k\mathcal{A}} &= \left(I + \frac{1}{2}k\mathcal{A} + \frac{1}{8}k^2\mathcal{A}^2 + \cdots\right)\left(I + k\mathcal{B} + \frac{1}{2}k^2\mathcal{B}^2 + \cdots\right) \\ &\quad \left(I + \frac{1}{2}k\mathcal{A} + \frac{1}{8}k^2\mathcal{A}^2 + \cdots\right) \qquad (5.20) \\ &= I + k(\mathcal{A}+\mathcal{B}) + \frac{1}{2}k^2(\mathcal{A}^2 + \mathcal{A}\mathcal{B} + \mathcal{B}\mathcal{A} + \mathcal{B}^2) + O(k^3)\,. \end{aligned}$$

Comparing with (5.15), we seee that the $O(k^2)$ term is now captured correctly. The $O(k^3)$ term is not correct in general, however, unless $\mathcal{A}\mathcal{B} = \mathcal{B}\mathcal{A}$.

Note that over several time steps we can simplify the expression obtained with the Strang splitting. After n steps we have

$$Q^n = \left(e^{\frac{1}{2}k\mathcal{A}}e^{k\mathcal{B}}e^{\frac{1}{2}k\mathcal{A}}\right)\left(e^{\frac{1}{2}k\mathcal{A}}e^{k\mathcal{B}}e^{\frac{1}{2}k\mathcal{A}}\right)\cdots\left(e^{\frac{1}{2}k\mathcal{A}}e^{k\mathcal{B}}e^{\frac{1}{2}k\mathcal{A}}\right)Q^0 \qquad (5.21)$$

repeated n times. Dropping the parentheses and noting that $e^{\frac{1}{2}k\mathcal{A}}e^{\frac{1}{2}k\mathcal{A}} = e^{k\mathcal{A}}$, we obtain

$$Q^n = e^{\frac{1}{2}k\mathcal{A}}e^{k\mathcal{B}}e^{k\mathcal{A}}e^{k\mathcal{B}}e^{k\mathcal{A}}\cdots e^{k\mathcal{B}}e^{\frac{1}{2}k\mathcal{A}}Q^0\,. \qquad (5.22)$$

This differs from the first-order splitting only in the fact that we start and end with a half time step on $\mathcal{A}$, rather than starting with a full step and ending with $\mathcal{B}$.

The fact that the Strang splitting is so similar to the first-order splitting suggests that the first-order splitting is not really so bad, and in fact it is not. While formally only first-order accurate, the errors made are analogous to evaluating the solution at slightly the wrong time, off by single time step, say. While this is enough to make the results formally only first-order accurate, relative to the exact solution at this particular time, the quality of the solution has not suffered at all. For problems of the sort we are considering here, where the solution involves discontinuities such as shock waves, the resolution achieved is every bit as good with the Godunov splitting as with Strang splitting. In practice the Godunov splitting is often easier to use in terms of specifying correct boundary conditions, and also when using variable time-step methods as is often done in practice.

5.4 Stiff Source Terms

Source terms sometimes model phenomena which occur on much faster time scales than we are attempting to resolve with our time step k, and perhaps also act over much smaller spatial scales than our grid can resolve. In this case the source terms are said to be **stiff**, by analogy with the stiff equations discussed in Sect. 4.1.6.

5.4.1 A Combustion Model. Stiff source terms that are not treated carefully can lead to serious numerical difficulties. One such example is combustion, where chemical reactions (or nuclear reactions in stars) may occur on much faster time scales than the gas flow, much faster even than high-speed detonation waves. Computations may produce waves that look reasonable at first glance and yet are propagating at nonphysical speeds due to purely numerical artifacts. This was observed in a simple model combustion problem studied by Colella, Majda, and Roytburd [58].

Although in practical problems there may be dozens of chemical species of importance, in this model there are only two chemical species: "unburnt gas" and "burnt gas". By using ϱ as one of our variables, we can eliminate the continuity equation for burnt gas and be left with the Euler equations for ϱ, ϱu and E together with an additional continuity equation for the unburnt gas. We let Z represent the mass fraction of the unburnt gas and assume that unburnt gas is converted to burnt gas via a simple decay process of the form

$$\text{unburnt gas} \stackrel{K(T)}{\longrightarrow} \text{burnt gas} ,$$

where $K(T)$ represents the reaction rate of the burning process. In general the reaction rate depends on the temperature T via some Arrhenius relation. Typically the reaction rate is very large when T is sufficiently high but negligible for small T. For simplicity we can approximate this by an "ignition temperature" kinetics model, in which the Arrhenius behavior is idealized to

$$K(T) = \begin{cases} 1/\tau & T \geq T_0 \\ 0 & T < T_0 \, . \end{cases} \tag{5.23}$$

Here T_0 is the ignition temperature and the rate constant for large T is expressed in the form $1/\tau$ where τ is the time scale of the chemical reaction. After ignition we expect exponential decay of the unburnt fraction like $\exp(-t/\tau)$ (ignoring the fluid dynamics).

When this kinetics model is combined with the Euler equations we obtain

$$\begin{aligned} (\varrho)_t + (\varrho u)_x &= 0 \, , \\ (\varrho u)_t + (\varrho u^2 + p)_x &= 0 \, , \\ E_t + (u(E+p))_x &= 0 \, , \\ (\varrho Z)_t + (\varrho u Z)_x &= -K(T)\varrho Z \, . \end{aligned} \tag{5.24}$$

For simplicity we assume that both the unburnt and burnt gases are ideal gases with the same ratio of specific heats γ and temperature $T = p/\varrho\mathcal{R}$. The equation of state is modified by the fact that the unburnt gas contains chemical energy that is released as heat in the process of burning. The total energy E takes the form

$$E = \frac{p}{\gamma - 1} + \frac{1}{2}\varrho u^2 + q_0 \varrho Z \, , \tag{5.25}$$

where q_0 is the heat release.

We consider the particular case of a *Chapman–Jouguet detonation wave.* See, *e.g.*, [53], [60], [97], [257] for more discussion of combustion in gas dyanamics, and also Müller's notes on nuclear burning and supernova explosions. For $\tau > 0$ we expect some region of finite width across which the reactions take place, but we expect this region to be very thin when τ is small. Independently, Zel'dovich, von Neumann, and Döring made the following assumption about the structure of a detonation wave with finite rate chemistry, which gives the so-called ZND structure. The detonation wave consists of an ordinary fluid dynamic shock followed by a deflagration wave. The shock raises the temperature of the unburned gas to a value greater than T_0. Through the deflagration wave, the gas burns so that Z decreases from 1 to 0 while the pressure and density decrease to their final equilibrium values. This ZND structure is illustrated in Fig. 5.1 for the particular form of the reactions given in (5.24) with the ignition temperature kinetics (5.23).

The width of the reaction zone is proportional to τ. In order to resolve this structure numerically it is necessary to use a sufficiently fine grid that several grid points are in the reaction zone, and a small enough time step that the rapid reaction can be tracked. In practice, however, τ is typically very small relative to the time scale of primary interest computationally, which is the time scale on which the reaction front moves (and is independent of τ). For very small τ we might be content to obtain the C-J limit computationally, a jump discontinuity linking the correct equilibrium values and traveling at the correct speed, but without the correct ZND structure.

Applying the fractional step approach to the equations (5.24), we first apply a conservative method to the homogeneous equations that result by setting $K(T) = 0$ in (5.24). This gives the Euler equations together with an advection equation for Z. To solve the Riemann problem for this system we compute pressures p_l and p_r from the conserved quantities and the equation of state (noting that Z_l and Z_r may be different). Based on p_l, p_r, u_l and u_r, we solve the Riemann problem as indicated in Sect. 3.4. We then set $Z = Z_l$ to the left of the contact discontinuity and $Z = Z_r$ to the right of the contact discontinuity, since Z is simply advected with the fluid velocity in the nonreacting case. This produces the solution to the Riemann problem for the system of 4 equations. Using this solution to the Riemann problem, we can apply Godunov's method or a high resolution method to produce U^* from U^n.

In the next step we solve the reaction equation in each grid cell. Since the source terms are zero for all equations except the Z equation, we obtain $\varrho^{n+1} = \varrho^*$, $(\varrho u)^{n+1} = (\varrho u)^*$ and $E^{n+1} = E^*$. The Z equation becomes

$$Z_t = -K(T)Z \,, \tag{5.26}$$

since ϱ is held constant in this stage of the splitting. Ordinarily we would apply a stable numerical method to the resulting kinetics equations, but here we have a sufficiently simple equation that we can in fact use the exact solution. We then obtain

$$Z_j^{n+1} = \exp(-K(T_j^*)k)\, Z_j^* \,, \tag{5.27}$$

where $T_j^* = p_j^*/(\varrho_j^* \mathcal{R})$ is the temperature in the jth cell after the first stage of the splitting. Note that we are assuming $K(T)$ remains constant during this stage of the splitting in order to obtain the true solution (5.27). Although ϱ, ϱu, and E remain constant, the pressure p and hence the temperature T will vary with Z. However, for an ignition kinetics model it is easy to check that T remains either above or below T_0 throughout this stage, so that $K(T)$ is constant. (If $T^* < T_0$ then $Z_t = 0$ so T is constant, while if $T^* \geq T_0$ then Z is decreasing which implies that p and hence T are increasing.) It follows that (5.27) does in fact solve (5.26) over the time step k.

5.4.2 Nonphysical Wave Speeds. Figures 5.1 and 5.2 show the results of two computations with this approach on the equations (5.24), using Godunov's method for the conservation law stage. In each case we used initial data corresponding to the ZND profile linking the states

$$\begin{aligned} \varrho_l &= 1.4 & \varrho_r &= 0.887564 \\ u_l &= 0 & u_r &= -0.577350 \\ p_l &= 1 & p_r &= 0.191709 \\ Z_l &= 0 & Z_r &= 1 \,, \end{aligned}$$

with $\gamma = 1.4$, $\mathcal{R} = 1$, $q_0 = 1$, and $T_0 = 0.22$. We used 200 grid points with $h = 0.01$ and took $k = 0.005$, giving a Courant number between 0.5 and 1.

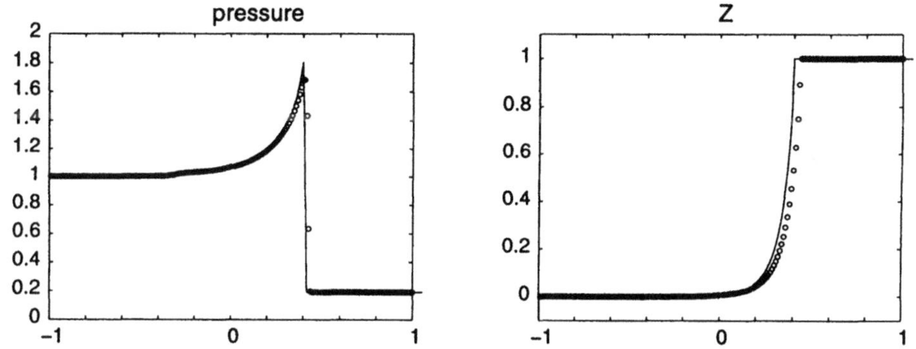

Fig. 5.1. True ZND structure (solid line) and numerical solution to the combustion problem with $\tau = 0.1$

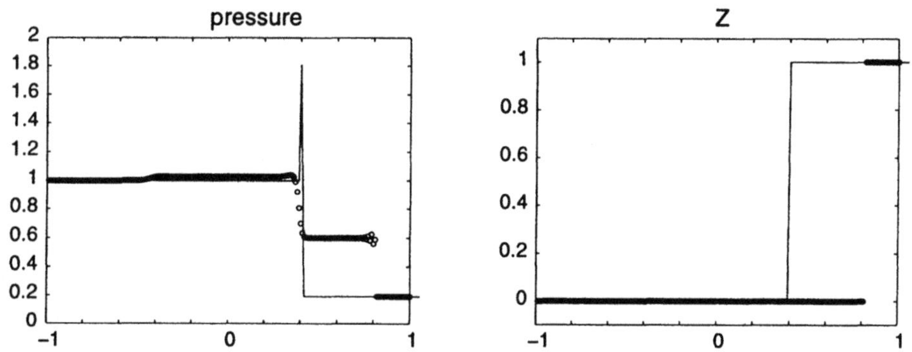

Fig. 5.2. True ZND structure (solid line) and numerical solution to the combustion problem with $\tau = 10^{-6}$

In Fig. 5.1 we use $\tau = 0.1$ so that the ZND profile is well resolved on the grid. The solution is shown at $t = 0.4$. Although there is some smearing of the pressure peak due to the dissipation of Godunov's method, the wave form is essentially correct and is moving at the correct speed $s = 1$.

In Fig. 5.2 we use $\tau = 10^{-6}$. The ZND profile can no longer be resolved on the grid. In the exact solution the reaction wave should still move at speed $s = 1$, but now the numerical solution behaves in a manner that is entirely nonphysical. There is a numerical wave traveling at speed 2 in which all of the chemical energy is released, followed by an ordinary shock at a slower speed.

Note that the fast wave seen here is moving at a speed that corresponds to one grid cell per time step, *i.e.*, at speed $2 = h/k$. If we change the mesh ratio then this wave speed changes to the new value h/k, showing that this wave is a purely numerical artifact. Similar experiments are reported in [58].

To understand what goes wrong here, consider Fig. 5.3, which indicates the solution to the Riemann problem arising at an initially sharp detonation wave (so $Z = 0$ to the left of this interface and $Z = 1$ to the right). Detonation results if the shock wave raises the temperature of the gas to a value above the ignition temperature. Physically it is the gas between the shock and the contact discontinuity that should then be burned in the second half of the fractional step method, since it is only in this region that there is both unburned gas and also a sufficiently high temperature.

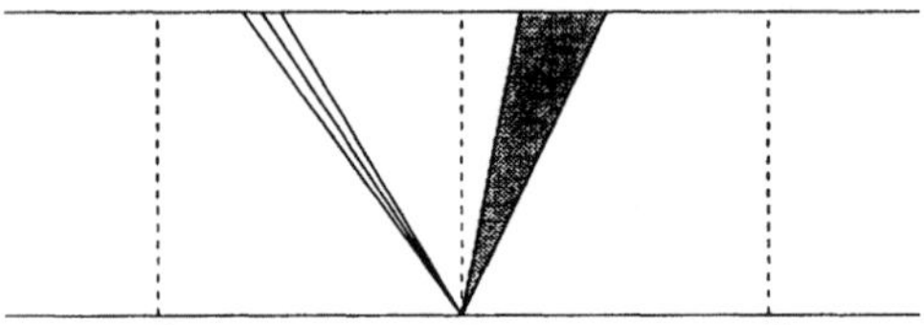

Fig. 5.3. Solution to the Riemann problem in the combustion or radiation examples. The source terms should be active only between the shock wave, marked $\mathcal{S}$, and the contact discontinuity, marked $\mathcal{C}$

In the standard fractional step method, however, we first update cell averages using this Riemann solution (and perhaps high-resolution corrections), and *then* apply the source terms based on the updated cell averages. The problem of incorrect waves speeds arises from the fact that if the new cell average yields a temperature greater then the ignition temperature, and if $k/\tau \gg 1$, then *all* of the unburnt gas in this cell will be burned in this time step, not just the gas that has been heated by the shock. The result is that the interface between burnt and unburnt gas moves over by one full grid cell, yielding a combustion wave that propagates at the speed of one cell per time step.

This phenomenon was discussed and analyzed in the context of an even simpler scalar equation in [148]. A more interesting scalar model is discussed in [140]. Fryxell, Müller, and Arnett [94] also demonstrate some of the difficulties one may encounter when computing detonations numerically. See also Müller's notes in this volume.

Although the potential difficulty with stiff source terms has been discussed in the context of fractional step methods, it should be stressed that similar problems can also arise with unsplit methods (which also have numerical viscosity) and it is not the fractional step approach that is at fault. This is demonstrated in [148] for the scalar model problem.

On the other hand, in order to properly model problems with troublesome stiff source terms it is necessary to better model the coupling between the source term and the fluid dynamics.

One possible solution to this problem would be to implement the fractional step method slightly differently. Rather than basing the source term update on the new cell averages computed after solving the conservation law, one could compute the source term contributions in the process of solving the Riemann problem, *e.g.*, by determining how much gas should actually burn in the above example. This could then be used to update Z after completing the conservation law updates. This approach has recently been explored by Christiane Helzel from Magdeburg, who has obtained some encouraging results (see Fig. 5.4 for an example.)

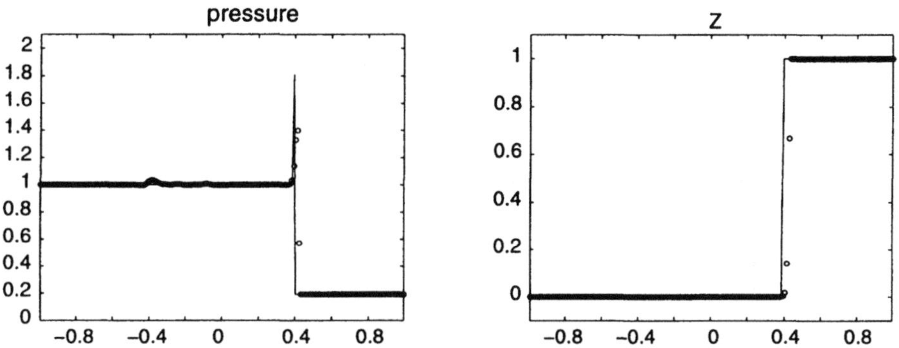

Fig. 5.4. True ZND structure (solid line) and numerical solution to the combustion problem with $\tau = 10^{-6}$, using a modified method in which the source term solver takes into account the structure of the Riemann solution.

Other approaches also involve solving some kind of generalized Riemann problem where the source terms are taken into account. A wide variety of different approaches to handling stiff source terms have been explored in the recent literature. A few examples are [24], [120], [184],

One can also avoid problems by making sure that the grid is sufficiently fine, and the time step small enough, that the source terms are fully resolved. For τ fixed, no matter how small, the fractional step method will converge as $k,\ h \to 0$. But this is what we typically want to avoid in the "stiff" case, if possible. If this is not possible to avoid, then *adaptive mesh refinement* will certainly be crucial to efficiently solve the problem (see Sect. 6.3 below). *Shock tracking* or front tracking (see Sect. 6.3.1) may also be a useful technique in this context, see [38], [39], [145].

5.4.3 Isothermal Equations and Radiation. Consider a molecular cloud that radiates heat as mentioned in Sect. 2.4. If a shock wave passes through such a cloud, then the molecules will be heated when passing through the shock. But this heat will then be so quickly radiated away relative to the fluid dynamic time scales, and the width of the radiation zone so narrow relative to the spatial domain of interest, that we can lump the shock wave and radiation zone together into a single isothermal shock and use the simpler isothermal equations to model the flow.

If, however, we do use the full Euler equations together with a source term for the thermal radiation, then the source terms will be very stiff, as in the combustion example above. However, this does not appear to lead to the same numerical difficulties of incorrect propagation speeds, and a simple splitting method can be successfully employed for this case. This is presumably not true for radiation problems in general (see the notes of Mihalas and Dorfi), but it is interesting to investigate this particular example since it sheds more light on the nature of the splitting error.

Again consider Fig. 5.3, but now suppose we are solving the Euler equations with radiation. Again it is only the region behind the shock where the source terms should be active, now by radiating heat, since only this gas has been heated by the shock. In the standard fractional step method we first average the energy and other conserved quantities over the entire grid cell and then radiate the heat, rather than radiating it from the smaller post-shock region. However, in this case we have not made a catastrophic error by doing the averaging first. The averaging spreads the heat over a larger region, but the total amount of energy to be radiated is the same either way, regardless of whether we have a whole cell with a high temperature or a thin zone with a very high temperature. In the combustion problem, by contrast, this makes a big difference since the gas burns equally well at any sufficiently high temperature, and so the larger region of high temperature leads to more gas being burned.

Again the issue of how well the splitting method works comes down to a question of whether certain processes commute — in this case we need to investigate whether the averaging process commutes with the process modeled by the source term. This is discussed by Pember [183] in a slightly different guise. He considers the conservation law with both a source term having time scale τ and also a viscous term with coefficient ε (a viscous regularization of the equation). His conclusion is that the stiff source term may cause a problem numerically if the limiting behavior of solutions to this equation as $\varepsilon,\ \tau \to 0$ depends on which parameter vanishes more rapidly. This is relevant because physically we are trying to model the case where $\varepsilon = 0$ and $\tau > 0$ (but very small), whereas numerically ε may be much larger than τ due to numerical viscosity. With Godunov's method this numerical viscosity arises from the averaging process, and so this is consistent with the analysis above.

5.5 Quasi-stationary Flow and Gravity

Another difficulty with source terms arises when the flow is nearly stationary, so $q_t \approx 0$. In one space dimension we then have

$$f(q)_x \approx \psi(q)\ , \tag{5.28}$$

but it may be that the two expressions in (5.28) are both large. The source term is roughly balanced by the divergence of the flux. As an example, consider the isothermal equations in a one-dimensional tube oriented vertically, with gravity acting in the negative x-direction. This force appears in the momentum equation and we have

$$
\begin{aligned}
\varrho_t + (\varrho u)_x &= 0\,, \\
(\varrho u)_t + (\varrho u^2 + a^2\varrho)_x &= -g\varrho\,,
\end{aligned}
\tag{5.29}
$$

where g is the gravitational constant. In equilibrium, the velocity is $u = 0$ while there is a gradient in the pressure $p = a^2\varrho$ that balances the force of gravity, yielding an exponentially varying density,

$$
\varrho(x) = \varrho_0 \exp(-gx/a^2)\,. \tag{5.30}
$$

Suppose we start with this state initially and apply a splitting method. In the first step we solve the homogeneous conservation law, ignoring gravity. The density variation will lead to an acceleration of the fluid. In the second step we modify the momentum using the source term, and we hope that this will restore the initial state in order for equilibrium to be maintained. We are relying on two large changes to cancel out, and clearly they won't exactly. The Strang splitting can be used to obtain a second order accurate method, but the error constant may be quite large.

This causes serious difficulties if we are attempting to model the propagation of small perturbations against the background of a nearly stationary flow. This is a common situation in modeling stellar atmosphere dynamics, where the self-gravity of a star leads to power-law variations in the density and pressure that serves as a background for smaller amplitude dynamics.

In the context of (5.29), suppose we wish to compute the propagation of a small amplitude acoustic wave through the exponentially stratified equilibrium. This signal will be an insignificant component in the solution of the Riemann problems in the homogeneous problem, since the density jumps will be dominated by the exponential variation. While the second step may nearly restore equilibrium, the signal can be totally corrupted in the process.

Various approaches are discussed in the literature for dealing with quasi-stationary problems of the type discussed above. One possibility is to view the cell averages as defining a piecewise stationary solution instead of a piecewise constant or piecewise linear function as we have done so far. The form of the solution in each grid cell is chosen so that (5.28) is roughly satisfied in the cell, a piecewise exponential function, for example. As a simpler approximation we might use a piecewise linear function with the slope q_x chosen so that

$$
f'(q)q_x \approx \psi(q)\,.
$$

This piecewise stationary solution is then used in conjunction with a **generalized Riemann solver** that resolves the jump at each interface taking into account the variation nearby. See [84], [95], [96], [250] for some methods based on this approach.

Here I will discuss a different approach that is related to this but easier to implement, and which also allows high-resolution correction terms to be computed that are based on variations in the perturbations from equilibrium.

Rather than determining a smooth function in each cell whose derivative matches the source term, we will reconstruct a piecewise constant function in the i'th cell that has a single discontinuity from Q_i^- to Q_i^+ at the middle of the cell. For simplicity we assume

$$Q_i^- = Q_i - \delta_i/2 , \qquad Q_i^+ = Q_i + \delta_i/2 , \tag{5.31}$$

for some vector δ_i, to be determined below. Now suppose we apply Godunov's method. The data is still piecewise constant and the jumps at cell interfaces, between Q_{i-1}^+ and Q_i^-, give rise to waves propagating into the two neighboring cells as usual. The Riemann problem in the middle of a grid cell gives a fan of waves which, for small time, remain entirely in the single cell. Hence we do not need to compute the detailed solution to this Riemann problem, since we easily compute that its contribution to the cell average is simply $\frac{k}{h}(f(Q_i^+) - f(Q_i^-))$. Rather than using a splitting method, we will also include the source term in an unsplit manner. For clarity suppose we are solving a *linear* hyperbolic problem with a source term ψ, so that $f(q) = Aq$ and the solution of the Riemann problem is easy to determine. Then the method takes the form

$$\begin{aligned} Q_i^{n+1} &= Q_i^n - \frac{k}{h}[A^+(Q_i^- - Q_{i-1}^+) + A^-(Q_{i+1}^- - Q_i^+)] \\ &\quad - \frac{k}{h}(AQ_i^+ - AQ_i^-) + k\psi(Q_i) . \end{aligned} \tag{5.32}$$

We have not yet discussed how to choose δ. The form of (5.32) suggests choosing it so that the term $\frac{k}{h}(AQ_i^+ - AQ_i^-)$ which arises from the discontinuity introduced at the cell center exactly cancels out the source term $k\psi(Q_i)$. Then the second line of (5.32) would vanish and the method reduces to what looks like Godunov's method based only on jumps at the interface, with no source term. But the jumps have been modified so the Riemann problem solved at each interface uses data Q_{i-1}^+ and Q_i^- rather than the original cell averages. High-resolution correction terms can now be based on the waves resulting from these Riemann problems. As Fig. 5.5 indicates, a large variation in Q that is primarily balanced by a source term will lead to small residual jumps $Q_i^- - Q_{i-1}^+$ at the cell interfaces.

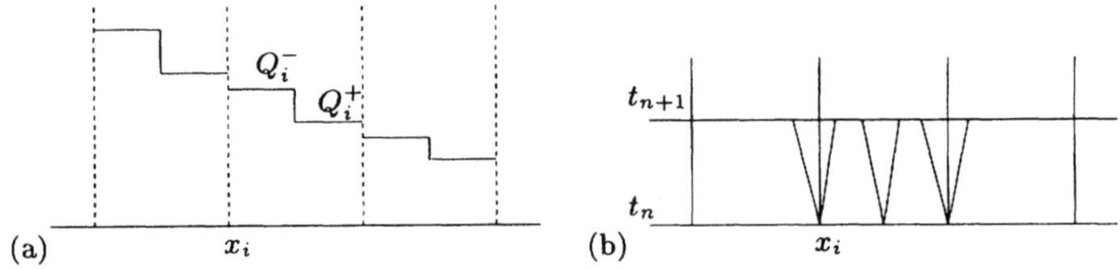

Fig. 5.5. (a) Introducing a jump of size δ at the center of each grid cell gives a new piecewise constant reconstruction. (b) Solving the Riemann problem at each jump. Three Riemann solutions affect the cell value Q_i. The flux difference from the Riemann problem at the cell center balances the source term

For a general nonlinear flux function, requiring that the second line of (5.32) vanish gives the condition

$$\frac{1}{h}(f(Q_i^+) - f(Q_i^-)) = \psi(Q_i) , \tag{5.33}$$

which mimics (5.28). Splitting Q_i in this manner thus gives a piecewise constant approximation to the equilibrium solution within the cell, while preserving the piecewise constant nature near the cell interfaces where Riemann problems must be solved. The resulting jumps at the interfaces correspond to perturbations from the equilibrium solution, as desired.

For a linear system, combining (5.33) and (5.31) and solving for δ gives

$$\delta = hA^{-1}\psi(Q_i) . \tag{5.34}$$

We use this to define the modified states in (5.31) and then the simplified form of (5.32) gives

$$Q_i^{n+1} = Q_i^n - \frac{k}{h}(A^+(Q_i^- - Q_{i-1}^+) + A^-(Q_{i+1}^- - Q_i^+)) . \tag{5.35}$$

This is the form used in the implementation since it avoids the source terms altogether. It is this form that can now be extended to a high-resolution method by using waves based on the modified states.

It is interesting, however, to rewrite (5.35) in the following form, obtained by using the expression (5.34) in (5.31),

$$\begin{aligned} Q_i^{n+1} &= Q_i^n - \frac{k}{h}(A^+(Q_i - Q_{i-1}) + A^-(Q_{i+1} - Q_i)) \\ &\quad + k(A^+A^{-1}\psi_{i-1/2} + A^-A^{-1}\psi_{i+1/2}) , \end{aligned} \tag{5.36}$$

where

$$\psi_{i-1/2} = \frac{1}{2}(\psi(Q_{i-1}) + \psi(Q_i)) .$$

In this form we can interpret the first-order method as being simply the standard Godunov method with a special form of the source term which consists of a certain combination of interface values $\psi_{i-1/2}$ and $\psi_{i+1/2}$ instead of $\psi(Q_i)$. Note that

$$A^+A^{-1} = R\Lambda^+\Lambda^{-1}R^{-1} = RI^+R^{-1} ,$$

where I^+ is the diagonal matrix with 1's on the diagonal in positions corresponding to positive eigenvalues of A and zeros elsewhere. Hence the decomposition

$$\psi_{i-1/2} = A^+A^{-1}\psi_{i-1/2} + A^-A^{-1}\psi_{i-1/2}$$

splits $\psi_{i-1/2}$ into components that are carried to the right and left respectively by the characteristics. In (5.36) we use the incoming portion of ψ from each cell edge to define the source term used in the cell.

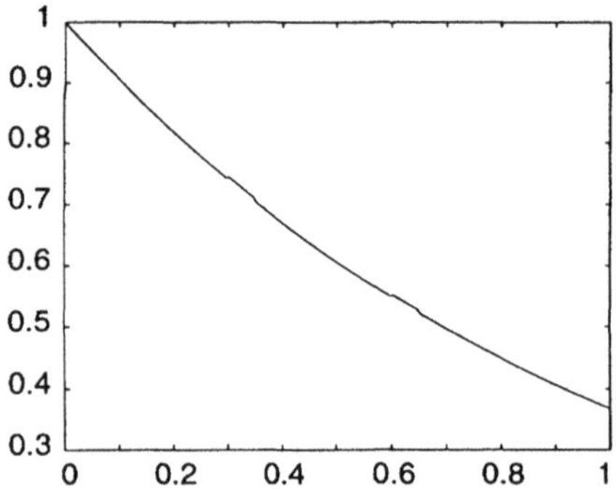

Fig. 5.6. Solution to the isothermal equations with gravity. Very weak acoustic waves are visible which are amplified in Fig. 5.7 below

This approach is frequently used in the literature (*e.g.*, [82], [164]) and has the additional advantage that the first-order accurate method (5.36) gives a second-order accurate steady state solution [198]. The derivation given here shows how this approach can be unified with the idea of a piecewise steady reconstruction. In adding the high-resolution terms it seems highly desirable to start with the form (5.35) rather than (5.36).

For a nonlinear problem it may not be so easy to determine δ_i so that (5.33) is satisfied. Consider the isothermal equations (5.29), for example. Let ϱ be the density and $m = \varrho u$ be the momentum in the i'th cell. We must take $m^- = m^+ = m$ since there is no source term in the conservation of mass equation, and then we need to choose $\varrho^{\pm} = \varrho \pm \delta/2$ so that

$$(m^2/\varrho^+ + a^2\varrho^+) - (m^2/\varrho^- + a^2\varrho^-) = -hg\varrho\,.$$

In equilibrium, $m = 0$ and this reduces to a linear equation for δ with solution $\delta = -hg\varrho/a^2$. In general clearing the denominator gives a cubic equation for δ, and near equilibrium this has a single real root which can be easily found by iteration using the equilibrium value as an initial guess.

Figure 5.7 shows some sample calculations on the isothermal system with solid walls at the two ends of a vertical tube of length 1, using $a = g = 1$ and the equlibrium state $\varrho(x) = \mathrm{e}^{-x}$. The equilibrium density is perturbed by 10^{-2} over $0.45 < x < 0.55$ initially, giving rise to weak acoustic waves propagating in both directions. These are barely visible in the plot of the density shown in Fig. 5.6. Figure 5.7 shows the true and computed density perturbations at time $t = 0.15$ with the two different approaches. Here the background exponential has been subtracted out. The solid line shows the "true" acoustic pulses as computed on a much finer grid.

In Fig. 5.7(a) the Strang splitting has been used. The perturbation is not very well captured. Near the perturbation we see overshoots and undershoots in spite of the fact that a flux-limiter method is used. The limiter is applied to the waves that arise in solving the homogeneous isothermal equations, which include the exponential variation as well as the perturbation.

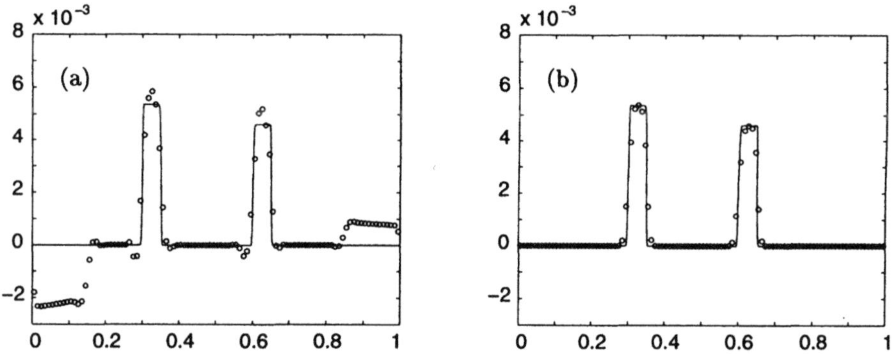

Fig. 5.7. Solution to the isothermal equations with gravity. The density perturbation from the equilibrium solution $\varrho(x) = \mathrm{e}^{-x}$ is plotted. Note these perturbations are of the order 10^{-3} relative to the equilibrium. (**a**) Using the fractional step method. (**b**) Using a method of the form (5.35)

We also see perturbations of a similar magnitude propagating in from the boundaries. These arise from the fact that the solid wall boundary conditions which are correct for the full problem are not the correct boundary conditions to impose for the homogeneous conservation law in the fractional step method. Developing appropriate boundary conditions for intermediate time steps in a fractional step method can be an important component in achieving full accuracy, see [139] for one example. In this problem these perturbations from the boundary, though small in relation to the exponentially varying equilibrium solution, will overwhelm the waves we are attempting to capture at later times.

In Fig. 5.7(b) the approach described above is used instead, and the waves are well captured, at later times as well as at the time shown. This approach is developed more fully in [143].

6. Multi-dimensional Problems

The derivation of multi-dimensional conservation laws was discussed in Sect. 2.5 and now we will look at the hyperbolic structure of such problems, and numerical methods for their solution. For a linear problem in two space dimensions, the system of equations takes the form

$$q_t + Aq_x + Bq_y = 0\,. \tag{6.1}$$

This problem is said to be *hyperbolic* if the matrix $\cos\theta A + \sin\theta B$ is diagonalizable with real eigenvalues for every choice of θ. Then taking plane-wave initial data which varies only in one direction (say the ξ-direction) at angle θ to the x-axis leads to a one-dimensional hyperbolic equation of the form

$$q_t + (\cos\theta A + \sin\theta B)q_\xi = 0\,.$$

The nonlinear conservation law $q_t+f(q)_x+g(q)_y=0$ is hyperbolic at some state q if the above condition is satisfied for the Jacobian matrices $A=f'(q)$ and $B=g'(q)$. The Euler equations (2.33), for example, are hyperbolic in 2 or 3 dimensions away from the vacuum state.

The two-dimensional advection equation takes the form

$$q_t + uq_x + vq_y = 0 \ , \tag{6.2}$$

where (u,v) is the velocity vector. This equation is hyperbolic and the solution simply translates along (u,v).

In one space dimension we can diagonalize a general linear hyperbolic equation using the matrix of eigenvectors, decoupling it into independent scalar advection equations for each characteristic variable. For a linear system in more dimensions, this can only be done if the coefficient matrices commute, e.g., if $AB = BA$ in (6.1), in which case the matrices have the *same* eigenvectors. Then A and B can be ***simultaneously diagonalized*** by the eigenvector matrix R:

$$A = R\Lambda R^{-1} \ , \qquad B = RMR^{-1} \ ,$$

where $\Lambda = \text{diag}(\lambda^1,\ldots,\lambda^m)$ and $M=\text{diag}(\mu^1,\ldots,\mu^m)$ contain the eigenvalues, which may be different. The system in (6.1) can then be diagonalized to obtain

$$\hat{q}_t + \Lambda\hat{q}_x + M\hat{q}_y = 0 \ ,$$

yielding independent advection equations. The p'th equation has the velocity vector (λ^p,μ^p).

If $AB \neq BA$, then there is no single transformation that will simultaneously diagonalize A and B. The equations are genuinely coupled. This is the usual situation physically. Consider linear acoustics, for example. This can be modelled by a linear system of 3 equations in two space dimensions:

$$q = \begin{bmatrix} p \\ u \\ v \end{bmatrix}, \qquad A = \begin{bmatrix} 0 & \gamma p_0 & 0 \\ 1/\varrho_0 & 0 & 0 \\ 0 & 0 & 0 \end{bmatrix}, \qquad B = \begin{bmatrix} 0 & 0 & \gamma p_0 \\ 0 & 0 & 0 \\ 1/\varrho_0 & 0 & 0 \end{bmatrix}.$$

Here p is the pressure perturbation and (u,v) the velocity, while ϱ_0 and p_0 are the background density and pressure. The matrix $\cos\theta A + \sin\theta B$ then has eigenvalues $0, \pm c$ for all values of θ, where $c = \sqrt{\gamma p/\varrho}$ is the sound speed, which is the same in all directions.

The multi-dimensional Euler equations have shock waves that travel isotropically in all directions. For some problems we may be able to reduce the complexity of the numerical problem substantially by taking advantage of symmetry. For example, if we are solving a problem where the solution is known to be radially symmetric, then we should be able to rewrite the equations in polar or spherical coordinates, obtaining a system that reduces to a problem in the single space dimension r. The transformed equations will typically involve *geometric source terms*.

Even if the real problems of interest must be studied multidimensionally, radially-symmetric solutions are very valuable in testing and validating numerical codes. A highly accurate solution to the one-dimensional problem can be computed on a fine grid and used to test solutions computed with the multi-dimensional solver. This is useful not only in checking that the code gives essentially the correct answer in at least some special cases, but also helps to determine whether the numerical method is isotropic or suffers from *grid-orientation effects* that lead to the results being better resolved in some directions than in others. See [134], [144], [147] for some examples of such numerical tests.

6.1 Dimensional Splitting

The simplest way to extend the finite volume methods developed in one dimension to more dimensions is to use **dimensional splitting**. The fractional step method described in Sect. 5. is applied to the multi-dimensional problem by splitting it into one-dimensional subproblems. For example, the two-dimensional linear problem (6.1) would be split into

$$\begin{aligned} x\text{-sweeps}: \qquad & q_t + Aq_x = 0\,, \\ y\text{-sweeps}: \qquad & q_t + Bq_y = 0\,. \end{aligned}$$

In the x-sweeps we would start with cell averages Q^n_{ij} at time t_n and solve one-dimensional problems $q_t + Aq_x = 0$ along each row with j fixed, updating Q^n_{ij} to Q^*_{ij}. In the y-sweeps we then use these Q^*_{ij} values as data for solving $q_t + Bq_y = 0$ along each row with i fixed, which results in Q^{n+1}_{ij}.

Note that there will generally be a splitting error unless the operators $\mathcal{A} = A\partial_x$ and $\mathcal{B} = B\partial_y$ commute, *i.e.*, unless $AB = BA$. Only in the case where the multi-dimensional problem decouples into scalar advection equations can we use dimensional splitting with no splitting error.

However, the splitting error is often no worse than the errors introduced by the numerical methods in each sweep, and dimensional splitting can be a very effective approach. It gives a simple and relatively inexpensive way to extend one-dimensional high-resolution methods to 2 or 3 dimensions.

6.2 Multi-dimensional Finite Volume Methods

Alternatively, finite-volume methods can be developed that are based directly on the multi-dimensional integral form of conservation law, *e.g.*(2.28) in three dimensions. The cell averages are then updated due to fluxes through all faces of the grid cell, and the fluxes are all computed based on the current cell averages rather than first updating by sweeps in one direction and then the other. In two space dimensions a conservative flux-differencing formulation would have the form

$$Q_{ij}^{n+1} = Q_{ij}^{n} - \frac{\Delta t}{\Delta x}(F_{i+1,j} - F_{ij}) - \frac{\Delta t}{\Delta y}(G_{i,j+1} - G_{ij}) .$$

The simplest Godunov method of this form consists in solving the one-dimensional Riemann problem normal to each cell face in order to compute the flux. Extending this method to second-order accuracy requires incorporating not only slopes in the normal direction, as in the one-dimensional algorithm, but also *transverse derivative* terms that arise from cross-derivatives in a Taylor series expansion. With the Strang splitting these terms would arise naturally from the fractional step procedure, but with an unsplit method they must be explicitly modeled. A wide variety of unsplit methods have been developed, *e.g.*, [22], [56], [85], [200], [205], [219], [220], [243], [266],

A multi-dimensional version of the wave-propagation algorithms described above has recently been developed, yielding a class of algorithms that follow fairly simply from the one-dimensional ideas and yet appear to be quite broadly applicable, fairly robust, and stable up to Courant numbers of 1 relative to the maximum wave speed in any direction. These algorithms were described in the course, but this material will not be repeated here since a complete description (in two space dimensions) has recently appeared in [144].

These algorithms are also implemented in the CLAWPACK software [138], in a manner that I hope can be comprehended and will aid in understanding the algorithms as well as providing a code that can be applied on many hyperbolic systems. This package also contains three-dimensional extensions as described in [133], [134].

6.3 Grids and Adaptive Refinement

In many technical fields where computational fluid dynamics is used, a major computational difficulty arises from the fact that the physical geometry can be very complex in 2 or 3 space dimensions. Examples include aerodynamical calculations of the flow around an airplane or inside a gas turbine. One must first determine a suitable grid on which to perform the calculation, and various issues arise in generating suitable grids and then solving the equations efficiently and accurately on such grids. Many books on CFD and grid generation discuss this issues in detail.

In astrophysical calculations the geometry is often quite simple by comparison, and many intersting calculations can be done in a rectangular domain using Cartesian coordinates, or perhaps with simple spherical coordinates. In these notes I will not discuss the grid generation problem at all.

There are, however, often a wide range of spatial and temporal scales present, so that efficient calculation generally requires the use of a nonuniform grid even if the geometry is simple. To resolve a shock wave cleanly may require a fairly fine grid near the shock, though one should remember that a shock from a hyperbolic system is an idealization and typically it will not be resolved exactly as a discontinuity no matter how fine the grid is, unless perhaps we employ "shock tracking" as described below. If we add viscous terms to the equations, or have reactions or radiation effects that are active over very thin regions, then we may need to resolve the physically correct width of these transition regions in order to obtain correct solutions, as discussed in Sect. 5.4.

6.3.1 Moving Grids and Front Tracking. One possibility is to adjust the computational grid in each time step so that it conforms to the features of interest, by moving a fixed set of grid points. One approach is to use a *Lagrangian formulation* of the equations of motion, in which the computational grid moves with the fluid. Müller has discussed this to some extent in his lectures. This can lead to grid tangling problems when there are large defomations of the fluid, and often such approaches are combined with fixed Eulerian grids through some form of remapping procedure.

Another approach is to move the grid points in such a way that they are clustered near the regions of rapid variation, without necessarily moving them at the fluid velocity. Dorfi's lectures contain a discussion of one such method for 1-dimensional (spherically symmetric) radiation hydrodynamics. See also [74], [105].

If the feature of interest is a shock wave that should be a sharp discontinuity, then instead of clustering many points near the shock in order to resolve it well with a standard finite-volume method, another approach is to keep track of the location of the shock itself and deal with the motion of the shock in some special manner. This is called **shock tracking** or more generally **front tracking** since this approach can also be applied to more general discontinuities in q that may not be shocks, *e.g.*, contact discontinuities or phase boundaries.

In one space dimension this requires introducing just one additional grid point for the shock location. In solving the Riemann problem at this point we can determine the exact speed of the shock and adjust this grid point accordingly. In more than one space dimension, front tracking can get quite complicated, particularly when the shocks develop interesting structure or begin to interact with one another, or when two shocks collide.

A variety of front-tracking methods of have been developed. See [145] [146] for one approach related to the wave propagation algorithms described here. The references in these papers point to many more examples.

By contrast, methods of the type developed in these lectures are generally called **shock capturing** methods, in that they attempt to capture shocks and other discontinuities nicely on a fixed grid, rather than adjusting the grid to track the shock. Shock-capturing methods are much simpler to implement than shock-tracking methods, particularly in more than one space dimension. However, in order to capture features such as shock waves very sharply it is often necessary to have a rather fine grid.

6.3.2 Adaptive Mesh Refinement. Rather than allowing grid points to move, another approach to clustering grid points where they are needed is to refine a uniform grid in regions where more resolution is needed. This approach can be used both for capturing shocks well and for introducing fine grids in regions of rapid smooth transition. Such an approach should be automatically adaptive in that it chooses the regions of refinement based on the behavior of the solution, and for time-dependent problems the regions of refinement must change with time.

I will discuss adaptive mesh refinement (AMR) in the context of Cartesian grids, though the same approaches can be applied on logically-rectangular curvilinear grids.

There are two main strategies that have been used to handle the data structures involved when refining a rectangular grid. One approach is to *refine individual grid cells* as needed, typically by splitting a single cell in two dimensions into four pieces. Each of these pieces may be further subdivided recursively, depending on how many levels of refinement are allowed. A convenient way to keep track of these cells in two dimensions is to use a **quad-tree structure**, a tree in which each node corresponds to a grid cell and those which are subdivided have four leaves corresponding to the subcells. In three dimensions cells would be subdivided into 8 pieces, leading to an oct-tree data structure.

The paper [99] shows an example of such a grid in 3D, used to solve the MHD equations for the interaction of the solar wind with a comet. This calculation illustrates another reason why adaptive mesh refinement can be essential when dealing with a variety of spatial scales. The comet in this calculation is 10km across, while the full computational domain is $50 \times 50 \times 75$ millions of kilometers. 16 levels of refinement have been used so that the smallest grid cells, near the comet, are roughly 48km on a side. Even at this resolution, the comet itself is a point source within one grid cell, but this is adequate to give the correct global solution. It would be absolutely impossible, however, to use this resolution over the entire computational domain. See [68] for another application of this adaptive MHD code, in calculating the magnetic field around Venus.

Another approach is to *refine rectangular patches* of the computational domain rather than individual cells, again recursively applying this idea to allow increased resolution in some regions. This approach, illustrated in Fig. 6.1, was pioneered by Berger and Oliger [27] and has been developed further by Berger and coworkers [19] [25] [28] [26].

Refinement on rectangular patches has the advantage that the data structures remain relatively simple, and consist of a nested set of grids. On each grid a standard finite-volume method is used to sweep over the grid, though a certain amount of additional work must then be done at the interface between grids to maintain accuracy and global conservation. Refinement of individual cells requires more work per cell in advancing the solution, but has the

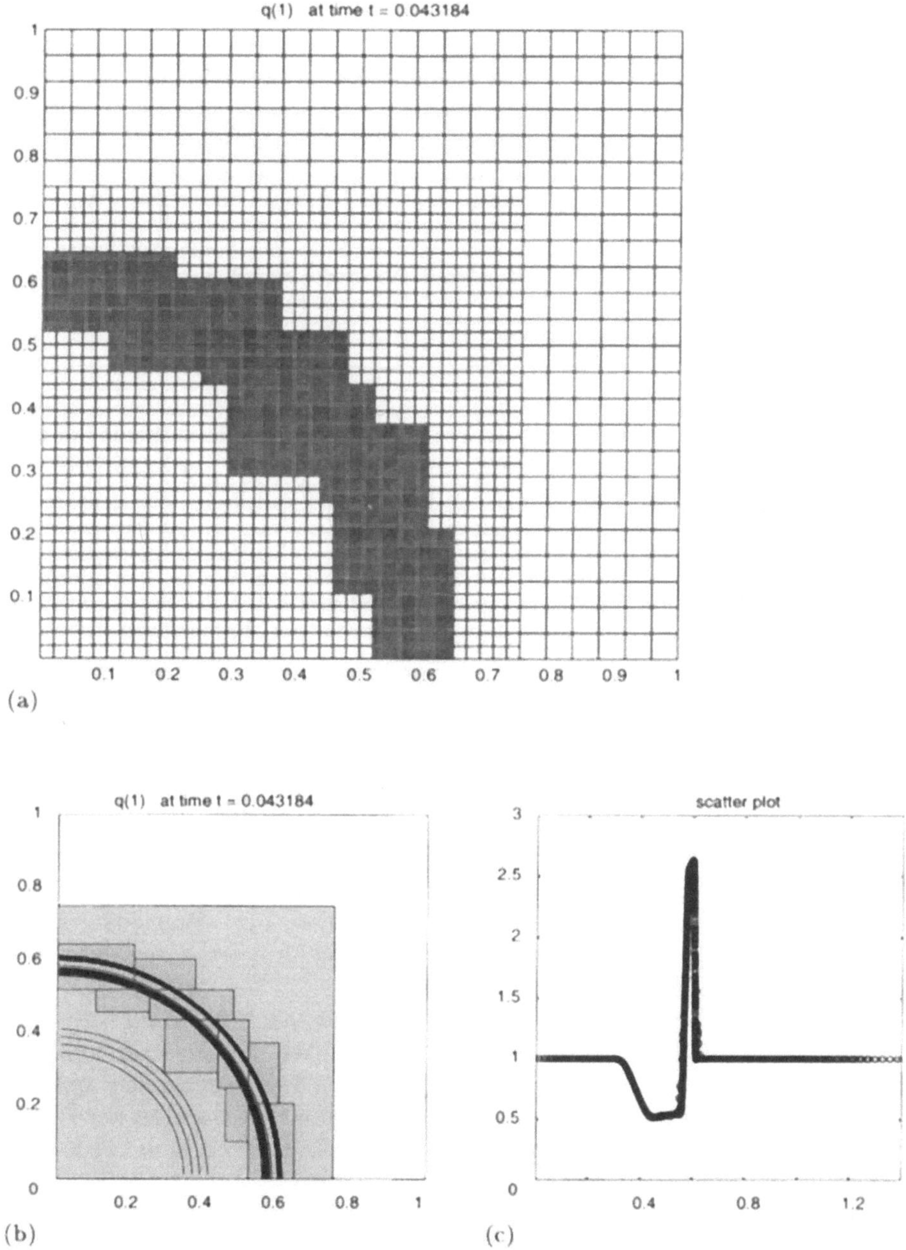

Fig. 6.1. Adaptive mesh refinement on rectangular patches for a spherical Riemann problem. Initially the pressure is 10 inside a sphere of radius 0.5 and 1 outside, with constant density $\varrho = 1$ and zero velocity. (**a**) shows the adaptively refined grid. Refinement by a factor of 2 is used in Level 2 and by an additional factor 4 in Level 3. (**b**) shows contours of density

advantage that fewer grid cells need to be refined since the refinement can really be focused where it is most needed. When refining on patches, the cells which are flagged as needing refinement must be clustered into rectangular patches. See [19] for more details on the refinement and clustering algorithms. Powell [186] gives some additional comparison of these approaches.

When refining in space and then solving the hyperbolic equation, we also generally want to refine in time by the same factor. If we refine by a factor of 4, for example, then we should take 4 time steps on the refined patch for each time step on the coarse grid. This insures that we are using a comparable Courant number on each grid and avoids the need for very small time steps on the coarsest grids.

At an interface between coarse and fine grids, we must also insure that the formulas used to update the solution on each grid are consistent with one another. In particular, when a conservation law is being solved we must preserve global conservation. See [28], [30] for details on how this is done.

Berger's work has recently been combined with the CLAWPACK code to create AMRCLAW [29], [30], which can be used on quite general hyperbolic systems. Figure 6.1 shows a sample calculation using AMRCLAW. A spherical Riemann problem is solved in which the gas is initially stationary with uniform density but with higher pressure (and temperature) inside a sphere of radius 0.5. The data agrees with that used for the one-dimensional Riemann problem shown in Fig. 2.2 and for short times roughly the same structure is seen in the radial direction. The outward-moving shock and contact discontinuity are separated by a thin shell of high-density gas that can only be resolved on a fairly fine grid. The computation shown was done in two space dimensions using the AMRCLAW package, including source terms for cylindrical symmetry about the y-axis (handled by the fractional step approach).

Figure 6.1(a) shows the grid structure at time $t = 0.043$. In this computation two levels of refined grids have been introduced, refining first by a factor of 2 and then by a factor of 4 in going to the finest level. Regridding every few time steps based on Richardson extrapolation error estimates allows the region of refinement to follow the thin structure.

Figure 6.1(b) shows contours of the density at this time, along with outlines of the individual rectangular grid patches. We can see that both the shock and contact discontinuity are well resolved. The solution also appears to be radially symmetric, with no "grid orientation" effects that sometimes plague less accurate methods (Sect. 7.6). This is also clearly seen in Fig. 6.1(c), which shows a scatter plot of the density ϱ_{ij} in every grid cell plotted against r_{ij}, the distance from the center of the cell to the origin. The density profile is similar to what is seen in Fig. 2.2 for the one-dimensional Riemann problem with the same data.

Berger's code has previously been adapted to astrophysical problems by Rolf Walder, who has studied colliding flows in binary-star systems [177], [253]. Figure 6.2 shows one such calculation in two space dimensions with

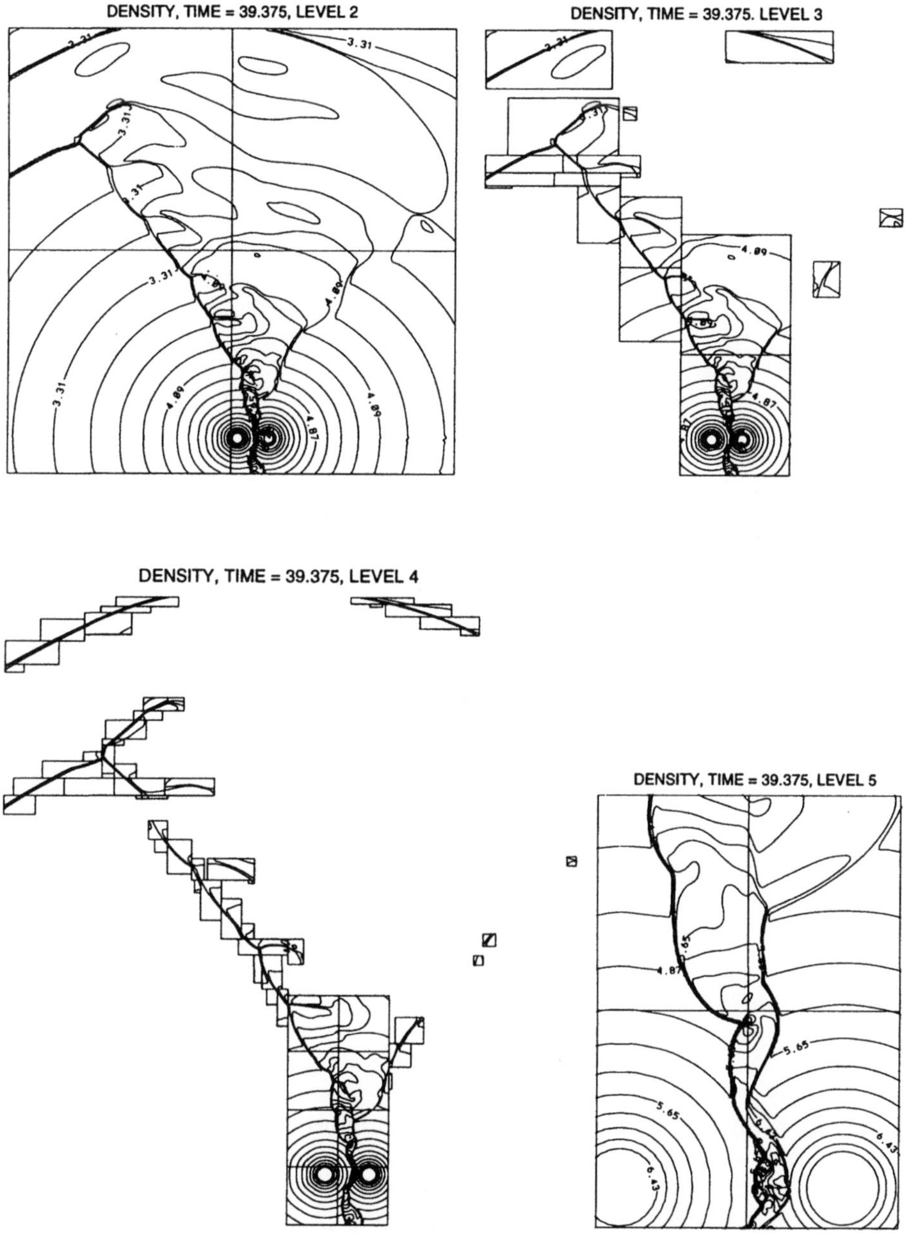

Fig. 6.2. Adaptive mesh refinement calculation of colliding winds between binary stars. Five levels of mesh refinement are used and the solutions on the finest four levels are shown. Note that Level 5 is used only in the interaction region directly between the stars. Calculation by R. Walder

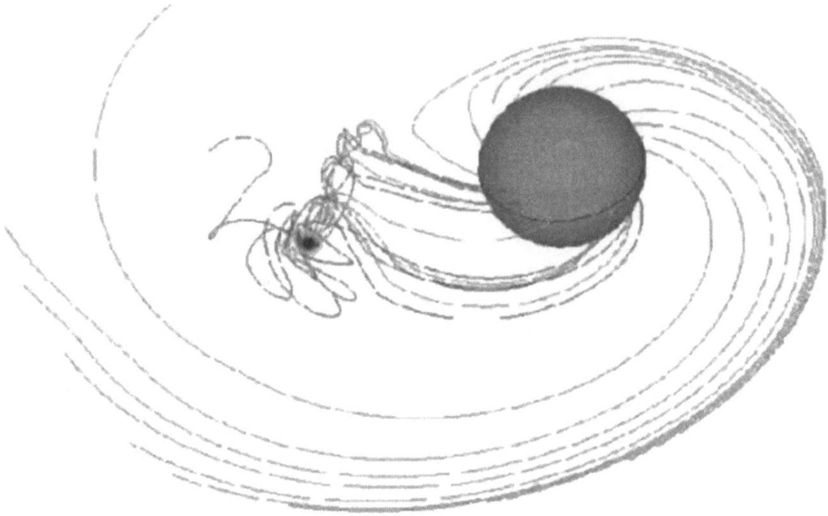

Fig. 6.3. Several particle paths are shown to indicate mass accreting from a red giant star to a companion white dwarf. Adaptive refinement was used to concentrate the grid in the region of interest in this three-dimensional calculation by R. Walder

five levels of refinement. Figure 2.5 shows results from a three-dimensional calculation.

Figure 6.3 shows the results of another three-dimensional calculation by Walder where AMR was a crucial ingredient. A Red Giant / White Dwarf pair is modeled, in which the wind velocity for the red giant is comparable to the angular velocity of the rotating system. This results in turbulent flow near the white dwarf, where mass is accreting. Figure 6.3 shows several particle paths from the three-dimensional simulation. AMR is used to resolve the flow in the neighborhood of the white dwarf without the need for an excessively fine grid everywhere. Five levels of refinement were used, with refinement by factors of 2 or 4. A total of 861,184 cells were used. A uniform grid calculation with the same resolution as the finest grid would have required $2560^3 = 1.68 \times 10^{10}$ cells and 64 times as many time steps in some regions. A uniform calculation might be more accurate, but this calculation is sufficient to give a rough estimate of what might happen in such an accretion process. For a deeper insight, one would also need to add viscosity and a turbulence model.

Adaptive mesh refinement is a powerful technique for solving astrophysical problems efficiently, especially for problems in three space dimensions. Related AMR methods have been used elsewhere in computational astrophysics, *e.g.*, [48], [75], [163], [228].

7. Computational Difficulties

The high-resolution finite methods discussed in these lectures can be used very successfully for a wide range of problems. They are not foolproof, however, and one should never accept computed results without a critical study of their accuracy. This is often hard to assess for complex problems where the exact solution is not known, but the following techniques can help:

- Investigate simple cases where exact solutions might be known, or at least the correct qualitative behavior of the solution is well understood.
- Reduce the number of space dimensions by considering radially symmetric solutions, for example. Then a fine-grid solution in one space dimension can be used as a reference solution for the multi-dimensional solution.
- Perform grid refinement studies on the real problem of interest. If you refine the grid does the solution remain basically the same? If not, then you probably cannot trust either solution. (If so, both solutions may still be completely incorrect. An error in the code that changes the equations might lead to a method that converges very nicely to a solution of the wrong equation.)

A number of specific difficulties that can arise in solving the Euler equations have already been mentioned, such as

- The use of a nonconservative method can give shocks that look reasonable but which travel at the wrong speed (Sect. 4.3).
- Stiff source terms can lead to similar results (Sect. 5.4).
- The computed solution may not satisfy the entropy condition, leading to discontinuities where there should be a smooth rarefaction (Sect. 4.6.4).

A number of other problems are frequently encountered in solving the Euler equations using high-resolution methods, and in the following sections a few of these are briefly discussed. (See Quirk [190] for some additional difficulties.) Along with alerting the newcomer to a few potential pitfalls, an exploration of these difficulties should offer further proof that a good understanding of the mathematics and physics of these equations is essential in diagnosing and improving computational algorithms.

7.1 Low-Density Flows

In conservative formulations of the Euler equations, the total energy is one of the conserved quantities. The internal energy and pressure are computed from this after subtracting out the kinetic energy determined from the momentum and density. In regions where the internal energy is very small compared with the kinetic energy, e.g., in high-speed, low-density flow, inaccuracies in the conserved quantities can easily lead to the kinetic energy exceeding the total energy, which results in the computed pressure being negative. This generally causes the code to crash, for example in the square root required to compute the sound speed. This problem is often dealt with by simply resetting negative pressures to some slightly positive value in the code, but it is preferable to use more robust methods that do not suffer this deficiency.

Linearized Riemann solvers, such as the Roe solver discussed in Sect. 4.6.4, may also fail near low densities or pressures by returning intermediate states that are nonphysical. Einfeldt, Munz, Roe and Sjogreen [79] give an analysis of some of these problems. They call a method "positively conservative" if the internal energy always remains positive for any physical data, and show that a variant of the HLLE method (Sect. 4.6.3) is positively conservative.

7.2 Discrete Shocks and Viscous Profiles

A number of difficulties are associated with the fact that the numerical approximation to a shock wave is not a sharp discontinuity (unless shock tracking is used) but rather is smeared over one or more grid cells. A single isolated shock in the p'th family consists, in principle, of only two values q_l and q_r separated by a discontinuity. In phase space, q_r lies on the p'th branch of the Hugoniot locus of q_l, and vice versa. The solution to the Riemann problem between these states consists of this one shock wave, with zero strength waves in the other families. A shock-capturing method, however, will introduce new values of q in the smeared region, and solving the Riemann problems involving these new states will, for a nonlinear system, introduce spurious waves in the other families. In spite of this, and somewhat remarkably in view of this interpretation, high-resolution shock-capturing methods often produce discrete shocks that propagate as quite sharp representations of the correct shock, generating very little visible noise in the other families. This is largely because the dissipation in the numerical methods mimics the physical dissipation that is natural for such systems, e.g., viscosity and heat conduction. A shock wave is a mathematical idealization of a thin transition, which results in a smooth curve of states between q_l and q_r in phase space (the **viscous profile**). As long as the numerical viscosity is of sufficient magnitude and of a reasonable form, the discrete shock may behave similarly to the physical shock, without generating noise in the other families.

In certain circumstances, however, this noise can be very noticeable and gives rise to various numerical phenomena that are frequently observed with the Euler equations and other systems. In particular, a wave that is moving at a slow speed s, so that sk/h is much less than 1 for this wave, experiences little numerical dissipation with many high-resolution methods. The next three sections discuss some situations where this lack of dissipation can lead to spurious noise since the numerical method is no longer modeling the dissipative aspect of the physics properly. All of the phenomena mentioned in the next few sections can be minimized by using a more dissipative method, though perhaps at the expense of some sharpness in the resolution of shocks and particularly contact discontinuities. Extra dissipation must be added judiciously where needed. For example, Marquina's flux-vector splitting method described in Sect. 4.8.3 puts in extra dissipation when the eigenvalue changes sign from one cell to the next, which signals that the shock speed may be close to zero. Donat and Marquina [73] show some comparisons for the type of problems discussed below.

For conservation laws other than the Euler equations, there may be additional trouble if the form of numerical dissipation introduced by a given method is not modeling a reasonable physical dissipation. For some nonlinear systems, different forms of dissipation can lead to different weak solutions in the limit of vanishing dissipation, in which case numerical methods must be used with particular caution.

7.3 Start-Up Errors

Numerical noise of the type described in the previous section is often observed in studying the propagation of an isolated shock numerically. If the data is chosen as a sharp discontinuity, then relatively large-magnitude waves in the other families are initially generated, before the discrete shock settles down into a viscous traveling wave.

Consider one step of Godunov's method, for example, on data

$$Q_i^0 = \begin{cases} q_l & \text{if } i \le 0 \\ q_r & \text{if } i \ge 1 \,. \end{cases}$$

Solving the Riemann problem at interface $i = 1$ gives the single shock wave propagating at some speed $s > 0$, say. Then

$$Q_i^1 = \begin{cases} q_l & \text{if } i \le 0 \\ q_m & \text{if } i = 1 \\ q_r & \text{if } i > 1 \,, \end{cases}$$

where

$$q_m = q_r - \frac{sk}{h}(q_r - q_l) = (1-\nu)q_r + \nu q_l \,,$$

with $\nu = sk/h$. The CFL condition requires $\nu \le 1$ and in fact $\nu < 1$ in general since the corresponding wave speed λ will be larger than s near state q_l or q_r. Hence q_m is a convex combination of q_l and q_r which lies on a straight line connecting these two states in phase space. In the next time step, we must solve Riemann problems between q_l and q_m and between q_m and q_r.

For a linear system, the Hugoniot loci are straight lines, so no waves in other families will be generated. The shock will become smeared, but will not

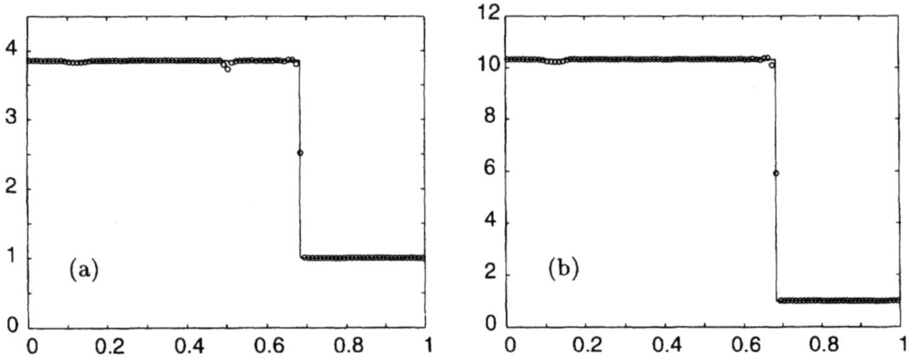

Fig. 7.1. Start-up error resulting from a Riemann problem with an isolated shock initially at $x = 0.5$. (**a**) Density (**b**) Pressure

generate noise in the other families. For a nonlinear problem, on the other hand, the Hugoniot loci are not straight lines and so q_m lies off of the locus connecting q_l and q_r. Solving these new Riemann problems then results in waves in the other characteristic families along with new approximations to parts of the original shock.

As a result, starting a shock-wave computation for the Euler equations with piecewise constant data of this sort will typically result in a spurious acoustic wave in the other nonlinear family and a spurious entropy wave. The magnitude of these waves depends on the strength of the initial shock, since they result from the nonlinearity of the Hugoniot locus. The spurious acoustic wave typically dissipates quickly and may not be noticed at all. The entropy wave, however, is often observed, particularly if the fluid velocity behind the shock is small. The entropy wave is simply a variation in density moving at this velocity, which gives rise to a contact discontinuity in the solution to any Riemann problem. If the velocity is small then this wave has little effect on the neighboring cells and experiences little numerical dissipation. Any variations in density remain essentially stationary, an artifact of the initial conditions which is often called **start-up error**.

Figure 7.1 shows an example when the high-resolution method from Sect. 4.7 with the superbee limiter is used on the data

$$\begin{aligned} &\varrho_l = 3.86, \quad &u_l = 0.0, \quad &p_l = 10.33, \quad &\text{for } x < 0.5\,, \\ &\varrho_r = 1, \quad &u_r = -2.63, \quad &p_r = 1, \quad &\text{for } x > 0.5\,, \end{aligned} \tag{7.1}$$

which gives only a 3-wave moving at velocity $s = 0.92$. Since the velocity behind the shock is zero, the density variation remains stationary at $x = 0.5$. There is also a weak acoustic 1-wave visible in the figure at about $x = 0.1$.

7.4 Wall Heating

This same sort of artifact is also often seen when two shocks collide or when a shock reflects off a solid wall. Consider two identical shocks approaching each other with zero velocity in between (which also models reflection at a wall halfway between the shocks — see Sect. 4.9.3). Each shock may have settled down to some numerical traveling wave that does not appear to generate any noise. But when the shocks collide, the result is two new out-going shocks with a different state in between than before the collision, with higher density and pressure but still zero velocity. During the interaction phase considerable noise will be generated in the other families, and in particular a spurious entropy wave will be generated which is then stationary in the zero-velocity region between the out-going shocks. This wave yields a dip in the density. The pressure, however, is nearly constant and so this dip in density results in an increase in the temperature $T = p/\mathcal{R}\varrho$. The gas appears to have been heated at the point where the collision occurs.

In particular, in computing the reflection of a shock off a solid wall, this spurious temperature rise occurs at the wall itself. This phenomenon is frequently observed in numerical simulations where shocks reflect off physical boundaries, and is known as **wall heating** in the literature. See, for example, [73], [175].

7.5 Slow-Moving Shocks

Figure 7.2 shows the numerical solution to the same shock propagation problem as in Fig. 7.1, but with the velocity shifted on each side so that the shock speed in this reference frame is much closer to zero. The data is the same as (7.1) but with 0.81 subtracted from each velocity. Otherwise, exactly the same method has been used as in Fig. 7.1, again with Courant number near 0.9. The solution is shown at time $t = 0.5$, after 260 time steps. In this case the propagating shock does not settle down into a smooth traveling wave, but instead continuously generates strong acoustic and entropy waves, giving rise to some oscillations in the pressure as well as the density. Discussion and analysis of this problem can be found in many papers. Arora and Roe [7] show nice plots of this behavior in phase space, showing that the numerical waves can deviate substantially from the correct Hugoniot locus. Karni and Canic [125] develop a modified equation analysis that illustrates the lack of numerical dissipation in this situation. See also [121], [193].

Note that the high-resolution method used here, with the superbee limiter, is a TVD method (see Sect. 4.7.5) on a scalar conservation law. Scalar shocks must be captured in a nonoscillatory fashion. For systems of equations there is no such guarantee in general. Developing high-resolution methods by extending the scalar methods as described in these notes has proved remarkably effective, but is not entirely foolproof as these examples show.

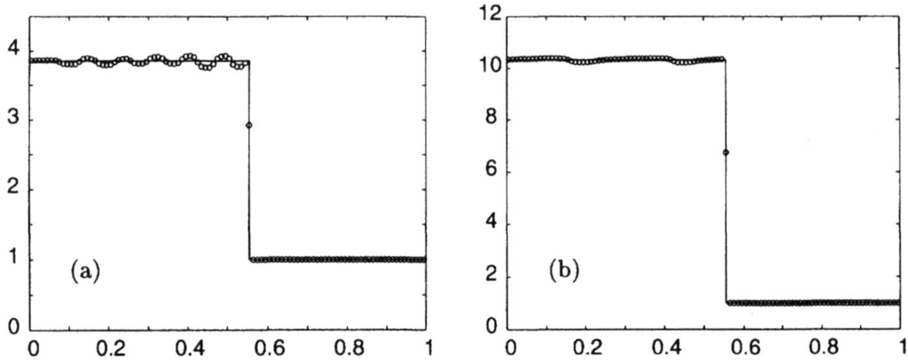

Fig. 7.2. Oscillations generated in a slowly-moving shock. The shock was initially at $x = 0.5$ and has moved to $x = 0.555$ at time $t = 0.5$. **(a)** Density **(b)** Pressure

7.6 Grid Orientation Effects

When a shock-capturing method is applied in more than one space dimension, shocks are generally not exactly aligned with the computational grid. Instead they cut through the grid at some angle. The methods which have been discussed here, based on solving one-dimensional Riemann problems normal to the cell interfaces, are using a series of Riemann problems in the coordinate directions to approximate a single shock at some angle to the grid. It is somewhat surprising that this works at all, but in fact it is remarkably successful in general. There are, however, some situations in which the orientation of the computational grid can be directly observed in computational results. See [147] for some examples. However, using a higher-order method which includes appropriate cross-derivative terms from the Taylor series expansion typically eliminates much of this grid-orientation effect. This is generally not a serious problem except perhaps with very strong shocks or when a curved structure is very poorly resolved on the grid.

7.7 Grid-Aligned Shocks

In a multi-dimensional calculation it may seem advantageous to have a shock aligned with the grid so that Riemann problems normal to the cell edges are also in the physically correct direction. However, a shock that is nearly aligned with the grid can also suffer certain numerical instabilities that have no analog in one-dimensional calculations. Figure 7.3 shows density contours at a sequence of times for a colliding flow problem. Initially $\varrho \equiv p \equiv 1$, $v \equiv 0$ everywhere, while the x-velocity is $u = +20$ on the left half of the domain and $u = -20$ on the right. This colliding flow should give rise to two symmetric shock waves propagating outwards. If the initial data is exactly uniform then the calculation will yield a reasonable approximation to this. For the calculation in Fig. 7.3, however, the density was initially perturbed

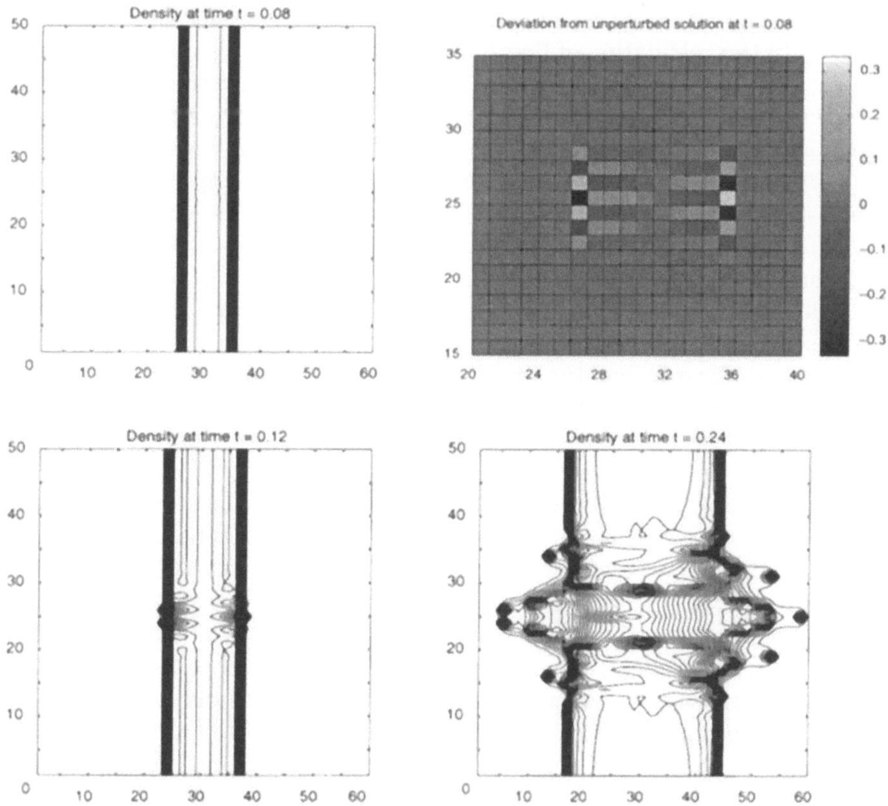

Fig. 7.3. Colliding winds give rise to two outgoing shock waves. An initial density perturbation of magnitude 10^{-6} grows due to numerical instability

to $\varrho = 1 + 10^{-6}$ in a single grid cell in the center. At time $t = 0.08$ the effect of this perturbation is just barely visible in the contour plot, though examining the deviation from an unperturbed calculation shows that it has grown to a magnitude of 0.3 and now covers a larger set of grid cells. By time $t = 0.12$ its effect is clearly visible in the contour plot and there are order 1 perturbations in the density that spread out from the center and completely destroy the shock structure.

In this calculation, $\gamma = 1.05$ and the simplest Godunov method has been used with the Roe solver and Courant number 0.45. The shocks are moving quite slowly; roughly 1000 time steps have been taken in Fig. 7.3. This type of numerical instability is seen only for strong shocks that are aligned with the grid, or nearly so. The instability develops as an oscillation on the grid scale in the y-direction, as seen in Fig. 7.3. Quirk [190] refers to this as **odd-even decoupling** and gives some analysis of this. It is in solving the Riemann problems in the y-direction, where nothing should be happening, that small amplitude oscillations in density and pressure on the grid scale are magnified unstably. In the calculation of Fig. 7.3, these were excited by a perturbation at a single point (on one side of the initial discontinuity, hence the asymmetry in x). In an actual calculation with a strong shock nearly aligned with the grid, perturbations due to other flow features or even to rounding errors can lead to this instability developing simultaneously along the length of the shock, leading to a more clearly visible oscillatory structure at later times.

The choice of Riemann solver affects this stability. Since the density perturbations should be propagating at velocity $v \approx 0$ in the y-direction, methods which use the Roe solver or even an exact solver will add very little numerical dissipation, as mentioned in Sect. 7.5, allowing the instability to grow. According to Quirk, the problem is eliminated using the HLLE solver, for example, and he discusses a hybrid approach where one switches to such a method near strong shocks. The Marquina solver (Sect. 4.8.3) also appears to be more stable in this situation.

A shock propagating at an angle to the grid does not suffer from this same instability. Shock fronts are often curved, however, and will be tangent to the grid lines at some point. Where the shock is locally aligned with the grid, an instability of this same form can arise which remains confined to this region. This resembles the early stages of the instability shown in Fig. 7.3, giving a small protrusion ahead of the shock that is sometimes called a **carbuncle** [190]. This has been observed in some calculations of radially expanding shocks, *e.g.*, in supernova explosions, where instabilities can develop along the coordinate axes in a Cartesian calculation. This type of instability can also plague colliding wind calculations in binaries [253]. It is also seen in supersonic flow around blunt bodies, which produces a bow shock. If the bow shock is aligned with the grid along the stagnation line, where there is no transverse flow, then a carbuncle can form with some methods [190], [258].

Switching to a more diffusive method can alleviate this problem, but care must be taken to insure that physically correct instabilities are not also damped. For example, the high-density region which develops between the shock waves in the colliding flow of Fig. 7.3 is subject to the **thin shell instability**, which is physically relevant in many astrophysical problems, particularly for radiating shocks. See, *e.g.*, [32], [70], [230], [251].

8. Magnetohydrodynamics

In astrophysical problems one must often study the flow of ionized gases subject to magnetic fields. The field exerts a force on the gas while the motion of the gas perturbs the magnetic field. To model such flows it is necessary to use the equations of magnetohydrodynamics (MHD) to couple equations for the magnetic field to the hydrodynamic (actually gas dynamic) equations. This uses a simplified version of Maxwell's equations to model the coupling. A simplified set is used because we only seek to model the motion of the particles, not the propagation of electromagnetic waves on much faster time scales.

In this section these equations will be briefly reviewed, in the context of hyperbolic theory, *i.e.*, with emphasis on the characteristic structure, wave speeds, solution to the Riemann problem, *etc.* I will concentrate on pointing out how this system differs from the simpler Euler equations and what new difficulties arise compuationally. The presentation of the equations given here owes much to other sources, particularly [124], [199]. The MHD equations are discussed in many texts, *e.g.*, [118], [129], [218].

8.1 The MHD Equations

In MHD we assume that the flow is nonrelativistic and that we are interested in variations on a slow time scale. Then in Maxwell's equations we can ignore the displacement current and approximate

$$\boldsymbol{j} \approx \frac{1}{4\pi}\nabla \times \boldsymbol{B}\,, \tag{8.1}$$

$$\boldsymbol{E} \approx \boldsymbol{j}/\sigma - \boldsymbol{u} \times \boldsymbol{B}\,, \tag{8.2}$$

and hence

$$\frac{\partial \boldsymbol{B}}{\partial t} = -\nabla \times \boldsymbol{E} \tag{8.3}$$

$$\approx \nabla \times (\boldsymbol{u} \times \boldsymbol{B}) + \eta \nabla^2 \boldsymbol{B}\,. \tag{8.4}$$

Explicit dependence on the electric field has been eliminated and we have an evolution equation for the magnetic field $\boldsymbol{B}$ that depends only on $\boldsymbol{B}$ and $\boldsymbol{u}$, the velocity. Here η is the resistivity and for ideal MHD we assume $\eta = 0$ so that this dissipative term vanishes.

These equations are also coupled with the condition

$$\nabla \cdot \boldsymbol{B} = 0 \tag{8.5}$$

from Maxwell's equations: the divergence of the magnetic field must always be zero. This is a physical law that is assumed to always hold (barring the existence of "magnetic monopoles", which have never been observed). Solutions to the MHD equations will automatically maintain (8.5) for $t > 0$ provided the initial data satisfies (8.5) at $t = 0$. (The numerical solution, however, may not. This constraint must then generally be imposed separately on the numerical solution, as discussed in Sect. 8.6.)

Setting $\eta = 0$, we can rewrite (8.3) as

$$\frac{\partial \boldsymbol{B}}{\partial t} + \nabla \cdot (\boldsymbol{u}\boldsymbol{B} - \boldsymbol{B}\boldsymbol{u}) = 0\,. \tag{8.6}$$

This is coupled with the Euler equations for the evolution of $\boldsymbol{u}$, with additional coupling coming from the fact that the magnetic field exerts a force on the fluid, which appears in the momentum equation. This force is

$$\begin{aligned} j \times \boldsymbol{B} &= \frac{1}{4\pi}(\nabla \times \boldsymbol{B}) \times \boldsymbol{B} \\ &= \frac{1}{4\pi}\left(\nabla \cdot \left(\boldsymbol{BB} - \frac{1}{2}(B^2)I\right) - \boldsymbol{B}\nabla \cdot \boldsymbol{B}\right) , \end{aligned} \tag{8.7}$$

where $B^2 = \boldsymbol{B} \cdot \boldsymbol{B}$. In view of (8.5), the last term in (8.7) drops out and this is simply the divergence of the **Maxwell stress tensor**. Using this, the momentum equation can be written in conservation form as

$$(\varrho \boldsymbol{u})_t + \nabla \cdot \left(\varrho \boldsymbol{uu} + I\left(p + \frac{1}{8\pi}B^2\right) - \frac{1}{4\pi}\boldsymbol{BB}\right) = 0 . \tag{8.8}$$

The isotropic force $\frac{1}{8\pi}B^2$ enters in a form similar to the pressure and is called the **magnetic pressure**, while the tensor $\boldsymbol{BB}$ is the **magnetic tension**.

The magnetic field also affects the energy equation. The total energy is now defined as

$$E = \varepsilon + \frac{1}{2}\varrho(\boldsymbol{u} \cdot \boldsymbol{u}) + \frac{1}{8\pi}B^2 ,$$

where the internal energy ε is still given by

$$\varepsilon = \frac{p}{\gamma - 1}$$

for a gamma-law gas. The energy flux is given by

$$\left(E + p + \frac{1}{8\pi}B^2\right)\boldsymbol{u} - \frac{1}{4\pi}\boldsymbol{B}(\boldsymbol{u} \cdot \boldsymbol{B}) .$$

The new terms relative to the Euler equations result from the work done on the gas by the magnetic field.

At this point it is convenient to rescale $\boldsymbol{B}$ to incorporate a factor of $1/\sqrt{\mu_0} = 1/\sqrt{4\pi}$ into the definition of $\boldsymbol{B}$. This eliminates all of these factors in the above equations and the ideal MHD equations are thus

$$\frac{\partial}{\partial t}\begin{bmatrix} \varrho \\ \varrho \boldsymbol{u} \\ \boldsymbol{B} \\ E \end{bmatrix} + \nabla \cdot \begin{bmatrix} \varrho \boldsymbol{u} \\ \varrho \boldsymbol{uu} + I\left((p + \frac{1}{2}B^2) - \boldsymbol{BB}\right) \\ \boldsymbol{uB} - \boldsymbol{Bu} \\ (E + p + \frac{1}{2}B^2)\,\boldsymbol{u} - \boldsymbol{B}(\boldsymbol{u} \cdot \boldsymbol{B}) \end{bmatrix} = 0 . \tag{8.9}$$

This is in the conservation form (2.27). Writing out all of the components, we obtain

$$q = \begin{bmatrix} \varrho \\ \varrho u \\ \varrho v \\ \varrho w \\ B_x \\ B_y \\ B_z \\ E \end{bmatrix} , \quad f(q) = \begin{bmatrix} \varrho u \\ \varrho u^2 + p + \frac{1}{2}B^2 - B_x^2 \\ \varrho uv - B_x B_y \\ \varrho uw - B_x B_z \\ 0 \\ B_y u - v B_x \\ B_z u - w B_x \\ u\left(E + p + \frac{1}{2}B^2\right) - B_x(uB_x + vB_y + wB_z) \end{bmatrix} ,$$

and similar expressions for the fluxes g and h in the y- and z-directions. Here B_x, B_y, and B_z are the three components of $\boldsymbol{B}$, and subscripts do not indicate derivatives.

8.2 One-Dimensional MHD

Consider a plane-wave solution in which $q(x,y,z,t)$ varies only with x and t. Then all y- and z-derivatives are zero and we are left with the equations of one-dimensional MHD,

$$q_t + f(q)_x = 0 , \tag{8.10}$$

with q and $f(q)$ as above. One thing to note immediately is that the fifth component of $f(q)$ is zero, and hence the fifth equation of (8.10) is simply

$$\frac{\partial}{\partial t} B_x = 0 . \tag{8.11}$$

Any spatial variation in B_x remains stationary in time. This should seem nonphysical; one would expect all wave speeds to be relative to the speed u of the gas, by Galilean invariance. However, for purely one-dimensional MHD the condition (8.5) requires that $\frac{\partial}{\partial x} B_x = 0$ since $\frac{\partial}{\partial y} B_y = \frac{\partial}{\partial z} B_z = 0$. Hence B_x must be spatially constant as well as temporally constant and so the wave speed is irrelevant. We can in fact drop the equation for B_x altogether, and view the constant value of B_x as a parameter in the remaining system of seven equations for

$$\tilde{q} = (\varrho, \varrho u, \varrho v, \varrho w, B_y, B_z, E) .$$

Linearizing these equations allows us to determine the characteristic wave speeds. Computing the eigenvalues of the Jacobian matrix about some constant state yields

$$\begin{aligned} &\lambda_1 = u - c_f , \quad \lambda_2 = u - c_A , \quad \lambda_3 = u - c_s , \quad \lambda_4 = u , \\ &\qquad \lambda_5 = u + c_s , \quad \lambda_6 = u + c_A , \quad \lambda_7 = u + c_f . \end{aligned} \tag{8.12}$$

Waves propagate at speeds c_f, c_A, c_s relative to the gas. These speeds are defined and desribed below. The middle wave with $\lambda_4 = u$ travels at the fluid velocity and corresponds to the **contact discontinuity** as in the Euler equations. This wave can carry an arbitrary variation in density while all other variables are constant, *i.e.*, the corresponding eigenvector is

$$r^4 = (1,0,0,0,0,0,0) .$$

Compare this with the three-dimensional Euler equations in the case of a plane wave, which gives a system of 5 equations with wave speeds

$$\lambda_1 = u - a , \quad \lambda_2 = \lambda_3 = \lambda_4 = u , \quad \lambda_5 = u + a ,$$

where a is the sound speed. In this case there are three waves with speed u since the contact discontinuity can also carry arbitrary variation in the transverse velocities v and w as well as in ϱ. This shear velocity, assuming it varies only with x and not with y or z, is simply advected with the flow and has no effect on the fluid dynamics. (This is only true for an ideal fluid where we ignore viscosity and hence there is no resistence to shear motion.)

In the MHD equations this is no longer true. In ideal MHD the field lines are "frozen into" the flow since the equation (8.3) implies that $\boldsymbol{B}$ is transported with the fluid. Hence a shear motion will cause a distortion of the field lines and will result in a force on the fluid. In this sense MHD is more similar to the equations of elasticity from solid mechanics than to fluid mechanics. An elastic solid supports two different types of waves: compressional waves (pressure waves, or P-waves) that are acoustic waves as in a fluid, and transverse waves (shear waves, or S-waves) that arise from the fact that a transverse displacement results in a restoring force from bonds between molecules that are absent in a fluid. These two types of waves propagate at distinct speeds, with the P-waves (primary waves) propagating faster than the S-waves (secondary waves).

In MHD, the magnetic field resists transverse motion and hence there exist purely transverse waves in which ϱ, u, B_x, and p are constant but the transverse velocities and magnetic fields vary. These are known as **Alfvén waves**, which propagate at the Alfvén velocity

$$c_A = \sqrt{B_x^2/\mu_0\varrho}$$

relative to the fluid velocity u. These are the 2-wave and 6-wave in the ordering (8.12). Like the contact discontinuity, these waves are linearly degenerate since B_x, and hence the wave speed, is constant throughout the wave.

There are two sets of waves remaining to discuss, propagating with speeds c_f and c_s relative to the gas. These are the **magnetosonic waves** which are nonlinear waves in the general Riemann problem that will typically have shock or rarefaction wave solutions. The two sets are called the **slow** and **fast** magnetosonic waves since $c_s \leq c_A \leq c_f$. These waves involve compression of the gas. In certain cases one set of waves or the other reduces to simple acoustic waves, but in general there is a more complicated coupling of each with the magnetic field.

Following [199], set

$$b_{x,y,z} = B_{x,y,z}/\sqrt{\varrho}\,, \qquad b^2 = b_x^2 + b_y^2 + b_z^2$$

and again let $a = \sqrt{\gamma p/\varrho}$ be the sound speed. Then computing the eigenvalues of the Jacobian matrix, the fast and slow wave speeds can be found as the larger and smaller values of c^2 satisfying the equation

$$c^4 - (a^2 + b^2)c^2 + a^2 b_x^2 = 0\,.$$

This yields

$$c_{f,s}^2 = \frac{1}{2}\left(a^2 + b^2 \pm \sqrt{(a^2 + b^2)^2 - 4a^2 b_x^2}\right) \tag{8.13}$$

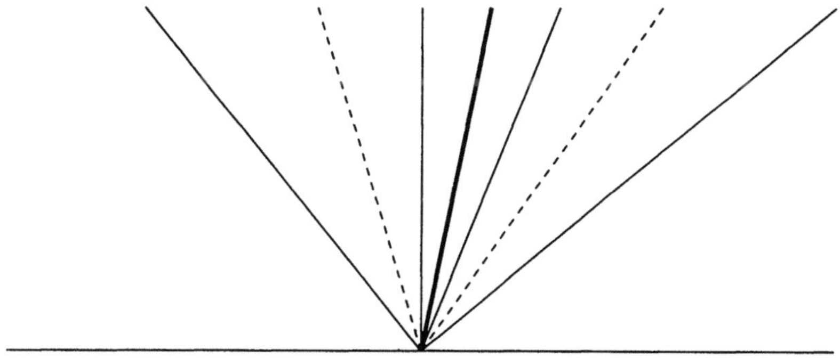

Fig. 8.1. Wave structure for the MHD equations. The symbols f, A, and s denote fast, Alfvén, slow waves, and u the contact discontinuity

with f, s corresponding to $+, -$ respectively. Since in general $c_s \leq c_A \leq c_f$, the Alfvén wave is also often called the **intermediate wave**. The wave structure is illustrated in Fig. 8.1.

The magnetic field B in the state we are linearizing about lies at some angle to the x-axis. If this angle is 0 or 90 degrees (*i.e.*, $b_\perp$ or b_x is 0, respectively), then certain degeneracies arise as discussed below. In general $b_\perp \neq 0$ and these two directions define a unique third direction, orthogonal to both. It is in this direction that the magnetic field and velocity variations in an Alfvén wave occur. If $b_\perp = 0$ then such transverse motions have 2 degrees of freedom in the entire y-z plane and one would expect 2 linearly independent waves traveling at the Alfvén speed. In fact, in this case either $c_s = c_A$ or $c_f = c_A$, as discussed further below, and one of the magnetosonic waves degenerates to a second Alfvén wave.

Except in this degenerate case, the fast and slow waves each have some compressional component. In ordinary gas dynamics a compressional wave travels at the sound speed determined by $a^2 = \mathrm{d}p/\mathrm{d}\varrho$ at constant entropy. Fast and slow waves propagate similary, but the magnetic pressure must now be included in this expression and the magnetosonic speeds can be described by

$$c^2 = \frac{\mathrm{d}}{\mathrm{d}\varrho}(p + B^2) \,. \tag{8.14}$$

The fluid pressure p increases with density, $\mathrm{d}p/\mathrm{d}\varrho > 0$, but the magnetic pressure B^2 may either increase or decrease, depending on the orientation of $\boldsymbol{B}$ and the changes in velocity across the wave. This means that the magnetosonic speed c given by (8.14) will either be greater than a or less than a, depending on whether changes in the fluid and magnetic pressure complement one another or partially cancel out. These two cases yield fast and slow waves, respectively.

The degenerate cases $b_x = 0$ and $b_\perp = 0$ are worth discussing in detail for two reasons. First, they give some indication of how these waves relate to gas-dynamic waves in the Euler equations. Second, in these special cases several waves coalesce and the system is not strictly hyperbolic, which can create numerical difficulties as discussed in Sect. 8.4. Genuine nonlinearity can also be lost at these points, leading to nonconvexity.

If $b_x = 0$ but $b^2 = b_y^2 + b_z^2 \equiv b_\perp^2$ is nonzero, so the magnetic field is completely in the transverse direction, then the resistance to transverse motion described above disappears. We then have

$$c_f^2 = a^2 + b_\perp^2 \,, \qquad c_s = c_A = 0 \,.$$

In this case the Alfvén waves, slow waves, and contact discontinuity all collapse into a contact discontinuity and shear waves as in the Euler equations. The fast magnetosonic wave is similar to an acoustic wave but the transverse magnetic field affects the wave speed and structure. If we also let $b_\perp = 0$, *i.e.*, impose no magnetic field, then $c_f = a$ and we recover the ordinary acoustic waves. This is the **hydrodynamic limit** in which MHD reduces to the Euler equations.

On the other hand, if $b_\perp = 0$ while $b_x \neq 0$, so that the magnetic field is entirely in the x-direction, then we expect two sets of Alfvén waves and only one set of nonlinear waves. These should in fact be ordinary acoustic waves, which are unaffected by the magnetic field when $\boldsymbol{B}$ is aligned with the direction of compression. We can check this by computing the wave speeds from (8.13). If $b_\perp = 0$ then $b^2 = b_x^2$ and so

$$c_{f,s}^2 = \frac{1}{2}\left(a^2 + b_x^2 \pm \sqrt{(a^2 - b_x^2)^2}\right) \,.$$

Now there are two cases to consider depending on how the Alfvén speed $c_A = |b_x|$ is related to the sound speed a, *i.e.*, how strong the magnetic field is.

If $b_\perp = 0$ and $b_x^2 < a^2$, then

$$c_f = a \,, \qquad c_s = c_A = |b_x| \,.$$

In this case the slow magnetosonic wave degenerates to an Alfvén wave, while the fast wave propagates at the acoustic speed, and is in fact an ordinary acoustic wave. Again note that if we also set $b_x = 0$ we recover the structure of the Euler solution.

If $b_\perp = 0$ and $b_x^2 > a^2$ then

$$c_f = c_A = |b_x| \,, \qquad c_s = a \,.$$

In this case the slow wave is an ordinary acoustic wave, while the fast wave becomes an Alfvén wave.

The most interesting case arises when $b_\perp = 0$ and $b_x^2 = a^2$, in which case

$$c_f = c_s = c_A = a \,.$$

The two-dimensional families of Alfvén waves propagate at the same speed as the acoustic waves. This is called the **triple umbilic point** (see Sect. 8.4).

8.3 Solving the Riemann Problem

The eigenvalues and eigenvectors of the Jacobian matrix derived above are useful in studying the propagation of small-amplitude disturbances about some fixed state. Solving the Riemann problem with a large-amplitude discontinuity between the states is a different matter, however. For a strictly hyperbolic system, such as the Euler equations, we expect the solution to the $m \times m$ Riemann problem to consist of m waves, one corresponding to each eigenvalue of the linearized matrix. For the 2×2 isothermal system studied in Sect. 3.3, for example, the eigenvalues are $\lambda^1 = u - a$, $\lambda^2 = u + a$, and the solution to the general Riemann problem consists of two waves that are easily identified with these two families. At each point in the relevant part of phase space ($\varrho > 0$) there are two distinct eigendirections and two distinct curves corresponding to 1-waves and 2-waves, determined by using the entropy condition to select the relevant portions of the integral curves and Hugoniot loci. These curves form a coordinate system for this portion of phase space, so that any two states can be connected in a unique way by a 1-wave followed by a 2-wave. A nonstrictly hyperbolic system can have a richer structure in solutions to the Riemann problem.

8.4 Nonstrict Hyperbolicity

If a system is not strictly hyperbolic then there are at least some points q in phase space where two or more eigenvalues of the Jacobian matrix are equal. If the problem is hyperbolic, then the matrix must still be diagonalizable and hence an eigenvalue of multiplicity r must have an r-dimensional eigenspace associated with it. At such a point there are infinitely many eigen-directions in phase space, leading to a possible breakdown in the structure mentioned above.

The multi-dimensional Euler equations fail to be strictly hyperbolic everywhere in phase space since $\lambda = u$ is an eigenvalue of multiplicity N in N space dimensions. The entropy wave and shear waves all travel at the fluid velocity. However, these are all linearly degenerate waves that behave completely independently of one another and no complication is introduced by this.

For nonlinear fields a multi-dimensional eigenspace can result in the number of integral curves through such a point being different from the m we normally expect for an $m \times m$ system at a strictly hyperbolic point. This singularity in the vector field can lead to a global change in the structure of the integral curves and Hugoniot loci, and difficulties in solving the Riemann problem.

In some systems there are isolated points in phase space where strict hyperbolicity fails, called **umbilic points** [207]. In systems such as MHD there are curves or surfaces in phase space along which there are multiple eigenvalues, *e.g.*, the surface $b_x = 0$ or $b_\perp = 0$.

A simple 2×2 nonstrictly hyperbolic system illustrates some of the potential difficulties. The model problem discussed here is a special case of a system studied by Schaeffer and Shearer [207]. Brio and Rosenau [43] developed a 3×3 extension of this as a model of intermediate waves and the triple umbilic point in MHD equations, which has also been studied by Myong [173].

Consider the 2×2 system

$$\begin{aligned} u_t + \frac{1}{2}(u^2 + v^2)_x &= 0\,, \\ v_t + (uv)_x &= 0\,. \end{aligned} \tag{8.15}$$

The Jacobian matrix is

$$f'(q) = \begin{bmatrix} u & v \\ v & u \end{bmatrix}$$

with eigenvalues $u \pm v$. For $v \neq 0$ the systems is strictly hyperbolic and has distinct eigen-directions $(1,1)$ and $(1,-1)$. Along the u-axis (where $v = 0$) in phase space the eigenvalues are both equal to u, the system fails to be strictly hyperbolic, and every vector is an eigenvector of $f'(q) = uI$.

For the simple system (8.15), the integral curves and Hugoniot loci can be shown to agree, and are simply the lines with slope ± 1 in the phase plane. These appear to behave well as we cross the u-axis. However, note that for $v > 0$ we have $u + v > u - v$, while for $v < 0$ we have $u + v < u - v$, so our convention that $\lambda^1 < \lambda^2$ forces us to define

$$\lambda^1 = \begin{cases} u + v & \text{if } v \le 0 \\ u - v & \text{if } v > 0 \end{cases}, \qquad \lambda^2 = \begin{cases} u - v & \text{if } v \le 0 \\ u + v & \text{if } v > 0 \end{cases}$$

with freedom in the labeling along $v = 0$. As a result, each integral curve takes a 90 degree turn as we cross the u-axis. This can happen only because every direction is an eigendirection along the u-axis.

Now consider a Riemann problem with states on opposite sides of the u-axis, say $q_l = (\alpha, \alpha)$, $q_r = (\beta, \beta)$ with $\alpha > 0 > \beta$. The Rankine–Hugoniot jump conditions are satisfied between these states if we take $s = \alpha - \beta$. This jump, propagating at speed s, is a weak solution to the conservation laws. It is not so clear, however, whether it should be associated with 1-waves or 2-waves, or whether such a wave is physically realizable.

The situation with MHD is similar, though more complicated, and has a rich structure of "intermediate shocks" that are not easily identified with a particular family. In some cases the number of characteristics impinging on such a shock as time advances in the x-t plane is different from the $m + 1$ characteristics that are expected in the $m \times m$ strictly hyperbolic case (see Sect. 3.3.4), and the shocks are said to be *overcompressive* or *undercompressive* if the number is greater or less than the number we expect.

This is related to the fact that MHD is further complicated by a failure of genuine nonlinearity (Sect. 3.3.3) near points where strict hyperbolicity fails, leading to an even richer structure of compound waves that involve both shock and rarefaction waves in the same family traveling together. The existence of such waves was first discovered computationally by Brio and Wu [44], [259] who used one of the earliest Riemann-based methods for MHD.

The analysis of intermediate waves in MHD is currently an active area of research, and there is still some controversy about what the correct admissibility criteria are for such waves, and which of these waves are physically observable. In this analysis one must remember that the ideal MHD equations ignore a variety of physical effects, and the hyperbolic system is only an approximation to more correct systems of equations. In the Euler equations for an ideal gas, fluid viscosity and heat conduction are ignored. In MHD we have also ignored the dissipative effect of resistivity, and also dispersive effects due to the Hall current.

There is a large literature on the topic of nonstrictly hyperbolic systems, for the MHD equations, in other applications such as oil recovery and nonlinear elasticity where similar problems arise, and in more general mathematical investigations. Along with the papers mentioned above, a small sample of interesting recent work includes [17], [42], [43], [91], [92], [114], [127], [154], [173], [207], [208], [209], [229], [261], [262]. Zachary, Malagoli, and Colella [265] discuss a numerical approach to dealing with nonstrict hyperbolicity, based on ideas in [21].

8.4.1 Artificial Resistivity. We obtain a hyperbolic problem in MHD by assuming the fluid is perfectly conducting, *i.e.*, we ignore the dissipative effect of resistivity along with viscosity and heat conduction. The resulting hyperbolic model is only useful if weak solutions obtained are in fact "vanishing resistivity" solutions to the more correct equations. For MHD the exact form of dissipation added can affect the wave structure. Numerical methods add some **numerical resistivity** and one might worry about whether the physically correct solutions are obtained on problems where physical resistivity is important. The analysis of the correct structure is complicated by the loss of strict hyperbolicity and genuine nonlinearity, and some of the papers mentioned in the previous section discuss issues of correct viscosity criteria.

8.5 Stiffness

In some contexts the MHD equations are stiff in the sense that there are widely varying wave speeds in different parts of the computational domain. In modeling the solar corona, for example, the density varies exponentially and so the waves speeds, which are proportional to $1/\sqrt{\varrho}$, will be orders of magnitude larger in some regions than others. If we use an explicit method with the same size time step everywhere, then the time step will be limited by the largest wave speed over the entire domain. In other parts of the domain we will then be using a tiny Courant number. This is not only inefficient, but can lead to loss of accuracy in these regions. For such problems it may be necessary to develop an implicit method for solving the hyperbolic system, *e.g.*, [123], [151], [210].

8.6 The Divergence of B

Numerical solutions to the MHD equations may not satisfy the condition (8.5) that is automatically satisfied by the true solution. It should be maintained numerically to the level of the truncation error, but this can still have a severe impact on the numerical solution, as observed in early numerical work by Brackbill and Barnes [40]. A variety of approaches have been suggested to deal with this problem.

One common approach (*e.g.*, [191], [264]) is to use some form of **divergence cleaning**, which consists of projecting the numerical $\boldsymbol{B}$ field onto the space of divergence-free vector fields every few time steps, thus removing any nonzero divergence from the field.

An arbitrary vector field $\boldsymbol{V}$ can be decomposed as the sum of a gradient and a curl,

$$\boldsymbol{V} = \nabla\phi + \nabla \times \boldsymbol{W} \tag{8.16}$$

for some ϕ and $\boldsymbol{W}$. If we have such an orthogonal decomposition then $\boldsymbol{V} - \nabla\phi$ has zero divergence and is the orthogonal projection of $\boldsymbol{V}$ onto the space of divergence-free vector fields, which are also called *solenoidal*. This is the **Hodge projection**. We can compute ϕ by taking the divergence of (8.16), which results in a Poisson problem for ϕ,

$$\nabla \cdot \boldsymbol{V} = \nabla^2\phi\,. \tag{8.17}$$

The numerical algorithm for MHD with divergence-cleaning proceeds as follows. Given Q^n at the start of a time step, which includes magnetic field components $\boldsymbol{B}^n$ that are divergence-free, first update this by solving the hyperbolic system over the time step Δt, resulting in some new values Q^*. This includes the updated field $\boldsymbol{B}^*$ which will not in general be divergence-free. Now solve the elliptic equation

$$\nabla^2\phi^{n+1} = \nabla \cdot \boldsymbol{B}^* \tag{8.18}$$

for ϕ^{n+1} and then set

$$\boldsymbol{B}^{n+1} = \boldsymbol{B}^* - \nabla\phi^{n+1}$$

and use these values in place of $\boldsymbol{B}^*$ to define Q^{n+1} from Q^*. In practice this projection may be done every few time steps rather than every step, since solving the elliptic equation can be expensive. Note that multi-dimensional elliptic equations are typically solved with iterative methods (see Sect. 1.3.2) which require an initial guess at the solution ϕ^{n+1}. We should take advantage of the fact that ϕ^{n+1} should be close to ϕ^n, so that the solution from the previous time step (or whenever divergence-cleaning was last performed) can be used as a good starting guess.

This elliptic equation must be solved together with some boundary conditions for ϕ on the boundary of the computational domain. Specifying appropriate boundary conditions can be a difficulty and careful thought must be given to this issue depending on the specific problem being solved. In studying fundamental phenomena, such as the behavior of certain instabilities in MHD (*e.g.*, [203]), it may be possible to solve a simplified problem with **periodic boundary conditions**. In this case periodic boundary conditions could be used for ϕ which also allows fast Poisson solvers based on fast Fourier transforms to be used in place of iterative methods [233].

8.6.1 Incompressible Flow. Similar projection methods are commonly used in numerical algorithms for the incompressible Navier–Stokes equations, which are frequently used in studying the flow of liquids. Hence this literature is a good source of information on such methods. See [185], for example, for some references. A brief discussion of incompressible flow is worth a detour here since this also introduces some other new ideas that may be useful in astrophysical flow calculations.

For an incompressible fluid the density ϱ is constant and the continuity equation $\varrho_t + \nabla \cdot (\varrho \boldsymbol{u}) = 0$ becomes

$$\nabla \cdot \boldsymbol{u} = 0 . \tag{8.19}$$

Hence the velocity field must be divergence-free. The conservation of momentum equation has the form

$$\boldsymbol{u}_t + \boldsymbol{u} \cdot \nabla \boldsymbol{u} + \nabla p = \mu \nabla^2 \boldsymbol{u} , \tag{8.20}$$

where μ is the viscosity (which typically cannot be ignored in modeling the flow of liquids). The **incompressible Navier–Stokes equations** (INS) consist of (8.20) together with the constraint (8.19). There is no energy equation and no equation of state. Instead the pressure p must be determined in each time step in such a way that (8.19) remains satisfied. There is no local evolution equation for p; it must be determined using this global constraint.

One common approach is the **projection method** introduced in [51] (see also, *e.g.*, [23], [20], [185]). In each time step one first advances (8.20) without the ∇p term to obtain $\boldsymbol{u}^*$, *i.e.*, we solve

$$\boldsymbol{u}_t + \boldsymbol{u} \cdot \nabla \boldsymbol{u} = \mu \nabla^2 \boldsymbol{u} \tag{8.21}$$

over time Δt. The resulting $\boldsymbol{u}^*$ will not be divergence-free because of the nonlinear terms.

Projecting $\boldsymbol{u}^*$ onto the divergence-free space gives $\boldsymbol{u}^{n+1}$. We solve

$$\Delta t \nabla^2 p^{n+1} = \nabla \cdot \boldsymbol{u}^*$$

and then set

$$\boldsymbol{u}^{n+1} = \boldsymbol{u}^* - \Delta t \nabla p^{n+1} . \tag{8.22}$$

The scalar factor Δt is introduced into the potential function ($\phi = \Delta t\, p$) so that (8.22) can be viewed as an approximation to the equation

$$u_t + \nabla p = 0 .$$

Solving this equation after solving (8.21) can be viewed as a fractional step approach (Sect. 5.) to the full INS equations, and the p obtained in this manner approximates the pressure.

Note that for INS it is not fundamentally unphysical to have a velocity field that fails to satisfy (8.19) — such velocity fields arise in compressible flow all the time and a real liquid is slightly compressible. One could model a liquid more properly by using the compressible Navier–Stokes equations with a suitable equation of state. However, for many applications the acoustic waves that would then arise are of no interest and all the motion of interest is on the much slower fluid time scale. Using the compressible equations would require taking much smaller time steps than desirable to resolve the uninteresting acoustic waves — again a problem of stiffness.

While the individual acoustic waves may not be of interest, their net effect is important. It is the small amplitude, rapidly moving, acoustic waves that mediate pressure changes throughout the liquid and cause a force exerted at one point in the fluid to be felt nearly instantly (relative to the fluid time scale) far away. In the incompressible equations we have filtered out the acoustic waves, but maintained their effect by introducing an elliptic equation for the pressure. This can be viewed as imposing a global "equation of state" which requires that the fluid behave incompressibly. The elliptic equation couples all points together and hence allows information to propagate infinitely quickly throughout the entire domain. The pressure can also be viewed formally as a **Lagrange multiplier** which imposes the constraint (8.19) in a variational formulation of the INS equations.

In the case of the MHD equations, there does not appear to be an analogous interpretation of the divergence-cleaning procedure, since in theory $\nabla \cdot \boldsymbol{B} = 0$ should be exactly satisfied. The numerical method, however, may introduce nonphysical magnetic monopoles which require a nonphysical potential ϕ for their removal.

8.7 Riemann Problems in Multi-dimensional MHD

As discussed in Sect. 8.2, the one-dimensional MHD equations contain the equation $\frac{\partial}{\partial t} B_x = 0$, which can be eliminated to give the standard system of seven equations (8.10) with seven physical waves. In a true one-dimensional plane wave there can also be no spatial variations in B_x, so it is reasonable to drop this variable and the corresponding equation.

However, in applying a multi-dimensional numerical algorithm of the type described in Sect. 6., we must solve one-dimensional Riemann problems at each cell interface with data which comes from the two cell averages on either side. We wish to solve the one-dimensional MHD equations in the direction

$\boldsymbol{n}$ normal to this interface, but there is no reason that the two cell averages should have the same value of $\boldsymbol{B} \cdot \boldsymbol{n}$, when these values come from multi-dimensional data.

Hence at the cell interface we have two sets of data q_l and q_r for the one-dimensional Riemann problem that do not have the same value for the normal component of the magnetic field, say B_x for x-sweeps. The one-dimensional Riemann problem does not formally have a solution, then, as a strictly one-dimensional problem. The data does not satisfy $\nabla \cdot \boldsymbol{B} = 0$ but rather has a singularity along the interface. To compute a solution using the 7-wave one-dimensional solution described above, we must first decide on some average value of B_x to use as the parameter in this solution.

Typically in practice an approximate Riemann solver is used, such as a Roe solver (Sect. 4.6.4) which is based on the eigenstructure of an approximate Jacobian matrix, determined by some averaged state based on q_l and q_r. In this case it is quite natural to think of averaging the two values of B_x in the definition of this matrix along with the other variables.

Brio and Wu [44] discuss a Roe solver for MHD in the case $\gamma = 2$ which is based on $\sqrt{\varrho}$-weighted averages as in the Euler equations. For $\gamma \neq 2$ they used a simpler averaging since their approach did not appear to yield an average Jacobian satisfying all the conditions of a Roe matrix in the general case. Other approximate Riemann solvers for MHD are discussed in [11], [15], [64].

8.8 Staggered Grids

Another approach to maintaining the divergence-free condition in MHD is to use a staggered grid in which the magnetic field variables are stored at cell edges rather than cell centers. In two dimensions, for example, the values of B_x would be stored at interfaces in the x-direction while B_y is stored at interfaces in the y-direction. Denote these values by

$$B_{x,i\pm 1/2,j} \quad \text{and} \quad B_{y,i,j\pm 1/2} \,.$$

Then the discrete divergence-free condition amounts to requiring that

$$\frac{1}{\Delta x}(B_{x,i+1/2,j} - B_{x,i-1/2,j}) + \frac{1}{\Delta y}(B_{y,i,j+1/2} - B_{y,i,j-1/2}) = 0 \,. \tag{8.23}$$

A variety of algorithms have been developed that update these edge values in a manner that exactly preserves the condition (8.23) from one time step to the next. See, for example, [63], [69], [228].

A similar staggered-grid approach is often used in incompressible flow, where the velocities $u_{i\pm 1/2,j}$ and $v_{i,j\pm 1/2}$ are maintained at cell edges while the pressure is a cell-centered value. This is often called a MAC grid, since an early application of this approach was in the Marker-and-Cell method [102].

8.9 The 8-Wave Riemann Solver

The standard derivation of the MHD equations uses the fact that $\nabla \cdot \boldsymbol{B} = 0$ to eliminate certain terms along the way, as in dropping the last term of (8.7), for example. If one does not make this assumption at the start, then it is possible to derive more general equations of the form

$$\frac{\partial}{\partial t}\begin{bmatrix} \varrho \\ \varrho \boldsymbol{u} \\ \boldsymbol{B} \\ E \end{bmatrix} + \nabla \cdot \begin{bmatrix} \varrho \boldsymbol{u} \\ \varrho \boldsymbol{u}\boldsymbol{u} + I\left(\left(p + \frac{1}{2}B^2\right) - \boldsymbol{B}\boldsymbol{B}\right) \\ \boldsymbol{u}\boldsymbol{B} - \boldsymbol{B}\boldsymbol{u} \\ \left(E + p + \frac{1}{2}B^2\right)\boldsymbol{u} - \boldsymbol{B}(\boldsymbol{u}\cdot\boldsymbol{B}) \end{bmatrix} = \nabla \cdot \boldsymbol{B} \begin{bmatrix} 0 \\ \boldsymbol{B} \\ \boldsymbol{u} \\ \boldsymbol{u}\cdot\boldsymbol{B} \end{bmatrix} \qquad (8.24)$$

in place of (8.9). A derivation due to M. Vinokur is given in [186]. This looks like the MHD equations with a source term on the right-hand side which vanishes if $\nabla \cdot \boldsymbol{B} = 0$. These equations can be rewritten as a quasilinear system

$$q_t + A^x(q)q_x + A^y(q)q_y + A^z(q)q_z = 0 .$$

The matrix $A^x(q)$ has essentially the same structure as the Jacobian matrix $f'(q)$ from the conservative form, but with the column corresponding to B_x modified in such a way that the eigenvalues consist of the seven values from (8.12) together with $\lambda = u$ in place of the eighth eigenvalue $\lambda = 0$ of $f'(q)$. The quasilinear equation has Galilean invariance and the eighth characteristic field can be interpreted as an advection equation for $\nabla \cdot \boldsymbol{B}$. This modification to $f'(q)$ was developed by Powell [187] and advocated as an approach to controlling the accumulation of errors in $\nabla \cdot \boldsymbol{B}$, at least for problems with open boundaries so that any $\nabla \cdot \boldsymbol{B}$ produced by the numerical method can advect out of the domain instead of remaing stationary. This approach has also been used in conjunction with occasional divergence cleaning. See [8], [187], [188], [186] for more discussion of this approach.

9. Relativistic Hydrodynamics

In modeling flow at speeds where relativistic effects become important, space and time become intrinsically coupled and the Euler equations of gas dynamics become more complicated. Nonetheless, it is still possible to write the relativistic Euler equations as a first-order hyperbolic system that can be advanced forward in time in some fixed reference frame. The reference frame will be called the lab frame since this is typically the frame from which we are observing.

Here we will only consider the simplest case of flat spacetime. To compute flows near massive objects it would be necessary to modify the equations further to include the local curvature of spacetime. This can be done without essential new difficulties arising by including the appropriate metric tensor in place of the Minkowski metric η below, provided the geometry of spacetime is assumed to be fixed or to vary in some known manner. This is what is generally refered to as **relativistic hydrodynamics** (RHD).

To model the evolution of spacetime as mass is redistributed requires the Einstein equations of general relativity, which are significantly more difficult to solve (see Sect. 9.7).

9.1 Conservation Laws in Spacetime

There is a symmetry between space and time in any system of conservation laws such as (1.7) or (2.27) that may not be apparent from the derivation of Sect. 2., but which is worth observing when starting to think about relativistic flows. The three-dimensional continuity equation for the conservation of mass, for example, can be written as

$$\bar{\nabla} \cdot \bar{F}(\varrho) = 0\,, \tag{9.1}$$

where

$$\bar{\nabla} = \begin{bmatrix} \partial_t \\ \partial_x \\ \partial_y \\ \partial_z \end{bmatrix}, \qquad \bar{F}(\varrho) = \begin{bmatrix} \varrho \\ \varrho u \\ \varrho v \\ \varrho w \end{bmatrix}.$$

If we call $\bar{F}(\varrho)$ the 4-flux of ϱ, then this equation states that the divergence of the 4-flux is zero. The flux functions ϱu, ϱv, and ϱw represent the flux of ϱ through surfaces with constant x, y and z, respectively. So we see that ϱ itself can be thought of as the flux of ϱ through constant t.

If we let $\bar{\Omega}$ be an arbitrary 4-dimensional volume in ordinary spacetime, then (9.1) results from the requirement that integrating the normal 4-flux over the boundary of $\bar{\Omega}$ gives zero. Equation (2.28) is a special case of this where the volume is a 4-cube. In (2.32) we generalized this to a "cylinder" in time with arbitrary cross-section Ω, but we now recognize that this can be generalized further to any arbitrary 4-dimensional volume.

Figure 9.1 illustrates the concept of flux-in-time for the continuity equation in one space dimension. Here the spacetime volume is taken to be a rectangle. Several particle paths are drawn. Suppose the particles each have unit mass. Then the mass flux $f = \varrho u$ can be interpreted as the number of particles crossing a fixed point x per unit time, and the integral of this flux over the left or right side of the rectangle in Fig. 9.1 gives the flux through that side, which is simply the number of particles crossing it.

The density ϱ is the number of particles per unit length at a fixed time, and hence integrating the density over the top or bottom of the rectangle simply gives the number of particles crossing that side. Hence we see that the density function can be interpreted as mass-flux through a constant-time surface.

The conservation law $\partial_t \varrho + \partial_x f = 0$ says that the divergence of the spacetime flux is zero, which simply reflects the fact that particles are conserved

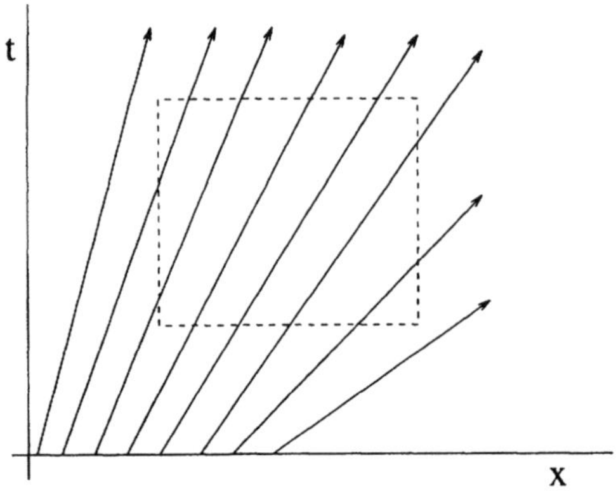

Fig. 9.1. Particle paths through a spacetime volume

— the same number of particles must be leaving the rectangle in Fig. 9.1 as are entering it.

Furthermore, we can combine the continuity equation with the conservation of momentum equations by introducing the tensor

$$M = \begin{bmatrix} \varrho & \varrho u & \varrho v & \varrho w \\ \varrho u & \varrho u^2 + p & \varrho uv & \varrho uw \\ \varrho v & \varrho uv & \varrho v^2 + p & \varrho vw \\ \varrho w & \varrho uw & \varrho vw & \varrho w^2 + p \end{bmatrix} . \tag{9.2}$$

This is an extension of the stress tensor (2.34) which includes density and the mass fluxes, so it might be called the stress–mass tensor. The conservation laws for mass and momentum can then be summarized by stating that the 4-divergence of M must vanish. This tensorial viewpoint might be valuable if we wanted to use an arbitrary curvilinear grid in spacetime, as the conservation laws would then be given by requiring that the covariant derivative of M be zero. Ordinarily a curvilinear grid is only introduced in space, however, combined with time stepping in the usual way. The tensorial nature of the stress tensor (9.2) is important in transforming the momentum equations to an arbitrary spatial grid, but the tensor M is generally not introduced (but see Section 5.10 of [167]).

At relativistic velocities, however, the coupling between space and time requires that we extend the ordinary momentum to the 4-momentum, which includes a temporal component and introduces a 4×4 tensor for its conservation. It is not, however, the continuity equation that gets incorporated into this tensor, as one might expect from the above discussion. Instead it is the conservation of energy equation that is used, and the resulting tensor T is called the *stress–energy* tensor. However, because of the interchange between mass and energy in relativity theory, this is related to conservation of mass. The continuity equation stands alone and enforces the conservation of particles, though the perceived partitioning between mass and energy in each particle depends on its relative speed.

9.2 The Continuity Equation

The relativistic Euler equations will be briefly derived using the notation of [162], for example. We first extend the continuity equation to the relativistic case. It is still true that particles are conserved, just as in Fig. 9.1. This is illustrated in Fig. 9.2 for a flow at relativistic speed u hitting a wall that is stationary in the lab frame (the x-t axes). The x'-ct' axes show the Lorentz frame of the moving gas. A relativistic shock wave from the wall decelerates the gas to zero velocity in the lab frame. Several particle paths are shown and again the net number of particles passing through the surface of any spacetime rectangle must be zero. Now, however, the perceived mass of each particle changes with its velocity, so it is **not** true that the net mass flux, as measured in the lab frame, will be zero. If m_0 is the rest mass of a particle, then the perceived mass is Wm_0, where

$$W = \frac{1}{\sqrt{1-(u/c)^2}}$$

is the *Lorentz factor*. If ϱ_0 is the rest density of the gas, *i.e.*, mass per unit distance in the local rest frame of the gas at each point, then the density measured in the lab frame (perceived mass per unit distance along the x-axis) is

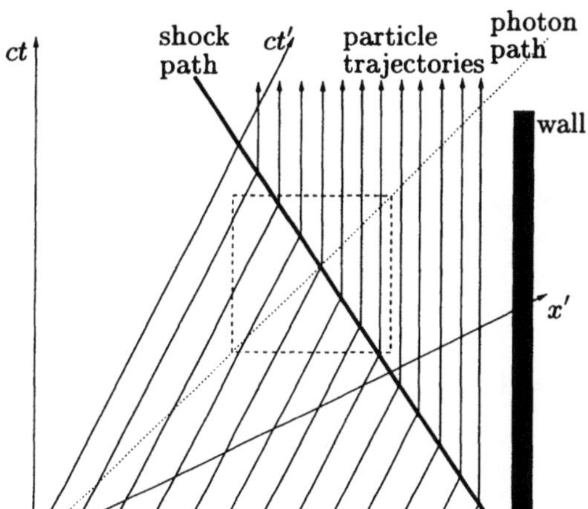

Fig. 9.2. Minkowski diagram for relativistic flow hitting a wall, generating a shock wave which brings the gas to a rest in the lab frame. Particles are still conserved when integrating over any spacetime region, such as the dashed rectangle

$$D \equiv W \varrho_0 \; .$$

Then the continuity equation for the relativistic gas can be written as

$$D_t + (Du)_x = 0 \; . \tag{9.3}$$

Note that W depends on u (and hence varies with x and t in general) so that the factor W cannot be cancelled out of this equation. It is *not* true that $(\varrho_0)_t + (\varrho_0 u)_x = 0$.

9.3 The 4-Momentum of a Particle

We now study the generalization of the momentum to the 4-momentum. First consider a single particle moving along an arbitary path in spacetime. Denote a point in spacetime, and the resulting covariant and contravariant gradient operators, by

$$\boldsymbol{x} = \begin{bmatrix} ct \\ x^1 \\ x^2 \\ x^3 \end{bmatrix}, \qquad \nabla_\alpha = \begin{bmatrix} \frac{1}{c}\partial_t \\ \partial_x \\ \partial_y \\ \partial_z \end{bmatrix}, \qquad \nabla^\alpha = \begin{bmatrix} -\frac{1}{c}\partial_t \\ \partial_x \\ \partial_y \\ \partial_z \end{bmatrix},$$

respectively, with the convention $x^0 = ct$ (and as usual Greek indices range from 0 to 3 while latin indices range from 1 to 3). We use the Minkowski metric

$$\eta = \begin{bmatrix} -1 & & & \\ & 1 & & \\ & & 1 & \\ & & & 1 \end{bmatrix} . \tag{9.4}$$

We can then parametrize the particle path by

$$\boldsymbol{x}(\hat{t}) \; ,$$

where $\hat{t}$ is proper time, *i.e.*, time as measured by the particle. It is then natural to define the 4-velocity of the particle as

$$\boldsymbol{u} = \frac{\mathrm{d}\boldsymbol{x}(\hat{t})}{\mathrm{d}\hat{t}} \; .$$

In the lab frame we have $\mathrm{d}\hat{t} = \frac{1}{W}\mathrm{d}t$ and hence

$$\boldsymbol{u} = W \begin{bmatrix} c \\ u^1 \\ u^2 \\ u^3 \end{bmatrix} .$$

The 4-momentum of the particle is then defined as

$$\boldsymbol{P} = m_0 \boldsymbol{u} \; ,$$

where m_0 is the rest mass. The 0'th component of this is

$$P^0 = Wm_0c = WE_0/c \,,$$

where

$$E_0 = m_0c^2$$

is the rest energy of the particle.

9.4 The Stress–Energy Tensor

Generalizing the stress–mass tensor M of (9.2) to the relativistic stress–energy tensor is complicated by the need to properly model pressure. (Ignoring pressure gives "relativistic dust" as discussed in [211].) In the nonrelativistic Euler equations, the energy for an ideal gas is defined by

$$E = \varrho\varepsilon + \frac{1}{2}\varrho u^2 \,,$$

where ε is the specific internal energy. Note that

$$E + p = \varrho(\varepsilon + p/\varrho) + \frac{1}{2}\varrho u^2 = \varrho h + \frac{1}{2}\varrho u^2 \,,$$

where $h = \varepsilon + p/\varrho$ is the *specific enthalpy*. The energy flux $u(E+p)$ contains both an advective term uE and a term up that corresponds to work done compressing the gas.

In the relativistic case, we must incorporate the relativistic enthalpy into the momentum in order to obtain a tensor conservation law. The **relativistic specific enthalpy** is defined by

$$h = c^2 + \varepsilon + p/\varrho_0 \,.$$

Note that

$$\varrho_0 h = \varrho_0 c^2 + \varrho_0\varepsilon + p$$

is the ordinary enthalpy plus a new term corresponding to the rest-mass energy which must be included with the internal energy in the relativistic case. In some works (e.g., [14]) the quantity h/c^2 is used instead, and is called the *mass-equivalent enthalpy density*.

The **stress–energy tensor** then takes the form

$$T^{\mu\nu} = (\varrho_0 h/c^2)u^\mu u^\nu + p\eta^{\mu\nu} \,, \tag{9.5}$$

where $\eta^{\mu\nu}$ is the Minkowski metric. With this definition of T, the conservation laws for momentum and energy take the desired form, and simply state that the divergence of this tensor must vanish:

$$\nabla_\nu T^{\mu\nu} = 0 \,.$$

In the lab frame we have

$$\begin{aligned} T^{00} &= E = \varrho_0 hW^2 - p = W^2(c^2\varrho_0 + \varrho_0\varepsilon + p) - p \\ &= W(c^2D + D\varepsilon + Dp/\varrho_0) - p \end{aligned} \tag{9.6}$$

while the momentum is defined by

$$T^{i0} = S^i = \varrho_0 h W^2 u^i / c$$

and the stress–energy tensor is

$$T = \begin{bmatrix} E & S^1 & S^2 & S^3 \\ S^1 & S^1u^1 + p & S^1u^2 & S^1u^3 \\ S^2 & S^2u^1 & S^2u^2 + p & S^2u^3 \\ S^3 & S^3u^1 & S^3u^2 & S^3u^3 + p \end{bmatrix} . \tag{9.7}$$

In one space dimension, for example, we have the conservation laws

$$S_t^1 + (S^1u^1 + p)_x = 0 \tag{9.8}$$

$$E_t + S_x^1 = 0 \tag{9.9}$$

together with the continuity equation (9.3).

9.4.1 Energy Perturbation. Computationally, the fact that the energy E contains the rest-mass energy can cause difficulties, since this term can overwhelm the internal energy due to the presence of the c^2 factor. This can lead to a loss of all accuracy on a finite-precision computer. For this reason it is necessary to solve for the energy perturbation rather than the energy itself. Note that expanding

$$W = \frac{1}{\sqrt{1-(u/c)^2}} = 1 + \frac{1}{2}(u/c)^2 + O((u/c)^4)$$

in (9.6) gives

$$\begin{aligned} E &= \left(1 + \tfrac{1}{2}\left(\tfrac{u}{c}\right)^2 + \cdots\right)\left(c^2 D + D\varepsilon + Dp/\varrho)0\right) - p \\ &= c^2 D + \tfrac{1}{2}u^2 D + D\varepsilon + (W-1)p + O((u/c)^2) . \end{aligned} \tag{9.10}$$

In the nonrelativistic limit ($u/c \to 0$ and $W \to 1$, $D \to \varrho_0$), we see that this approaches

$$E = \varrho_0 c^2 + \frac{1}{2}\varrho_0 u^2 + \varrho\varepsilon .$$

This gives the usual kinetic and internal energy, plus the rest-mass energy. (Note that kinetic energy arises from the rest-mass energy and the expansion of W.) The expansion (9.10) suggests that computationally the appropriate energy perturbation to consider is thus

$$\tau \equiv E - c^2 D .$$

Subtracting c times the continuity equation (9.3) from the conservation of energy equation (9.9) then gives the equation

$$\frac{1}{c}\tau_t + (S^1 - cDu^1)_x = 0 .$$

adiapThis equation, together with the continuity equation (9.3) and the conservation of momentum equation (9.8), gives the **relativistic Euler equations in one dimension**:

$$\begin{aligned} D_t + (Du^1)_x &= 0\,, \\ S^1_t + (S^1u^1 + cp)_x &= 0\,, \\ \tau_t + (cS^1 - c^2Du^1)_x &= 0\,. \end{aligned} \tag{9.11}$$

In the literature, units are often chosen so that $c = 1$ and does not appear in these equations.

9.5 Finite Volume Methods

The finite volume methods discussed in Sections 4.2, 4.5, and 4.7 can be extended to the relativistic case fairly easily. Solving the one-dimensional Riemann problem at each cell interface can be done in the same way as outlined in Sect. 3.4 for the Newtonian case, with a few minor changes. Taub [236] first developed the theory of simple waves and shocks for relativistic flow. Further work was done by [116], [150], [241], and other researchers. The complete solution to the relativistic Riemann problem is clearly presented by Marti and Müller [160]. An all-shock approximate Riemann solver (see Sect. 4.6.1) was developed by Balsara [14]. See also [66].

The basic approach is the same as in Sect. 3.4, once the functions $U^1(p^*; q_l)$ and $U^3(p^*; q_r)$ have been appropriately modified. Figure 9.3 shows the solution to the Riemann problem in the u-p plane in the relativistic case, for a case where the initial velocities are zero but there is an enormous pressure jump that gives rise to motion at relativistic speeds. This figure also shows the solution that would be obtained with the nonrelativistic Riemann solver. Note that this would give a superluminal velocity in the intermediate state!

One additional complication in the relativistic case is that it is more difficult to switch from the conserved quantities (D, S, τ) to the primitive variables (ϱ, u, p) needed in this solution process. In the Newtonian case it is trivial to compute u from ϱ and ϱu, for example, but in the relativistic case we have

$$D = \varrho W\,, \qquad S = \varrho h W^2 u/c$$

and the value of W depends on u, while the enthalpy h also depends on the internal energy ε and the pressure. Hence we must solve a nonlinear system of equations to recover the primitive variables from the conserved quantities, as described in [160].

Figure 9.4 shows a relativistic shock tube computation performed by D. Bale with the one-dimensional CLAWPACK code. The initial data is taken from Problem 2 of [162], and the results agree favorably with what is shown in Fig. 4 of that paper, as computed with a relativistic PPM method on the same grid (400 points). The superbee limiter has been used in this computation. The structure of this Riemann solution is similar to what was seen in Fig. 2.2 for a nonrelativistic flow with analogous initial data (constant density but higher temperature on the left). Again a thin zone of high-density gas is swept up by the leading shock. In the relativistic case, however, this structure can be exceedingly thin, leading to computational challenges.

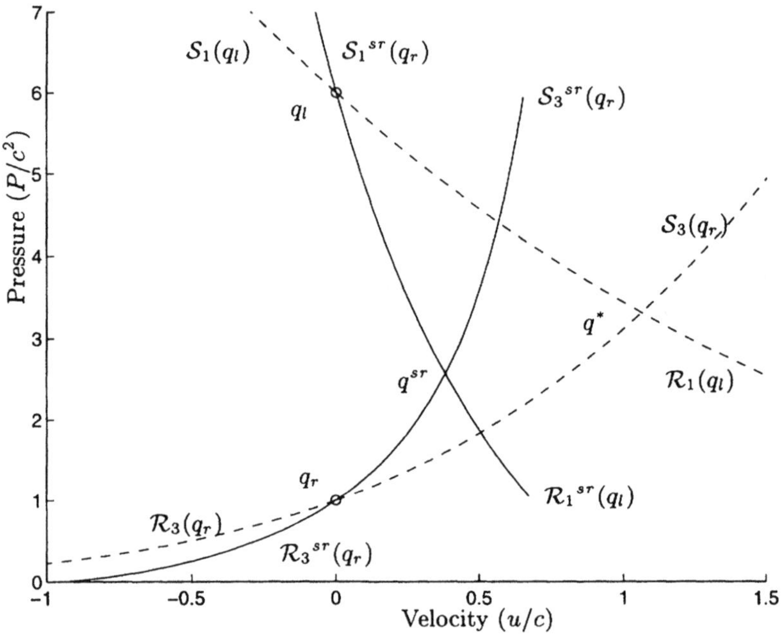

Fig. 9.3. Comparison of Hugoniot adiabats for the flow that is assumed to be non-relativistic (dashed lines) with the Taub adiabats of relativistic flow (solid lines). $\mathcal{S}$ and $\mathcal{R}$ refer to shock and rarefaction curves and the superscript sr refers to the special relativistic case. With data q_l and q_r the nonrelativistic Riemann solver would predict the state q^* in solving the Riemann problem, which has a superluminal velocity. The correct intermediate state in the solution to this relativistic Riemann problem is denoted by q^{sr}. Note that u is in units of c while the pressure is in units of c^2

It is easy to see why this structure should be very thin at ultrarelativistic speeds. The high temperature gas creates a shock moving into the cooler gas which accelerates this gas to a very high velocity v. The contact discontinuity moves at this fluid velocity. The shock is moving at some greater velocity s, but must still be subluminal, and so $v < s < c$. As we raise the temperature on the left, $v \to c$ and s must lie in between, so from the lab frame it appears that the shock and contact are moving at nearly the same speed, with a very thin zone in between. (From the reference frame of the gas in this zone, on the other hand, the shock would be moving ahead at enormous speed and this zone would appear to widen very rapidly.)

Note that in this Riemann problem the initial velocity is zero everywhere but the enormous left-state pressure gives rise to macroscopic flow at relativistic speeds. But even the left-state gas which is still "at rest" is in fact moving microscopically at relativistic speeds; it must be to create such pressure. Such a gas is refered to as "relativistically hot". By contrast a "cold" relativistic gas is one which is moving at relativistic speed in the chosen reference frame,

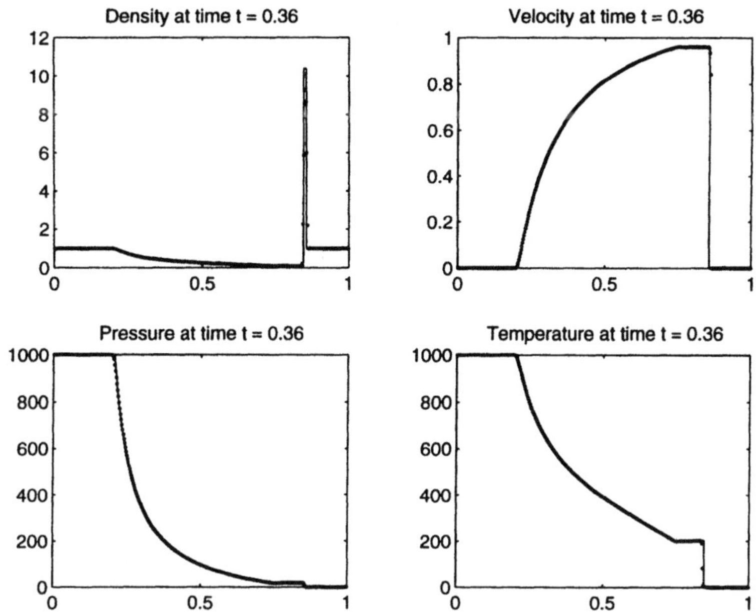

Fig. 9.4. Relativistic shock tube computation with initial states $(\varrho_l, u_l, p_l) = (1, 0, 1000.0)$ and $(\varrho_r, u_r, p_r) = (1, 0, 0.01)$. Velocity is in units of c and pressure in units of c^2, while the temperature plotted is p/ϱ. The solid line shows the true solution

but for which there exists a reference frame in which all the molecules are moving at nonrelativistic speeds. (See also Section 5.5 of Müller's notes.)

9.6 Multi-dimensional Relativistic Flow

A further complication arises in multi-dimensional flow. For a multi-dimensional finite volume method we must solve a one-dimensional Riemann problem normal to each interface between grid cells. For the Newtonian Euler equations this is easily done with a trivial extension of the one-dimensional Riemann solver. The structure of the solution is completely determined by the density, pressure, and normal velocity in the two neighboring grid cells. There are also transverse velocities defined in each grid cell, but these values have no effect on the solution to the one-dimensional problem. The solution to the three-dimensional problem consists of 5 waves, the 3 waves from the one-dimensional solution plus two shear waves that carry the jumps in the transverse velocities. These shear waves travel at the velocity u^* of the contact discontinuity, and an arbitrary jump in transverse velocity is allowed across these waves.

At relativistic speeds, however, the magnitude of the transverse velocity does have an impact on the solution of the Riemann problem in the normal direction. This is because the Lorentz factor W depends on the *total* velocity,

$$W = \left(1 - (u^2 + v^2 + w^2)/c^2\right)^{-1/2} ,$$

and so changing the transverse velocities affects the structure of the one-dimensional solution. The transverse velocities are constant across a shock or through a rarefaction fan in the Newtonian case, but are not constant at relativistic speeds. For shock waves, the presence of transverse flow only complicates the algebraic expressions. For rarefaction waves, however, the solution process now requires solving a coupled system of two ordinary differential equations to integrate the two transverse velocities through the rarefaction fan [160].. Using the 2-shock approximation avoids this complication.

9.7 Gravitation and General Relativity

So far we have only considered the case of flat spacetime, using the Minkowski metric (9.4). In the neighborhood of massive objects such as black holes or neutron stars, it is necessary to take into account the curvature of spacetime using a more general metric g in place of η, with distance measured locally by

$$\mathrm{d}s^2 = g_{\mu\nu}\mathrm{d}x^\mu \mathrm{d}x^\nu .$$

9.7.1 Test Fluid Approximation. In the simplest case, we assume that this metric is known and we wish to solve the hydrodynamic equations for the motion of a fluid whose mass has negligible effect on the metric. This is sometimes called a *test fluid.* An example might be the study of accretion into a black hole, taking g to be the the radially symmetric Schwarzschild metric ([167], [211], [252]).

In this case the hydrodynamics equations are modified simply by replacing the stress–energy tensor (9.5) by

$$T^{\mu\nu} = (\varrho_0 h/c^2)u^\mu u^\nu + p g^{\mu\nu} , \tag{9.12}$$

and the conservation laws still take the form

$$\nabla_\nu T^{\mu\nu} = 0 .$$

When this is written out in the lab frame we obtain conservation laws similar to (9.11) (in one space dimension) but the equations now also have source terms which arise from the covariant differentiation of the metric term.

An approximate Riemann solver for general-relativistic hydrodynamics was developed by [82]. Related methods are developed in [16], [201].

9.7.2 The Einstein Equations. If we wish to study the motion of massive objects, say the collision of two black holes or neutron stars or the core collapse of a neutron star in the formation of a black hole, then we must determine the metric g as part of the solution process. Such events give rise to gravitational waves which propagate outward at the speed of light as ripples in the spacetime metric. Gravitational radiation of this form should be constantly bombarding the earth from interesting astronomical events. In principle this can be detected, for example by accurately measuring distances in two orthogonal directions and monitoring how the distances compare to one another. See [1], [238], [240] for surveys. See also [239] for an excellent introductory overview of this topic and others in gravitational astrophysics.

Unfortunately, such events are sufficiently far away that oscillations in the metric observed on earth are expected to be of relative magnitude 10^{-21} or less. Thus the entire earth distorts by on 10^{-14} meter (10 times the diameter of an atomic nucleus), and laboratory instruments of more modest size have so far been unable to detect gravitational waves. However, there are currently several highly-sensitive instruments under construction which may be capable of measuring these waves within a few years. For this reason, there is currently considerable interest among numerical relativists in predicting the gravitational wave signature of various possible events, in order to help interpret any data received and separate the signal from noise.

Astronomy has been revolutionized in the past by developing the ability to observe electromagnetic radiation in new wavelength regimes (*e.g.*, by X-ray and radio telescopes). If successful, the development of gravitational-wave observatories will surely lead to many new surprises.

To compute the evolution of the metric, we must solve the Einstein equations

$$G = 8\pi T \tag{9.13}$$

which are described in detail in many sources, *e.g.*, [88], [167], [211], [252].

Here T is the stress–energy tensor (9.12) and G is the Einstein tensor, which is a function of g and its first and second derivatives which gives the proper measure of the curvature of spacetime. The structure of spacetime affects the hydrodynamics, just as in the previous section, but the hydrodynamics also affects the structure of spacetime when the motion of massive objects is being modeled.

Since G and T are symmetric tensors, the Einstein equations (9.13) are a system of 10 equations in 4-dimensional spacetime. One might imagine trying to find solutions to these equations over some region of spacetime with given boundary conditions, but it is not clear what sort of boundary conditions are required since there should be causality in time-like directions.

9.7.3 The 3+1 Formalism. Instead of viewing (9.13) as a boundary-value problem over a region of 4-dimensional spacetime, the modern approach is to view spacetime as a foliation of 3-dimensional space-like manifolds with a spatial metric γ (obtained from g) defined on each sheet. The equations are then solved by marching forward in the time-like direction that links these manifolds together, advancing γ forward in "time". This approach was originally proposed by Arnowitt, Deser, and Misner [6] and is also known as the *ADM formalism.* The equations they proposed have a Hamiltonian structure. More recently, other $3+1$ formulations of the Einstein equations have been proposed that give first-order hyperbolic systems of conservation laws. These equations are amenable to solution by high-resolution finite-volume methods of the sort discussed in these lectures.

The Einstein equations involve second derivatives of the metric g. In order to write these equations as a first-order hyperbolic system, it is necessary to introduce a larger set of variables corresponding to various derivatives or combinations of derivatives of g. There are several ways in which this can be done, and it is not clear at this time which formulation is best. The number of variables and equations varies between different formulations, ranging from 30 to 58, so these systems are quite large. Some possibilities are discussed in [35], [36], [50], [93].

Hyperbolic formulations have received recent attention for theoretical as well as numerical work. The well-developed mathematical theory of hyperbolic conservation laws can be used to gain new insights in general relativity. For example, Smoller and Temple [224] have found new solutions to the Einstein equations which allow the possibility that the visible universe may only locally resemble an expanding 3-sphere, being separated from the remainder of the universe by a shock wave. See also [222], [223].

9.7.4 Elliptic Constraints. In addition to the first-order hyperbolic equations, there are also various constraint equations which should be satisfied between these variables. These constraints typically take the form of elliptic equations. In principle it is enough to impose these constraints at the initial time, known as the *initialization problem* (e.g., [170]). The time evolution of the Einstein equations guarantees that the constraints will be satisfied also at all later times. Numerically, however, they may fail to be satisfied at later times in the same way that the $\nabla\cdot \boldsymbol{B}=0$ condition in the MHD equations can fail numerically (see Sect. 8.6). It may be necessary to use similar projection procedures to maintain these constraints. In current practice, however, these constraints are often imposed only at the initial time in the process of preparing the initial data. It is then hoped that an accurate numerical method will maintain them sufficiently well at later times, avoiding a costly elliptic solve in each time step. Choptuik [47] gives some justification for this for smooth solutions, where one can argue that the constraints should remain satisfied to the order of accuracy of the numerical method. For problems involving shock waves this will not generally be true.

9.7.5 The Vacuum Einstein Equations. The simplest case of the Einstein equations arises from setting $T \equiv 0$ so that the equations reduce to $G = 0$. This corresponds to assuming that there is no mass in the computational domain. This simplifies the equations since we then do not need to worry about the hydrodynamics and associated complications such as fluid dynamical shocks. Such an approach would be valid for modeling the interaction of two black holes, for example, if the domain does not include the black holes themselves.

As noted above, the first-order hyperbolic system that we obtain is quite large, with at least 30 variables and equations. Developing and using a Riemann solver for such a system would appear to be a daunting task. However, these systems typically decouple into sets of linear hyperbolic systems for which the eigenstructure can be relatively easily determined.

9.7.6 Choice of Time Slicing. To solve the Einstein equations numerically in the 3 + 1 formulation, let $x^\mu = (t, x^i)$ represent the computational coordinates, so that we march forward in t from one spacelike manifold $\mathcal{M}(t)$ to the next $\mathcal{M}(t + \Delta t)$. The goal is to construct the metric $g_{\mu\nu}$ as we go. There is a great deal of flexibility in this construction since there are many different foliations of spacetime that we could choose, and many different choices of spatial coordinates within a given foliation. This flexibility must be pinned down and also exploited in order to avoid singularities. We may need to avoid physical singularities such as black holes, but we must also avoid artificial singularities that can arise even in computing flat spacetime from a poor choice of coordinates, as illustrated at the end of this section.

The metric g is generally decomposed as

$$\begin{aligned} \mathrm{d}s^2 &= g_{\mu\nu}\mathrm{d}x^\mu \mathrm{d}x^\nu \\ &= -\alpha^2 \mathrm{d}t^2 + g_{ij}(\mathrm{d}x^i + \beta^i \mathrm{d}t)(\mathrm{d}x^j + \beta^j \mathrm{d}t) , \end{aligned}$$

where g_{ij} is the spatial part of the metric, α is the **lapse** and β^i the **shift vector**. The lapse measures the elapsed proper time between two manifolds $\mathcal{M}(t)$ and $\mathcal{M}(t + \Delta t)$ (note that α, β^i, and g_{ij} all vary with x^μ in general, so the lapse can be different at different locations). An observer starting at location x^i on $\mathcal{M}(t)$ who moves normal to this manifold (*i.e.*, remains at rest in his comoving frame) for proper time $\alpha \mathrm{d}t$, will find himself on the manifold $\mathcal{M}(t + \Delta t)$ at location $x^i + \beta^i \mathrm{d}t$. This shift in coordinate location is due only to a change in the labeling of coordinates from one manifold to the next, and defines the shift vector.

The particular foliation of spacetime is determined by the choice of α together with the initial metric g_{ij} on $\mathcal{M}(0)$ and its derivatives needed to initialize q in the hyperbolic conservation law. In particular, the lapse α determines how rapidly proper time advances in different regions of the simulation, and hence can be used to avoid certain physical singularities. Figure 9.5(a) shows a schematic diagram of a time slicing that might be useful in modeling gravitational collapse and the formation of a black hole. The physical singularity forms on the t-axis at time t^* and the time slicing is chosen in such a way that it is possible to compute out to larger times away from the

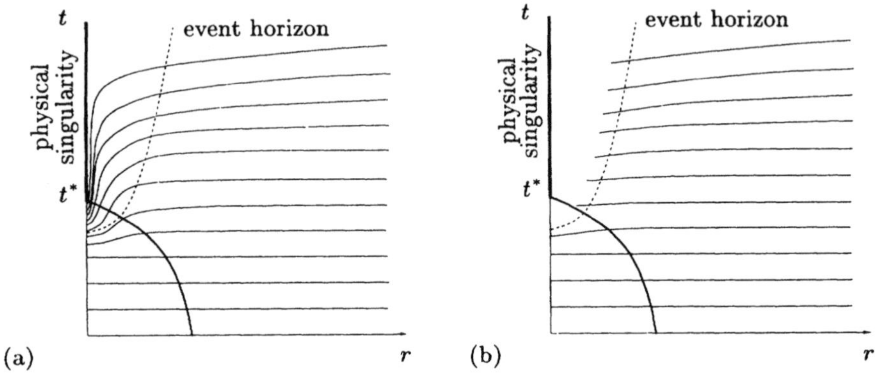

Fig. 9.5. Possible time slicing for gravitational collapse of matter (within the heavy curve) to a black hole along the t-axis. (**a**) Time is retarded to avoid the physical singularity by choosing a small lapse near the t-axis. (**b**) Horizon boundary conditions could also be used to avoid computing within the apparent horizon

singularity by slowing down the progression of time (taking α smaller) near the origin.

Another possibility for this particular calculation would be to use *apparent horizon boundary conditions* as discussed in [4] and [213]. Since information cannot escape from within the event horizon, the computational domain can be truncated at some point within the apparent horizon and nonreflecting outflow boundary conditions then used.

Once the time-evolution of the lapse and shift have been specified, one obtains a hyperbolic system for certain derivatives of the spatial metric g_{ij}.

One simple choice is $\alpha \equiv 1$, $\beta^i \equiv 0$, which gives **geodesic slicing**. This would not be suitable if we want to slow down the progression of time as in the previous example. But even if physical singularities are not a problem, geodesic slicing often leads to the formation of singularities in the coordinate system.

Figure 9.6 shows two examples of geodesic time slicing of flat Minkowski spacetime. In Fig. 9.6(a) the initial manifold was chosen to be a plane of simultaneity for an observer at constant velocity. Geodesic time slicing leads to the coordinates shown in Fig. 9.6(a), which are simply Cartesian coordinates in this Lorentz frame. In Fig. 9.6(b), the initial manifold was chosen to be a curved spacelike manifold. Now geodesic time slicing results in a focusing of the coordinate lines into a caustic, or "coordinate shock".

The appropriate choice of time slicing is a major concern in numerical relativity, and a variety of different approaches have been used. Some examples are given in [35], [36].

9.7.7 Adaptive Refinement. Some spherically symmetric problems in numerical relativity can be solved in $1+1$ dimensions, such as spherical stellar collapse and black hole formation. Spherically symmetric solutions cannot give rise to gravitational radiation, however, and the most interesting prob-

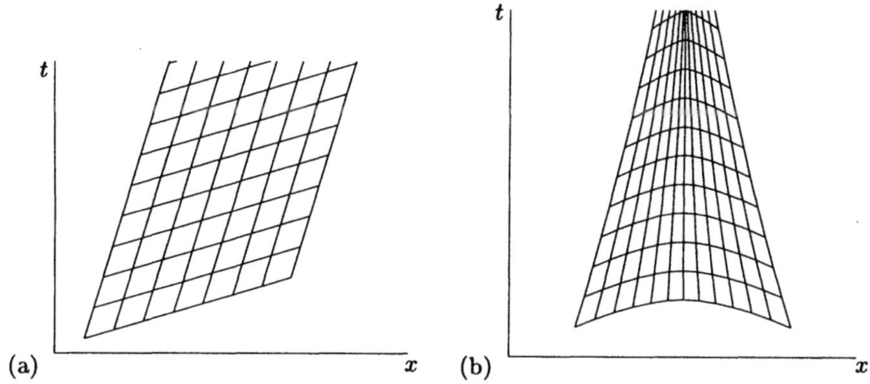

Fig. 9.6. Two examples of geodesic time slicing of flat Minkowski spacetime. (**a**) The initial manifold is a plane of simultaneity. (**b**) A different initial manifold leads to a coordinate singularity

lems will ultimately require solving the full $3+1$ equations. Given the size of the hyperbolic systems involved and the large domains required, this is a challenging task for which adaptive mesh refinement, as discussed in Sect. 6.3.2, will almost surely be needed for efficient computation.

Even for $1+1$ calculations, adaptive refinement can be crucial. For example, Choptiuk [48], [49] discovered an interesting new effect via careful high-resolution adaptive numerical simulation. Argyres [5] gives a concise overview of this and related work.

9.7.8 The Full Einstein Equations. If we wish to include the effect of mass in the Einstein equations with a time-dependent density function, then we must also solve the hydrodynamic equations for the fluid motion in curved spacetime in conjunction with advancing the metric that determines the geometry. This would be necessary, for example, in modeling gravitational waves resulting from core collapse supernovae, as discussed in Section 3.8 of Müller's notes. Presumably a full treatment would require coupling general relativity with magnetohydrodynamics, radiative transfer, nuclear burning, and a host of other physical effects.

One possible approach is to use a fractional step method, in which we alternate between freezing the geometry and solving the hydrodynamics equations as discussed in Sect. 9.7.1, and then freezing the mass distribution and using the Einstein equations to update the metric. Researchers are just beginning to explore such methods. Even with the on-going exponential growth of available computing power, it is clear that the further development of even more accurate and efficient numerical algorithms will play a crucial rôle.

References

1. A. Abramovici et al. Graviational wave astrophysics. In E. W. Kolb and R. D. Peccei, editors, *Particle and Nuclear Astrophysics and Cosmology in the Next Millennium, Proc. 1994 Snowmass Summer School*, pages 398–425. World Scientific, 1994.
2. D. A. Anderson, J. C. Tannehill, and R. H. Pletcher. *Computational Fluid Mechanics and Heat Transfer*. McGraw-Hill, 1984.
3. A. M. Anile. *Relativistic Fluids and Magneto-Fluids*. Cambridge University Press, Cambridge, 1989.
4. P. Anninos, G. Daues, J. Massó, E. Seidel, and W.-M. Suen. Horizon boundary conditions in numerical relativity. *Physical Review D*, 51:5562–5578, 1995.
5. P. C. Argyres. Universality and scaling in black hole formation. In E. W. Kolb and R. D. Peccei, editors, *Particle and Nuclear Astrophysics and Cosmology in the Next Millennium, Proc. 1994 Snowmass Summer School*, pages 447–449. World Scientific, 1994.
6. R. Arnowitt, S. Deser, and C. W. Misner. The dynamics of general relativity. In L. Witten, editor, *Gravitation: An Introduction to Current Work*, pages 227–265, New York, 1962. Wiley.
7. M. Arora and P. L. Roe. On postshock oscillations due to shock capturing schemes in unsteady flow. *J. Comput. Phys.*, 130:1–24, 1997.
8. N. Aslan. *Computational investigations of ideal MHD plasmas with discontinuities*. PhD thesis, University of Michigan, Nuclear Eng. Dept., 1993.
9. N. Aslan. Numerical solutions of one-dimensional MHD equations by a fluctuation approach. *Int. J. Numer. Meth. Fluids*, 22:569–580, 1996.
10. N. Aslan. Two dimensional solutions of MHD equations with a modified Roe's method. *Int. J. Numer. Meth. Fluids*, in print, 1996.
11. N. Aslan and T. Kammash. A Riemann solver for two dimensional MHD equations. *Int. J. Numer. Meth. Fluids*, to appear, 1997.
12. Martin Bailyn. *A Survey of Thermodynamics*. American Institute of Phyisics Press, Woodbury, NY, 1994.
13. D. Balsara. Modern schemes for solving hyperbolic conservation laws of interest in computational astrophysics on parallel machines. In D. A. Clarke, editor, *Halifax Conference on Computational Astrophysics*, 1996.
14. D. S. Balsara. Riemann solver for relativistic hydrodynamics. *J. Comput. Phys.*, 114:284–297, 1994.
15. D. S. Balsara. Linearized formulation of the Riemann problem for adiabatic and isothermal magnetohydrodynamics. *Astrophys. J.*, to appear.
16. F. Banyuls, J. A. Font, J. M. Ibáñez, J. M. Martí, and J. A. Miralles. Numerical $3+1$ general-relativistic hydrodynamics: a local characteristic approach. *Astrophysical J.*, pages 221–231, 1997.
17. A. A. Barmin, A. G. Kulikovskiy, and N. V. Pogorelov. Shock-capturing approach and nonevolutionary solutions in magnetohydrodynamics. *J. Comput. Phys.*, 126:77–90, 1996.
18. E. Battaner. *Astrophysical Fluid Dynamics*. Cambridge University Press, Cambridge, 1996.
19. J. Bell, M. J. Berger, J. Saltzman, and M. Welcome. Three dimensional adaptive mesh refinement for hyperbolic conservation laws. *SIAM J. Sci. Stat.*, 15:127–138, January 1994.
20. J. B. Bell, P. Colella, and H. M. Glaz. A second-order projection method for the incompressible Navier-Stokes equations. *J. Comput. Phys.*, 85:257–283, 1989.

21. J. B. Bell, P. Colella, and J. R. Trangenstein. Higher order Godunov methods for general systems of hyperbolic conservation laws. *J. Comput. Phys.*, 82:362, 1989.
22. J. B. Bell, C. N. Dawson, and G. R. Shubin. An unsplit, higher order Godunov method for scalar conservation laws in multiple dimensions. *J. Comput. Phys.*, 74:1–24, 1988.
23. J. B. Bell and D. L. Marcus. A second-order projection method for variable-density flows. *J. Comput. Phys.*, 101:334–348, 1992.
24. M. Ben-Artzi. The generalized Riemann problem for reactive flows. *J. Comput. Phys.*, 81:70–101, 1989.
25. M. Berger. *Adaptive mesh refinement for hyperbolic partial differential equations.* PhD thesis, Computer Science Deptartment, Stanford University, 1982.
26. M. Berger and A. Jameson. Automatic adaptive grid refinement for the Euler equations. *AIAA J.*, 23:561–568, 1985.
27. M. Berger and J. Oliger. Adaptive mesh refinement for hyperbolic partial differential equations. *J. Comput. Phys.*, 53:484–512, 1984.
28. M. J. Berger and P. Colella. Local adaptive mesh refinement for shock hydrodynamics. *J. Comput. Phys.*, 82:64–84, 1989.
29. M. J. Berger and R. J. LeVeque. AMRCLAW software. A test version is available at `http://www.amath.washington.edu/~rjl/amrclaw/`.
30. M. J. Berger and R. J. LeVeque. Adaptive mesh refinement using wave-propagation algorithms for hyperbolic systems. *SIAM J. Numer. Anal.*, to appear.
31. P. L. Bhatnagar, E. P. Gross, and M. Krook. A model for collision processes in gases I. small amplitude processes in charged and neutral one-component systems. *Phys. Rev.*, 94:511, 1954.
32. J. M. Blondin and B. S. Marks. Evolution of cold shock-bounded slabs. *New Astronomy*, 1:235–244, 1996.
33. C. Bona, J.M. Ibáñez, J.M. Martí, and J. Massó. Shock capturing methods in 1D numerical relativity. In F. Chinea, editor, *Gravitation and General Relativity: rotating bodies and other topics, Lecture Notes in Physics, Vol. 423.* Springer-Verlag, 1993.
34. C. Bona and J. Massó. A hyperbolic evolution system for numerical relativity. *Phys. Rev. Letters*, 68:1097, 1992.
35. C. Bona and J. Massó. Numerical relativity: Evolving spacetime. *International Journal of Modern Physics C: Phys. and Computers*, 4:883, 1993.
36. C. Bona, J. Massó, E. Seidel, and J. Stela. A new formalism for numerical relativity. *Phys. Rev. Lett.*, 75:600, 1995.
37. J. P. Boris and D. L. Book. Flux corrected transport I, SHASTA, a fluid transport algorithm that works. *J. Comput. Phys.*, 11:38–69, 1973.
38. A. Bourlioux. *Numerical Study of Unstable Detonations.* PhD thesis, Princeton University, June, 1991.
39. A. Bourlioux and A. J. Majda. Theoretical and numerical structure of unstable detonations. *Phil. Trans. R. Soc. Lond. A*, 350:29–68, 1995.
40. J. U. Brackbill and D. C. Barnes. The effect of nonzero $\nabla \cdot \boldsymbol{B}$ on the numerical solution of the magnetohydrodynamic equations. *J. Comput. Phys.*, 35:426–430, 1980.
41. W. L. Briggs. *A Multigrid Tutorial.* SIAM, 1987.
42. M. Brio and P. Rosenau. Stability of shock waves for 3×3 model MHD equations. Fourth Intl. Conf. Hyperbolic Prob., Taormina, 1992.
43. M. Brio and P. Rosenau. Evolution of the fast-intermediate shock wave in an MHD model problem. preprint, 1997.

44. M. Brio and C. C. Wu. An upwind differencing scheme for the equations of ideal magnetohydrodynamics. *J. Comput. Phys.*, 75:400–422, 1988.
45. Tung Chang and Ling Hsiao. *The Riemann Problem and Interaction of Waves in Gas Dynamics.* John Wiley and Sons Inc., New York, 1989.
46. M. Choptuik. Experiences with an adaptive mesh refinement algorithm in numerical relativity. In C. R. Evans, L. S. Finn, and D. W. Hobill, editors, *Frontiers in Numerical Relativity.* Cambridge University Press, 1989.
47. M. Choptuik. Consistency of finite-difference solutions of Einstein's equations. *Phys. Rev. D*, 44:3124–3135, 1991.
48. M. Choptuik. Universality and scaling in gravitational collapse of a massless scalar field. *Phys. Rev. Lett.*, 70:9–12, 1993.
49. M. W. Choptuik. Critical behaviour in massless scalar field collapse. In R. d'Inverno, editor, *Approaches to Numerical Relativity*, page 202. Cambridge University Press, 1992.
50. Y. Choquet-Bruhat and J. W. York, Jr. Geometrical well-posed systems for the Einstein equations. *C. R. Acad. Sci. Paris*, A321:1089–1095, 1995.
51. A. J. Chorin. Numerical solution of the Navier-Stokes equations. *Math. Comp.*, 22:745–762, 1968.
52. A. J. Chorin. Random choice solution of hyperbolic systems. *J. Comput. Phys.*, 22:517–533, 1976.
53. A. J. Chorin and J. E. Marsden. *A Mathematical Introduction to Fluid Mechanics.* Springer-Verlag, 1979.
54. P. Colella. Glimm's method for gas dynamics. *SIAM J. Sci. Stat. Comput.*, 3:76–110, 1982.
55. P. Colella. A direct Eulerian MUSCL scheme for gas dynamics. *SIAM J. Sci. Stat. Comput.*, 6:104–117, 1985.
56. P. Colella. Multidimensional upwind methods for hyperbolic conservation laws. *J. Comput. Phys.*, 87:171–200, 1990.
57. P. Colella and H. M. Glaz. Efficient solution algorithms for the Riemann problem for real gases. *J. Comput. Phys.*, 59:264–284, 1985.
58. P. Colella, A. Majda, and V. Roytburd. Theoretical and numerical structure for reacting shock waves. *SIAM J. Sci. Stat. Comput.*, 7:1059–1080, 1986.
59. P. Colella and P. Woodward. The piecewise-parabolic method (PPM) for gas-dynamical simulations. *J. Comput. Phys.*, 54:174–201, 1984.
60. R. Courant and K. O. Friedrichs. *Supersonic Flow and Shock Waves.* Springer, 1948.
61. R. Courant, K. O. Friedrichs, and H. Lewy. Über die partiellen Differenzengleichungen der mathematischen Physik. *Math. Ann.*, 100:32–74, 1928.
62. T. G. Cowling. *Magnetohydrodynamics.* Adam Hilger, The Institute of Physics, Bristol, England, 1976.
63. W. Dai and P. R. Woodward. On the divergence-free condition and conservation laws in numerical simulations for supersonic magnetohydrodynamical flows. preprint.
64. W. Dai and P. R. Woodward. An approximate Riemann solver for ideal magnetohydrodynamics. *J. Comput. Phys.*, 111:354–372, 1994.
65. W. Dai and P. R. Woodward. A high-order Godunov-type scheme for shock interactions in ideal magnetohydrodynamics. *SIAM J. Sci. Comput.*, 18:957–981, 1997.
66. W. Dai and P. R. Woodward. An iterative Riemann solver for relativistic hydrodynamics. *SIAM J. Sci. Comput.*, 18:982–995, 1997.
67. M. B. Davies, M. Ruffert, W. Benz, and E. Müller. A comparison between SPH and PPM: simulations of stellar collisions. *Astron. Astrophys.*, 272:430–441, 1993.

68. D. L. De Zeeuw, A. F. Nagy, T. I. Gombosi, K. G. Powell, and J. G. Luhmann. A new axisymmetric MHD model of the interaction of the solar wind with Venus. *J. Geophys. Res.*, 101:4547–4556, 1996.
69. R. DeVore. Transport techniques for multidimensional compressible magnetohydrodynamics. *J. Comput. Phys.*, 92:142, 1991.
70. Ruth Dgani, Rolf Walder, and Harry Nussbaumer. Stability analysis of colliding winds in a double star system. *Astron. Astrophys.*, 278:209–225, 1993.
71. W. G. Dixon. *Special Relativity.* Cambridge University Press, Cambridge, 1978.
72. R. Donat. Studies on error propagation for certain nonlinear approximations to hyperbolic equations: discontinuities in derivatives. *SIAM J. Numer. Anal.*, 31:655–679, 1994.
73. R. Donat and A. Marquina. Capturing shock reflections: an improved flux formula. *J. Comput. Phys.*, 125:42–58, 1996.
74. E. A. Dorfi and M. U. Feuchtinger. Adaptive radiation hydrodynamics of pulsating stars. *Computer Phys. Comm.*, 89:69–90, 1995.
75. G. C. Duncan and P. A. Hughes. Simulations of relativistic extragalactic jets. *Astrophysical J.*, 436:L119–L122, 1994.
76. V. Dwarkadas, R. Chevalier, and J. Blondin. The shaping of planetary nebulae: Asymmetry in the external wind. *Astrophys. J.*, 457:773, 1996.
77. Carl Eckart. The thermodynamics of irreversible processes. *Phys. Rev.*, 58:919–924, 1940.
78. B. Einfeldt. On Godunov-type methods for gas dynamics. *SIAM J. Num. Anal.*, 25:294–318, 1988.
79. B. Einfeldt, C. D. Munz, P. L. Roe, and B. Sjogreen. On Godunov type methods near low densities. *J. Comput. Phys.*, 92:273–295, 1991.
80. P. G. Eltgroth. Similarity analysis for relativistic flow in one dimension. *Phys. Fluids*, 14:2631, 1971.
81. B. Engquist and S. Osher. Stable and entropy satisfying approximations for transonic flow calculations. *Math. Comp.*, 34:45–75, 1980.
82. R. Eulderink and G. Mellema. General relativistic hydrodynamics with a Roe solver. *Astron. Astrophys. Suppl. Ser.*, 110:587–623, 1995.
83. T. E. Faber. *Fluid Dynamics for Physicists.* Cambridge University Press, Cambridge, 1995.
84. J. Falcovitz and M. Ben-Artzi. Recent developments of the GRP method. *JSME International J. Series B*, 38:497–517, 1995.
85. M. Fey, R. Jeltsch, and A.-T. Morel. Multidimensional schemes for nonlinear systems of hyperbolic conservation laws. In D. F. Griffiths and G. A. Watson, editors, *16th Biannual Dundee Conference on Numerical Analysis, 1995.* Longman, 1996.
86. C. A. J. Fletcher. *Computational Techniques for Fluid Dynamics.* Springer-Verlag, 1988.
87. J. A. Font, J. Ma. Ibáñez, A. Marquina, and J. Ma. Martí. Multidimensional relativistic hydrodynamics: characteristic fields and modern high-resolution shock-capturing schemes. *Astron. Astrophys*, 282:304–314, 1994.
88. J. Foster and J. D. Nightingale. *A Short Course in General Relativity.* Springer-Verlag, New York, 1995.
89. A. Frank, T. W. Jones, and D. Ryu. Time dependent simulation of oblique MHD cosmic-ray shocks using the two-fluid model. *Astrophys. J.*, 441:629, 1995.
90. A. Frank and G. Mellema. From the owl to the eskimo: Radiation gasdynamics of planetary nebulae IV. *Astrophys. J.*, 430:800, 1994.

91. H. Freistühler. Some remarks on the structure of intermediate magnetohydrodynamic shocks. *J. Geophys. Res.*, 96:3825–3827, 1991.
92. H. Freistühler and E. B. Pitman. A numerical study of a rotationally degenerate hyperbolic system, Part I: the Riemann problem. *J. Comput. Phys.*, 100:306, 1992.
93. H. Friedrich. Hyperbolic reductions for Einstein's equations. *Class. Quantum Grav.*, 13:1451–1469, 1996.
94. Fryxell, E. Müller, and Arnett. Hydrodynamics and nuclear burning. MPA Report 449, 1989.
95. H. M. Glaz and T.-P. Liu. The asymptotic analysis of wave interactions and numerical calculations of transonic nozzle flow. *Advances Appl. Math.*, 5:111–146, 1984.
96. J. Glimm, G. Marshall, and B. Plohr. A generalized Riemann problem for quasi-one-dimensional gas flows. *Advances Appl. Math.*, 5:1–30, 1984.
97. E. Godlewski and P.-A. Raviart. *Numerical Approximation of Hyperbolic Systems of Conservation Laws.* Springer-Verlag, New York, 1996.
98. S. K. Godunov. *Mat. Sb.*, 47:271, 1959.
99. T. I Gombosi, K. G. Powell, and D. L. De Zeeuw. Axisymmetric modelling of cometary mass loading on an adaptively refined grid. *J. Geophys. Res.*, 99, A11:21525–21539, 1994.
100. J. B. Goodman and R. J. LeVeque. A geometric approach to high resolution TVD schemes. *SIAM J. Num. Anal.*, 25:268–284, 1988.
101. A. Greenbaum. *Iterative Methods for Solving Linear Systems.* SIAM, Philadelphia, 1997.
102. F. H. Harlow and J. E. Welch. The MAC method: a computing technique for solving viscous, incompressible, transient fluid flow problems involving free surfaces. *Phys. Fluids*, 8:2182–2189, 1965.
103. A. Harten. High resolution schemes for hyperbolic conservation laws. *J. Comput. Phys.*, 49:357–393, 1983.
104. A. Harten, B. Engquist, S. Osher, and S. Chakravarthy. Uniformly high order accurate essentially nonoscillatory schemes, III. *J. Comput. Phys.*, 71:231, 1987.
105. A. Harten and J. M. Hyman. Self-adjusting grid methods for one-dimensional hyperbolic conservation laws. *J. Comput. Phys.*, 50:235–269, 1983.
106. A. Harten, J. M. Hyman, and P. D. Lax. On finite-difference approximations and entropy conditions for shocks. *Comm. Pure Appl. Math.*, 29:297–322 (with appendix by Barbara Keyfitz), 1976.
107. A. Harten and P. D. Lax. A random choice finite difference scheme for hyperbolic conservation laws. *SIAM J. Num. Anal.*, 18:289–315, 1981.
108. A. Harten, P. D. Lax, and B. van Leer. On upstream differencing and Godunov-type schemes for hyperbolic conservation laws. *SIAM Review*, 25:35–61, 1983.
109. A. Harten and S. Osher. Uniformly high-order accurate nonoscillatory schemes. I. *SIAM J. Num. Anal.*, 24:279–309, 1987.
110. A. Harten and G. Zwas. Self-adjusting hybrid schemes for shock computations. *J. Comput. Phys.*, 9:568, 1972.
111. C. Hirsch. *Numerical Computation of Internal and External Flows.* Wiley, 1988.
112. H. T. Huynh. Accurate upwind methods for the Euler equations. *SIAM J. Numer. Anal.*, 32:1565–1619, 1995.
113. V. Icke, B. Balick, and A. Frank. The hydrodynamics of aspherical planetary nebulae. *Astron. Astrophys.*, 253:224–243, 1991.

114. E. Isaacson, D. Marchesin, and B. Plohr. Transitional waves for conservation laws. *SIAM J. Math. Anal.*, 21:837–866, 1990.
115. A. Iserles. *Numerical Analysis of Differential Equations.* Cambridge University Press, Cambridge, 1996.
116. W. Israel. Relativistic theory of shock waves. *Proc. Roy. Soc. London*, A259:129, 1960.
117. J. D. Jackson. *Classical Electrodynamics.* John Wiley and Sons, Inc., New York, 1975.
118. A. Jeffrey and T. Taniuti. *Non-linear Wave Propagation.* Academic Press, 1964.
119. G. Jiang and C. W. Shu. Efficient implementation of weighted ENO schemes. *J. Comput. Phys.*, 126:202–228, 1996.
120. S. Jin. Implicit numerical schemes for hyperbolic conservation laws with stiff relaxation terms. *J. Comput. Phys.*, to appear.
121. S. Jin and J.-G. Liu. The effects of numerical viscosities: I. Slowly moving shocks. *J. Comput. Phys.*, 126:373–389, 1996.
122. Montgomery H. Johnson and Christopher F. McKee. Relativistic hydrodynamics in one dimension. *Phys. Rev. D*, 3:858–863, 1971.
123. O. S. Jones, U. Shumlak, and D. S. Eberhardt. An implicit scheme for nonideal magnetohydrodynamics. *J. Comput. Phys.*, 130:231–242, 1997.
124. A. Kantrowitz and H. E. Petschek. MHD characteristics and shock waves. In W. B. Kunkel, editor, *Plasma Physics in Theory and Application*, pages 148–206. McGraw-Hill, 1966.
125. S. Karni and S. Canic. Computation of slowly moving shocks. *J. Comput. Phys.*, to appear, 1997.
126. J. Kevorkian. *Partial Differential Equations.* Wadsworth & Brooks/Cole, 1990.
127. B. L. Keyfitz and H. C. Kranzer. A system of hyperbolic conservation laws arising in elasticity theory. *Arch. Rat. Mech. Anal.*, 72:219–241, 1980.
128. Arieh Königl. Relativistic gasdynamics in two dimensions. *Phys. Fluids*, 23:1083, 1980.
129. N. A. Krall and A. W. Trievelpiece. *Principles of Plasma Physics.* McGraw-Hill, New York, 1973.
130. D. Kröner. *Numerical Schemes for Conservation Laws.* Wiley-Teubner series, 1997.
131. J. D. Lambert. *Computational Methods in Ordinary Differential Equations.* Wiley, 1973.
132. L. D. Landau and E. M. Lifshitz. *Fluid Mechanics.* Pergamon Press Inc., Elmsford, New York, 1979.
133. J. O. Langseth and R. J. LeVeque. Three-dimensional Euler computations using CLAWPACK. In P. Arminjon, editor, *Conf. on Numer. Meth. for Euler and Navier-Stokes Eq., Montreal*, 1995. to appear.
134. J. O. Langseth and R. J. LeVeque. A wave-propagation method for three-dimensional hyperbolic conservation laws. preprint, 1997.
135. P. D. Lax. Hyperbolic systems of conservation laws, II. *Comm. Pure Appl. Math.*, 10:537–566, 1957.
136. P. D. Lax. *Hyperbolic Systems of Conservation Laws and the Mathematical Theory of Shock Waves.* SIAM Regional Conference Series in Applied Mathematics, #11, 1972.
137. B. P. Leonard, A. P. Lock, and M. K. MacVean. Conservative explicit unrestricted-time-step multidimensional constancy-preserving advection schemes. *Monthly Weather Rev.*, 124:2588–2606, 1996.

138. R. J. LeVeque. CLAWPACK software. available from `netlib.att.com` in `netlib/pdes/claw` or on the Web at the URL `http://www.amath.washington.edu/~rjl/clawpack.html`.
139. R. J. LeVeque. Intermediate boundary conditions for time-split methods applied to hyperbolic partial differential equations. *Math. Comput.*, 47:37–54, 1986.
140. R. J. LeVeque. Hyperbolic conservation laws and numerical methods. Von Karman Institute for Fluid Dynamics Lecture Series, 90-03, 1990.
141. R. J. LeVeque. *Numerical Methods for Conservation Laws.* Birkhäuser-Verlag, 1990.
142. R. J. LeVeque. Finite difference methods for differential equations. Class Notes, `http://www.amath.washington.edu/~rjl`, 1996.
143. R. J. LeVeque. Balancing source terms and flux gradients in high-resolution Godunov methods. in preparation, 1997.
144. R. J. LeVeque. Wave propagation algorithms for multi-dimensional hyperbolic systems. *J. Comput. Phys.*, 131:327–353, 1997.
145. R. J. LeVeque and K.-M. Shyue. One-dimensional front tracking based on high resolution wave propagation methods. *SIAM J. Sci. Comput.*, 16:348–377, 1995.
146. R. J. LeVeque and K.-M. Shyue. Two-dimensional front tracking based on high resolution wave propagation methods. *J. Comput. Phys.*, 123:354–368, 1996.
147. R. J. LeVeque and R. Walder. Grid alignment effects and rotated methods for computing complex flows in astrophysics. GAMM Conf. on Comput. Fluid Dyn., Lausanne, 1991.
148. R. J. LeVeque and H. C. Yee. A study of numerical methods for hyperbolic conservation laws with stiff source terms. *J. Comput. Phys.*, 86:187–210, 1990.
149. E. P. T. Liang. Relativistic simple waves: Shock damping and entropy production. *Astrophys. J.*, 211:361–376, 1977.
150. A. Lichnerowicz. *Relativistic Hydrodynamics and Magnetohydrodynamics.* Benjamin, New York, 1967.
151. I. Lindemuth and J. Killeen. Alternating direction implicit techniques for two-dimensional magnetohydrodynamic calculations. *J. Comput. Phys.*, 13:181, 1973.
152. M.-S. Liou. A sequel to AUSM: AUSM+. *J. Comput. Phys.*, to appear.
153. M.-S. Liou and C. J. Steffen Jr. A new flux splitting scheme. *J. Comput. Phys.*, 107:107, 1993.
154. T. P. Liu. On the viscosity criterion for hyperbolic conservation laws. In M. Shearer, editor, *Viscous Profiles and Numerical Methods for Shock Waves*, page 105, Philadelphia, 1991. SIAM.
155. X.-D. Liu, S. Osher, and T. Chan. Weighted essentially non-oscillatory schemes. *J. Comput. Phys.*, 115:200–212, 1994.
156. M.-M. Mac Low and M. Normann. Nonlinear growth of dynamical overstabilities in blast waves. *Astrophys. J.*, 407:207–218, 1993.
157. A. Marquina. Local piecewise hyperbolic reconstruction of numerical fluxes for nonlinear scalar conservation laws. *SIAM J. Sci. Comput.*, 15:892, 1994.
158. A. Marquina, J. Ma. Martí, J. Ma. Ibáñez, J. A. Miralles, and R. Donat. Ultrarelativistic hydrodynamics: high-resolution shock-capturing methods. *Astron. Astrophys.*, 258:566–571, 1992.
159. R. L. Marsa and M. W. Choptuik. Black hole–scalar field interactions in spherical symmetry. preprint, 1996.
160. J. M. Martí and E. Müller. The analytical solution of the Riemann problem in relativistic hydrodynamics. *J. Fluid Mech.*, 258:317–333, 1994.

161. J. Ma. Martí, J. Ma. Ibáñez, and J. A. Miralles. Numerical relativistic hydrodynamics: Local characteristic approach. *Phys. Rev. D*, 43:3794–3801, 1991.
162. J. Ma. Martí and E. Müller. Extension of the piecewise parabolic method to one-dimensional relativistic hydrodynamics. *J. Comput. Phys.*, 123:1–14, 1996.
163. Joan Massó, Edward Seidel, and Paul Walker. Adaptive mesh refinement in numerical relativity. To appear in General Relativity(MG7 Proceedings),R. Ruffini and M. Keiser (eds.), World Scientific, 1995.
164. G. Mellema, F. Eulderink, and V. Icke. Hydrodynamical models of aspherical planetary nebulae. *Astron. Astrophys.*, 252:718–732, 1991.
165. L. Mestel and N. O. Weiss. *Magnetohydrodynamics.* Fourth Saas-Fee Course, 1974.
166. D. Mihalas and B. W. Mihalas. *Foundations of Radiation Hydrodynamics.* Oxford Univesity Press, 1984.
167. C. W. Misner, K. S. Thorne, and J. A. Wheeler. *Gravitation.* W. H. Freeman and Co., New York, 1973.
168. C. Moller. *The Theory of Relativity, 2nd ed.* Oxford, University Press, Oxford, 1972.
169. K. W. Morton and D. F. Mayers. *Numerical Solution of Partial Differential Equations.* Cambridge University Press, Cambridge, 1994.
170. A. Mukherjee. *An adaptive finite element code for elliptic boundary value problems in three dimensions with applications in numerical relativity.* PhD thesis, Mathematics, Pennsylvania State University, 1996.
171. W. A. Mulder. Computation of a quasi-steady gas flow in a spiral galaxy by means of a multigrid method. *Astron. Astrophys.*, 156:354–380, 1986.
172. E. Müller and Steinmetz. Simulating self-gravitating hydrodynamics. *Comp. Phys. Comm.*, 89:45–58, 1995.
173. R. S. Myong. *Theoretical and computational investigations of nonlinear waves in magnetohydrodynamics.* PhD thesis, Aerospace Engineering, University of Michigan, 1996.
174. Gregory L. Naber. *The Geometry of Minkowski Spacetime.* Springer-Verlag, New York, 1992.
175. W. F. Noh. Errors for calculations of strong shocks using an artificial viscosity and an artificial heat flux. *J. Comput. Phys.*, 72:78, 1987.
176. M. L. Norman and K-H.A. Winkler. *Astrophysical Radiation Hydrodynamics.* Reidel, Norwell, MA, 1986.
177. H. Nussbaumer and R. Walder. Modification of the nebular environment in symbiotic systems due to colliding winds. *Astronomy and Astrophysics*, 278:209–225, 1993.
178. E. S. Oran and J. P. Boris. *Numerical Simulation of Reactive Flow.* Elsevier, 1987.
179. S. Osher. Riemann solvers, the entropy condition, and difference approximations. *SIAM J. Num. Anal.*, 21:217–235, 1984.
180. S. Osher and S. Chakravarthy. High resolution schemes and the entropy condition. *SIAM J. Num. Anal.*, 21:995–984, 1984.
181. S. Osher and F. Solomon. Upwind difference schemes for hyperbolic systems of conservation laws. *Math. Comp.*, 38:339–374, 1982.
182. W. Pauli. *Theory of Relativity.* Dover Publications, Inc., New York, 1981.
183. R. B. Pember. Numerical methods for hyperbolic conservation laws with stiff relaxation, I. Spurious solutions. *SIAM J. Appl. Math.*, 53:1293–1330, 1993.
184. R. B. Pember. Numerical methods for hyperbolic conservation laws with stiff relaxation, II. Higher order Godunov methods. *SIAM J. Sci. Comput.*, 14, 1993.

185. R. Peyret and T. D. Taylor. *Computational Methods for Fluid Flow*. Springer, 1983.
186. K. Powell. Solution of the Euler and Magnetohydrodynamic Equations on Solution-Adaptive Cartesian Grids. Von Karman Institute for Fluid Dynamics Lecture Series, 1996.
187. K. G. Powell. An approximate Riemann solver for magnetohydrodynamics (that works in more than one dimension). ICASE Report No. 94-24, NASA Langley Research Center, 1994.
188. K. G. Powell, P. L. Roe, R. S. Myong, T. Gombosi, and D. De Zeeuw. An upwind scheme for magnetohydrodynamics. AIAA-95-1704-CP, 1995.
189. K. H. Prendergast and K. Xu. Numerical hydrodynamics from gas-kinetic theory. *J. Comput. Phys.*, 109:53–66, 1993.
190. J. J. Quirk. A contribution to the great Riemann solver debate. *Internat. J. Numer. Methods Fluids*, 18:555–574, 1994.
191. J. Ramshaw. A method for enforcing the solenoidal condition on magnetic fields in numerical calculations. *J. Comput. Phys.*, 52:592–596, 1983.
192. R. D. Richtmyer and K. W. Morton. *Difference Methods for Initial-value Problems*. Wiley-Interscience, 1967.
193. T. W. Roberts. The behavior of flux difference splitting schemes near slowly moving shock waves. *J. Comput. Phys.*, 90:141–160, 1990.
194. P. Roe. The Harten memorial lecture — new applications of upwinding. preprint, 1997.
195. P. L. Roe. Approximate Riemann solvers, parameter vectors, and difference schemes. *J. Comput. Phys.*, 43:357–372, 1981.
196. P. L. Roe. Numerical algorithms for the linear wave equation. Royal Aircraft Establishment Technical Report 81047, 1981.
197. P. L. Roe. Some contributions to the modeling of discontinuous flows. *Lect. Notes Appl. Math.*, 22:163–193, 1985.
198. P. L. Roe. Upwind differencing schemes for hyperbolic conservation laws with source terms. In C. Carraso, P.-A. Raviart, and D. Serre, editors, *Nonlinear Hyperbolic Problems*, pages 41–51. Springer-Verlag, Lecture Notes in Mathematics 1270, 1986.
199. P. L. Roe and D. S. Balsara. Notes on the eigensystem of magnetohydrodynamics. *SIAM J. Appl. Math.*, 1996.
200. P. L. Roe and D. Sidilkover. Optimum positive linear schemes for advection in two and three dimensions. *SIAM J. Num. Anal.*, 29:1542–1568, 1992.
201. J. Romero, J. Ma. Ibáñez, J. Ma Martí, and J. A. Miralles. A new spherically symmetric general relativistic hydrodynamical code. *Astrophys. J.*, 462:836, 1996.
202. D. Ryu and T. W. Jones. Numerical magnetohydrodynamics in astrophysics: Algorithm and tests for one-dimensional flow. *Astrophysical J.*, 442:228–258, 1995.
203. D. Ryu, T. W. Jones, and A. Frank. Numerical magnetohydrodynamics in astrophysics: Algorithm and tests for multi-dimensional flow. *Astrophysical J.*, 452:785–796, 1995.
204. D. Ryu and E.T. Vishniac. The dynamic instability of adiabatic blast waves. *Astrophys. J.*, 368:411–425, 1991.
205. J. Saltzman. An unsplit 3-D upwind method for hyperbolic conservation laws. *J. Comput. Phys.*, pages 153–168, 1994.
206. R. H. Sanders and K. H. Prendergast. The possible relation of the 3-kiloparsec arm to explosions in the galactic nucleus. *Astrophys. J.*, 188:489, 1974.

207. D. G. Schaeffer and M. Shearer. The classification of 2×2 systems of nonstrictly hyperbolic conservation laws, with application to oil recovery. *Comm. Pure Appl. Math.*, 40:141–178, 1987.
208. D. G. Schaeffer and M. Shearer. Riemann problems for nonstrictly hyperbolic 2×2 systems of conservation laws. *Trans. Amer. Math. Soc.*, 304:267–306, 1987.
209. S. Schecter and M. Shearer. Transversality for undercompressive shocks in Riemann problems. In M. Shearer, editor, *Viscous Profiles and Numerical Methods for Shock Waves*, pages 142–154, Philadelphia, 1991. SIAM.
210. D. Schnack and J. Killeen. Nonlinear, two-dimensional magnetohydrodynamic calculations. *J. Comput. Phys.*, 35:110, 1980.
211. Bernard F. Schutz. *A first Course in General Relativity*. Cambridge University Press, Cambridge, 1990.
212. E. Seidel and W.-M. Suen. Numerical relativity. preprint.
213. E. Seidel and W.-M. Suen. Towards a singularity proof in numerical relativity. *Phys. Rev. Let.*, 69:1845–1848, 1992.
214. C.-W. Shu. TVB uniformly high-order schemes for conservation laws. *Math. Comp.*, 49:105–121, 1987.
215. C.-W. Shu and S. Osher. Efficient implementation of essentially nonoscillatory shock capturing schemes. *J. Comput. Phys.*, 77:439–471, 1988.
216. C.-W. Shu and S. Osher. Efficient implementation of essentially nonoscillatory shock capturing schemes II. *J. Comput. Phys.*, 83:32, 1989.
217. F. H. Shu. *The Physics of Astrophysics: Volume 1, Radiation.* University Science Books, Mill Valley, CA, 1991.
218. F. H. Shu. *The Physics of Astrophysics: Volume 2, Gas Dynamics.* University Science Books, Mill Valley, CA, 1991.
219. D. Sidilkover. A genuinely multidimensional upwind scheme for the compressible Euler equations. In J. Glimm et al., editors, *Proceedings of the Fifth International Conference on Hyperbolic Problems: Theory, Numerics, Applications.* World Scientific, June 1994.
220. P. K. Smolarkeiwicz and W. W. Grabowski. The multidimensional positive definite advection transport algorithm: nonoscillatory option. *J. Comput. Phys.*, 86:355–375, 1990.
221. J. Smoller. *Shock Waves and Reaction-Diffusion Equations.* Springer, 1983.
222. J. Smoller and B. Temple. Global solutions of the relativistic Euler equations. *Comm. Math. Phys.*, 156:67–99, 1993.
223. J. Smoller and B. Temple. Shock-waves and irreversibility in Einstein's theory of gravity. In J. Glimm et al., editors, *Proceedings of the Fifth International Conference on Hyperbolic Problems: Theory, Numerics, Applications*, pages 81–90, Singapore, June 1994. World Scientific.
224. J. Smoller and B. Temple. An astrophysical shock-wave solution of the Einstein equations. *Phys. Rev. D*, 51:2733–2743, 1996.
225. G. Sod. A survey of several finite diffeence methods for systems of nonlinear hyperbolic conservation laws. *J. Comput. Phys.*, 27:1–31, 1978.
226. G. Sod. *Numerical Methods for Fluid Dynamics.* Cambridge University Press, 1985.
227. J. L. Steger and R. F. Warming. Flux vector splitting of the inviscid gasdynamic equations with applications to finite-difference methods. *J. Comput. Phys.*, 40:263, 1981.
228. O. Steiner, M. Knölker, and M. Schüssler. Dynamic interaction of convection with magnetic flux sheets: first results of a new MHD code. In R. J. Rutten and C. J. Schrijver, editors, *Solar Surface Magnetism, NATO Advanced Research Workshop*, Dordrecht, 1993. Kluwer Academic Publishers.

229. R. S. Steinolfson and A. J. Hundhausen. MHD intermediate shocks in coronal mass ejections. *J. Geophys. Res.*, 95:6389, 1990.
230. I. R. Stevens, J. M. Blondin, and A. M. T. Pollock. Colliding winds from early–type stars in binary systems. *Ap.J.*, 386:265–287, 1992.
231. G. Strang. On the construction and comparison of difference schemes. *SIAM J. Num. Anal.*, 5:506–517, 1968.
232. J. C. Strikwerda. *Finite Difference Schemes and Partial Differential Equations.* Wadsworth & Brooks/Cole, 1989.
233. Paul N. Swarztrauber. Fast Poisson solvers. In Gene H. Golub, editor, *Studies in Numerical Analysis*, volume 24, pages 319–370, Washington, D.C., 1984. The Mathematical Association of America.
234. P. K. Sweby. High resolution schemes using flux limiters for hyperbolic conservation laws. *SIAM J. Num. Anal.*, 21:995–1011, 1984.
235. J. L. Synge. *The Relativistic Gas.* North-Holland Publishing Company, Amsterdam, 1957.
236. A. H. Taub. Relativistic Rankine-Hugoniot equations. *Phys. Rev.*, 74:328, 1948.
237. Kevin W. Thompson. The special relativistic shock tube. *J. Fluid Mech.*, 171:365–375, 1986.
238. K. S. Thorne. Gravitational radiation. In S. W. Hawking and W. Israel, editors, *300 Years of Gravitation.* Cambridge University Press, 1987.
239. K. S. Thorne. *Black Holes and Time Warps.* W. W. Norton & Co., New York, 1994.
240. K. S. Thorne. Gravitational waves. In E. W. Kolb and R. D. Peccei, editors, *Particle and Nuclear Astrophysics and Cosmology in the Next Millennium, Proc. 1994 Snowmass Summer School*, pages 160–184. World Scientific, 1994.
241. K.S. Thorne. Relativistic shock: The Taub adiabat. *Astrophys. J.*, 179:897, 1973.
242. E. F. Toro. *Riemann solvers and numerical methods for fluid dynamics.* Springer-Verlag, Berlin, Heidelberg, 1997.
243. J. A. Trangenstein. An unsplit Godunov method for three-dimensional polymer flooding. preprint, 1993.
244. L. N. Trefethen and D. Bau. *Numerical Linear Algebra.* SIAM, Philadelphia, 1997.
245. G. D. van Albada, B. van Leer, and W. W. Roberts, Jr. A comparative study of computational methods in cosmic gas dynamics. *Astron. Astrophys.*, 108:76–84, 1982.
246. B. van Leer. Towards the ultimate conservative difference scheme I. The quest of monotonicity. *Springer Lecture Notes in Physics*, 18:163–168, 1973.
247. B. van Leer. Towards the ultimate conservative difference scheme IV. A new approach to numerical convection. *J. Comput. Phys.*, 23:276–299, 1977.
248. B. van Leer. Towards the ultimate conservative difference scheme V. A second order sequel to Godunov's method. *J. Comput. Phys.*, 32:101–136, 1979.
249. B. van Leer. Flux-vector splitting for the Euler equations. In E. Krause, editor, *Lecture Notes in Physics*, volume 170, page 507. Springer-Verlag, 1982.
250. B. van Leer. On the relation between the upwind-differencing schemes of Godunov, Engquist-Osher, and Roe. *SIAM J. Sci. Stat. Comput.*, 5:1–20, 1984.
251. E. T. Vishniac. Nonlinear instabilities in shock-bounded slabs. *Astrophys. J.*, 428:186–208, 1994.
252. R. M. Wald. *General Relativity.* Univesity of Chicago Press, 1984.

253. R. Walder. *Some Aspects of the Computational Dynamics of Colliding Flows in Astrophysical Nebulae.* PhD thesis, Astronomy Institute, ETH-Zürich, No. 10302, 1993.
254. R. Walder and D. Folini. Radiative cooling instability in 1D colliding flows. *Astronomy and Astrophysics*, 315:265–284, 1996.
255. P. Wesseling. *An Introduction to Multigrid Methods.* Wiley, New York, 1992.
256. G. Whitham. *Linear and Nonlinear Waves.* Wiley-Interscience, 1974.
257. F. A. Williams. *Combustion Theory.* Benjamin/Cummings, 1985.
258. P. Woodward and P. Colella. The numerical simulation of two-dimensional fluid flow with strong shocks. *J. Comput. Phys.*, 54:115–173, 1984.
259. C. C. Wu. On MHD intermediate shocks. *Geophys. Res. Letters*, 14:668–671, 1987.
260. C. C. Wu. Formation, structure, and stability of MHD intermediate shocks. *J. Geophys. Res.*, 95:8149, 1990.
261. C. C. Wu. New theory of MHD shock waves. In M. Shearer, editor, *Viscous Profiles and Numerical Methods for Shock Waves*, pages 231–235, Philadelphia, 1991. SIAM.
262. C. C. Wu and C. F. Kennel. Evolution of small amplitude intermediate shocks in a dissipative and dispersive system. *J. Plasma Phys.*, 47:85, 1992.
263. H. C. Yee. Upwind and symmetric shock-capturing schemes. NASA Ames Technical Memorandum 89464, 1987.
264. A. L. Zachary and P. Colella. A higher order Godunov method for the equations of ideal magnetohydrodynamics. *J. Comput. Phys.*, 99:341–347, 1992.
265. A. L. Zachary, A. Malagoli, and P. Colella. A higher-order Godunov method for multidimensional ideal magnetohydrodynamics. *SIAM J. Sci. Comput.*, 15:263–284, 1994.
266. S. T. Zalesak. Fully multidimensional flux corrected transport algorithms for fluids. *J. Comput. Phys.*, 31:335–362, 1979.

Dimitri Mihalas, who had to stay in hospital, missed his chance for lecturing, sun, and snow

Radiation Hydrodynamics

Dimitri Mihalas

Department of Astronomy, University of Illinois, 1002 West Green Street, Urbana, IL 61801, U.S.A.

1. Basic Radiation Theory

References: [1, 5, 22, 29, 35]. *General Reference*: [24].

1.1 Specific Intensity

Define the *specific intensity* such that

$$\delta\mathcal{E} = I(\boldsymbol{x}, t; \boldsymbol{n}, \nu)\, \mathrm{d}S \cos\alpha\, \mathrm{d}\nu\, \mathrm{d}t \tag{1.1}$$

is the radiant energy in a range $(\nu, \nu + \mathrm{d}\nu)$ crossing $\mathrm{d}S$, whose normal makes an angle α with $\boldsymbol{n}$, the direction of propagation of the radiation, into a solid angle $\mathrm{d}\omega$, in a time $\mathrm{d}t$. The units of specific intensity are

$$[I] = \text{ergs cm}^{-2}\ \text{s}^{-1}\ \text{hz}^{-1}\ \text{sr}^{-1}. \tag{1.2}$$

In one-dimensional plane-parallel or spherically symmetric media I is independent of Φ, the azimuthal angle around the outward normal; it can depend only on one space coordinate (z or r) and on the polar angle Θ between the outward normal and the direction of photon propagation $\boldsymbol{n}$. We can then write $I = I(z, t; \mu, \nu)$ or $I(r, t; \mu, \nu)$ where $\mu \equiv \boldsymbol{n} \cdot \boldsymbol{k}$ or $\mu \equiv \boldsymbol{n} \cdot \hat{\boldsymbol{r}}$.

1.2 Photon Number Density

I is appropriate for the macroscopic view. To think in terms of photons, define the *photon number density* ψ such that $\psi(\boldsymbol{x}, t; \boldsymbol{n}, \nu)\, \mathrm{d}\nu\, \mathrm{d}\omega$ is the number of photons, per unit volume, at $(\boldsymbol{x}, t)$, with frequency $(\nu, \nu + \mathrm{d}\nu)$, traveling with speed c into solid angle $\mathrm{d}\omega$ around $\boldsymbol{n}$. Then the number crossing area $\mathbf{d}\boldsymbol{S}$ in time $\mathrm{d}t$ into $\mathrm{d}\omega$ is $\psi\ \boldsymbol{n} \cdot \mathbf{d}\boldsymbol{S}\ (c\, \mathrm{d}t)\, \mathrm{d}\nu\, \mathrm{d}\omega$. Each photon has energy $h\nu$, hence

$$\delta\mathcal{E} = c\, h\nu\, \psi(\boldsymbol{x}, t; \boldsymbol{n}, \nu)\, \mathrm{d}S \cos\alpha\, \mathrm{d}\omega\, \mathrm{d}\nu\, \mathrm{d}t\ . \tag{1.3}$$

Therefore

$$I(\boldsymbol{x}, t; \boldsymbol{n}, \nu) = c\, h\nu\ \psi(\boldsymbol{x}, t; \boldsymbol{n}, \nu)\ . \tag{1.4}$$

1.3 Photon Distribution Function

Define the photon distribution function such that $f_{\mathrm{R}}(\boldsymbol{x},t;\boldsymbol{n},p)\,\mathrm{d}^3p$ is the number of photons, per unit volume, at $(\boldsymbol{x},t)$, with momenta $(\boldsymbol{p},\boldsymbol{p}+\mathbf{d}\boldsymbol{p}) = (h\nu/c)(\boldsymbol{n},\boldsymbol{n}+\mathbf{d}\boldsymbol{n})$. Using $\mathrm{d}^3p = p^2\mathrm{d}p\,\mathrm{d}\omega = (h/c)^3\nu\,\mathrm{d}\nu\,\mathrm{d}\omega$ we find

$$(h^3\nu^2/c^3)f_{\mathrm{R}}\,\mathrm{d}\nu\,\mathrm{d}\omega = \psi\,\mathrm{d}\nu\,\mathrm{d}\omega\ , \tag{1.5}$$

which implies

$$\psi = \frac{h^3\nu^2}{c^3}f_{\mathrm{R}}\ , \tag{1.6}$$

and

$$I = \frac{h^4\nu^3}{c^2}f_{\mathrm{R}}\ . \tag{1.7}$$

f_{R} is the direct analogue of the particle distribution function for material particles. By exactly the same analysis as used for material particles, one can show that f_{R} *is a relativistic invariant* under Lorentz transformation.

1.4 Mean Intensity

The *mean intensity* is the average of the specific intensity over all solid angles,

$$J(\boldsymbol{x},t;\nu) = \frac{1}{4\pi}\oint I(\boldsymbol{x},t;\boldsymbol{n},\nu)\,\mathrm{d}\omega\ . \tag{1.8}$$

The units of the mean intensity are the same as the specific intensity. In one-dimensional planar or spherically-symmetric media, the integral over the azimuthal angle Φ is trivial (because I is independent of Φ). Therefore using $\mathrm{d}\omega = \sin\Theta\,\mathrm{d}\Theta\,\mathrm{d}\Phi = -\mathrm{d}\mu\,\mathrm{d}\Phi$ we have

$$J_\nu = \frac{1}{2}\int_{-1}^{1} I_\nu \mathrm{d}\mu\ . \tag{1.9}$$

Thus, the mean intensity is the zeroth angular moment of the specific intensity.

1.5 Radiation Energy Density

The *monochromatic radiation energy density* at frequency ν equals (the number density of photons) $\times(h\nu)$, summed over all solid angles, that is

$$E(\boldsymbol{x},t;\nu) = h\nu\oint \psi(\boldsymbol{x},t;\boldsymbol{n},\nu)\,\mathrm{d}\omega\ , \tag{1.10}$$

which implies

$$E_\nu = \frac{1}{c}\oint I(\boldsymbol{x},t;\boldsymbol{n},\nu)\,\mathrm{d}\omega = \frac{4\pi}{c}J_\nu\ . \tag{1.11}$$

The total radiation density is $E = \int_0^\infty E_\nu \mathrm{d}\nu = (4\pi/c)J$.

1.6 Radiation Energy Flux

Define the *monochromatic radiation energy flux* $\boldsymbol{F}(\boldsymbol{x},t;\nu)$ $(= \boldsymbol{F}_\nu)$ such that $\boldsymbol{F} \cdot \mathbf{d}\boldsymbol{S}$ is the *net* rate of radiant energy flow, per unit frequency interval, through $\mathbf{d}\boldsymbol{S}$. The *number flux* through $\mathbf{d}\boldsymbol{S}$ is

$$N = \left(\oint \psi(\boldsymbol{x},t;\boldsymbol{n},\nu)\, c\,\boldsymbol{n}\, \mathrm{d}\omega \right) \cdot \mathbf{d}\boldsymbol{S} \,. \tag{1.12}$$

To get the energy flux, multiply by $h\nu$. Then convert ψ to I, obtaining

$$\delta\mathcal{E} = \left(\oint I(\boldsymbol{x},t;\boldsymbol{n},\nu)\, \boldsymbol{n}\, \mathrm{d}\omega \right) \cdot \mathbf{d}\boldsymbol{S} \,, \tag{1.13}$$

hence

$$\boldsymbol{F}(\boldsymbol{x},t;\nu) = \oint I(\boldsymbol{x},t;\boldsymbol{n},\nu)\, \boldsymbol{n}\, \mathrm{d}\omega \,. \tag{1.14}$$

The units of $\boldsymbol{F}_\nu$ are $[F_\nu] = \text{ergs cm}^{-2}\ \text{s}^{-1}\ \text{hz}^{-1}$. The total radiant energy flux is $\boldsymbol{F} = \int_0^\infty \boldsymbol{F}_\nu\, \mathrm{d}\nu$. Note that in one-dimensional planar or spherical geometry, azimuthal symmetry of I implies only *one* nonzero component of $\boldsymbol{F}$ (along $\hat{k}$ in planar geometry and $\hat{r}$ in spherical geometry. This is easy to see by taking rays in cancelling pairs in the (x, y) plane. People often call this one component "the" flux *as if* it were a scalar:

$$F(z,t;\nu) = 2\pi \int_{-1}^{1} I(z,t;\mu,\nu)\mu\, \mathrm{d}\mu \equiv F_\nu \,. \tag{1.15}$$

But don't forget that $\boldsymbol{F}$ is really a vector! Following Eddington, it is useful to define

$$H_\nu \equiv \frac{1}{4\pi} F_\nu = \frac{1}{2} \int_{-1}^{1} I(z,t;\mu,\nu)\mu\, \mathrm{d}\mu \,. \tag{1.16}$$

Note that H_ν is the first angular moment of the specific intensity.

1.7 Radiation Momentum Density

The momentum of a photon traveling in direction $\boldsymbol{n}$ is $(h\nu/c)\,\boldsymbol{n}$. Therefore the *net* rate of radiation momentum transport through $\mathbf{d}\boldsymbol{S}$ is $(1/c)$ times the energy flux through $\mathbf{d}\boldsymbol{S}$, that is, $(1/c)\boldsymbol{F}_\nu \cdot \mathbf{d}\boldsymbol{S}$. Thus, the *monochromatic momentum flux* is $(1/c)\boldsymbol{F}_\nu$. This flux is transported by particles moving with speed c. Therefore the *monochromatic momentum density* is $(1/c) \cdot$ momentum flux,

$$\boldsymbol{G}_\nu \equiv \frac{1}{c^2} \boldsymbol{F}_\nu \,, \tag{1.17}$$

and the total radiation momentum density is $\boldsymbol{G} \equiv (1/c^2)\boldsymbol{F}$. The units of $\boldsymbol{G}_\nu$ are $[G_\nu] = \text{g cm}^{-2}\ \text{s}^{-1}\ \text{hz}^{-1}$.

1.8 Radiation Stress Tensor (Radiation Pressure Tensor)

Definition. The radiation stress tensor is most easily derived in terms of the photon picture. Let $P^{ij} \equiv$ net rate of momentum transport in the ith direction per unit area of a surface perpendicular to the jth coordinate axis. The number of photons of frequency ν moving in the jth direction and crossing a unit area per unit time is $\psi_\nu c n^j$. Each has momentum $(h\nu/c)\, n^i$ in the ith direction. Thus,

$$P^{ij}(\boldsymbol{x},t;\nu) = \oint \psi(\boldsymbol{x},t;\boldsymbol{n},\nu)(h\nu\, n^i/c)(c n^j)\,\mathrm{d}\omega\,, \tag{1.18}$$

or

$$P^{ij}(\boldsymbol{x},t;\nu) = \frac{1}{c}\oint I(\boldsymbol{x},t;\boldsymbol{n},\nu) n^i n^j \mathrm{d}\omega\,, \tag{1.19}$$

which implies

$$\mathsf{P}(\boldsymbol{x},t;\nu) = \frac{1}{c}\oint I(\boldsymbol{x},t;\boldsymbol{n},\nu)\boldsymbol{n}\,\boldsymbol{n}\,\mathrm{d}\omega\,. \tag{1.20}$$

P is the *monochromatic radiation stress tensor*. Note that P is manifestly symmetric, i.e., $P^{ij} \equiv P^{ji}$. The total radiation stress tensor is

$$\mathsf{P}(\boldsymbol{x},t) = \int_0^\infty \mathsf{P}_\nu\,\mathrm{d}\nu\,. \tag{1.21}$$

The units of P_ν and P are: $[\mathsf{P}_\nu]$ = dynes cm^{-2} hz^{-1} and $[\mathsf{P}]$ = dynes cm^{-2}.

One-Dimensional Media. In general one needs six distinct components to specify the (symmetric) radiation stress tensor in three dimensional media: three along the diagonal, and three off-diagonal. But for one dimensional symmetry (planar or spherical) there is substantial simplification.

1) The axisymmetry of I_ν in a one dimensional medium implies that the off-diagonal elements are identically zero (ray by ray cancellation).
2) We can evaluate the diagonal elements easily. The integrands are: $\sin^2\Theta\ \cos^2\Phi$, $\sin^2\Theta\ \sin^2\Phi$, and $\cos^2\Theta \equiv \mu^2$.

We find

$$\begin{aligned}\mathsf{P}_\nu &= \begin{pmatrix} \frac{1}{2}(E_\nu - P_\nu) & 0 & 0 \\ 0 & \frac{1}{2}(E_\nu - P_\nu) & 0 \\ 0 & 0 & P_\nu \end{pmatrix} \\ &= \begin{pmatrix} P_\nu & 0 & 0 \\ 0 & P_\nu & 0 \\ 0 & 0 & P_\nu \end{pmatrix} - \frac{1}{2}\begin{pmatrix} 3P_\nu - E_\nu & 0 & 0 \\ 0 & 3P_\nu - E_\nu & 0 \\ 0 & 0 & 0 \end{pmatrix}.\end{aligned} \tag{1.22}$$

Here E_ν is the radiation energy density and P_ν is the scalar

$$P_\nu = \frac{2\pi}{c}\int_{-1}^{1} I(\boldsymbol{x},t;\boldsymbol{n},\nu)\mu^2 d\mu = (\frac{4\pi}{c})\frac{1}{2}\int_{-1}^{1} I_\nu \mu^2 d\mu \equiv \frac{4\pi}{c}K_\nu \,. \tag{1.23}$$

K_ν is the second angular moment of the specific intensity. The scalar P_ν is often called "the" radiation pressure, *as if* P_ν were a scalar because: 1) in one-dimensional symmetry the operators $(\partial/\partial x)$ and $(\partial/\partial y) \equiv 0$ so only P_{33} (which equals P_{zz} in planar geometry, and P_{rr} in spherical geometry) matters, and 2) as we shall see later, when the radiation field is *isotropic* (e.g., many mean free paths from any boundary – as in the interior of a star) then $P_\nu \to \frac{1}{3}E_\nu$, so the second matrix in Eq. (1.22) vanishes, and P becomes diagonal, isotropic, and is fully specified by a single number, P_ν.

In short, in one-dimensional symmetry we need only two numbers, E_ν and P_ν to specify P_ν, which is diagonal.

Isotropy. The trace of P_ν is invariant:

$$P_\nu^{ii} = P_\nu^{xx} + P_\nu^{yy} + P_\nu^{zz} = \frac{1}{c}\oint I_\nu (n_x^2 + n_y^2 + n_z^2)\, d\omega \equiv E_\nu \,. \tag{1.24}$$

Therefore we can in general define a *mean pressure*

$$\bar{P}_\nu = \frac{1}{3}P_\nu^{ii} = \frac{1}{3}E_\nu \tag{1.25}$$

as was true for material particles. However, unlike material particles where the distribution function $f(\boldsymbol{x},\boldsymbol{u},t)$ *is* nearly isotropic on the small scale, so that all diagonal elements of P very nearly equal $\bar{P}$, for radiation the elements P_{zz} and P_{xx} (or P_{yy}) can be markedly different, especially near boundaries. Thus, for radiation, $\bar{P}_\nu$ has no useful dynamical significance.

But if we simply *assume* that $I(\boldsymbol{n})$ is isotropic, then $\oint I n_i^2\, d\omega \equiv \frac{1}{3}\oint I d\omega$, $(i = 1,2,3)$. Therefore in this case $P_\nu = \frac{1}{3}E_\nu$ and

$$\mathsf{P}_\nu = \begin{pmatrix} P_\nu & 0 & 0 \\ 0 & P_\nu & 0 \\ 0 & 0 & P_\nu \end{pmatrix} . \tag{1.26}$$

This case is of great importance because it applies with very high accuracy in $99^{+}\%$ of a star. Only one number is needed to specify P_ν in this limit.

Streaming Limit. To study the effect of a boundary, consider the free streaming limit. Take a plane wave traveling along the outward normal, $\mu = 1$, so $I = I_0\,\delta(\mu - 1)$. Example: the radiation field at very large distance from a point source (e.g., light from a distant star). Then it is trivial to show that $J_\nu \equiv H_\nu \equiv K_\nu$, so the pressure tensor becomes

$$\mathsf{P}_\nu = \begin{pmatrix} 0 & 0 & 0 \\ 0 & 0 & 0 \\ 0 & 0 & P_\nu \end{pmatrix} = \begin{pmatrix} 0 & 0 & 0 \\ 0 & 0 & 0 \\ 0 & 0 & E_\nu \end{pmatrix} . \tag{1.27}$$

Variable Eddington Factors. It is useful to introduce the *variable Eddington factor* [11, 22]

$$f_\nu \equiv P_\nu/E_\nu = \int_{-1}^{1} I_\nu(\mu)\mu^2 \mathrm{d}\mu \Big/ \int_{-1}^{1} I_\nu(\mu)\, \mathrm{d}\mu \,. \tag{1.28}$$

In one-dimensional symmetry this one scalar is sufficient to specify the asymmetry of the radiation field.

The variable Eddington factor provides a *closure relation* between P_ν and E_ν in the radiation moment equations discussed later. In a material gas one can always specify the closure relation [$p/\hat{e} = \frac{2}{3}$ (N.R.) or $\frac{1}{3}$ (E.R.)] because the particle mean free path is $\ll$ the characteristic system size. Even near boundaries it can be derived from the Chapman–Enskog solution. Radiation theory differs fundamentally from gas dynamics: one must *construct the closure by calculating f_ν from a solution of the transfer equation.* The reason is that a photon mean free path can be about the size of, or larger than, the characteristic system size.

In a one dimensional medium we can expect $\frac{1}{3} \le f_\nu \le 1$ in most cases. In some cases one finds $0 \le f_\nu \le 1$; values less than $\frac{1}{3}$ come from a "pancake" shaped distribution of radiation where the lateral radiation field is much stronger than the radial radiation field. An important example is an optically thin (radial direction) planar layer, which is representative of a *shock*.

In two- and three-dimensional media we have, more generally, the *Eddington tensor*

$$\mathsf{f}_\nu \equiv \mathsf{P}_\nu/E_\nu \,. \tag{1.29}$$

Allowing for symmetry, the Eddington tensor has one, three, and six components in one-dimensional, two-dimensional, and three-dimensional geometry, respectively. Ad hoc prescriptions [25, 20] are sometimes used to estimate the Eddington tensor, but often satisfactory results are obtained only from direct calculation.

1.9 Thermal Radiation

Consider the important case of radiation in thermal equilibrium. Then[1]

$$I_\nu^* \equiv B_\nu(T) = \frac{2h\nu^3}{c^2}/(\mathrm{e}^{h\nu/kT} - 1) \qquad \text{(Planck function)} \,. \tag{1.30}$$

Notice that I_ν^* is isotropic. Therefore $\boldsymbol{F}_\nu^* \equiv 0$ inside the hohlraum. A nonzero $\boldsymbol{F}_\nu$ imples an anisotropic radiation field; in a stellar interior a small temperature gradient produces the flux which transports the stellar luminosity. The monochromatic energy density in equilibrium is

[1] The superscript * designates quantities in thermal equilibrium.

$$J_\nu^* = I_\nu^* \Rightarrow E_\nu^* = \frac{4\pi}{c} B_\nu = \frac{8\pi h\nu^3}{c^3} / (\mathrm{e}^{h\nu/kT} - 1) \ . \tag{1.31}$$

Then the total energy density is:

$$E^*(T) = \int_0^\infty E_\nu^* \mathrm{d}\nu = \frac{8\pi k^4 T^4}{c^3 h^3} \int_0^\infty \frac{x^3 \mathrm{d}x}{\mathrm{e}^x - 1} = a_\mathrm{R} T^4 \ , \tag{1.32}$$

which is *Stefan's law*. The integrated Planck functon is

$$B(T) = \frac{c}{4\pi} E^*(T) = \frac{a_\mathrm{R} c}{4\pi} T^4 = \frac{\sigma_\mathrm{R}}{\pi} T^4 \ , \tag{1.33}$$

where $\sigma_\mathrm{R} \equiv (a_\mathrm{R} c/4)$ is the *Stefan–Boltzmann constant*. To calculate the flux *emerging* from a hohlraum, note that I^* is isotropic. Thus,

$$F_\mathrm{BB}(\nu) = 2\pi \int_0^1 B_\nu(T) \mu \mathrm{d}\mu = \pi B_\nu(T) \ , \tag{1.34}$$

hence

$$F_\mathrm{BB} = \pi B(T) = \sigma_\mathrm{R} T^4 \ , \tag{1.35}$$

where BB stands for "black body". Within the hohlraum the flux in the sense of (1.15) is zero. Again recalling that thermal radiation is isotropic its pressure is

$$P^* = \frac{1}{3} E^* = \frac{1}{3} a_\mathrm{R} T^4 \ . \tag{1.36}$$

$$T\mathrm{d}S = \mathrm{d}(E^* V) + P^* \mathrm{d}V \ , \tag{1.37}$$

or

$$\mathrm{d}S = 4 a_\mathrm{R} T^2 V \mathrm{d}T + \frac{4}{3} a_\mathrm{R} T^3 \mathrm{d}V = \mathrm{d}(\frac{4}{3} a_\mathrm{R} T^3 V) \ , \tag{1.38}$$

which implies

$$S_\mathrm{rad} = \frac{4}{3} a_\mathrm{R} T^3 V \ . \tag{1.39}$$

Suppose the enclosure is adiabatic. Then $\mathrm{d}S \equiv 0 \Rightarrow \frac{\mathrm{d}T}{T} + \frac{1}{3}\frac{\mathrm{d}V}{V} = 0 \Rightarrow TV^{1/3} =$ constant or $P^* V^{4/3} =$ constant. In fact, for pure radiation $\Gamma_1 = \Gamma_2 = \Gamma_3 = 4/3$. Therefore thermal radiation behaves like a polytrope with $\gamma = 4/3$. The specific heat at constant volume of thermal radiation is

$$c_v = \frac{\partial}{\partial T} (a_\mathrm{R} T^4 V)_V = 4 a_\mathrm{R} T^3 V \ . \tag{1.40}$$

In contrast $c_p \to \infty$. The reason is that if P^* is constant, T must be constant, which implies that $\mathrm{d}T \equiv 0$ even for finite $\mathrm{d}Q$.

The sound speed of pure radiation can be obtained from the general formula

$$a = c\left(\frac{\Gamma_1 p}{\hat{e}+p}\right)^{1/2} . \tag{1.41}$$

For pure radiation $\Gamma_1 = \frac{4}{3}$ and $p = \frac{1}{3}\hat{e}$. Therefore $a = c/\sqrt{3}$ is the "sound speed" of an "acoustic disturbance" in pure thermal radiation.

1.10 Thermodynamics of Thermal Radiation and a Perfect Gas

It is easy to find the thermodynamic properties of a radiating fluid comprising both thermal radiation and a material perfect gas. Thus,

$$p = p_{\mathrm{g}} + P^* = \frac{NkT}{V} + \frac{1}{3}a_{\mathrm{R}}T^4 \tag{1.42}$$

and

$$e = \frac{3}{2}RT + \frac{a_{\mathrm{R}}T^4}{\varrho} . \tag{1.43}$$

Further, $T\mathrm{d}S = \mathrm{d}(\varrho eV) + p\,\mathrm{d}V = \mathrm{d}(\frac{3}{2}NkT + a_{\mathrm{R}}T^4V) + (\frac{NkT}{V} + \frac{1}{3}a_{\mathrm{R}}T^4)\,\mathrm{d}V$ which implies that $\mathrm{d}S = Nk\,\mathrm{d}[\ln(T^{3/2}V)] + \mathrm{d}(\frac{4}{3}a_{\mathrm{R}}T^3V)$ or

$$S = Nk\ln(T^{3/2}V) + \frac{4}{3}a_{\mathrm{R}}T^3V . \tag{1.44}$$

Define $\alpha \equiv P^*/p_{\mathrm{g}}$. Then one can show that

$$c_v = \frac{3}{2}\frac{k}{m}(1+8\alpha) , \tag{1.45}$$

$$c_p = \frac{5}{2}\frac{k}{m}(1+8\alpha+\frac{32}{5}\alpha^2) . \tag{1.46}$$

One can see by inspection that these formulae go to the correct limits as $\alpha \to 0$ and $\alpha \to \infty$. Further

$$\Gamma_1 \equiv \left(\frac{\partial \ln p}{\partial \ln \varrho}\right)_s = \frac{\frac{5}{2}+20\alpha+16\alpha^2}{(\frac{3}{2}+12\alpha)(1+\alpha)} , \tag{1.47}$$

$$\frac{\Gamma_2 - 1}{\Gamma_2} \equiv \left(\frac{\partial \ln p}{\partial \ln T}\right)_s = \frac{1+5\alpha+4\alpha^2}{\frac{5}{2}+20\alpha+16\alpha^2} , \tag{1.48}$$

$$\Gamma_3 - 1 \equiv \left(\frac{\partial \ln T}{\partial \ln \varrho}\right)_s = \frac{1+4\alpha}{(\frac{3}{2}+12\alpha)} . \tag{1.49}$$

One easily sees that for all three values of the generalized adiabatic exponent Γ defined above, $\Gamma \to \frac{4}{3}$ as $\alpha \to \infty$ and $\Gamma \to \frac{5}{3}$ as $\alpha \to 0$. With a modest amount of algebra these results can be extended to an ionizing material gas plus radiation.

It should be noted that all of the results in this section apply for *thermal* radiation only (LTE). Thus, they hold in, say, the interior of a star, but *not* in a stellar atmosphere where $I_\nu \neq B_\nu$, or indeed in any open system which is out of thermodynamic equilibrium.

2. The Transfer Equation

2.1 Absorption, Emission, and Scattering

Opacity. Define the *opacity* or *extinction coefficient* χ such that an element of material of length $\mathrm{d}l$, cross-section $\mathrm{d}S$ perpendicular to a beam moving in direction $\boldsymbol{n}$, into solid angle $\mathrm{d}\omega$, removes an amount of energy

$$\delta\mathcal{E} = \chi(\boldsymbol{x}, t; \boldsymbol{n}, \nu)\, I(\boldsymbol{x}, t; \boldsymbol{n}, \nu)\, \mathrm{d}l\, \mathrm{d}S\, \mathrm{d}\omega\, \mathrm{d}\nu\, \mathrm{d}t \tag{2.1}$$

in a time $\mathrm{d}t$. The opacity χ is obtained from the sum $\Sigma_i n_i \alpha_i$ where n_i is the number density of particle of type i which can absorb or scatter radiation at frequency ν with a cross-section $\alpha_i(\nu)$. The units of χ are $[\chi] = \mathrm{cm}^{-1}$. Thus, $\lambda_\nu \equiv 1/\chi_\nu$ is a *photon mean free path* at frequency ν.

If the material is at rest and is homogeneous on the microscale, then χ is *isotropic*. The same is true in the *comoving frame* of moving material. But in the laboratory frame the opacity of moving material is *anisotropic* because of the strong coupling of frequency to angles through the Doppler effect. Therefore in general $\chi = \chi(\boldsymbol{x}, t; \boldsymbol{n}, \nu)$. This coupled angle-frequency dependence is a tremendous complication in moving media, and one is highly motivated to find how to work in the comoving frame of the fluid where material properties are isotropic. This goal is accomplished by development of the Lagrangian transport equation, which we shall discuss later.

Emissivity. Define the emission coefficient η such that an element $\mathrm{d}l\, \mathrm{d}S$ of material emits an amount of energy

$$\delta\mathcal{E} = \eta(\boldsymbol{x}, t; \boldsymbol{n}, \nu) \mathrm{d}l\, \mathrm{d}S\, \mathrm{d}\omega\, \mathrm{d}\nu\, \mathrm{d}t \tag{2.2}$$

into solid angle $\mathrm{d}\omega$ in the frequency interval $(\nu, \nu + \mathrm{d}\nu)$ in a time $\mathrm{d}t$. The units of the emissivity are

$$[\eta] = \mathrm{ergs\ cm^{-3}\ s^{-1}\ hz^{-1}\ sr^{-1}}\ . \tag{2.3}$$

Again, the emissivity is isotropic in the rest frame of the fluid (the comoving frame), but in general will be anisotropic, *i.e.*, $\eta = \eta(\boldsymbol{x}, t; \boldsymbol{n}, \eta)$, in the lab frame for a moving fluid for the same reasons χ is.

Scattering. It is sometimes useful to distinguish between "true" or *thermal absorption-emission* processes and *scattering.* Thermal processes couple radiation to the thermal state of the material locally. For example, suppose an atom is excited from its ground state to an excited state by the absorption of a photon, and then is de-excited back to the ground state by a superelastic collision with a particle in the plasma such as an electron. In this case the original photon is *destroyed*, and the photon's energy is contributed to the thermal pool of the gas. Or suppose an atom is excited from its ground state to an excited state by an inelastic collision with a particle in the plasma, and then de-excited by the emission of a photon. In this case a photon is *created* at the expense of the thermal energy of the particles in the gas. In a scattering process a photon interacts with a scattering center, and emerges with an altered direction and/or frequency. In this case the emission rate is coupled to the ambient number of photons present, and not to the thermal state of the gas. An example is when a photon interacts with a free electron; for low energy photons the result is a conservative scattering process called *Thomson scattering*, but if the energy in the photon approaches the rest energy of the electron (X-rays), we have *Compton scattering* which admits large shifts in the photon energy. Another example is when an atom is excited from its ground state by a photon, sits briefly in the excited state, and then emits a photon in order to return to the ground state.

We will adopt the notation $\chi = \kappa+\sigma$, where κ is the absorption coefficient and σ is the scattering coefficient. For our purposes it will be adequate to consider only Thomson scattering, $\sigma = n_e\sigma_0$ which is frequency independent. Then in the comoving frame[2]

$$\chi_0(\boldsymbol{x},t;\nu_0) = \kappa_0(\boldsymbol{x},t;\nu_0) + \sigma_0(\boldsymbol{x},t) \,. \tag{2.4}$$

In a similar manner we can consider the emission in the comoving frame as consisting of thermal emission plus scattering

$$\eta(\boldsymbol{x},t;\nu_0) = \eta^t(\boldsymbol{x},t;\nu_0) + \eta^s(\boldsymbol{x},t;\nu_0). \tag{2.5}$$

Assume the scattering process is *conservative* (i.e., energy in = energy out), *coherent* ($\nu_{\rm in} = \nu_{\rm out}$), and *isotropic*, in the *comoving frame*. Then

$$\begin{aligned} \delta\mathcal{E} = \oint \sigma_0(\boldsymbol{x},t) I_0(\boldsymbol{x},t;\boldsymbol{n}_0,\nu_0)\mathrm{d}\omega &= 4\pi\sigma_0(\boldsymbol{x},t) J_0(\boldsymbol{x},t;\nu_0) \\ &= 4\pi\eta_0^s(\boldsymbol{x},t;\nu_0) \,. \end{aligned} \tag{2.6}$$

The last equality follows because we said the process is conservative. Therefore,

$$\eta^s(\boldsymbol{x},t;\nu_0) = \sigma_0(\boldsymbol{x},t) J_0(\boldsymbol{x},t;\nu_0) \,. \tag{2.7}$$

Thus, we see explicitly that η^s depends on J. This fact is important because J at $(\boldsymbol{x},t)$ may be determined by sources and sinks *elsewhere* and not at $(\boldsymbol{x},t)$. Therefore η^s is *decoupled* from local conditions. One can show that photons can scatter over large distances, through *many* mean free paths.

Kirchhoff–Planck Relation. In the important case of thermal equilibrium we must have

$$(\eta_\nu^t)^* = \kappa_\nu^* B_\nu(T) \,. \tag{2.8}$$

This relationship is often extended to the idea of *local thermodynamic equilibrium (LTE)*. Then, in the comoving frame

[2] The subscript 0 designates quantities measured in the comoving frame.

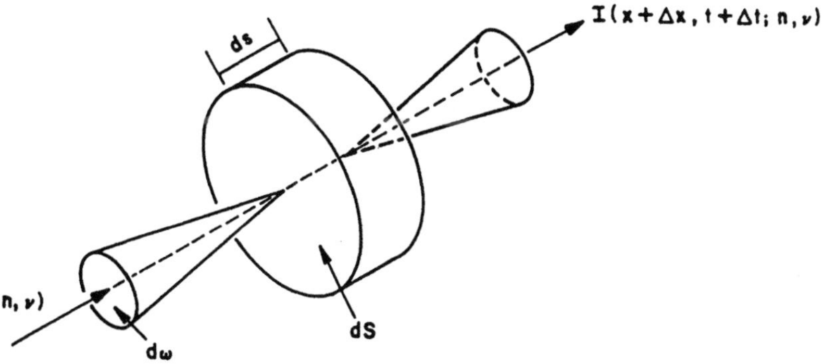

Fig. 2.1. Pencil of radiation passing through a material element

$$\eta_0^t(\boldsymbol{x},t;\nu_0) = \kappa_0^*(\boldsymbol{x},t;\nu_0) B[T(\boldsymbol{x},t),\nu_0] \, . \tag{2.9}$$

This result will be true when spatial gradients of material properties are small, *i.e.*, properties vary on scales $\ell \gg \lambda_{\mathrm{p}}$ (e.g., in a stellar interior where $| \nabla T | \sim 10^{-4}$). The Kirchhoff–Planck relation breaks down near boundary surfaces. In most radiation hydro problems we have little choice but to assume LTE (and the problem is hard even then!).

Opacity and Emission Coefficients. Accounting for bound-bound $(i\,j)$, bound-free $(i\,k)$, free-free $(k\,k)$, and scattering processes we have (*in* LTE)

$$\begin{aligned}\chi_0^*(\boldsymbol{x},t;\nu_0) &= \left\{ \sum_i n_i^* \left[\alpha_{ik}(\nu_0) + \sum_{j>i} \alpha_{ij}(\nu_0) \right] \right. \\ &\quad \left. + \sum_k n_{\mathrm{e}} n_k \alpha_{kk}(\nu_0,T) \right\} (1 - \mathrm{e}^{-h\nu/kT}) + n_{\mathrm{e}}\sigma_{\mathrm{e}} \\ &= \kappa_0^* + n_{\mathrm{e}}\sigma_{\mathrm{e}} \, . \end{aligned} \tag{2.10}$$

2.2 The Equation of Transfer

Derivation. In an element of material of length $\mathrm{d}s$, the difference between the energy that enters one end at $(\boldsymbol{x},t)$ and emerges from the other end at $(\boldsymbol{x}+\Delta\boldsymbol{x}, t+\Delta t)$ results from absorptions and emissions (see Fig. 2.1). Thus, in Cartesian coordinates

$$\begin{aligned}&[I(\boldsymbol{x}+\Delta\boldsymbol{x}, t+\Delta t;\boldsymbol{n},\nu) - I(\boldsymbol{x},t;\boldsymbol{n},\nu)]\, \mathrm{d}S\, \mathrm{d}\omega\, \mathrm{d}\nu\, \mathrm{d}t \\ &\quad = [\eta(\boldsymbol{x},t;\boldsymbol{n},\nu) - \chi(\boldsymbol{x},t;\boldsymbol{n},\nu)\, I(\boldsymbol{x},t;\boldsymbol{n},\nu)]\, \mathrm{d}s\, \mathrm{d}S\, \mathrm{d}\omega\, \mathrm{d}\nu\, \mathrm{d}t \, . \end{aligned} \tag{2.11}$$

Expanding I in a Taylor's series and noticing that $\Delta t = \mathrm{d}s/c$ we get

$$I(\boldsymbol{x} + \Delta\boldsymbol{x}, t + \Delta t; \boldsymbol{n}, \nu) = I(\boldsymbol{x}, t; \boldsymbol{n}, \nu) + \left(\frac{1}{c}\frac{\partial I}{\partial t} + \frac{\partial I}{\partial s}\right) \mathrm{d}s \,. \tag{2.12}$$

Using (2.12) in (2.11) we get the *transfer equation*

$$\left(\frac{1}{c}\frac{\partial}{\partial t} + \frac{\partial}{\partial s}\right) I(\boldsymbol{x}, t; \boldsymbol{n}, \nu) = \eta(\boldsymbol{x}, t; \boldsymbol{n}, \nu) - \chi(\boldsymbol{x}, t; \boldsymbol{n}, \nu)\, I(\boldsymbol{x}, t; \boldsymbol{n}, \nu) \,. \tag{2.13}$$

This is a purely phenomenological derivation. Approximations made here are discussed in *Equations of Radiation Hydrodynamics* by Pomraning [27, pp. 47–49]. For a quantum electrodynamical derivation see [15, 16, 17].

Because s is a coordinate-independent pathlength, the above equation applies in arbitrary coordinate systems, provided we use an appropriate expression in the evaluation of $(\partial/\partial s)$. For example, in Cartesian coordinates

$$\frac{\partial}{\partial s} = \frac{\partial x}{\partial s}\frac{\partial}{\partial x} + \frac{\partial y}{\partial s}\frac{\partial}{\partial y} + \frac{\partial z}{\partial s}\frac{\partial}{\partial z} = n_x\frac{\partial}{\partial x} + n_y\frac{\partial}{\partial y} + n_z\frac{\partial}{\partial z} = \boldsymbol{n} \cdot \boldsymbol{\nabla} \,. \tag{2.14}$$

Therefore in Cartesians we have

$$\begin{aligned} \left(\frac{1}{c}\frac{\partial}{\partial t} + \boldsymbol{n} \cdot \boldsymbol{\nabla}\right) I(\boldsymbol{x}, t; \boldsymbol{n}, \nu) \quad &= \quad \eta(\boldsymbol{x}, t; \boldsymbol{n}, \nu) \\ &\quad - \chi(\boldsymbol{x}, t; \boldsymbol{n}, \nu)\, I(\boldsymbol{x}, t; \boldsymbol{n}, \nu) \,. \end{aligned} \tag{2.15}$$

In one-dimensional planar media (2.15) reduces to

$$\frac{1}{c}\frac{\partial I}{\partial t} + \mu\frac{\partial I}{\partial z} = \eta(z, t; \mu, \nu) - \chi(z, t; \mu, \nu)\, I(z, t; \mu, \nu) \,. \tag{2.16}$$

For static media ($\boldsymbol{v} = 0$) or steady flow, we can drop $(\partial/\partial t)$ to obtain the "standard" one-dimensional transfer equation

$$\mu\frac{\partial I}{\partial z} = \eta(z, t; \mu, \nu) - \chi(z, t; \mu, \nu)\, I(z, t; \mu, \nu) \,. \tag{2.17}$$

Suppose η and χ are given. Then (2.17) is an ODE, and (2.16) is a PDE. Different techniques are needed to solve these two classes of equation. Suppose η contains a scattering term; then on the right-hand side we have a term containing $J = \int I(\mu)\mathrm{d}\mu$. Then the two equations just mentioned become (ordinary or partial) ***integrodifferential equations***, which are much harder to solve.

In curvilinear coordinates the basis vectors rotate with respect to a straight-line path determined by a fixed propagation vector $\boldsymbol{n}$. Therefore we must allow for changes in the components of $\boldsymbol{n}$ measured relative to those basis vectors. Thus,

$$\frac{\partial}{\partial s} \rightarrow \boldsymbol{n} \cdot \boldsymbol{\nabla} + \left(\frac{\mathrm{d}\boldsymbol{n}}{\mathrm{d}s}\right) \cdot \boldsymbol{\nabla}_{\boldsymbol{n}} \,. \tag{2.18}$$

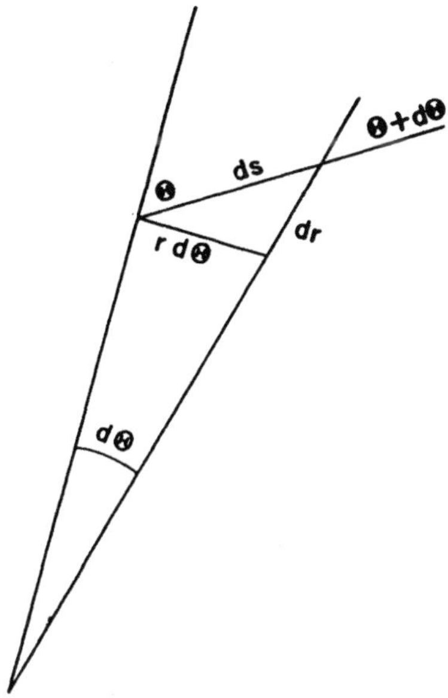

Fig. 2.2. Photon propagation angle in spherical symmetry

For example, in general spherical coordinates we will have derivatives in the space coordinates $\partial/\partial r, \partial/\partial\theta$, and $\partial/\partial\phi$, and the angles that specify the direction of the propagation vector $\boldsymbol{n}$, $\partial/\partial\Theta$ and $\partial/\partial\Phi$. In the special case of spherical symmetry, $\partial/\partial\theta = \partial/\partial\phi = \partial/\partial\Phi \equiv 0$. Furthermore, $\mathrm{d}r = \cos\Theta \mathrm{d}s = \mu \mathrm{d}s$, and $r\mathrm{d}\Theta = -\sin\Theta \mathrm{d}s = -(1-\mu^2)^{\frac{1}{2}}$ (see Fig. 2.2), hence

$$\frac{\partial}{\partial s} = \cos\Theta \frac{\partial}{\partial r} - \frac{\sin\Theta}{r}\frac{\partial}{\partial\Theta} = \mu\frac{\partial}{\partial r} + \frac{(1-\mu^2)}{r}\frac{\partial}{\partial\mu}\,, \tag{2.19}$$

which implies the transfer equation is

$$\frac{1}{c}\frac{\partial I}{\partial t} + \mu\frac{\partial I}{\partial r} + \frac{(1-\mu^2)}{r}\frac{\partial I}{\partial\mu} = \eta - \chi I\,. \tag{2.20}$$

Exercise: derive the transfer equation in *general* spherical coordinates.

Optical Depth. Define the *optical depth* as

$$\tau_\nu(\boldsymbol{x}, \boldsymbol{x}') \equiv \int_0^\ell \chi(\boldsymbol{x} + \boldsymbol{n}s; \boldsymbol{n}, \nu)\mathrm{d}s\,. \tag{2.21}$$

Physically, τ gives the number of photon mean free paths along the ray. In planar geometry it is customary to use the the *vertical* or *radial* optical depth, which is the number of mean free paths downward into the medium as seen by an outside observer. Thus,

$$\mathrm{d}\tau_\nu = -\chi_\nu \mathrm{d}z\,, \tag{2.22}$$

or

$$\tau_\nu(z) = \int_z^{z_{max}} \chi_\nu(z')\mathrm{d}z' \ . \tag{2.23}$$

Along a ray propagating such that $\boldsymbol{n} \cdot \boldsymbol{k} = \mu$, the *slant* optical depth is τ_ν/μ.

Source Function. The *source function* is defined as the ratio of emissivity to opacity:

$$S(\boldsymbol{x}, t; \boldsymbol{n}, \nu) \equiv \frac{\eta(\boldsymbol{x}, t; \boldsymbol{n}, \nu)}{\chi(\boldsymbol{x}, t; \boldsymbol{n}, \nu)} \ . \tag{2.24}$$

Then in static (or steady-flow) media we can write the transfer equation in the customary form

$$\mu \frac{\partial I_{\mu\nu}}{\partial \tau_{\mu\nu}} = I_{\mu\nu} - S_{\mu\nu} \ . \tag{2.25}$$

There are many ways of integrating this equation (and its generalizations). For certain special cases we can write explicit expressions for the source function. A few examples, in the *comoving* frame of the fluid are:
Strict LTE:

$$S(\nu_0) = B(\nu_0, T) \ . \tag{2.26}$$

Thermal emission/absorption plus isotropic coherent scattering:

$$\chi_0(\nu_0) = \kappa_0(\nu_0) + \sigma_0(\nu_0) \ , \tag{2.27}$$

$$\eta_0(\nu_0) = \kappa_0(\nu_0)B(\nu_0, T) + \sigma_0(\nu_0)J_0(\nu_0) \ , \tag{2.28}$$

$$S(\nu_0) = \frac{\kappa_0(\nu_0)B(\nu_0, T) + \sigma_0(\nu_0)J_0(\nu_0)}{\kappa_0(\nu_0) + \sigma_0(\nu_0)} \ . \tag{2.29}$$

2.3 Moments of the Transfer Equation

For simplicity work in Cartesian coordinates. The transfer equation is

$$\frac{1}{c}\frac{\partial I(\boldsymbol{x}, t; \boldsymbol{n}, \nu)}{\partial t} + n^i \frac{\partial I(\boldsymbol{x}, t; \boldsymbol{n}, \nu)}{\partial x^i} = \eta(\boldsymbol{x}, t; \boldsymbol{n}, \nu) - \chi(\boldsymbol{x}, t; \boldsymbol{n}, \nu)\, I(\boldsymbol{x}, t; \boldsymbol{n}, \nu) \ . \tag{2.30}$$

Take the zeroth moment of (2.30) over angles, i.e., multiply by $\mathrm{d}\omega/4\pi$ and integrate over solid angles. We find

$$\frac{1}{c}\frac{\partial J(\boldsymbol{x}, t; \nu)}{\partial t} + \boldsymbol{\nabla} \cdot \boldsymbol{H}(\boldsymbol{x}, t; \nu) = \frac{1}{4\pi} \oint [\eta(\boldsymbol{x}, t; \boldsymbol{n}, \nu) - \chi(\boldsymbol{x}, t; \boldsymbol{n}, \nu)\, I(\boldsymbol{x}, t; \boldsymbol{n}, \nu)]\, \mathrm{d}\omega \tag{2.31}$$

or, converting to physical units, we get the *monochromatic radiation energy equation*

$$\frac{\partial E_\nu}{\partial t} + \boldsymbol{\nabla} \cdot \boldsymbol{F}_\nu = \oint [\eta(\boldsymbol{x}, t; \boldsymbol{n}, \nu) - \chi(\boldsymbol{x}, t; \boldsymbol{n}, \nu)\, I(\boldsymbol{x}, t; \boldsymbol{n}, \nu)]\, \mathrm{d}\omega\,. \tag{2.32}$$

Now integrate over all frequencies to obtain the *radiation energy equation*

$$\frac{\partial E}{\partial t} + \boldsymbol{\nabla} \cdot \boldsymbol{F} = \int_0^\infty \mathrm{d}\nu \oint [\eta(\boldsymbol{x}, t; \boldsymbol{n}, \nu) - \chi(\boldsymbol{x}, t; \boldsymbol{n}, \nu)\, I(\boldsymbol{x}, t; \boldsymbol{n}, \nu)]\, \mathrm{d}\omega\,. \tag{2.33}$$

This equation states

Rate of change of the radiation energy density in a fixed volume
= the rate of emission from the material
− the rate energy is absorbed by material
− the net flow of radiant energy through the boundary surface of the volume .

Note that Eq. (2.33) is in tensor form so we can use it in other coordinate systems. For example, in spherical symmetry

$$\frac{\partial E}{\partial t} + \frac{1}{r^2}\frac{\partial (r^2 F)}{\partial r} = 2\pi \int_0^\infty \mathrm{d}\nu \int_{-1}^{1} [\eta(\boldsymbol{x}, t; \boldsymbol{n}, \nu) - \chi(\boldsymbol{x}, t; \boldsymbol{n}, \nu)\, I(\boldsymbol{x}, t; \boldsymbol{n}, \nu)]\, \mathrm{d}\mu\,. \tag{2.34}$$

If we define the *luminosity* to be $L(r,t) = 4\pi r^2 F(r,t)$ then we can write (2.34) as

$$\frac{\partial E}{\partial t} + \frac{1}{4\pi r^2}\frac{\partial L}{\partial r} = 2\pi \int_0^\infty \mathrm{d}\nu \int_{-1}^{1} [\eta(\boldsymbol{x}, t; \boldsymbol{n}, \nu) - \chi(\boldsymbol{x}, t; \boldsymbol{n}, \nu)\, I(\boldsymbol{x}, t; \boldsymbol{n}, \nu)]\, \mathrm{d}\mu\,. \tag{2.35}$$

In the limiting case of a *static* medium we have $\partial/\partial t \equiv 0$. Furthermore, in order to achive a steady state the material must emit exactly as much energy as it absorbs. Therefore

$$\int_0^\infty (\eta - \chi I)\mathrm{d}\nu \equiv 0\,, \tag{2.36}$$

which implies that $\boldsymbol{\nabla} \cdot \boldsymbol{F} \equiv 0$. This is the condition of *radiative equilibrium.* Note that (except for highly artificial situations) radiative equilibrium can occur *only* in absolutely static media. The instant the material moves or evolves in time, other terms come into play. This will become clearer later when we study moments of the Lagrangian transfer equation.

Now take the first angular moment of the transfer equation. Multiply by $\boldsymbol{n} d\omega/4\pi$. We find

$$\frac{1}{c}\frac{\partial H_\nu^i}{\partial t}+\frac{\partial K_\nu^{ij}}{\partial x^j}=\frac{1}{4\pi}\oint[\eta(\boldsymbol{x},t;\boldsymbol{n},\nu)-\chi(\boldsymbol{x},t;\boldsymbol{n},\nu)\,I(\boldsymbol{x},t;\boldsymbol{n},\nu)]n^i d\omega \quad (2.37)$$

or, converting to physical units, we get the *monochromatic radiation momentum equation*

$$\frac{1}{c^2}\frac{\partial \boldsymbol{F}_\nu}{\partial t}+\boldsymbol{\nabla}\cdot\mathsf{P}_\nu=\frac{1}{c}\oint[\eta(\boldsymbol{x},t;\boldsymbol{n},\nu)-\chi(\boldsymbol{x},t;\boldsymbol{n},\nu)\,I(\boldsymbol{x},t;\boldsymbol{n},\nu)]\boldsymbol{n} d\omega\ . \quad (2.38)$$

Integrating over frequency we get the *radiation momentum equation*

$$\frac{1}{c^2}\frac{\partial \boldsymbol{F}}{\partial t}+\boldsymbol{\nabla}\cdot\mathsf{P} =\frac{1}{c}\int_0^\infty d\nu\oint[\eta(\boldsymbol{x},t;\boldsymbol{n},\nu)-\chi(\boldsymbol{x},t;\boldsymbol{n},\nu)\,I(\boldsymbol{x},t;\boldsymbol{n},\nu)]\boldsymbol{n}\, d\omega\ . \quad (2.39)$$

Recall that $\boldsymbol{F}/c^2$ equals the radiation momentum density, P is the radiation momentum flux density, and on the right-hand side one can write $(\eta/h\nu)\times(h\nu/c)\boldsymbol{n}$ = (number of photons) × (momentum/photon) emitted. And similarly for the number absorbed. Thus,

Rate of change of the radiation momentum density
in a fixed volume
= the rate of momentum input to the radiation field
from the material
− the rate of momentum removal from the radiation field
by the material
− the net rate of radiant momentum through
the boundary surface of the volume .

Note in passing that

$$\boldsymbol{f}_{\mathrm{R}}=\frac{1}{c}\int_0^\infty d\nu\oint\chi(\boldsymbol{x},t;\boldsymbol{n},\nu)\,I(\boldsymbol{x},t;\boldsymbol{n},\nu)\boldsymbol{n}\, d\omega \quad (2.40)$$

is the *radiation force* on the material, per unit volume.

Again, Eq. (2.39) is in tensorial form, hence it can be used in any coordinate system, e.g., in spherical symmetry:

$$\frac{1}{c^2}\frac{\partial \boldsymbol{F}}{\partial t}+\frac{\partial P}{\partial r}+\frac{3P-E}{r} =\frac{2\pi}{c}\int_0^\infty d\nu\int_{-1}^{1}[\eta(\boldsymbol{x},t;\boldsymbol{n},\nu)-\chi(\boldsymbol{x},t;\boldsymbol{n},\nu)\,I(\boldsymbol{x},t;\boldsymbol{n},\nu)]\mu\, d\mu\ . \quad (2.41)$$

Static Moment Equations. In a *static medium*, we can take $v \equiv 0$, and $\partial/\partial t \equiv 0$. Further if we assume planar geometry, we get the standard moment equations used in the theory of stellar atmospheres:

$$\frac{\partial F_\nu}{\partial z} = 4\pi\eta_\nu - c\chi_\nu E_\nu \tag{2.42}$$

and

$$\frac{\partial P_\nu}{\partial z} = \frac{\partial(f_\nu E_\nu)}{\partial z} = 0 - \frac{1}{c}\chi_\nu F_\nu \,. \tag{2.43}$$

Here we have noted that the *net* contribution to the momentum balance by isotropic emission is exactly zero.

The Closure Problem. In solving problems of radiation hydrodynamics we undertake to calculate the evolution of the *total* (*i.e.*, material plus radiative) energy and momentum density in the flow. The radiative contributions to these quantities follow from the radiation energy and momentum equations. In order to close the system of radiation momentum equations we must know the Eddington factor f_ν (or, in general, the Eddington tensor f_ν which allows to express P_ν in terms of E_ν.

We must therefore be able, somehow, to *calculate* f_ν (or the components of f_ν). For ordinary material fluids we can either impose a closure by assuming isotropy of the microscopic velocity distribution (equilibrium flow) or calculate the departure from isotropy *analytically* (*e.g.*, the Chapman–Enskog solution) because $\lambda/\ell \ll 1$. The same is true for radiation when $\lambda_\mathrm{p}/\ell \ll 1$ (that is, in an opaque medium such as in the interior of a star). In this case we can develop the *diffusion limit* solution for radiation, which is similar to the Chapman–Enskog solution for matter. But when $\lambda_\mathrm{p}/\ell \geq 1$, *i.e.*, near a boundary surface or in a very transparent medium, *we must compute the closure from a full angle (and frequency, if relevant) dependent solution of the equation of transfer.* We shall not have time here to discuss this important problem other than to say that methods for making this closure exist.

Notice that the fluid flow responds only to the *total* (*i.e.*, frequency-integrated as opposed to monochromatic) radiative momentum and energy input into the material. Thus, one usually replaces the detailed frequency-variation of the opacity of the material with appropriate frequency averages over the spectrum. Again, we shall not have time here to discuss these issues. Suffice it to say that in the past 5–10 years very impressive advances in our knowledge of the opacity of stellar material have been made.

3. Lorentz Transformation of the Transfer Equation

Let us now begin the study of the behavior of radiation in *fluid flows*, *i.e.*, in moving media. The motivation for this study is that radiation can dominate both the energy *content* and the energy *transport* in a radiating fluid. We can get an intuitive feeling for when this happens by use of a couple of dimensionless numbers.

Energy Content. Define R to be

$$R = \frac{\text{energy density in the material}}{\text{energy density in the radiation}} = \frac{\hat{e}}{E} = \frac{\frac{3}{2}NkT}{a_R T^4} = \frac{0.028N}{T^3} . \tag{3.1}$$

One can easily show that $R \approx 1$, so that $\hat{e} \approx E$, when $T_{\text{KeV}} \approx 2\varrho^{1/3}$, where the density ϱ is measured in g cm^{-3} and the temperature corresponding to 1 KeV is approximately 10^7 K. For example:

> Center of the Sun, $\varrho \sim 100$, $T \sim 10^7$, and $R \approx 500$, so that radiation is *unimportant.*
> Center of an O-star, $\varrho \sim 3$, $T \sim 4 \times 10^7$, and $R \approx 1$, so that radiation is *important.*

Energy Transport. The importance of radiant energy transport is measured by the *Boltzmann number*

$$\text{Bo} = \frac{\varrho c_p T v}{\sigma_R T^4} \sim \frac{\text{material enthalpy flux}}{\text{radiation flux}} . \tag{3.2}$$

Bo is essentially $(v/c) \times R$ as defined above. Usually $(v/c) \ll 1$, therefore the radiation almost always dominates the energy transport (counterexample: convection in a star). In optically thick media one should replace $\sigma_R T^4$ in (3.2) with $\sigma_R T_{\text{eff}}^4$ where T_{eff} gives a measure of the net radiant flux in the medium.

In the analysis of the radiation equations one must carefully account for terms that are $O(v/c)$. These are *required* for consistency between frames (Eulerian $\leftrightarrow$ Lagrangian). In some cases these seemingly small terms can dominate all others in the equations! [Remark: the frame dependence of material terms is $O(v^2/c^2)$ while for radiation it is $O(v/c)$. Clearly the effect is much larger for radiation.] *References*: [31, 4].

3.1 Lorentz Transformation of the Photon 4-Momentum

The photon 4-momentum is

$$M^\alpha = \left(\frac{h\nu}{c}\right)(1, \boldsymbol{n}) , \tag{3.3}$$

where ν and $\boldsymbol{n}$ are the photon's frequency and direction of propagation in the laboratory frame. If we boost (Lorentz transform) to another frame moving with velocity $\boldsymbol{v}$ with respect to the laboratory frame, and in particular choose this frame to be the comoving frame of the fluid, then applying a Lorentz transformation to M^α we find:

$$\nu_0 = \gamma\nu\left(1 - \frac{\boldsymbol{n}\cdot\boldsymbol{v}}{c}\right) , \tag{3.4}$$

$$\boldsymbol{n}_0 = \left(\frac{\nu}{\nu_0}\right)\left\{\boldsymbol{n} - \gamma\left(\frac{\boldsymbol{v}}{c}\right)\left[1 - \frac{(\gamma\boldsymbol{n}\cdot\boldsymbol{v}/c)}{\gamma+1}\right]\right\} , \tag{3.5}$$

where $\gamma = 1/\sqrt{1-v^2/c^2}$. These four equations, when expanded to $O(v/c)$ yield the effects described classically as "aberration" and "Doppler shift". For one-dimensional motion along the z-axis, equations (3.4) and (3.5) simplify to:

$$(\nu_0, \nu_0 n_{x0}, \nu_0 n_{y0}, \nu_0 n_{z0}) = [\gamma\nu(1 - n_z\beta), \nu n_x, \nu n_y, \gamma\nu(n_z - \beta)] , \tag{3.6}$$

where $\beta = v/c$. Note that the photon direction in the (x, y) plane is unchanged. We can rewrite the above expressions in planar or spherical geometry as:

$$\left[\nu_0, \mu_0, (1-\mu_0^2)^{1/2}, \Phi_0\right] = \left[\gamma\nu(1-\beta\mu), \frac{(\mu-\beta)}{(1-\beta\mu)}, \frac{(1-\mu^2)^{1/2}}{\gamma(1-\beta\mu)}, \Phi\right] . \tag{3.7}$$

We can obtain the inverse transformations from the comoving frame to the laboratory frame by interchanging ν (*etc.*) with ν_0 (*etc.*) and also changing $\boldsymbol{v} \to -\boldsymbol{v}$. From the above general transformations we can easily show that

$$\mathrm{d}\nu_0 = \left(\frac{\nu_0}{\nu}\right)\mathrm{d}\nu , \tag{3.8}$$

$$\mathrm{d}\mu_0 = \left(\frac{\nu}{\nu_0}\right)^2 \mathrm{d}\mu , \tag{3.9}$$

and

$$\mathrm{d}\Phi_0 = \mathrm{d}\Phi . \tag{3.10}$$

Recall that $\mathrm{d}\omega = \mathrm{d}\mu\mathrm{d}\Phi$. Then from (3.8–10) we can show that the quantity $\nu\,\mathrm{d}\nu\,\mathrm{d}\omega$ is a *Lorentz invariant*, i.e.,

$$\nu\,\mathrm{d}\nu\,\mathrm{d}\omega = \nu_0\,\mathrm{d}\nu_0\,\mathrm{d}\omega_0 , \tag{3.11}$$

a result which we will use repeatedly. The invariance of $\nu\,\mathrm{d}\nu\,\mathrm{d}\omega$ is actually a special case, for photons, of the general result that for any kind of particle

$$\frac{\mathrm{d}^3 p}{\tilde{e}} = \frac{p^2\mathrm{d}p\,\mathrm{d}\omega}{\tilde{e}} \tag{3.12}$$

is a Lorentz invariant, where $\tilde{e}$ = energy per particle. This is a standard result of relativistic kinetic theory. For photons, $p = h\nu/c, \tilde{e} = h\nu$; then the connection between (3.11) and (3.12) is obvious.

3.2 Lorentz Transformation of the Specific Intensity, Opacity, and Emissivity

A very simple and elegant way of obtaining the Lorentz transformation properties of the physical quantities in the macroscopic transfer equation was devised by [31], who used thought-experiments based on *counting photons*, which is invariant between frames, to connect comoving-frame and lab-frame quantities.

Specific Intensity. Have an observer in the lab frame count the number of photons passing through a surface $\mathrm{d}S$ stationary in that frame. The result will be

$$N = \frac{I(\mu,\nu)}{h\nu}\mu\,\mathrm{d}S\;\mathrm{d}\nu\,\mathrm{d}\omega\,\mathrm{d}t\;. \tag{3.13}$$

Now put the observer on a fluid element moving with a velocity v to the right in the laboratory frame, and have him again count the number of photons passing through the *same* surface $\mathrm{d}S$. He now sees the surface moving to the left with a velocity v with respect to his frame. So he now finds

$$N_0 = \frac{I_0(\mu_0,\nu_0)}{h\nu_0}\mathrm{d}\nu_0\mathrm{d}\omega_0\left(\mu_0\mathrm{d}S\mathrm{d}t_0 + \frac{v}{c}\mathrm{d}S\mathrm{d}t_0\right)\;. \tag{3.14}$$

In (3.14) the first term is the number counted if $\mathrm{d}S$ were *stationary* in this frame, and the second terms is the photon number density $(I_0/h\nu_0 c)\times$ the volume $(v\,\mathrm{d}S\mathrm{d}t)$ swept up by the moving surface. But because we have counted the same photons in both frames, N must be identical to N_0. Furthermore, we know that $\mathrm{d}t_0 = \gamma\mathrm{d}t, \nu = \gamma\nu_0(1+\beta\mu_0), \nu\mathrm{d}\nu\mathrm{d}\omega = \nu_0\mathrm{d}\nu_0\mathrm{d}\omega_0$, and $\mu = (\mu_0+\beta)/(1+\beta\mu_0)$. Using these relationships to connect (3.13) with (3.14) we get $(I/\nu)\mu\,\mathrm{d}\nu\mathrm{d}\omega\mathrm{d}t = (I_0/\nu_0)\mathrm{d}\nu_0\mathrm{d}\omega_0(\mu_0+\beta)\gamma\mathrm{d}t = (I_0/\nu_0)\mathrm{d}\nu_0\mathrm{d}\omega_0\mu(\nu/\nu_0)\mathrm{d}t$ which implies

$$\frac{I(\mu,\nu)}{\nu^3} = \frac{I_0(\mu_0,\nu_0)}{\nu_0^3}\;. \tag{3.15}$$

Thus,

$$\mathcal{I}(\mu,\nu) \equiv \frac{I(\mu,\nu)}{\nu^3} \tag{3.16}$$

is a Lorentz invariant, known as the *invariant intensity.* Recalling that the photon distribution function $f_\mathrm{R} \propto I/\nu^3$, we see that we could have gotten the same result by invoking the Lorentz invariance of the distribution function.

Emissivity. Have an observer in the laboratory frame count the number of photons emitted from a definite element of material in $\mathrm{d}V$ into $\mathrm{d}\omega\mathrm{d}\nu$ in $\mathrm{d}t$. The result will be

$$N = \frac{\eta(\mu,\nu)\mathrm{d}\omega\mathrm{d}\nu\mathrm{d}V\mathrm{d}t}{h\nu} \,. \tag{3.17}$$

Now have the observer count again in the comoving frame of the fluid (the *same* material). Now the result will be

$$N_0 = \frac{\eta_0(\nu_0)\mathrm{d}\omega_0\mathrm{d}\nu_0\mathrm{d}V_0\mathrm{d}t_0}{h\nu_0} \,. \tag{3.18}$$

Here we have assumed isotropy of the emission coefficient in the fluid frame. Now recall that $\mathrm{d}V\mathrm{d}t$ is invariant, and $\nu\mathrm{d}\nu\mathrm{d}\omega$ is invariant. Therefore we have

$$\eta(\mu,\nu) = \left(\frac{\nu}{\nu_0}\right)^2 \eta_0(\nu_0) \,. \tag{3.19}$$

Thus,

$$\mathcal{E} \equiv \frac{\eta}{\nu^2} \tag{3.20}$$

is a Lorentz invariant, the ***invariant emissivity.***

Opacity. Now count the number of photons absorbed in a definite element of material. In the laboratory frame the result will be

$$N = \frac{\chi(\mu,\nu)I(\mu,\nu)}{h\nu}\mathrm{d}\omega\mathrm{d}\nu\mathrm{d}V\mathrm{d}t \tag{3.21}$$

and in the comoving frame we get

$$N_0 = \frac{\chi_0(\nu_0)I_0(\mu_0,\nu_0)}{h\nu_0}\mathrm{d}\omega_0\mathrm{d}\nu_0\mathrm{d}V_0\mathrm{d}t_0 \,. \tag{3.22}$$

Again demand $N \equiv N_0$, and use the invariants identified above, to get

$$\chi(\mu,\nu) = \left(\frac{\nu_0}{\nu}\right) \chi_0(\nu_0) \,. \tag{3.23}$$

Thus,

$$\mathcal{A} \equiv \nu\chi \tag{3.24}$$

is a Lorentz invariant, the ***invariant absorptivity.***

3.3 Lorentz Transformation of the Radiation Stress Energy Tensor

One can guess the form of R, the full radiation stress energy tensor by using the same technique that works for the material stress energy tensor M. Thus, we can demand that the space components give back the the radiation pressure tensor derived before. Then

$$R^{ij} \to P^{ij} = \int f_{\mathrm{R}} M^i c n^j \mathrm{d}^3 M \ , \tag{3.25}$$

which equals (number of photons/cm^3/phase volume)× (ith component of momentum) × (jth component of velocity) integrated over all space, which is correct.

Now for a photon, $cn^j = c^2\ (h\nu n^j/c)\ (1/h\nu) = c^2 M^j/\tilde{e}$. So try[3]

$$R^{\alpha\beta} = c^2 \int f_{\mathrm{R}} M^\alpha M^\beta \frac{\mathrm{d}^3 M}{\tilde{e}} \ . \tag{3.26}$$

By inspection we see that this choice has all the desired properties: It is manifestly a 4-tensor, being the outer product of two 4-vectors M^α and M^β, times the 4-invariants f_{R} and $\mathrm{d}^3 M/\tilde{e}$.

Now examine the other components of R:

$$\begin{aligned} R^{00} = \int f_{\mathrm{R}} h\nu \mathrm{d}^3 M \quad &= \quad \text{(number of photons)} \times \text{(energy/photon)} \\ &= \quad \text{radiation energy density} \ , \end{aligned} \tag{3.27}$$

$$R^{0i} = \frac{1}{c} \int f_{\mathrm{R}} h\nu c n^i \mathrm{d}^3 M = \frac{1}{c} \times \text{(radiation flux density)} \ , \tag{3.28}$$

$$\begin{aligned} R^{i0} \quad &= \quad c \int f_{\mathrm{R}} \frac{h\nu}{c} n^i \mathrm{d}^3 M \\ &= \quad c \times \text{(radiation momentum density in } i\text{th direction)} \ . \end{aligned} \tag{3.29}$$

Therefore we have exactly the right interpretations. One can, in fact, apply Eq. (3.26) to any particles, not just photons. In order to shift to the macroscopic picture, use the identities (see Sect. 1.3)

$$f_{\mathrm{R}} \equiv \frac{c^2}{h^4 \nu^3} I \tag{3.30}$$

and

$$\mathrm{d}^3 M \equiv \frac{h^3 \nu^2}{c^3} \mathrm{d}\nu\, \mathrm{d}\omega \ . \tag{3.31}$$

---

[3] We use Greek indices to denote components in the four dimensional space time, while Latin indices refer to components in the three dimensional Euclidian space.

Then

$$R^{\alpha\beta} = \frac{1}{c} \int_0^\infty \mathrm{d}\nu \oint \mathrm{d}\omega \, I(\boldsymbol{n}, \nu) \, n^\alpha n^\beta \, . \tag{3.32}$$

This tensor is also manifestly covariant: it is composed of the invariant $(I/\nu^3)\times$ the outer product of the photon 4-momentum with itself $\times$ the invariant $(\nu \, \mathrm{d}\nu \, \mathrm{d}\omega)$. Alternatively we can write

$$\mathsf{R} = \begin{pmatrix} E & \boldsymbol{F}/c \\ \boldsymbol{F}/c & \mathsf{P} \end{pmatrix} . \tag{3.33}$$

To express the radiation stress–energy tensor in the comoving frame, we merely append the subscript"0" to all radiation quantities ($I, \mu, \nu, \boldsymbol{n}, E, \boldsymbol{F}$, and P). To obtain the transformations laws for the components of R from one frame to another, boost according to the standard Lorentz transformation rules. In general 3D the results are rather complicated. But for 1D flows, they are simple:

$$E = \gamma^2 (E_0 + \frac{2\beta}{c} F_0 + \beta^2 P_0) \, , \tag{3.34}$$

$$F = \gamma^2 [(1 + \beta^2) F_0 + v E_0 + v P_0] \, , \tag{3.35}$$

$$P = \gamma^2 (P_0 + \frac{2\beta}{c} F_0 + \beta^2 E_0) \, . \tag{3.36}$$

To $O(v/c)$, useful for many applications,

$$E = E_0 + \frac{2\beta}{c} F_0 \, , \tag{3.37}$$

$$F = F_0 + v E_0 + v P_0 \, , \tag{3.38}$$

$$P = P_0 + \frac{2\beta}{c} F_0 \, . \tag{3.39}$$

As we shall see later, Eq. (3.36) has important consequences. It should be noted that we can apply standard Lorentz transformations only to angle-frequency integrated quantities, as only such quantities can truly be tensors on a 4 manifold $(\boldsymbol{r}, t)$. Transformations for E_ν, $\boldsymbol{F}_\nu$, and P_ν *can* also be derived, but are much more complicated; they also follow directly from the transformation laws for I/ν^3.

It is a worthwhile exercise to derive equations (3.37–39) by using the usual definitions of E, $\boldsymbol{F}$, and P and first-order transformations for (μ, ν). For example:

$$E = \frac{1}{c} \iint I \, \mathrm{d}\omega \, \mathrm{d}\nu \, , \tag{3.40}$$

$$I \mathrm{d}\nu \mathrm{d}\omega = \frac{I}{\nu} \nu \mathrm{d}\nu \mathrm{d}\omega = \frac{1}{\nu} \frac{\nu^3 I_0}{\nu_0^3} \nu_0 \mathrm{d}\nu_0 \mathrm{d}\omega = \left(\frac{\nu}{\nu_0}\right)^2 I_0 \mathrm{d}\nu_0 \mathrm{d}\omega_0 \, , \tag{3.41}$$

$$\frac{\nu}{\nu_0} \approx 1 + \beta\mu_0 \Rightarrow \left(\frac{\nu}{\nu_0}\right)^2 = 1 + 2\beta\mu_0 \,, \tag{3.42}$$

therefore,

$$E = \frac{1}{c}\int\int I_0(1 + 2\beta\mu_0)\mathrm{d}\nu_0\mathrm{d}\omega_0 = E_0 + \frac{2\beta}{c}F_0 \tag{3.43}$$

and similarly for $\boldsymbol{F}$ and P.

3.4 The Radiation 4-Force Density Vector

We expect the dynamical equations for the radiation field, as for matter, to have the general form

$$R^{\alpha\beta}_{;\beta} = -G^\alpha \,, \tag{3.44}$$

where G^α is the radiation force density acting on the material. Thus, G^0 must be $\frac{1}{c}\times$ (the net rate of radiative energy input to matter), while G^i must be the (net rate of radiative momentum input into the matter). Therefore

$$G^0 = \frac{1}{c}\int_0^\infty \mathrm{d}\nu \oint \mathrm{d}\omega[\chi(\boldsymbol{n},\nu)I(\boldsymbol{n},\nu) - \eta(\boldsymbol{n},\nu)] \tag{3.45}$$

and

$$G^i = \frac{1}{c}\int_0^\infty \mathrm{d}\nu \oint \mathrm{d}\omega[\chi(\boldsymbol{n},\nu)I(\boldsymbol{n},\nu) - \eta(\boldsymbol{n},\nu)]\, n^i \,. \tag{3.46}$$

The above vector is manifestly covariant: it is the 4-vector $\nu(1,\boldsymbol{n})$ times the invariants $\nu\mathrm{d}\nu\mathrm{d}\omega, \chi I/\nu^2$, and η/ν^2 summed over phase space.

With this choice for G^α and our previous expression for $R^{\alpha\beta}$, (3.44) is indeed a covariant conservation relation for the radiation field.[4] Transformation relations for G^α are obtained by boosting. These are simple for one-dimensional flow:

$$\begin{pmatrix} G^0 \\ G^1 \end{pmatrix} = \gamma \begin{pmatrix} G^0_0 + \beta G^1_0 \\ G^1_0 + \beta G^0_0 \end{pmatrix} \quad \text{or} \quad \begin{pmatrix} G^0_0 \\ G^1_0 \end{pmatrix} = \gamma \begin{pmatrix} G^0 - \beta G^1 \\ G^1 - \beta G^0 \end{pmatrix} . \tag{3.47}$$

Here, in the laboratory frame,

$$G^0 = \frac{2\pi}{c}\int_0^\infty \mathrm{d}\nu \int_{-1}^1 \mathrm{d}\mu[\chi(\mu,\nu)I(\mu,\nu) - \eta(\mu,\nu)] \,, \tag{3.48}$$

$$G^1 = \frac{2\pi}{c}\int_0^\infty \mathrm{d}\nu \int_{-1}^1 \mathrm{d}\mu\, \mu[\chi(\mu,\nu)I(\mu,\nu) - \eta(\mu,\nu)] \,. \tag{3.49}$$

[4] A relation (physical law) that preserves its mathematical form under a transformation between inertial frames is called "covariant".

In this frame the double integrals are irreducible because of the anisotropy of the material coefficients. However, in the comoving frame

$$G_0^0 = \frac{1}{c}\int_0^\infty [c\chi_0(\nu_0)E_0(\nu_0) - 4\pi\eta_0(\nu_0)]\mathrm{d}\nu_0 \ , \tag{3.50}$$

$$G_0^1 = \frac{1}{c}\int_0^\infty \chi_0(\nu_0)F_0(\nu_0)\mathrm{d}\nu_0 \ . \tag{3.51}$$

In the comoving frame we have single integrals because the material coefficients are isotropic in this frame. Isotropy of photon emissions in the comoving frame also implies the net momentum transfer from material to radiation is exactly zero, hence there is no term from the emission coefficient in (3.51).

3.5 Covariant Form of the Transfer Equation

Is the transfer equation covariant, or not? This question has led to much confusion in the physical and astrophysical literature. Here we shall just sketch some of the approaches that have been made to answering it, most of which have serious deficiencies, leaving a definitive analysis to the next section. One very basic point is to remember that the specific intensity is a seven-dimensional distribution function, not a 4-vector or 4-tensor. Thus, we cannot expect simple Lorentz transformation to apply, but must account for the behavior of the *tangent spaces* (i.e., momentum phase space) attached to each point of the four-dimensional spacetime manifold.

Consider transformation from one inertial frame (say the laboratory frame, at rest) to another inertial (*i.e.*, *uniformly* moving) frame.

1) We can derive the transformation properties of a *4-gradient*, $(\partial/c\,\partial t, \partial/\partial x, \partial/\partial y, \partial/\partial z)$, by expanding with a chain rule:

$$\frac{\partial}{\partial t} \to \frac{\partial t'}{\partial t}\frac{\partial}{\partial t'} + \frac{\partial z'}{\partial t}\frac{\partial}{\partial z'} \ , \tag{3.52}$$

$$\frac{\partial}{\partial z} \to \frac{\partial t'}{\partial z}\frac{\partial}{\partial t'} + \frac{\partial z'}{\partial z}\frac{\partial}{\partial z'} \ , \tag{3.53}$$

and using the Lorentz transformation to evaluate $\partial\alpha/\partial t$ and $\partial\alpha/\partial z$, where $\alpha = t'$ or z'. We obtain

$$\left(\frac{1}{c}\frac{\partial}{\partial t}, \frac{\partial}{\partial x}, \frac{\partial}{\partial y}, \frac{\partial}{\partial z}\right) = \left[\gamma\left(\frac{1}{c}\frac{\partial}{\partial t'} - \beta\frac{\partial}{\partial z'}\right), \frac{\partial}{\partial x'}, \frac{\partial}{\partial y'}, \gamma\left(\frac{\partial}{\partial z'} - \frac{\beta}{c}\frac{\partial}{\partial t'}\right)\right] \ . \tag{3.54}$$

Rearranging, we find

$$\frac{1}{c}\frac{\partial}{\partial t} + \boldsymbol{n}\cdot\boldsymbol{\nabla} = \left(\frac{\nu'}{\nu}\right)\left[\frac{1}{c}\frac{\partial}{\partial t'} + \boldsymbol{n}'\cdot\boldsymbol{\nabla}'\right] , \tag{3.55}$$

where we used the results obtained earlier for the transformation of ν and $\boldsymbol{n}$. Therefore, $(1/c)(\partial I/\partial t) + \boldsymbol{n}\cdot\boldsymbol{\nabla}I = \eta - \chi I$ transforms to

$$\left(\frac{\nu'}{\nu}\right)\left(\frac{1}{c}\frac{\partial}{\partial t'} + \boldsymbol{n}'\cdot\boldsymbol{\nabla}'\right)\left[\left(\frac{\nu}{\nu'}\right)^3 I'(\boldsymbol{n}',\nu')\right]$$
$$= \left(\frac{\nu}{\nu'}\right)^2\eta' - \left(\frac{\nu'\chi'}{\nu}\right)\left(\frac{\nu}{\nu'}\right)^3 I' . \tag{3.56}$$

Now between two *uniformly* moving frames the ratio (ν/ν') is independent of $(\boldsymbol{x},t)$, hence it can be taken through the derivatives. Therefore we get

$$\frac{1}{c}\frac{\partial I'}{\partial t'} + \boldsymbol{n}'\cdot\boldsymbol{\nabla}'I' = \eta' - \chi' I' . \tag{3.57}$$

Equation (3.57) implies that the transfer equation is covariant in the sense that it has the same algebraic form in all inertial frames. But in a sense this is a trivial result because we know from experience that we can write the same transfer equation in a uniformly moving medium (say a star) as in the laboratory frame simply by taking into account the Doppler shift $\Delta\nu_{\text{Doppler}}$ induced by the motion of the star relative to the laboratory frame.

2) Alternatively we could argue that $\mathcal{I}_{,\alpha}$, the 4-gradient of the 4-invariant $\mathcal{I}$, should transform like a covariant 4-vector, so that the inner product $M^\alpha\mathcal{I}_{,\alpha}$ should be an invariant. Thus,

$$\frac{1}{\nu^2}\left[\eta(\boldsymbol{n},\nu) - \chi(\boldsymbol{n},\nu)\, I(\boldsymbol{n},\nu)\right] = \frac{1}{\nu^2}\left(\frac{1}{c}\frac{\partial}{\partial t} + \boldsymbol{n}\cdot\boldsymbol{\nabla}\right) I = M^\alpha\mathcal{I}_{,\alpha}$$
$$\equiv M'^\alpha\mathcal{I}'_{,\alpha} = \frac{1}{\nu'^2}\left(\frac{1}{c}\frac{\partial}{\partial t'} + \boldsymbol{n}'\cdot\boldsymbol{\nabla}'\right) I' . \tag{3.58}$$

By converting the material properties in the first expression of (3.58) to the comoving frame and equating the result to the last expression, we again find

$$\frac{1}{c}\frac{\partial I'}{\partial t'} + \boldsymbol{n}'\cdot\boldsymbol{\nabla}'I' = \eta' - \chi' I' . \tag{3.59}$$

Although this argument looks plausible, it misses the point: $\mathcal{I}$ is a *distribution function*, *not* a *scalar*, so we have no way of knowing that $\mathcal{I}_{,\alpha}$ is a covariant vector (in general it is not). Thus, we cannot actually claim that $M^\alpha\mathcal{I}_{,\alpha}$ is an invariant. Furthermore, this "derivation" doesn't apply if the frames are not moving *uniformly* with respect to one another because then we can't take the factor $(\nu/\nu')^3$ through derivatives. And, of course, *nonuniform flow* where $\boldsymbol{v} = \boldsymbol{v}(\boldsymbol{x},t)$, i.e., there are gradients in space and accelerations in time, is precisely the case we wish to study.

So how should we proceed? Recall that the particle Boltzmann equation is

$$\frac{\partial f}{\partial t} + v^i \frac{\partial f}{\partial x^i} + \dot{p}^i \frac{\partial f}{\partial p^i} = \left(\frac{Df}{Dt}\right)_{\text{coll}} . \tag{3.60}$$

Here (D/Dt) is the standard Lagrangian time derivative. The covariant generalization of (3.60) is

$$\left(\frac{\mathrm{d}x^\alpha}{\mathrm{d}\tau}\right) \frac{\partial f}{\partial x^\alpha} + \left(\frac{\mathrm{d}p^\alpha}{\mathrm{d}\tau}\right) \frac{\partial f}{\partial p^\alpha} = \left(\frac{\delta f}{\delta t}\right)_{\text{coll}} \tag{3.61}$$

Here $(\delta/\delta t)$ is the intrinsic derivative with respect to time, calculated along the path followed in phase space.

We can apply Eq. (3.61) to photons. For convenience use units in which $c = h = 1$. A technical problem to be faced is that photon world lines lie on the null cone, so $\Delta\tau \equiv 0$. To get around this problem we replace τ with some other *affine pathlength* ℓ. Define ℓ such that

$$\frac{\mathrm{d}x^\alpha}{\mathrm{d}\ell} \equiv p^\alpha . \tag{3.62}$$

Then rewrite the Boltzmann equation as

$$p^\alpha \frac{\partial f}{\partial x^\alpha} + \dot{p}^\alpha \frac{\partial f}{\partial p^\alpha} = \left(\frac{\delta f}{\delta \ell}\right)_{\text{coll}} , \tag{3.63}$$

where

$$\dot{p}^\alpha \equiv \frac{\mathrm{d}p^\alpha}{\mathrm{d}\ell} . \tag{3.64}$$

For photons, $f \to f_{\mathrm{R}}, p^\alpha \to M^\alpha$, the source term $= \varepsilon$ (emission), and the sink term $= -\alpha f_{\mathrm{R}}$ (absorption), so that the *photon Boltzmann equation* is

$$M^\alpha \frac{\partial f_{\mathrm{R}}}{\partial x^\alpha} + \dot{M}^\alpha \frac{\partial f_{\mathrm{R}}}{\partial M^\alpha} = \varepsilon - \alpha f_{\mathrm{R}} . \tag{3.65}$$

Or, in terms of the invariant intensity, emissivity, and absorptivity defined earlier,

$$M^\alpha \frac{\partial \mathcal{I}}{\partial x^\alpha} + \dot{M}^\alpha \frac{\partial \mathcal{I}}{\partial M^\alpha} = \mathcal{E} - \mathcal{A}\mathcal{I} . \tag{3.66}$$

Equation (3.66) applies in *all* frames. In particular it applies in the *inertial* frame. In such a frame photons move on straight lines with constant frequency. Therefore, ignoring general relativity effects, $\dot{M}^\alpha \equiv 0$. It is *then* we get

$$M^\alpha \mathcal{I}_{,\alpha} = \mathcal{E} - \mathcal{A}\mathcal{I} , \tag{3.67}$$

which is the equation studied above in connection with Lorentz invariance of the transfer equation. As we have just seen, this equation is valid if and *only* if

$\dot{M}^\alpha \equiv 0$, *i.e.*, only for *uniformly* moving frames. In a *noninertial* frame, e.g., a moving fluid, photons still move on *geodesics*, but now the geodesics aren't Euclidian straight lines, and the photon frequency isn't constant because, as seen by observers attached to different fluid elements, each having its own $\boldsymbol{v}(\boldsymbol{r},t)$, a photon with fixed lab-frame $(\nu, \boldsymbol{n})$ has a *different* $(\nu_0, \boldsymbol{n}_0)$ at each fluid element.

We can write Eq. (3.66) in a form which is similar to Eq. (3.67), namely

$$M^\alpha \mathcal{I}_{|\alpha} = \mathcal{E} - \mathcal{A}\mathcal{I} \,, \tag{3.68}$$

where $_{|\alpha}$ is an operator which denotes derivatives with respect to x^α taken subject to the constraint that photons remain on the null cone as seen in the (noninertial) fluid frame. The physical content of (3.66) and (3.68) are identical, whereas (3.67) is quite different from either of the others. When we can evaluate either (3.66) or (3.68) for a specified geometry and flow, we will have derived the *Lagrangian transfer equation.*

To prevent misunderstandings characteristic of early work, it is essential to bear in mind that I or $\mathcal{I}$ or f_R are *distribution functions.* They exist, and describe the behavior of photons, not only in the four-dimensional spacetime $(\boldsymbol{r},t)$ but also in attached three-dimensional *tangent spaces* $\boldsymbol{p}$ or $(\nu,\boldsymbol{n})$. All seven variables are independent, therefore when calculating the variation of $\mathcal{I}$ or f_R along a path, one must remember, and differentiate with respect to, *all* the independent variables.

4. Inertial-Frame Equations of Radiation Hydrodynamics

4.1 Inertial-Frame Radiation Equations

Let us work in the laboratory frame where the fluid moves with velocity $\boldsymbol{v}(\boldsymbol{r},t)$. We already wrote the correct radiation energy and momentum equations:

$$\begin{aligned}\frac{\partial E}{\partial t} + \frac{\partial F^i}{\partial x^i} &= -cG^0 \\ &= \int_0^\infty \mathrm{d}\nu \oint \left[\eta(\boldsymbol{x},t;\boldsymbol{n},\nu) - \chi(\boldsymbol{x},t;\boldsymbol{n},\nu)\, I(\boldsymbol{x},t;\boldsymbol{n},\nu)\right] \mathrm{d}\omega \,,\end{aligned} \tag{4.1}$$

$$\begin{aligned}\frac{1}{c^2}\frac{\partial F}{\partial t} + \frac{P^{ij}}{x^j} &= -G^i \\ &= \frac{1}{c}\int_0^\infty \mathrm{d}\nu \oint \left[\eta(\boldsymbol{x},t;\boldsymbol{n},\nu) - \chi(\boldsymbol{x},t;\boldsymbol{n},\nu)\, I(\boldsymbol{x},t;\boldsymbol{n},\nu)\,\right] n^i \mathrm{d}\omega \,.\end{aligned} \tag{4.2}$$

These are just $R^{0\beta}_{,\beta} = -G^0$ and $R^{i\beta}_{,\beta} = -G^i$. So what is new here? Actually nothing. Then what is the problem here? The right-hand sides. They are messy because they contain hidden in them a host of velocity-dependent terms resulting from Doppler shifts. In most problems it would be essentially impossible to do the work (usually interpolation in tables) to account for these effects. Do we *really* need to worry about Doppler-shift-induced anisotropies of χ and η? After all, $v/c \ll 1$. The usual response in the literature has been to *ignore* the $\boldsymbol{v}$-dependent terms and to evaluate G^0 and G^i *as if* $\eta \equiv \eta_0(\nu)$ and $\chi \equiv \chi_0(\nu)$ at the laboratory frame frequency. However, this is a bad idea. To see why, we must examine the effects of $O(v/c)$ terms. Start with the transfer equation.

Mixed-Frame Equations. Choose $(\nu, \boldsymbol{n})$ in the lab frame. Then evaluate χ and ν at ν_0, where

$$\nu_0 \approx \nu\left(1 - \frac{\boldsymbol{n}\cdot\boldsymbol{v}}{c}\right) . \tag{4.3}$$

Then, using (3.23) and (3.19) and expanding to $O(v/c)$ we get

$$\chi(\mu,\nu) = \left(\frac{\nu_0}{\nu}\right)\chi_0(\nu_0) = \chi_0(\nu) - \left(\frac{\boldsymbol{n}\cdot\boldsymbol{v}}{c}\right)\left[\chi_0(\nu) + \nu\frac{\partial\chi_0}{\partial\nu}\right] \tag{4.4}$$

and similarly

$$\eta(\mu,\nu) = \eta 0(\nu) + \left(\frac{\boldsymbol{n}\cdot\boldsymbol{v}}{c}\right)\left[2\eta_0(\nu) - \nu\frac{\partial\eta_0}{\partial\nu}\right] . \tag{4.5}$$

Hence the transfer equation is

$$\begin{aligned}\frac{1}{c}\frac{\partial I(\boldsymbol{n},\nu)}{\partial t} &+ \boldsymbol{n}\cdot\boldsymbol{\nabla} I(\boldsymbol{n},\nu) = \eta_0(\nu) - \chi_0(\nu)I(\boldsymbol{n},\nu) \\ &+ \left(\frac{\boldsymbol{n}\cdot\boldsymbol{v}}{c}\right)\left\{2\eta_0(\nu) - \nu\frac{\partial\eta_0}{\partial\nu} + \left[\chi_0(\nu) + \nu\frac{\partial\chi_0}{\partial\nu}\right]I(\boldsymbol{n},\nu)\right\} .\end{aligned} \tag{4.6}$$

This is a *mixed-frame* transfer equation: I, ν, and $\boldsymbol{n}$ are measured in the laboratory frame, whereas χ_0 and η_0 are measured in the fluid frame, which is advantageous because these quantities are isotropic in that frame. Integrating over angle, we get the *monochromatic radiation energy equation*

$$\frac{\partial E_\nu}{\partial t} + \frac{\partial F^i_\nu}{\partial x^i} = 4\pi\eta_0(\nu) - c\chi_0(\nu)E_\nu + \frac{v_i F^i_\nu}{c}\left[\chi_0(\nu) + \nu\frac{\partial\chi_0}{\partial\nu}\right] \tag{4.7}$$

and the *monochromatic radiation momentum equation*

$$\begin{aligned}\frac{1}{c^2}\frac{\partial F^i_\nu}{\partial t} + \frac{\partial P^{ij}_\nu}{\partial x^j} &= -\frac{1}{c}\chi_0(\nu)F^i_\nu + \frac{v_j P^{ij}_\nu}{c}\left[\chi_0(\nu) + \nu\frac{\partial\chi_0}{\partial\nu}\right] \\ &+ \left(\frac{4\pi v^i}{3c^2}\right)\left[2\eta_0(\nu) - \nu\frac{\partial\eta_0}{\partial\nu}\right] ,\end{aligned} \tag{4.8}$$

where we used the result that $\oint n^i n^j \mathrm{d}\omega = (4\pi/3)\delta^{ij}$. Integrating over frequency we get the *radiation energy equation*

$$\begin{aligned}\frac{\partial E}{\partial t}+\frac{\partial F^i}{\partial x^i} &= \int_0^\infty [4\pi\eta_0(\nu) - c\chi_0(\nu)E_\nu]\,\mathrm{d}\nu \\ &+ \frac{v_i}{c}\int_0^\infty \left[\chi_0(\nu)+\nu\frac{\partial\chi_0}{\partial\nu}\right] F^i_\nu\,\mathrm{d}\nu \\ &= -cG^0 \,. \end{aligned} \tag{4.9}$$

Note that the absorption-emission term on the right-hand side is *not* $-cG^0_0$: both ν and E_ν are in the *laboratory* frame. Similarly, taking the first angular moment we get the *radiation momentum equation*:

$$\begin{aligned}\frac{1}{c^2}\frac{\partial F^i}{\partial t}+\frac{\partial P^{ij}}{\partial x^j} &= -\frac{1}{c}\int_0^\infty \chi_0(\nu)F^i_\nu \mathrm{d}\nu \\ +\frac{v_j}{c}\int_0^\infty &\left[\chi_0(\nu)+\nu\frac{\partial\chi_0}{\partial\nu}\right]P^{ij}\mathrm{d}\nu + \frac{4\pi v^i}{c^2}\int_0^\infty \eta_0(\nu)\mathrm{d}\nu \\ &= -cG^i \,. \end{aligned} \tag{4.10}$$

Again note that the first term on the right-hand side is *not* $-cG^i_0$: both ν and F_ν are in the *laboratory* frame.

On the Importance of (v/c)*Terms.* What are the above equations trying to tell us? To extract the physical flavor, let us make a simple analysis for *grey* material. Thus drop the terms in $\partial\chi_0/\partial\nu$ and $\partial\eta_0/\partial\nu$. We then have

$$\frac{\partial E}{\partial t}+\frac{\partial F^i}{\partial x^i} = \kappa(4\pi B - cE) + \frac{\kappa}{c}v_i F^i \tag{4.11}$$

and

$$\frac{1}{c^2}\frac{\partial F^i}{\partial t}+\frac{\partial P^{ij}}{\partial x^j} = \frac{\kappa}{c}\left[-F^i + v_j P^{ij} + v^i\left(\frac{4\pi B}{c}\right)\right] . \tag{4.12}$$

Consider the energy equation first:

1) We instantly see that if we drop the $O(v/c)$ term $\boldsymbol{v}\cdot\boldsymbol{F}/c$, we have omitted a term equal to the rate of work done by the radiation on the material, which can be quite large when the radiation field is very strong. Clearly this is not a good choice if we want to do radiation hydrodynamics.

2) Next, examine the diffusion regime. From the Lorentz transformation

$$cE = cE^0 + 2\boldsymbol{v}\cdot\boldsymbol{F}^0/c \to 4\pi B + 2\boldsymbol{v}\cdot\boldsymbol{F}/c + O(v^2/c^2) \,. \tag{4.13}$$

Using this result in Eq. (4.11) we have

$$\frac{\partial E}{\partial t}+\frac{\partial F^i}{\partial x^i} \to -\frac{\kappa}{c}v_i F^i \,. \tag{4.14}$$

Equation (4.14) makes the physically correct statement: (the rate of change of the radiation energy density in a fixed volume) plus (the rate of work done by the radiation on the material) equals the net rate of radiant energy inflow into the volume. We thus reach two important conclusions: (a) omission of $O(v/c)$ terms leads to an error of the order of the net absorption-emission terms, and (b) we get the correct first law of thermodynamics for radiation only if we keep the $O(v/c)$ terms.

3) From a simple dimensional analysis one can show that $(\kappa \boldsymbol{v} \cdot \boldsymbol{F}/c)$ is $O(\ell v/\lambda_{\mathrm{p}} c) \ll 1$ relative to $\nabla \cdot F$. Hence in the limit of dynamic diffusion, $(v/c) \geq (\lambda_{\mathrm{p}}/\ell)$, the velocity-dependent terms may actually *dominate* the energy balance.

Now consider the momentum equation. Examine the diffusion regime. In that limit one can drop the term $c^{-2}(\partial F/\partial t)$ because by dimensional analysis one finds that it is $O(\lambda_{\mathrm{p}} v/\ell c) \ll 1$ in comparison with $\kappa F/c$. Thus we get

$$F^i = -\left(\frac{c}{\kappa}\right) P^{ij}_{,j} + v_j P^{ij} + v^i \left(\frac{4\pi B}{c}\right) . \tag{4.15}$$

In the diffusion limit

$$E_0 \to \frac{4\pi B}{c} + O\left(\frac{\lambda_{\mathrm{p}} v}{\ell c}\right) , \tag{4.16}$$

$$P^{ij} \to P_0^{ij} + O\left(\frac{\lambda_{\mathrm{p}} v}{\ell c}\right) = \frac{1}{3} E_0 \delta^{ij} + O\left(\frac{\lambda_{\mathrm{p}} v}{\ell c}\right) , \tag{4.17}$$

$$\boldsymbol{F}_0 \to -\frac{c}{\kappa} \boldsymbol{\nabla} \cdot \mathsf{P}_0 . \tag{4.18}$$

Thus (4.15) simply states

$$F^i = F_0^i + v^i E_0 + v_j P_0^{ij} , \tag{4.19}$$

which is the standard Lorentz transformation of the flux, simplified to $O(v/c)$. Therefore, unless we keep $O(v/c)$ terms in the transfer equation we fail to discriminate $\boldsymbol{F}$ from $\boldsymbol{F}_0$. Is this important? Consider the diffusion limit: there we have $(vE_0/F) \to (v\sigma T^4/c)/(\sigma T_{\mathrm{eff}}^4) \to (v/c)(T/T_{\mathrm{eff}})^4$. Recall that in the Sun, the ratio of the central temperature to the effective temperature $(T_{\mathrm{c}}/T_{\mathrm{eff}}) \sim 10^3$. So even if (v/c) is only 10^{-12}, the *advected* flux can exceed the *diffusive* flux. (Note: this problem doesn't arise in the comoving frame.) Thus it is clear that in the laboratory frame formulation it is important to analyze (v/c) terms.

Dimensional Analysis.

Rules:

1. $(\partial/\partial x) \to (1/\ell)$
2. $(\partial/\partial t) \to 1/(\ell/v) = (v/\ell)$
3. We can drop any term that is *always* at most $O(v/c)$ in comparison with the dominant term.
4. But we must keep in *all* regimes any term found to be comparable to the dominant term in *any* regime. (Otherwise we won't get the transition from one regime to another right.)
5. Grey material.

Regimes:

1. Streaming limit: $\lambda_p/\ell \geq 1; E \sim P; F \sim cE$
2. Diffusion limit: $\lambda_p/\ell \ll 1; P \to \frac{1}{3}E$

a) *Static diffusion:* $v/c \ll \lambda_p/\ell$

$$F \to F_0, \quad \text{therefore,} \quad F \to \left(\frac{c}{\chi}\frac{\partial E}{\partial x}\right) \to O\left(\frac{c\lambda_p}{\ell}\right) E$$

$$\kappa(cE - 4\pi B) \to \frac{\partial}{\partial x}\left(\frac{c}{\chi}\frac{\partial E}{\partial x}\right) \to O\left(\frac{c\lambda_p}{\ell^2}\right) E$$

b) *Dynamic diffusion:* $v/c \geq \lambda_p/\ell$

$$F \to vE_0, \quad \text{therefore,} \quad \left(\frac{F}{cE}\right) \sim O\left(\frac{v}{c}\right)$$

$$\kappa(cE - 4\pi B) \to \frac{\partial}{\partial x}(vE) \sim O\left(\frac{v}{\ell}\right) E$$

Begin with an analysis of the transfer equation following the rules stated above.

	A	B	C	D
	$\frac{1}{c}\frac{\partial I}{\partial t}$	$+\,\boldsymbol{n}\cdot\boldsymbol{\nabla} I$	$= \eta - \chi I$	$+\left(\frac{\boldsymbol{n}\cdot\boldsymbol{v}}{c}\right)(2\eta + \chi I)$
Streaming:	(v/c)	1	(ℓ/λ_p)	$(v/c)(\ell/\lambda_p)$
Static diffusion:	(v/c)	1	(λ_p/ℓ)	$(v/c)(\ell/\lambda_p)$
Dynamic diffusion:	(v/c)	1	(v/c)	$(v/c)(\ell/\lambda_p)$

Thus in the streaming limit, we need only the spatial gradient term B and the source-sink term C, which yields the standard static transfer equation. Term A is needed only in a radiation front. For static diffusion we need terms B and C, and drop A. Normally we can also drop term D. But if (λ_p/ℓ) is so small that $(v/c) \geq (\lambda_p/\ell)^2$, then the velocity-dependent term may become the dominant source-sink term on the right-hand side. In the dynamic diffusion regime we need terms B and D.

In summary, in order to guarantee a correct solution of the mixed-frame transfer equation we must retain the spatial operator on the left-hand side, and all terms on the right-hand side.

Next, consider the energy equation.

	A	B	C	D
	$\frac{\partial E}{\partial t}$	$+\nabla \cdot \boldsymbol{F}$	$= \kappa(4\pi B - cE)$	$+\left(\frac{\kappa}{c}\right)\boldsymbol{v}\cdot\boldsymbol{F}$
Streaming:	(v/c)	1	(ℓ/λ_p)	$(v/c)(\ell/\lambda_p)$
Static diffusion:	$(v/c)(\ell/\lambda_p)$	1	1	$(v/c)(\ell/\lambda_p)$
Dynamic diffusion:	1	1	1	$(v/c)(\ell/\lambda_p)$

In the streaming limit we normally need terms B and C, but can drop the $(\partial/\partial t)$ term A and the velocity-dependent term D. We need to include term A if we try to follow a radiation front. For static diffusion we normally need only terms B and C and can drop terms A and D. But note that as $(v/c) \sim (\lambda_p/\ell)$, *all* terms are of the same order and all must be retained. In the limit of dynamic diffusion, we need *all* terms, and in fact the velocity-dependent terms may dominate all others.

In summary, to guarantee the correct solution of the mixed-frame energy equation in all regimes we must retain *all* terms.

Finally, consider the momentum equation.

	A	B	C	D	E
	$\frac{1}{c^2}\frac{\partial \boldsymbol{F}}{\partial t}$	$+\nabla\cdot\mathsf{P}$	$= -\frac{\kappa}{c}\boldsymbol{F}$	$+\left(\frac{\kappa}{c}\right)\left(\frac{4\pi B}{c}\boldsymbol{v}\right)$	$+\left(\frac{\kappa}{c}\right)(\boldsymbol{v}\cdot\mathsf{P})$
Streaming:	(v/c)	1	(ℓ/λ_p)	$(v/c)(\ell/\lambda_p)$	$(v/c)(\ell/\lambda_p)$
Static diffusion:	$(v/c)(\lambda_p/\ell)$	1	1	$(v/c)(\ell/\lambda_p)$	$(v/c)(\ell/\lambda_p)$
Dynamic diffusion:	$(v/c)^2$	1	$(v/c)(\ell/\lambda_p)$	$(v/c)(\ell/\lambda_p)$	$(v/c)(\ell/\lambda_p)$

In the streaming limit we must retain the basic terms B and C. We can drop the time-dependent term A compared to B *unless* we wish to track a radiation front. We can drop the velocity-dependent terms D and E in comparison with C. In the static diffusion limit we can normally drop the time- and velocity-dependent terms A, D, and E in comparison with B and C. *However*, note that if v/c becomes comparable to λ_p/ℓ, then we must retain *all* terms on the right hand side. In the dynamic diffusion limit we can drop the time derivative term A, but we must keep all terms on the right-hand side.

In summary, in solving the mixed-frame radiation momentum equation, we can guarantee the correct solution in all regimes only if we retain all terms, except $\partial/\partial t$ (which will also be needed if we follow radiation fronts). Thus, in the mixed frame there are no simplifications, no short-cuts. Basically all the terms in the equations must be kept. We will find a different result when we work with Lagrangian (comoving-frame) equations.

4.2 Inertial-Frame Equations of Radiation Hydrodynamics

The equations of radiation hydrodynamics can be written in covariant form as

Radiation:

$$R^{\alpha\beta}_{;\beta} = -G^{\alpha} \ . \tag{4.20}$$

Matter:

$$\text{view 1:} \quad M^{\alpha\beta}_{;\beta} = F^{\alpha} + G^{\alpha} \ , \tag{4.21}$$

$$\text{view 2:} \quad (M^{\alpha\beta} + R^{\alpha\beta})_{;\beta} = F^{\alpha} \ . \tag{4.22}$$

Here $M^{\alpha\beta}$ is the material stress–energy tensor, $R^{\alpha\beta}$ is the radiation stress–energy tensor, F^{α} is an external 4-force, and G^{α} is the radiation-matter 4-force. Equations (4.21) and (4.22) are based on two different, but equivalent, views. In (4.21) we have an external force (e.g., electromagnetic, gravitational, *etc.*) plus a radiation force acting on the material. In (4.22) we have an external body force acting on a *radiating fluid.* In view of (4.20), (4.21) and (4.22) are mathematically identical; nevertheless "view 1" is intuitively more natural in the streaming limit, whereas "view 2" is more natural in the diffusion limit.

Recalling the L.H. Thomas expressions for $M^{\alpha\beta}_{;\beta}$ and combining them with the radiation terms, we get the relativistically correct equations:

$$(\varrho_1 c^2 - p)_{,t} + (\varrho_1 c^2 v^j)_{,j} = v_j f^j + cG^0 \tag{4.23}$$

and

$$(\varrho_1 v_i)_{,t} + (\varrho_1 v_i v^j)_{,j} = f_i - p_{,i} + G_i \ , \tag{4.24}$$

or

$$(\varrho_1 c^2 - p + E)_{,t} + (\varrho_1 c^2 v^j + F^j)_{,j} = v_j f^j \tag{4.25}$$

and

$$(\varrho_1 v_i + c^{-2} F^i)_{,t} + (\varrho_1 v_i v^j + P^{ij})_{,j} = f^i - \delta^{ij} p_{,j} \ . \tag{4.26}$$

Here, using Thomas' notation, $\varrho_1 = \gamma^2 \varrho_{000}$, where $\varrho_{000} = \varrho_0 \left(1 + \frac{e}{c^2}\right) + \frac{p}{c2}$. To cancel the large rest mass terms from the energy equations above, subtract c^2 times the continuity equation from them, to obtain:

$$\begin{aligned} &[(\gamma - 1)\varrho c^2 + \gamma \varrho e + (\gamma^2 - 1)p]_{,t} \\ &\quad + \{[(\gamma - 1)\varrho c^2 + \gamma \varrho e + \gamma^2 p] v^i\}_{,i} = v_i f^i + c\, G^0 \end{aligned} \tag{4.27}$$

or

$$[(\gamma-1)\varrho c^2+\gamma\varrho e+(\gamma^2-1)p+E]_{,t} \\ +\{[(\gamma-1)\varrho c^2+\gamma\varrho e+\gamma^2 p]v^i+F^i\}_{,i}=v_i f^i \,. \tag{4.28}$$

In (4.27) and (4.28) the term $(\gamma-1)\varrho c^2$ is the kinetic energy of the fluid and the term $\gamma\varrho e$ is its internal energy. Let us now reduce these equations to $O(v/c)$ to derive equations that can be applied in Newtonian radiation hydrodynamics.

Momentum Equation. Thus defining $\varrho_*=\gamma\varrho_{000}$, we can manipulate (4.24) and (4.26) into the relativistically correct momentum equation:

$$\varrho_*\frac{D\boldsymbol{v}}{D\tau}=\boldsymbol{f}-\nabla p-\frac{\boldsymbol{v}}{c^2}\left(\frac{\partial p}{\partial t}+\boldsymbol{v}\cdot\boldsymbol{f}\right)+\boldsymbol{G}-\frac{\boldsymbol{v}}{c}G^0 \tag{4.29}$$

or

$$\varrho_*\frac{D\boldsymbol{v}}{D\tau}=\boldsymbol{f}-\nabla p-\frac{\boldsymbol{v}}{c^2}\left(\frac{\partial p}{\partial t}+\boldsymbol{v}\cdot\boldsymbol{f}\right) \\ -\left(\frac{1}{c^2}\frac{\partial \boldsymbol{F}}{\partial t}+\nabla\cdot\mathsf{P}\right)+\frac{\boldsymbol{v}}{c^2}\left(\frac{\partial E}{\partial t}+\nabla\cdot\boldsymbol{F}\right)\,. \tag{4.30}$$

On the right-hand sides of (4.29) and (4.30), the first term is the externally imposed force, the second term is the force from a pressure gradient, the third term is the radiation force, and the fourth term accounts for the change in the equivalent mass density of the material resulting from its interaction with radiation (remember: energy ⇒ mass ⇒ inertia ⇒ dynamical effect).

The velocity-dependent terms in equations (4.27–30) are *logically* essential, for they guarantee consistent equations spanning the optically thick/thin transition, and between the laboratory frame and the Lagrangian frame. Therefore, in discussion of such points one must retain all of them. But for practical applications in Newtonian flow (small v/c) we can omit any terms that are *at most* $O(v/c)$ in all regimes. It is easy to show that $(\boldsymbol{v}/c)G^0$ is at most $O(v/c)$ compared to $\boldsymbol{G}$; therefore we can drop it. Likewise, in (4.30) one can show:

	A	B	C	D
	$-\frac{1}{c^2}\frac{\partial \boldsymbol{F}}{\partial t}$	$-\nabla\cdot\mathsf{P}$	$+\frac{\boldsymbol{v}}{c^2}\left(\frac{\partial E}{\partial t}\right)$	$+\frac{\boldsymbol{v}}{c^2}(\nabla\cdot\boldsymbol{F})$
Streaming:	(v/c)	1	$(v/c)^2$	(v/c)
Static diffusion:	$(v/c)(\lambda_p/\ell)$	1	$(v/c)^2$	$(v/c)(\lambda_p/\ell)$
Dynamic diffusion:	$(v/c)^2$	1	$(v/c)^2$	$(v/c)^2$

One sees that all the radiative terms are *at most* $O(v/c)$ compared to $\nabla\cdot\mathsf{P}$, so that only P need be retained for practical work in the Newtonian limit. Thus the Newtonian momentum equation is

$$\varrho \frac{D\boldsymbol{v}}{Dt} = \boldsymbol{f} - \boldsymbol{\nabla} p + \boldsymbol{G} , \tag{4.31}$$

or

$$\varrho \frac{D\boldsymbol{v}}{Dt} = \boldsymbol{f} - \boldsymbol{\nabla} p - \boldsymbol{\nabla} \cdot \mathsf{P} . \tag{4.32}$$

These are just the standard Newtonian momentum equations for gasdynamics in the presence of a radiation force or, equivalently, a divergence of the radiation pressure. What is new here? At first sight, nothing. But recall that when we *evaluate* $\boldsymbol{G}$ (which is *not* $\boldsymbol{G}_0$) or $\boldsymbol{\nabla} \cdot \mathsf{P}$ (which is *not* $\boldsymbol{\nabla} \cdot \mathsf{P}_0$) we *must* carry along $O(v/c)$ terms, *especially in the diffusion limit.* We see here, and will see again later, it is the radiation terms, not the material terms, that cause trouble!

Total Energy Equation. Let $\gamma \to 1$, so that $\gamma - 1 \to \frac{1}{2}(v^2/c^2)$. Then the total energy equations (4.27) and (4.28) become

$$\left(\varrho e + \frac{1}{2}\varrho v^2\right)_{,t} + \left\{\left[\varrho\left(e + \frac{1}{2}v^2\right) + p\right] v^i\right\}_{,i} = v_i f^i + cG^0 , \tag{4.33}$$

or

$$\left(\varrho e + \frac{1}{2}\varrho v^2 + E\right)_{,t} + \left\{\left[\varrho\left(e + \frac{1}{2}v^2\right) + p\right] v^i + F^i\right\}_{,i} = v_i f^i . \tag{4.34}$$

Note that the kinetic energy $\frac{1}{2}\varrho v^2$ is really a *relativistic* effect arising from boosting the rest mass energy $\varrho_0 c^2$ into the laboratory frame as the fluid moves. Equation (4.33) is the standard Newtonian energy equation with the addition of the net radiative energy input into the fluid as a source term on the right-hand side. Likewise Eq. (4.34) is the standard Newtonian energy equation for the radiating fluid, written to account for the total energy density (material + radiation) and the total energy flux (material + radiation). Is there anything new here? Yes: the caveat that when evaluating the radiative terms one *must* account for $O(v/c)$ terms. Note in particular that the energy exchange term G^0 on the right-hand side is *not* G_0^0. Note in passing that (4.33) can be written in quasi-Lagrangian form

$$\varrho \frac{D}{Dt}(e + \frac{1}{2}\varrho v^2) + \boldsymbol{\nabla} \cdot (p\boldsymbol{v}) = \boldsymbol{v} \cdot \boldsymbol{f} + G^0 . \tag{4.35}$$

Mechanical Energy Equation. If we take the inner product of $\boldsymbol{v}$ with the momentum equation and expand to $O(v/c)$ we have

$$\varrho \frac{D}{Dt}\left(\frac{1}{2}v^2\right) = -\boldsymbol{v} \cdot \boldsymbol{\nabla} p + \boldsymbol{v} \cdot (\boldsymbol{f} + \boldsymbol{G}) , \tag{4.36}$$

or

$$\varrho \frac{D}{Dt}\left(\frac{1}{2}v^2\right) = -\boldsymbol{v} \cdot (\boldsymbol{\nabla} p + \boldsymbol{\nabla} \cdot \mathsf{P}) + \boldsymbol{v} \cdot \boldsymbol{f} . \tag{4.37}$$

These again are standard Newtonian equations which we could have written intuitively (*provided*, of course, we remember to account for the velocity-dependent terms in $\boldsymbol{G}$ and P. They state that the rate of change of the kinetic energy equals (a) the rate of work done by the gas pressure gradient plus the rate of work done by the total (external plus radiation) force, or, as in (4.37), the rate of work done by the total (gas plus radiation) pressure gradient plus the rate of work done by external body forces.

Gas Energy Equation. Following L.H. Thomas again, one can reduce (4.24) to the relativistically correct *gas energy equation*

$$\varrho_0\left[\frac{De}{D\tau}+p\frac{D}{D\tau}\left(\frac{1}{\varrho_0}\right)\right] = -V_\alpha F^\alpha - V_\alpha G^\alpha \,. \tag{4.38}$$

For ordinary body forces, $V_\alpha F^\alpha \equiv 0$. If thermonuclear reactions occur, then in the comoving frame $-V_\alpha F^\alpha = \varrho_0\varepsilon$ and $-V_\alpha G^\alpha = \gamma(cG^0 a - \boldsymbol{v}\cdot\boldsymbol{G}) \equiv cG_0^0$. Equation (4.38) is clearly just the covariant generalization of the first law of thermodynamics. It states that: the rate of change of material energy plus the rate of work done by material equal the rate of energy input into material from radiation and thermonuclear sources, both measured in the comoving frame. Now to O(v/c), $\varrho_0 \to \varrho, \tau \to t$, and $\gamma \to 1$. We then obtain

$$\varrho\left[\frac{De}{Dt}+p\frac{D}{Dt}\left(\frac{1}{\varrho}\right)\right] = \varrho\varepsilon + cG^0 - \boldsymbol{v}\cdot\boldsymbol{G} = \varrho\varepsilon + cG_0^0 \,, \tag{4.39}$$

or

$$\varrho\left[\frac{De}{Dt}+p\frac{D}{Dt}\left(\frac{1}{\varrho}\right)\right] = \varrho\varepsilon - \boldsymbol{\nabla}\cdot\boldsymbol{F} - \frac{\partial E}{\partial t} + \boldsymbol{v}\cdot(\boldsymbol{\nabla}\cdot\mathsf{P}) \,. \tag{4.40}$$

To see what terms on the right hand side of (3.39) need to be retained for practical applications, one can perform a dimensional analysis. One finds that the ratio $(\boldsymbol{v}\cdot\boldsymbol{G}/cG^0)$ is < 1 in the static diffusion limit, ≥ 1 in the dynamic diffusion limit, and ≥ 1 in the streaming limit if the material is nearly in radiative equilibrium. Therefore we should retain all terms on the right hand side. One comes to similar conclusions for (4.40); in particular all terms are of the same order in dynamic diffusion. We can rewrite (4.40) as

$$\varrho\left[\frac{D}{Dt}\left(e+\frac{E}{\varrho}\right)+p\frac{D}{Dt}\left(\frac{1}{\varrho}\right)\right] + \boldsymbol{\nabla}\cdot(\boldsymbol{F}-\boldsymbol{v}E) = \boldsymbol{v}\cdot\boldsymbol{\nabla}\cdot\mathsf{P} + \varrho\varepsilon \,. \tag{4.41}$$

Note that

$$\boldsymbol{v}\cdot\boldsymbol{\nabla}\cdot\mathsf{P} = v_i P^{ij}_{\ ,j} = (v_i P^{ij})_{,j} - P^{ij}v_{i,j} = \boldsymbol{\nabla}\cdot(\boldsymbol{v}\cdot\mathsf{P}) - \mathsf{P}:\boldsymbol{\nabla}\boldsymbol{v} \,, \tag{4.42}$$

hence

$$\varrho\left[\frac{D}{Dt}\left(e+\frac{E}{\varrho}\right)+p\frac{D}{Dt}\left(\frac{1}{\varrho}\right)+\mathsf{P}:\boldsymbol{\nabla}\boldsymbol{v}\right] + \boldsymbol{\nabla}\cdot(\boldsymbol{F}-\boldsymbol{v}E-\boldsymbol{v}\cdot\mathsf{P}) = \varrho\varepsilon \,. \tag{4.43}$$

Recall $\boldsymbol{F} - \boldsymbol{v}E - \boldsymbol{v} \cdot \mathsf{P} = \boldsymbol{F}_0$, the fluid-frame flux. Also, $\mathsf{P} : \boldsymbol{\nabla v}$ is the rate of work done by the (anisotropic) pressure tensor. Equation (4.43) is a quasi-Lagrangian first law of thermodynamics for the radiating fluid, with $\boldsymbol{F}_0$ playing the role of an equivalent heat-flux vector. We will rederive this result from the Lagrangian view later.

Synopsis. We have shown that for practical Newtonian calculations the dynamical equations for the radiating fluid are just what we would have expected intuitively from purely Newtonian arguments by adding an appropriate radiation counterpart for each material term in the equation. Most previous formulations have penetrated this far. However it must be remembered that the radiation equations from which the radiation terms in the fluid equations are to be obtained contain terms, formally of $O(v/c)$, which are *essential* to keep. Without these terms one fails to discriminate between, say, the lab frame flux $\boldsymbol{F}$ and the comoving frame flux $\boldsymbol{F}_0$ (purely diffusive). It is here that most previous formulations have gone astray, and potentially serious errors were made by ignoring $\boldsymbol{v}$-dependent terms in solving for the radiation field, and by evaluating quantities like G^0 and G^i *as if* they were G_0^0 and G_0^i
The mixed-frame equations have some significant disadvantages:

1. We would need opacity and emissivity derivatives $(\partial/\partial\nu)$ in order to evaluate the right-hand side of the transfer equation or its moments. It is almost impossible to obtain this information for a complex spectrum.

2. One may encounter problems with the centering of the $O(v/c)$ terms in the difference equations. Yet these terms can be crucial (e.g., in discriminating between $\boldsymbol{F}$ and $\boldsymbol{F}_0$).

We shall see that these difficulties are mitigated in the comoving-frame approach.

5. Comoving-Frame Equation of Transfer

References: [4, 21, 23, 24, 3].

Our goal in this section is to obtain a transfer equation (and its moments) in which *all* radiation and material properties $(I, \mu, \nu, E, \boldsymbol{F}, \mathsf{P}, \chi, \eta)$ are expressed in the comoving fluid frame (the proper frame). The complication to be faced is that the comoving frame is noninertial; photons move on geodesics but these aren't Euclidian straight lines; both the direction and frequency of a photon as measured by comoving observers change along its path.

In our discussion it is important to distinguish the *frame* in which the *physical variables* (*dependent* variables) are measured, and the *coordinate grid* (*independent* variables) used to specify locations in spacetime. We want the former to be the comoving frame, but the latter can be *any* grid. For example:

Lagrangian grid	(Castor):	grid fixed in the *fluid*.
Inertial grid	(Mihalas, Buchler):	grid fixed in the *lab*.
Adaptive grid	(Winkler):	grid that tracks *features* in the flow, and *floats* with respect to both fluid and lab.

I will call *all* such formulations "Lagrangian" or "comoving frame" because that is where the *physics* is being done, even though only Castor's work is 100% Lagrangian in the narrowest sense.

The choice of coordinates is important. Castor's derivation (historically the first) is somewhat harder to follow than the others because it employs a full GR formalism using concepts from differential geometry (even though the physics requires only SR). To get a start into the problem, let's work first with lab-frame coordinates.

5.1 Special Relativistic Derivation (D. Mihalas)

Choose an inertial spacetime; the laboratory coordinates will do. Proceed exactly as one does in deriving relativistic material fluid equations, *i.e.*, leave the coordinate system in the lab frame but boost all fluid (radiation) quantities from the lab frame to the proper frame. Assume 1D spherically-symmetric flow. In the lab frame we know that

$$\frac{1}{c}\frac{\partial I(\mu,\nu)}{\partial t} + \mu\frac{\partial I(\mu,\nu)}{\partial r} + \frac{(1-\mu^2)}{r}\frac{\partial I(\mu,\nu)}{\partial \mu} = \eta(\mu,\nu) - \chi(\mu,\nu)I(\mu,\nu)\,. \tag{5.1}$$

Recall that

$$I(\mu,\nu) = \left(\frac{\nu}{\nu_0}\right)^3 I_0(\mu_0,\nu_0) \qquad \eta(\mu,\nu) = \left(\frac{\nu}{\nu_0}\right)^2 \eta_0(\nu_0)\,, \tag{5.2}$$

$$\chi(\mu,\nu) = \left(\frac{\nu_0}{\nu}\right) \chi_0(\nu_0)$$

hence

$$\begin{aligned}\left(\frac{\nu}{\nu_0}\right) & \left[\frac{1}{c}\frac{\partial I_0(\mu_0,\nu_0)}{\partial t} + \mu\frac{\partial I_0(\mu_0,\nu_0)}{\partial r} + \frac{(1-\mu^2)}{r}\frac{\partial I_0(\mu_0,\nu_0)}{\partial \mu}\right] \\ & - 3\left(\frac{\nu}{\nu_0^2}\right)\left[\frac{1}{c}\frac{\partial \nu_0}{\partial t} + \mu\frac{\partial \mu_0}{\partial r} + \frac{(1-\mu_0^2)}{r}\frac{\partial \nu_0}{\partial \mu}\right] I_0(\mu_0,\nu_0) \\ & = \eta_0(\nu_0) - \chi_0(\nu_0) I_0(\mu_0,\nu_0)\,. \end{aligned} \tag{5.3}$$

We must now express all quantities in the coefficients in the comoving frame. Likewise we convert all derivatives with respect to lab frame variables to derivatives with respect to comoving frame variables by use of the chain rule. These conversions account for the fact that all angles and frequencies are in the comoving frame. For example, in (5.3), in the lab frame

$$\frac{\partial}{\partial r} \Rightarrow \left.\frac{\partial}{\partial r}\right|_{t\mu\nu}\,,$$

that is, a partial derivative with respect to r is taken, holding t, μ, and ν fixed. What we want is to shift the derivative to fixed (μ_0, ν_0). Thus,

$$\left.\frac{\partial}{\partial r}\right|_{t\mu\nu} = \left.\frac{\partial}{\partial r}\right|_{t\mu_0\nu_0} + \left.\frac{\partial \mu_0}{\partial r}\right|_{t\mu\nu}\frac{\partial}{\partial \mu_0} + \left.\frac{\partial \nu_0}{\partial r}\right|_{t\mu\nu}\frac{\partial}{\partial \nu_0}\,, \tag{5.4}$$

$$\left.\frac{\partial}{\partial t}\right|_{r\mu\nu} = \left.\frac{\partial}{\partial t}\right|_{r\mu_0\nu_0} + \left.\frac{\partial \mu_0}{\partial t}\right|_{r\mu\nu}\frac{\partial}{\partial \mu_0} + \left.\frac{\partial \nu_0}{\partial t}\right|_{r\mu\nu}\frac{\partial}{\partial \nu_0}\,, \tag{5.5}$$

$$\left.\frac{\partial}{\partial \mu}\right|_{rt\nu} = \left.\frac{\partial \mu_0}{\partial \mu}\right|_{rt\nu}\frac{\partial}{\partial \mu_0} + \left.\frac{\partial \nu_0}{\partial \mu}\right|_{rt\nu}\frac{\partial}{\partial \nu_0}\,. \tag{5.6}$$

Now use the relations (from (3.7))

$$\nu_0 = \gamma\nu(1-\beta\mu) \tag{5.7}$$

$$\mu_0 = \frac{\mu-\beta}{1-\beta\mu} \tag{5.8}$$

$$1-\mu_0^2 = \frac{1-\mu^2}{\gamma^2(1-\beta\mu)^2} \tag{5.9}$$

to derive (after some algebra)

$$\frac{\partial \mu_0}{\partial t} = -\gamma^2(1-\mu_0^2)\frac{\partial \beta}{\partial t} \quad (5.10) \qquad \frac{\partial \nu_0}{\partial r} = -\gamma^2\mu_0\nu_0\frac{\partial \beta}{\partial r} \quad (5.13)$$

$$\frac{\partial \nu_0}{\partial t} = -\gamma^2\mu_0\nu_0\frac{\partial \beta}{\partial t} \quad (5.11) \qquad \frac{\partial \mu_0}{\partial \mu} = \gamma^2(1+\beta\mu_0)^2 \quad (5.14)$$

$$\frac{\partial \mu_0}{\partial r} = -\gamma^2(1-\mu_0^2)\frac{\partial \beta}{\partial r} \quad (5.12) \qquad \frac{\partial \nu_0}{\partial \mu} = -\beta\gamma^2(1+\beta\mu_0)\nu_0 \quad (5.15)$$

which are in the desired form because all physical quantities are in the comoving frame.

Using these expressions in the chain rules (5.4–6), and then using those expressions in the transfer equation we find, after a moderate amount of algebra, the *comoving frame transfer equation*:

$$\begin{aligned}
&\frac{\gamma}{c}(1+\beta\mu_0)\frac{\partial I_0(\mu_0,\nu_0)}{\partial t} + \gamma(\mu_0+\beta)\frac{\partial I_0(\mu_0,\nu_0)}{\partial r} \\
&\quad + \frac{\partial}{\partial \mu_0}\left\{\gamma(1-\mu_0^2)\left[\frac{(1+\beta\mu_0)}{r} - \gamma^2(\mu_0+\beta)\frac{\partial \beta}{\partial r}\right.\right. \\
&\qquad\qquad \left.\left. - \frac{\gamma^2}{c}(1+\beta\mu_0)\frac{\partial \beta}{\partial t}\right] I_0(\mu_0,\nu_0)\right\} \\
&\quad - \frac{\partial}{\partial \nu_0}\left\{\gamma\nu_0\left[\frac{\beta(1-\mu_0^2)}{r} + \gamma^2\mu_0(\mu_0+\beta)\frac{\partial \beta}{\partial r}\right.\right. \\
&\qquad\qquad \left.\left. + \frac{\gamma^2}{c}\mu_0(1+\beta\mu_0)\frac{\partial \beta}{\partial t}\right] I_0(\mu_0,\nu_0)\right\} \\
&\quad + \gamma\left\{\frac{2\mu_0+\beta(3-\mu_0^2)}{r} + \gamma^2(1+\mu_0^2+2\beta\mu_0)\frac{\partial \beta}{\partial r}\right. \\
&\qquad\qquad \left. + \frac{\gamma^2}{c}\left[2\mu_0+\beta(1+\mu_0^2)\right]\frac{\partial \beta}{\partial t}\right\} I_0(\mu_0,\nu_0) \\
&\quad = \eta_0(\nu_0) - \chi_0(\nu_0)I_0(\mu_0,\nu_0)\,.
\end{aligned} \tag{5.16}$$

A few remarks about (5.16) are in order:

- This transfer equation is relativistically correct; it is accurate to all orders of (v/c) $(0 \le \beta \le 1)$.
- The left-hand side contains all the mess (most of which was missed in earlier attempts to derive). But the material terms on the right-hand side are simple, and are the quantities we actually know (can compute from first principles).
- The left-hand side is written in conservation form for angles and frequencies.
- The left-hand side can be made conservative in space too, even on an adaptive grid.
- The $O(v/c)$ equations will be less formidable.

Now integrate over angle to obtain the (relativistic) *monochromatic radiation energy equation*:

$$\begin{aligned}
&\gamma\left[\frac{\partial E_0(\nu_0)}{\partial t}+\frac{v}{c^2}\frac{\partial F_0(\nu_0)}{\partial t}\right]+\gamma\left[\frac{\partial F_0(\nu_0)}{\partial r}+v\frac{\partial E_0(\nu_0)}{\partial r}\right]\\
&\quad-\frac{\partial}{\partial \nu_0}\left(\gamma\nu_0\left\{\frac{v}{r}\left[E_0(\nu_0)-P_0(\nu_0)\right]+\gamma^2\frac{\partial v}{\partial r}\left[P_0+\frac{v}{c^2}F_0(\nu_0)\right]\right.\right.\\
&\qquad\qquad\left.\left.+\frac{\gamma^2}{c^2}\frac{\partial v}{\partial t}\left[F_0(\nu_0)+vP_0(\nu_0)\right]\right\}\right)\\
&\quad+\gamma\left\{\frac{1}{r}\left[2F_0(\nu_0)+3vE_0(\nu_0)-vP_0(\nu_0)\right]\right.\\
&\qquad\qquad+\gamma^2\frac{\partial v}{\partial r}\left[E_0(\nu_0)+P_0(\nu_0)+\frac{2v}{c^2}F_0(\nu_0)\right]\\
&\qquad\qquad\left.+\frac{\gamma^2}{c^2}\frac{\partial v}{\partial t}\left[2F_0(\nu_0)+vE_0(\nu_0)+vP_0(\nu_0)\right]\right\}\\
&\quad=4\pi\eta_0(\nu_0)-c\chi_0(\nu_0)E_0(\nu_0)\,.
\end{aligned}\tag{5.17}$$

If we integrate the transfer equation against $\mu_0 \mathrm{d}\omega_0$ we obtain the (relativistic) *monochromatic radiation momentum equation*:

$$\begin{aligned}
&\frac{\gamma}{c^2}\left[\frac{\partial F_0(\nu_0)}{\partial t}+v\frac{\partial P_0(\nu_0)}{\partial t}\right]+\gamma\left[\frac{\partial P_0(\nu_0)}{\partial r}+\frac{v}{c^2}\frac{\partial F_0(\nu_0)}{\partial r}\right]\\
&\quad-\frac{\partial}{\partial \nu_0}\left(\gamma\nu_0\left\{\frac{v}{c^2 r}\left[F_0(\nu_0)-Q_0(\nu_0)\right]+\frac{\gamma^2}{c^2}\frac{\partial v}{\partial r}\left[Q_0+vP_0(\nu_0)\right]\right.\right.\\
&\qquad\qquad\left.\left.+\frac{\gamma^2}{c^2}\frac{\partial v}{\partial t}\left[P_0(\nu_0)+\frac{v}{c^2}Q_0(\nu_0)\right]\right\}\right)\\
&\quad+\gamma\left\{\frac{1}{r}\left[3P_0(\nu_0)-E_0(\nu_0)+\frac{2v}{c^2}F_0(\nu_0)\right]\right.\\
&\qquad\qquad+\frac{\gamma^2}{c^2}\frac{\partial v}{\partial r}\left[2F_0(\nu_0)+vE_0(\nu_0)+vP_0(\nu_0)\right]\\
&\qquad\qquad\left.+\frac{\gamma^2}{c^2}\frac{\partial v}{\partial t}\left[E_0(\nu_0)+P_0(\nu_0)+\frac{2v}{c^2}F_0(\nu_0)\right]\right\}\\
&\quad=-\frac{1}{c}\chi_0(\nu_0)F_0(\nu_0)\,.
\end{aligned}\tag{5.18}$$

Here

$$Q_0(\nu_0)\equiv 2\pi\int_{-1}^{1}\mu_0^3 I_0(\mu_0,\nu_0)\,\mathrm{d}\mu_0\,.\tag{5.19}$$

Note that these two equations contain *four* moments (E_0, F_0, P_0, Q_0), not three. Therefore to solve them we need *two* closure factors: f relating P_0 to

E_0, and g relating Q_0 to F_0. In fact, in general Q_0 is a three-index tensor Q_{ijk}; this tensor has four independent components in 2D, and ten in 3D.

Finally, integrate over frequency. We get the (relativistic) *radiation energy equation*:

$$\begin{aligned}
&\gamma\left(\frac{\partial E_0}{\partial t}+\frac{v}{c^2}\frac{\partial F_0}{\partial r}\right)+\gamma\left(\frac{\partial F_0}{\partial r}+v\frac{\partial E_0}{\partial r}\right)\\
&\quad+\gamma\left[\frac{1}{r}\left(2F_0+3vE_0-vP_0\right)+\gamma^2\frac{\partial v}{\partial r}\left(E_0+P_0+\frac{2v}{c^2}F_0\right)\right.\\
&\qquad\qquad\left.+\frac{\gamma^2}{c^2}\frac{\partial v}{\partial t}\left(2F_0+vE_0+vP_0\right)\right]\\
&\quad=\int_0^\infty\left[4\pi\eta_0(\nu_0)-c\chi_0(\nu_0)E_0(\nu_0)\right]\mathrm{d}\nu_0
\end{aligned}\tag{5.20}$$

and the (relativistic) *radiation momentum equation*:

$$\begin{aligned}
&\frac{\gamma}{c^2}\left(\frac{\partial F_0}{\partial t}+v\frac{\partial P_0}{\partial t}\right)+\gamma\left(\frac{\partial P_0}{\partial r}+\frac{v}{c^2}\frac{\partial F_0}{\partial r}\right)+\gamma\left[\frac{1}{r}\left(3P_0-E_0+\frac{2v}{c^2}F_0\right)\right.\\
&\quad\left.+\frac{\gamma^2}{c^2}\frac{\partial v}{\partial r}\left(2F_0+vE_0+vP_0\right)+\frac{\gamma^2}{c^2}\frac{\partial v}{\partial t}\left(E_0+P_0+\frac{2v}{c^2}F_0\right)\right]\\
&\quad=-\frac{1}{c}\int_0^\infty\chi_0(\nu_0)F_0(\nu_0)\,\mathrm{d}\nu_0\,.
\end{aligned}\tag{5.21}$$

These two equations contain only tensorial quantities $\boldsymbol{E}, \boldsymbol{F}$, and P. So we can check our derivation by starting with the laboratory-frame equations

$$\frac{\partial E}{\partial t}+\frac{1}{r^2}\frac{\partial}{\partial r}(r^2F)=-cG^0 \tag{5.22}$$

and

$$\frac{1}{c^2}\frac{\partial F}{\partial t}+\frac{\partial P}{\partial r}+\frac{3P-E}{r}=-cG^1 \tag{5.23}$$

and boosting E, F, P, and G into the comoving frame. After some algebra we find linear combinations of the comoving radiation energy and momentum equations written above. So it checks. The approach we took has the advantage of giving the monochromatic energy and momentum equations, as well as the full transfer equation. These allow us to determine the full angle-frequency variation of the radiation field. They have in fact been solved in some highly relativistic demonstration problems and in realistic supernova-envelope models.

For many purposes we need equations accurate only to $O(v/c)$. If we expand (5.16) to first order in β, we get the *Lagrangian transfer equation*

$$
\begin{aligned}
&\frac{1}{c}\frac{\mathrm{D}I_0(\mu_0,\nu_0)}{\mathrm{D}t} + \frac{\mu_0}{r^2}\frac{\partial}{\partial r}\left[r^2 I_0(\mu_0,\nu_0)\right] \\
&\quad + \frac{\partial}{\partial \mu_0}\left\{(1-\mu_0^2)\left[\frac{1}{r} + \frac{\mu_0}{c}\left(\frac{v}{r} - \frac{\partial v}{\partial r}\right) - \frac{a}{c^2}\right] I_0(\mu_0,\nu_0)\right\} \\
&\quad - \frac{\partial}{\partial \nu_0}\left\{\nu_0\left[(1-\mu_0^2)\frac{v}{cr} + \frac{\mu_0^2}{c}\frac{\partial v}{\partial r} + \frac{\mu_0 a}{c}\right] I_0(\mu_0,\nu_0)\right\} \\
&\quad + \left[(3-\mu_0^2)\frac{v}{cr} + \frac{(1+\mu_0^2)}{c}\frac{\partial v}{\partial r} + \frac{2\mu_0 a}{c^2}\right] I_0(\mu_0,\nu_0) \\
&\quad = \eta_0(\nu_0) - \chi_0(\nu_0) I_0(\mu_0,\nu_0)\,. \qquad (5.24)
\end{aligned}
$$

A few things about (5.24) should be noted:

- $a \equiv (\partial v/\partial t)$ = fluid acceleration
- Aside from terms in a (normally negligible) this equation agrees with the Lagrangian equation first derived by Castor.
- This equation looks possible to solve. In fact the full relativistic version has been solved, but only for steady flow [23].
- We get the planar limit by letting $(1/r) \to 0$.

If we integrate over $\mathrm{d}\omega_0$ we get the *monochromatic radiation energy equation*

$$
\begin{aligned}
&\frac{\mathrm{D}E_0(\nu_0)}{\mathrm{D}t} + \frac{1}{r^2}\frac{\partial}{\partial r}\left[r^2 F_0(\nu_0)\right] \\
&\quad - \frac{\partial}{\partial \nu_0}\left(\nu_0\left\{\frac{v}{r}\left[E_0(\nu_0) - P_0(\nu_0)\right] + \frac{\partial v}{\partial r}P_0 + \frac{a}{c^2}F_0(\nu_0)\right\}\right) \\
&\quad + \frac{v}{r}\left[3E_0(\nu_0) - P_0(\nu_0)\right] + \frac{\partial v}{\partial r}\left[E_0(\nu_0) + P_0(\nu_0)\right] + \frac{2a}{c^2}F_0(\nu_0) \\
&\quad = 4\pi\eta_0(\nu_0) - c\chi_0(\nu_0)E_0(\nu_0)\,. \qquad (5.25)
\end{aligned}
$$

If we integrate over $\mu_0 \mathrm{d}\omega_0$ we get the *monochromatic radiation momentum equation*

$$
\begin{aligned}
&\frac{1}{c^2}\frac{\mathrm{D}F_0(\nu_0)}{\mathrm{D}t} + \frac{\partial P_0(\nu_0)}{\partial r} + \frac{3P_0(\nu_0) - E_0(\nu_0)}{r} \\
&\quad - \frac{\partial}{\partial \nu_0}\left(\nu_0\left\{\frac{v}{c^2 r}\left[F_0(\nu_0) - Q_0(\nu_0)\right] + \frac{1}{c^2}\frac{\partial v}{\partial r}Q_0(\nu_0) + \frac{a}{c^2}P_0(\nu_0)\right\}\right) \\
&\quad + \frac{2}{c^2}\left(\frac{\partial v}{\partial r} + \frac{v}{r}\right)F_0(\nu_0) + \frac{a}{c^2}\left[E_0(\nu_0) + P_0(\nu_0)\right] \\
&\quad = -\frac{1}{c}\chi_0(\nu_0)F_0(\nu_0)\,. \qquad (5.26)
\end{aligned}
$$

Finally, by integrating over ν_0 we get the *radiation energy equation*

$$\frac{\mathrm{D}E_0}{\mathrm{D}t} + \frac{1}{r^2}\frac{\partial}{\partial r}(r^2 F_0) + \frac{v}{r}(3E_0 - P_0) + \frac{\partial v}{\partial r}(E_0 + P_0) + \frac{2a}{c^2}F_0$$
$$= \int_0^\infty [4\pi\eta_0(\nu_0) - c\chi_0(\nu_0)E_0(\nu_0)]\,\mathrm{d}\nu_0 \tag{5.27}$$

and the *radiation momentum equation*

$$\frac{1}{c^2}\frac{\mathrm{D}F_0}{\mathrm{D}t} + \frac{\partial P_0}{\partial r} + \frac{3P_0 - E_0}{r} + \frac{2}{c^2}\left(\frac{\partial v}{\partial r} + \frac{v}{r}\right)F_0 + \frac{a}{c^2}(E_0 + P_0)$$
$$= -\frac{1}{c}\int_0^\infty \chi_0(\nu_0)F_0(\nu_0)\,\mathrm{d}\nu_0\,. \tag{5.28}$$

As we shall see, these equations have a direct physical interpretation. Further, it is clear that they can be incoporated into a practical computational scheme.

5.2 Consistency Between Comoving-Frame and Inertial-Frame Equations

Consider a grey, planar, pure-absorbing medium in LTE. Then the comoving-frame energy and momentum equations (omitting acceleration terms) simplify to

$$\frac{\partial E_0}{\partial t} + \frac{\partial F_0}{\partial z} + v\frac{\partial E_0}{\partial z} + \left(\frac{\partial v}{\partial z}\right)(E_0 + P_0) = \kappa_0(4\pi B_0 - cE_0) \tag{5.29}$$

and

$$\frac{1}{c^2}\frac{\partial F_0}{\partial t} + \frac{\partial P_0}{\partial z} + \frac{v}{c^2}\frac{\partial F_0}{\partial z} + \frac{2}{c^2}\left(\frac{\partial v}{\partial z}\right)F_0 = -\frac{\kappa_0}{c}F_0\,. \tag{5.30}$$

Start from the inertial-frame energy equation (4.11)

$$\frac{\partial E}{\partial t} + \frac{\partial F}{\partial z} = \kappa_0(4\pi B_0 - cE) + \frac{v}{c}\kappa_0 F \tag{5.31}$$

and use the $O(v/c)$ transformations for E and F (3.37–38) to get

$$\frac{\partial E_0}{\partial t} + \frac{\partial F_0}{\partial z} + v\left(\frac{\partial E_0}{\partial z} + \frac{\partial P_0}{\partial z}\right) + \left(\frac{\partial v}{\partial z}\right)(E_0 + P_0)$$
$$= \kappa_0(4\pi B_0 - cE_0) - \frac{v}{c}\kappa_0 F_0 + O(v^2/c^2)\,. \tag{5.32}$$

Now rearrange terms and use (5.30) to get

$$\begin{aligned}\frac{\mathrm{D}E_0}{\mathrm{D}t} + \frac{\partial F_0}{\partial z} + \left(\frac{\partial v}{\partial z}\right)(E_0 + P_0) &= \kappa_0(4\pi B_0 - cE_0) \\ &\quad - v\left(\frac{\kappa_0 F_0}{c} + \frac{\partial P_0}{\partial z}\right) + O(v^2/c^2) \\ &= \kappa_0(4\pi B_0 - cE_0) + O(v^2/c^2)\end{aligned} \tag{5.33}$$

which is identical to (5.29). Many similar consistency checks are possible; they all work out. The upshot of the analysis is that we get consistency between the inertial-frame and comoving frame equations *if, and only if,* we retain the $O(v/c)$ terms in both frames.

5.3 Noninertial Frame Derivation (J.I. Castor)

We now work directly in the accelerated frame of a Lagrangian fluid element. The algebra is rather heavy, but the deepened conceptual basis makes the additional effort worthwhile. We can sketch only the main points here.
Start from the Boltzmann equation for photons:

$$M^\alpha \frac{\partial \mathcal{I}}{\partial x^\alpha} + \dot{M}^\alpha \frac{\partial \mathcal{I}}{\partial M^\alpha} = \mathcal{E} - \mathcal{A}\mathcal{I} = \left(\frac{\delta \mathcal{I}}{\delta \ell}\right)_{\text{coll}} . \tag{5.34}$$

Recalling that photon trajectories are geodesics even in the curved spacetime of the comoving frame, we can write, using Christoffel symbols,

$$\frac{\delta M^\alpha}{\delta \ell} \equiv 0 = \frac{\mathrm{d}M^\alpha}{\mathrm{d}\ell} + \left\{ {\alpha \atop \beta\,\gamma} \right\} M^\beta \frac{\mathrm{d}x^\gamma}{\mathrm{d}\ell} . \tag{5.35}$$

But $M^\alpha = \mathrm{d}x^\alpha/\mathrm{d}\ell$, and $\dot{M}^\alpha = \mathrm{d}M^\alpha/\mathrm{d}\ell$, so (5.35) can be rewritten

$$\dot{M}^\alpha = -\left\{ {\alpha \atop \beta\,\gamma} \right\} M^\beta M^\gamma . \tag{5.36}$$

Using (5.36) in the photon Boltzmann equation we have

$$M^\alpha \frac{\mathrm{D}\mathcal{I}}{\mathrm{D}x^\alpha} = \mathcal{E} - \mathcal{A}\mathcal{I} , \tag{5.37}$$

where

$$\frac{\mathrm{D}}{\mathrm{D}x^\alpha} \equiv \frac{\partial}{\partial x^\alpha} - \left\{ {\gamma \atop \alpha\,\beta} \right\} M^\beta \frac{\partial}{\partial M^\gamma} . \tag{5.38}$$

To evaluate the Christoffel symbols we need the metric of our Lagrangian spacetime. We return to this point later.
For purposes of calculation it is more convenient to not use contravariant components M^γ, but rather components measured in an *orthonormal tetrad,* which can be chosen to exploit any symmetries in the problem. Thus we use

$$M^a \equiv \epsilon^a_\alpha M^\alpha , \tag{5.39}$$

where ϵ^a_α is an appropriate transformation matrix. After some algebra one can show that the transfer equation in the tetrad frame is

$$M^\alpha \frac{\mathrm{D}\mathcal{I}}{\mathrm{D}x^a} = M^a \left(\partial_a - \Gamma^b_{ac} M^c \frac{\partial}{M^b}\right) \mathcal{I} = \mathcal{E} - \mathcal{A}\mathcal{I} = \left(\frac{\delta \mathcal{I}}{\delta \ell}\right)_{\text{coll}} , \tag{5.40}$$

where

$$\partial_a \equiv \epsilon_a^\alpha \frac{\partial}{\partial x^\alpha} \tag{5.41}$$

is called the *Pfaffian derivative*, and the connection coefficients

$$\Gamma^b_{ac} \equiv \epsilon_a^\alpha \, \epsilon_\beta^b \, \epsilon^\beta_{c;\alpha} \tag{5.42}$$

are the *Ricci rotation coefficients*. If we are given the metric we can evaluate the ϵ's and the Γ's; note that the Γ's are not symmetric in lower indices (Christoffel symbols are). In practice only three tetrad components of $\boldsymbol{M}$ are independent because $\boldsymbol{M}$ is null. Thus, we will use only space components in the tetrad frame.

Lindquist [21] considers the general metric

$$\mathrm{d}s^2 = -\mathrm{e}^{2\Psi}\mathrm{d}\tau^2 + \mathrm{e}^{2\Lambda}\mathrm{d}\mathcal{R}^2 + R^2(\mathrm{d}\theta^2 + \sin^2\theta \mathrm{d}\phi^2) , \tag{5.43}$$

where $\mathcal{R}$ is a generalized Lagrangian radial coordinate, and τ is "time" as measured in the Lagrangian frame. Now impose spherical symmetry; then $\partial/\partial\theta = \partial/\partial\phi \equiv 0$, and we need calculate only terms in $\mathcal{R}$ and τ.

This calculation is not difficult, but it is laborious. One finds 16 nonzero Christoffel symbols for the metric in (5.43). The metric also implies ϵ_a^α and its inverse ϵ_α^a. Given the $\epsilon's$ and the Christoffel symbols, one can calculate the $\Gamma's$; after a lot of algebra one gets ten nonzero values. Next, choosing tetrad components:

$$M^0 = \nu \tag{5.44}$$

$$M^1 = \nu \cos\Theta = \nu\mu \tag{5.45}$$

$$M^2 = \nu \sin\Theta \cos\Phi = \nu(1-\mu^2)^{1/2}\cos\Phi \tag{5.46}$$

$$M^3 = \nu \sin\Theta \sin\Phi = \nu(1-\mu^2)^{1/2}\sin\Phi \tag{5.47}$$

we can find the Jacobian $J\left(\dfrac{M^1, M^2, M^3}{\nu,\ \Theta,\ \Phi}\right)$ and its inverse $J\left(\dfrac{\nu,\ \Theta,\ \Phi}{M^1, M^2, M^3}\right)$. We can then expand

$$\frac{\partial}{\partial M^1} = \frac{\partial\nu}{\partial M^1}\frac{\partial}{\partial\nu} + \frac{\partial\mu}{\partial M^1}\frac{\partial}{\partial\mu} + \frac{\partial\Phi}{\partial M^1}\frac{\partial}{\partial\Phi} = \mu\frac{\partial}{\partial\nu} + \frac{1-\mu^2}{\nu}\frac{\partial}{\partial\mu} \tag{5.48}$$

and similarly for $\partial/\partial M^2$ and $\partial/\partial M^3$. Putting all the pieces together, with a large amount of algebra we can evaluate $M^\alpha \mathrm{D}\mathcal{I}/\mathrm{D}x^\alpha$, obtaining finally *Lindquist's transfer equation*

$$\begin{aligned} D_\tau\mathcal{I} + \mu\mathrm{D}_\mathcal{R}\mathcal{I} - \nu\left[\mu\mathrm{D}_\mathcal{R}\Psi + \mu^2\mathrm{D}_\tau\Lambda + (1-\mu^2)\frac{U}{R}\right]\frac{\partial\mathcal{I}}{\partial\nu} \\ + (1-\mu^2)\left\{\frac{\Gamma}{R} - \mathrm{D}_\mathcal{R}\Psi + \mu\left[\frac{U}{R} - \mathrm{D}_\tau\Lambda\right]\right\}\frac{\partial\mathcal{I}}{\partial\mu} = \frac{1}{\nu}(\mathcal{E} - \mathcal{A}\mathcal{I}) , \end{aligned} \tag{5.49}$$

where

$$D_\tau \equiv e^{-\Psi}\frac{\partial}{\partial\mathcal{T}} \qquad (5.50\text{-a}) \qquad\qquad \Gamma \equiv D_{\mathcal{R}}R \qquad (5.50\text{-c})$$

$$D_{\mathcal{R}} \equiv e^{-\Lambda}\frac{\partial}{\partial\mathcal{R}} \qquad (5.50\text{-b}) \qquad\qquad U \equiv D_\tau R\,. \qquad (5.50\text{-d})$$

So all we need is a metric!

Castor adopted the following coordinate transformation from the inertial-frame coordinates (t', r, θ, ϕ) to Lagrangian coordinates (t, M_r, θ, ϕ):

$$M_r(r,t') = \int_0^r 4\pi r'^2 \varrho(r',t')\,\mathrm{d}r' \tag{5.51}$$

and

$$t(r,t') = t' - \frac{1}{c^2}\int_0^r v(r',t')\,\mathrm{d}r' \tag{5.52}$$

which repesent an approximate Lorentz transformation locally in the neighborhood of (r, t'). It is then easy to show that

$$\mathrm{d}x \equiv \left(\frac{\mathrm{d}M_r}{4\pi r^2\varrho}\right) = \mathrm{d}r - v\mathrm{d}t'\ , \tag{5.53-a}$$

$$\mathrm{d}t = \left(1 - \frac{I}{c^2}\right)\mathrm{d}t' - \frac{v}{c^2}\mathrm{d}r\ , \tag{5.53-b}$$

where

$$I \equiv \int_0^r \frac{\partial v(r',t')}{\partial t'}\mathrm{d}r' \tag{5.54}$$

and we used the equation of continuity to show that $(\partial M_r/\partial t')_r = -4\pi r^2\varrho v$. Inverting equations (5.53) we get

$$\mathrm{d}r = \left[\frac{1-(I/c^2)}{D}\right]\mathrm{d}x + \left(\frac{v}{D}\right)\mathrm{d}t\ , \tag{5.55}$$

$$\mathrm{d}t' = \left(\frac{v}{c^2 D}\right)\mathrm{d}x + \left(\frac{1}{D}\right)\mathrm{d}t\ , \tag{5.56}$$

where

$$D \equiv 1 - \frac{1}{c^2}(I + v^2)\,. \tag{5.57}$$

Now plug equations (5.55) and (5.56) into the inertial-frame metric

$$\mathrm{d}s^2 = \mathrm{d}r^2 + r^2(\mathrm{d}\theta^2 + \sin^2\theta\mathrm{d}\phi^2) - c^2(\mathrm{d}t')^2 \tag{5.58}$$

to get

$$\mathrm{d}s^2 = F\left(\frac{\mathrm{d}M_r}{4\pi r^2 \varrho}\right)^2 + r^2\left(\mathrm{d}\theta^2 + \sin^2\theta \mathrm{d}\phi^2\right) - G\mathrm{d}t^2 - 2H\mathrm{d}M_r\mathrm{d}t\ , \qquad (5.59)$$

where

$$F \equiv \frac{1}{D^2}\left[\left(1 - \frac{I}{c^2}\right)^2 - \frac{v^2}{c^2}\right]\ , \qquad (5.60)$$

$$G \equiv (c^2 - v^2)/D^2\ , \qquad (5.61)$$

$$H \equiv vI/(4\pi r^2 \varrho c^2 D^2)\ . \qquad (5.62)$$

One can easily show that H is $O(v^2/c^2)$, hence it can be dropped. Similarly $F \to 1 + O(v^2/c^2)$ and $G \to c^2 + 2I + O(v^2/c^2)$.

We now have Ψ, Λ, and $\mathcal{R}$ in the Lindquist metric. Therefore we can plug these quantities into equations (5.50) and (5.49), and after considerable amount of algebra we get the comoving frame transfer equation

$$\begin{aligned}
&\frac{1}{c}\frac{\mathrm{D}I_0(\mu_0,\nu_0)}{\mathrm{D}t} + 4\pi\varrho\mu_0\frac{\partial}{\partial M_r}[r^2 I_0(\mu_0,\nu_0)] \\
&\quad + \frac{\partial}{\partial\mu_0}\left\{(1-\mu_0^2)\left[\frac{1}{r} + \frac{\mu_0}{c}\left(\frac{3v}{r} + \frac{\mathrm{D}\ln\varrho}{\mathrm{D}t}\right) - \frac{a}{c^2}\right] I_0(\mu_0,\nu_0)\right\} \\
&\quad - \frac{\partial}{\partial\nu_0}\left\{\nu_0\left[(1-3\mu_0^2)\frac{v}{cr} - \frac{\mu_0^2}{c}\frac{\mathrm{D}\ln\varrho}{\mathrm{D}t} + \frac{\mu_0 a}{c^2}\right] I_0(\mu_0,\nu_0)\right\} \\
&\quad + \left[(1-3\mu_0^2)\frac{v}{cr} - \frac{(1+\mu_0^2)}{c}\frac{\mathrm{D}\ln\varrho}{\mathrm{D}t} + \frac{2\mu_0 a}{c^2}\right] I_0(\mu_0,\nu_0) \\
&\quad = \eta_0(\nu_0) - \chi_0(\nu_0) I_0(\mu_0,\nu_0)\ . \qquad (5.63)
\end{aligned}$$

The following remarks are appropriate:

1. Equation (5.63) is a truly Lagrangian equation. Note the presence of $\partial/\partial M_r$ and $\mathrm{D}\ln\varrho/\mathrm{D}t$.
2. Of course Eq. (5.16) can be cast into this form by an appropriate grouping of terms. Indeed, using the equation of continuity, $\mathrm{D}\ln\varrho/\mathrm{D}t + (\partial v/\partial r) + (2v/r) = 0$, it is easy to show that Eq. (5.63) is identical to (5.16).
3. Taking angle moments we get the same radiation energy and momentum equations as before.

For future reference, the radiation energy equation becomes

$$\begin{aligned}
&\varrho\frac{\mathrm{D}}{\mathrm{D}t}\left(\frac{E_0}{\varrho}\right) + 4\pi\varrho\frac{\partial}{\partial M_r}\left(r^2 F_0\right) - \left(\frac{\mathrm{D}\ln\varrho}{\mathrm{D}t}\right)P_0 - \frac{v}{r}(3P_0 - E_0) + \frac{2a}{c^2}F_0 \\
&\quad = \int_0^\infty \left[4\pi\eta_0(\nu_0) - c\chi_0(\nu_0)E_0(\nu_0)\right]\mathrm{d}\nu_0 \qquad (5.64)
\end{aligned}$$

and the radiation momentum equation becomes

$$\frac{\varrho}{c^2}\frac{\mathrm{D}}{\mathrm{D}t}\left(\frac{F_0}{\varrho}\right) + 4\pi r^2 \varrho \frac{\partial P_0}{\partial M_r} + \frac{(3P_0 - E_0)}{r} - \frac{1}{c^2}\left(\frac{\mathrm{D}\ln\varrho}{\mathrm{D}t} + \frac{2v}{r}\right) F_0 \\ + \frac{a}{c^2}(E_0 + P_0) = -\frac{1}{c}\int_0^\infty \chi_0(\nu_0) F_0(\nu_0)\,\mathrm{d}\nu_0 \,. \tag{5.65}$$

General Geometry. From a different $O(v/c)$ analysis Buchler showed that the radiation energy and momentum equations can be rewritten in general tensorial form:

$$\varrho \frac{\mathrm{D}}{\mathrm{D}t}\left(\frac{E_0}{\varrho}\right) + \boldsymbol{\nabla}\cdot\boldsymbol{F}_0 + \mathsf{P}_0 : \boldsymbol{\nabla v} + \frac{2}{c^2}\boldsymbol{a}\cdot\boldsymbol{F}_0 + c\,G_0^0 = 0 \tag{5.66}$$

and

$$\frac{\varrho}{c^2}\frac{\mathrm{D}}{\mathrm{D}t}\left(\frac{\boldsymbol{F}_0}{\varrho}\right) + \boldsymbol{\nabla}\cdot\mathsf{P}_0 + \frac{1}{c^2}(\boldsymbol{F}_0\cdot\boldsymbol{\nabla})\boldsymbol{v} \\ + \frac{1}{c^2}(E_0\boldsymbol{a} + \boldsymbol{a}\cdot\mathsf{P}_0) + \boldsymbol{G}_0 = 0 \,. \tag{5.67}$$

These equations are useful for working in other geometries (e.g., cylindrical, 3D Cartesian). Note that the monochromatic forms and the full transfer equation in Buchler's paper contain numerous typos (and some alleged errors); more complete and accurate formulae have been published by [26].

5.4 Analysis of $O(v/c)$ Terms

We can play the same dimensional analysis game with these equations as we did with the inertial-frame equations. Consider first the energy equation (5.66); as before consider fluid-flow timescales only. Ignore the acceleration terms which are never larger than $O(v/c)$.

		Streaming:	Static diffusion:	Dynamic diffusion:
A	$\varrho\frac{\mathrm{D}}{\mathrm{D}t}\left(\frac{E_0}{\varrho}\right)$	(v/c)	$(v/c)(\ell/\lambda_\mathrm{p})$	1
B	$+\frac{1}{r^2}\frac{\partial}{\partial r}(r^2 F_0)$	1	1	$(c/v)(\lambda_\mathrm{p}/\ell)$
C	$-\frac{v}{r}(3P_0 - E_0)$	(v/c)	$(v/c)^2$	$(v/c)(\lambda_\mathrm{p}/\ell)$
D	$-\left(\frac{\mathrm{D}\ln\varrho}{\mathrm{D}t}\right)P_0$	(v/c)	$(v/c)(\ell/\lambda_\mathrm{p})$	1
E	$= (4\pi\eta - c\chi E)$	$(\ell/\lambda_\mathrm{p})$	1	1

In the streaming limit we need to retain the flux divergence B and the source-sink term E. We would need the time derivative term A if we tried to track radiation fronts. Also, if the medium is very nearly in radiative equilibrium, term E $\to 0$, and both DE_0/Dt and the velocity-dependent terms may be needed to balance (or drive) $\nabla \cdot \boldsymbol{F}$. In static diffusion we need only the divergence term B and the absorption-emission term E so long as $(v/c) < (\lambda_p/\ell)$. As $(v/c) \to (\lambda_p/\ell)$ we need to retain all terms except term C, $(3P_0 - E_0)$. In practice we would keep C anyway so that in spherical geometry we can identify all terms in the radiation work term $\mathsf{P}_0 : \nabla \boldsymbol{v}$. In dynamic diffusion we need the DE_0/Dt term A, the work term D, and the absorption-emission term E.

In short, to guarantee a correct solution of the comoving frame radiation energy equation in all regimes, we must retain all time- and velocity-dependent terms. Thus we reach the same conclusion as we did in the intertial frame (though the actual terms are, of course, quite different).
Similarly for the momentum equation (5.67)

	A	B	C	D	E
	$\frac{\varrho}{c^2}\frac{D}{Dt}\left(\frac{F_0}{\varrho}\right)$	$+\frac{\partial P_0}{\partial r}$	$+\ \frac{3P_0-E_0}{r}$	$-\ \frac{F_0}{c^2}\frac{\partial v}{\partial r}$	$= -\frac{\chi_0 F_0}{c}$
Streaming:	(v/c)	1	1	(v/c)	(ℓ/λ_p)
Diffusion regime:	$(v/c)(\lambda_p/\ell)$	1	$(v/c)(\lambda_p/\ell)$	$(v/c)(\lambda_p/\ell)$	1

In the streaming regime we need terms B and C (which together give $\nabla \cdot \mathsf{P}_0$), and the radiation force E. In the diffusion regime we also need only these terms. Thus to guarantee the correct solution of the radiation momentum equation in all regimes we need retain only

$$\frac{\partial P_0}{\partial r} + \frac{3P_0 - E_0}{r} = -\frac{1}{c}\chi_0 F_0 \,. \tag{5.68}$$

This result is much simpler than the inertial-frame equation. If we want to follow radiation flow, we would also need to retain the term in DF/Dt. Castor arrived at the same conclusions for the radiation energy and momentum equations by considering stellar pulsation with a frequency ω (which sets the characteristic value of D/Dt).

6. Lagrangian Equations of Radiation Hydrodynamics

In this section we shall derive radiation hydrodynamics equations in the comoving frame, ignoring acceleration terms, which are $O(v^2/c^2)$, on a fluid-flow timescale.

6.1 Momentum Equation

If we reduce the relativistically correct equation, derived earlier (4.29), to the comoving frame (in which $\boldsymbol{v} = 0$ instantaneously), we get to $O(v/c)$,

$$\varrho_{000}\frac{\mathrm{D}\boldsymbol{v}}{\mathrm{D}t} = \boldsymbol{f} - \boldsymbol{\nabla}p + \boldsymbol{G}_0 \,. \tag{6.1}$$

Here we ignored the $O(v^2/c^2)$ difference between the laboratory frame time t and proper time τ. If the fluid is not relativistic internally (*i.e.*, at extremely high temperature or degenerate), we can set $\varrho_{000} \to \varrho$. We then get (for 1D spherical flow)

$$\varrho\frac{\mathrm{D}v}{\mathrm{D}t} = -\frac{GM_r\varrho}{r^2} - \frac{\partial p}{\partial r} + \frac{1}{c}\int_0^\infty \chi_0(\nu_0)F_0(\nu_0)\,\mathrm{d}\nu_0 \,. \tag{6.2}$$

This equation is just what one would guess intuitively from Newtonian arguments; but note the specification of a particular (comoving) frame for the radiative term.

We can rewrite Eq. (6.2) by using the radiation momentum equation (5.65) and the continuity equation. We obtain

$$\varrho\frac{\mathrm{D}}{\mathrm{D}t}\left(v + \frac{F_0}{\varrho c^2}\right) = -\frac{GM_r\varrho}{r^2} - \frac{\partial(p+P_0)}{\partial r} - \frac{3P_0 - E_0}{r} - \frac{1}{c^2}\frac{\partial v}{\partial r}F_0 \,. \tag{6.3}$$

This equation states:

The rate of change of the total momentum density
of the radiating fluid
= the external force
− the divergence of the total pressure ($p\mathsf{I} + \mathsf{P}_0$) in the radiating fluid
− a relativistic term arising form the inertia for radiant energy .

On a fluid-flow timescale all terms in F_0 are at most $O(v/c)$, therefore we may drop them in nonrelativistic practical calculations. We obtain:

$$\varrho\frac{\mathrm{D}v}{\mathrm{D}t} = \boldsymbol{f} - \boldsymbol{\nabla}p - \boldsymbol{\nabla}\cdot\mathsf{P}_0 \,. \tag{6.4}$$

This result is again what one would guess from purely Newtonian arguments.

6.2 Gas Energy Equation

We have seen (in Sect. 4.2) that the relativistic gas energy equation is

$$\varrho_0\left[\frac{\mathrm{D}e}{\mathrm{D}\tau} + p\frac{\mathrm{D}}{\mathrm{D}\tau}\left(\frac{1}{\varrho_0}\right)\right] = -V_\alpha(F^\alpha + G^\alpha) \,. \tag{6.5}$$

In the comoving frame the right-hand side reduces to $c(F_0^0 + G_0^0)$. For ordinary body forces $F_0^0 \equiv 0$; if thermonuclear reactions occur, $V_\alpha F^\alpha = \varrho_0\varepsilon$ (*e.g.*, in stellar interiors or in a laser-fusion pellet). Then

$$\varrho_0 \left[\frac{\mathrm{D}e}{\mathrm{D}\tau} + p \frac{\mathrm{D}}{\mathrm{D}\tau} \left(\frac{1}{\varrho_0} \right) \right] = \int_0^\infty \left[c\chi_0(\nu_0) E_0(\nu_0) - 4\pi\eta_0(\nu_0) \right] \mathrm{d}\nu_0 + \varrho_0 \varepsilon \,. \quad (6.6)$$

We can now drop $O(v^2/c^2)$ terms, hence $\varrho_0 \to \varrho$, $\mathrm{D}/\mathrm{D}\tau \to \mathrm{D}/\mathrm{D}t$, obtaining

$$\varrho \left[\frac{\mathrm{D}e}{\mathrm{D}t} + p \frac{\mathrm{D}}{\mathrm{D}t} \left(\frac{1}{\varrho} \right) \right] = \int_0^\infty \left[c\chi_0(\nu_0) E_0(\nu_0) - 4\pi\eta_0(\nu_0) \right] \mathrm{d}\nu_0 + \varrho \varepsilon \,. \quad (6.7)$$

Clearly Eq. (6.7) is the first law of thermodynamics for the material.

6.3 First Law of Thermodynamics for the Radiation Field

To obtain the desired result, we can simply rewrite the Lagrangian radiation energy equation (5.64) as

$$\begin{aligned} &\varrho \left[\frac{\mathrm{D}}{\mathrm{D}t} \left(\frac{E_0}{\varrho} \right) + P_0 \frac{\mathrm{D}}{\mathrm{D}t} \left(\frac{1}{\varrho} \right) - \frac{v}{\varrho r} (3P_0 - E_0) \right] \\ &\quad = \int_0^\infty \left[4\pi\eta_0(\nu_0) - c\chi_0(\nu_0) E_0(\nu_0) \right] \mathrm{d}\nu_0 - \frac{1}{r^2} \frac{\partial}{\partial r} \left(r^2 F_0 \right) \,, \end{aligned} \quad (6.8)$$

The rate of change of the radiation energy density
per unit volume

+ the rate of work done by radiation stress ($\mathsf{P} : \boldsymbol{\nabla v}$)

= the net rate of "heat delivery" to the radiation field
(*i.e.*, energy emitted by gas – energy absorbed by gas)

– the divergence of the radiant "heat flux" out of the volume .

Equation (6.8) is just what one would expect for a nonviscous, heat-conducting fluid.

6.4 First Law of Thermodynamics for the Radiating Fluid

To obtain the first law of thermodynamics for the radiating fluid simply add the first law of thermodynamics for the material (6.7) to the first law of thermodynamics for the radiation (6.8). We find

$$\begin{aligned} &\frac{\mathrm{D}}{\mathrm{D}t} \left(e + \frac{E_0}{\varrho} \right) + p \frac{\mathrm{D}}{\mathrm{D}t} \left(\frac{1}{\varrho} \right) + \left[P_0 \frac{\mathrm{D}}{\mathrm{D}t} \left(\frac{1}{\varrho} \right) - \frac{v}{\varrho r} (3P_0 - E_0) \right] \\ &\quad = \varepsilon - \frac{\partial}{\partial M_r} \left(4\pi r^2 F_0 \right) \,, \end{aligned} \quad (6.9)$$

The rate of change of total radiating fluid energy density per unit volume
+ the rate of work done by total stress in radiating fluid
= the rate of energy input from irreversible processes
− the divergence of the total "heat flux" out of the volume (material conduction ignored here) .

In the diffusion limit, $3P_0 = E_0$ + high order terms, therefore we can simplify (6.9) to

$$\frac{\mathrm{D}}{\mathrm{D}t}\left(e+\frac{E_0}{\varrho}\right)+(p+P_0)\frac{\mathrm{D}}{\mathrm{D}t}\left(\frac{1}{\varrho}\right)=\varepsilon-\frac{\partial L_r}{\partial M_r}\,, \tag{6.10}$$

which is one of the standard equations of time-dependent stellar structure. Because the nuclear timescale t_{N} (of the order of 10^9 years) is so much greater than a characteristic dynamical time of a star (of the order of a few days at most), in following stellar evolution we can go to the quasistatic limit, hence

$$\frac{\partial L_r}{\partial M_r}=\varepsilon\,, \tag{6.11}$$

which is one of the standard equations of stellar structure and (quasistatic) evolution.

6.5 Mechanical Energy Equation

If we multiply the momentum equation (6.2) by v, we get

$$\varrho\frac{\mathrm{D}(\frac{1}{2}v^2)}{\mathrm{D}t}=-\frac{GM_r\varrho v}{r^2}-v\left(\frac{\partial p}{\partial r}\right)+\frac{v}{c}\int_0^\infty \chi_0(\nu_0)F_0(\nu_0)\,\mathrm{d}\nu_0\,, \tag{6.12}$$

The rate of change of the mechanical energy density
= the rate of work done by gravity
+ the rate of work done by the pressure gradient
+ the rate of work done by radiation forces .

6.6 Total Energy Equation

Using the equation of continuity, and the relation $\boldsymbol{\nabla}\cdot(p\boldsymbol{v})=p\boldsymbol{\nabla}\cdot\boldsymbol{v}+\boldsymbol{v}\cdot\boldsymbol{\nabla}p$, we can rewrite the mechanical energy equation as

$$\frac{\mathrm{D}}{\mathrm{D}t}\left(\frac{1}{2}v^2-\frac{GM_r}{r}\right)+\frac{\partial}{\partial M_r}\left(4\pi r^2 vp\right)$$
$$=p\frac{\mathrm{D}}{\mathrm{D}t}\left(\frac{1}{\varrho}\right)+\frac{v}{c\varrho}\int_0^\infty \chi_0(\nu_0)F_0(\nu_0)\,\mathrm{d}\nu_0\,. \tag{6.13}$$

Eliminating the radiation force on the right-hand side by using the radiation momentum equation (5.65) we get

$$\frac{\mathrm{D}}{\mathrm{D}t}\left(\frac{1}{2}v^2 - \frac{GM_r}{r}\right) + \frac{\partial}{\partial M_r}\left[4\pi r^2 v(p+P_0)\right] = (p+P_0)\frac{\mathrm{D}}{\mathrm{D}t}\left(\frac{1}{\varrho}\right) - \frac{v}{\varrho r}(3P_0 - E_0)\,. \tag{6.14}$$

Now add the first law of thermodynamics for the radiating fluid (6.9) to obtain the *total energy equation of the radiating fluid:*

$$\frac{\mathrm{D}}{\mathrm{D}t}\left(e + \frac{E_0}{\varrho} + \frac{1}{2}v^2 - \frac{GM_r}{r}\right) + \frac{\partial}{\partial M_r}\left\{4\pi r^2\left[F_0 + v(p+P_0)\right]\right\} = \varepsilon\,, \tag{6.15}$$

The rate of change of the total energy density in a fluid element
+ the divergence of radiant energy flux
+ the divergence of the rate of work done by stresses in the radiating fluid
= the rate of energy production by irreversible processes .

This equation is a direct analog of the total energy equation for a heat-conducting, but inviscid, ordinary material fluid. (Compare!) The energy density and stress now have both material and radiation components. In the diffusion limit we get the same equation as is used in dynamical stellar structure calculations. If we now integrate (6.15) over a volume containing a star, apply the standard boundary conditions of stellar structure work ($T \to 0, p \to 0$, as $r \to R$), and note that $\mathrm{D}/\mathrm{D}t$ and $\partial/\partial M_r$ commute, we get

$$\frac{\mathrm{D}}{\mathrm{D}t}\int_M\left(e + \frac{E_0}{\varrho} + \frac{1}{2}\varrho v^2\right)\mathrm{d}M_r + L(M) = -\frac{\mathrm{D}\Phi}{\mathrm{D}t} + \int_0^M \varepsilon\,\mathrm{d}M_r\,, \tag{6.16}$$

The rate of change of the internal, radiant, and kinetic energy in the entire star
+ the rate radiant energy emerges from stellar surface
= the total rate of release of gravitational potential energy
+ the rate of energy production by thermonuclear energy release .

We can express (6.15) in Eulerian coordinates:

$$\frac{\partial}{\partial t}\left(\varrho e + E_0 + \frac{1}{2}\varrho v^2 - \frac{GM_r\varrho}{r}\right) + \frac{1}{r^2}\frac{\partial}{\partial r}\left(r^2\left\{F_0 + v\left[E_0 + P_0 + p + \varrho\left(e + \frac{1}{2}v^2 - \frac{GM_r}{r}\right)\right]\right\}\right) = \varrho\varepsilon\,. \tag{6.17}$$

But $F_0 + vE_0 + vP_0 = F$, and $E_0 = E + O(v/c)$. Therefore

$$\frac{\partial}{\partial t}\left(\varrho e + E + \frac{1}{2}\varrho v^2 - \frac{GM_r\varrho}{r}\right) + \frac{1}{r^2}\frac{\partial}{\partial r}\left(r^2\left\{F + v\left[p + \varrho\left(e + \frac{1}{2}v^2 - \frac{GM_r}{r}\right)\right]\right\}\right) = \varrho\varepsilon \, . \tag{6.18}$$

6.7 Consistency of Different Forms of the Radiating-Fluid Energy and Momentum Equations

Let us now examine the consistency of various forms of the radiating fluid energy equation. In the optically thin limit the "natural" form of this equation is the gas energy equation.

$$\varrho\left[\frac{\mathrm{D}e}{\mathrm{D}t} + p\frac{\mathrm{D}}{\mathrm{D}t}\left(\frac{1}{\varrho}\right)\right] = \int_0^\infty \left[c\chi_0(\nu_0)E_0(\nu_0) - 4\pi\eta_0(\nu_0)\right]\mathrm{d}\nu_0 + \varrho\varepsilon \, , \tag{6.19}$$

which expresses energy gains and losses of the gas directly. In the diffusion limit, the right-hand side of (6.19) approaches zero by cancellation to many digits. It is then natural to use the first law of thermodynamics of a radiating fluid:

$$\frac{\mathrm{D}}{\mathrm{D}t}\left(e + \frac{E_0}{\varrho}\right) + p\frac{\mathrm{D}}{\mathrm{D}t}\left(\frac{1}{\varrho}\right) + \left[P_0\frac{\mathrm{D}}{\mathrm{D}t}\left(\frac{1}{\varrho}\right) - \frac{v}{\varrho r}(3P_0 - E_0)\right] = \varepsilon - \frac{\partial}{\partial M_r}\left(4\pi r^2 F_0\right) \, . \tag{6.20}$$

In practice we must choose only *one* of these equations in a real computation even though the flow spans *both* optically thick and thin regimes. We can guarantee (analytically) a correct transition from one regime to the other *if and only if*

1. *all* of the $\mathrm{D}/\mathrm{D}t$ and v-dependent terms are retained in whichever of (6.19) or (6.20) are chosen, and
2. *all* of the $\mathrm{D}/\mathrm{D}t$ and v-dependent terms we decided were necessary are retained in the radiation energy equation.

If we do not satisfy these requirements, then disaster follows. Castor shows that dropping the $O(v/c)$ terms causes an error of $O(P/p)$ in the regime for which the energy equation chosen is not well suited. This error may be acceptable if $P/p \ll 1$, but will be a serious error if $P/p \gg 1$ (high temperature material, interior of a star, *etc.*).

The situation for the momentum equation is totally different. Here we can drop $\mathrm{D}F_0/\mathrm{D}t$ and velocity-dependent terms multiplying F_0 because all are at most $O(v/c)$. Moreover these terms become $O(v^2/c^2)$ in the mechanical energy equation. Therefore we can ignore all $O(v/c)$ terms in the radiation momentum equation without affecting consistency among various forms of the energy or momentum equations.

6.8 Consistency of Inertial-Frame and Comoving-Frame Radiation Energy and Momentum Equations

We have already shown that we *must* keep $O(v)$ terms to assure consistency of inertial-frame and comoving-frame *radiation* energy and momentum equations. How about the *radiating fluid* equations?
Start from the inertial-frame energy equation (4.40):

$$\varrho\left[\frac{\mathrm{D}e}{\mathrm{D}t}+p\frac{\mathrm{D}}{\mathrm{D}t}\left(\frac{1}{\varrho}\right)\right]=-\left(\frac{\partial E}{\partial t}+\nabla\cdot\boldsymbol{F}\right)+\boldsymbol{v}\cdot\left(\frac{1}{c^2}\frac{\partial \boldsymbol{F}}{\partial t}+\nabla\cdot\mathsf{P}\right). \tag{6.21}$$

Convert this equation to the fluid frame to $O(v/c)$:

- On a fluid-flow timescale $c^{-2}(\partial F/\partial t)$ is $O(v^2/c^2)$ compared to $\nabla\cdot\boldsymbol{F}$. Therefore drop it.
- We don't need $O(v/c)$ terms in converting $E\to E_0$ and $P\to P_0$ because they produce $O(v^2/c^2)$ terms,
- We must retain all terms in $\boldsymbol{F}=\boldsymbol{F}_0+\boldsymbol{v}E_0+\boldsymbol{v}\cdot\mathsf{P}_0$.
- In spherical geometry:

$$\begin{aligned}\nabla\cdot\boldsymbol{F} &= \frac{1}{r^2}\frac{\partial}{\partial r}[r^2(F_0+vE_0+vP_0)]\\ &= \frac{\partial F_0}{\partial r}+\frac{2F_0}{r}+v\left(\frac{\partial E_0}{\partial r}+\frac{\partial P_0}{\partial r}\right)+\left(\frac{\partial v}{\partial r}+\frac{2v}{r}\right)(E_0+P_0)\ ,\\ \nabla\cdot\mathsf{P}_0 &= \frac{\partial P_0}{\partial r}+\frac{(3P_0-E_0)}{r}\ .\end{aligned}$$

Using these results in (6.21) we obtain:

$$\begin{aligned}&\varrho\left[\frac{\mathrm{D}e}{\mathrm{D}t}+p\frac{\mathrm{D}}{\mathrm{D}t}\left(\frac{1}{\varrho}\right)\right]\\ &\quad=-\left[\frac{\mathrm{D}E_0}{\mathrm{D}t}+\frac{1}{r^2}\frac{\partial(r^2F_0)}{\partial r}+\frac{v}{r}(3E_0-P_0)+(E_0+P_0)\frac{\partial v}{\partial r}\right]\ .\end{aligned} \tag{6.22}$$

Note that the right-hand side of (6.22) is identical to the left-hand side of the radiation energy equation (5.27). Replace it by the right-hand side of that equation. We then get an exact reduction to

$$\varrho\left[\frac{\mathrm{D}e}{\mathrm{D}t}+p\frac{\mathrm{D}}{\mathrm{D}t}\left(\frac{1}{\varrho}\right)\right]=\int_0^\infty[c\chi_0(\nu_0)E_0(\nu_0)-4\pi\eta_0(\nu_0)]\,\mathrm{d}\nu_0\ , \tag{6.23}$$

the comoving frame gas energy equation (6.7).
Alternatively we can start with the inertial-frame equation (4.39) for grey material,

$$\varrho\left[\frac{\mathrm{D}e}{\mathrm{D}t}+p\frac{\mathrm{D}}{\mathrm{D}t}\left(\frac{1}{\varrho}\right)\right]=\kappa_0\left(cE-4\pi B_0-\frac{2\boldsymbol{v}\cdot\boldsymbol{F}}{c}\right). \tag{6.24}$$

But from the $O(v/c)$ version of the Lorentz transformation, $E = E_0 + 2\boldsymbol{v}\cdot\boldsymbol{F}_0/c^2$, and $\boldsymbol{v}\cdot\boldsymbol{F} = \boldsymbol{v}\cdot\boldsymbol{F}_0 + O(v^2/c^2)$. Thus (6.24) reduces to

$$\varrho\left[\frac{\mathrm{D}e}{\mathrm{D}t}+p\frac{\mathrm{D}}{\mathrm{D}t}\left(\frac{1}{\varrho}\right)\right]=\kappa_0\left(cE_0-4\pi B_0\right) \tag{6.25}$$

again, the comoving-frame gas energy equation for grey material (6.7).
Now consider the inertial-frame momentum equation (4.30)

$$\varrho\frac{\mathrm{D}\boldsymbol{v}}{\mathrm{D}t}=\boldsymbol{f}-\boldsymbol{\nabla}p-\left(\frac{1}{c^2}\frac{\partial\boldsymbol{F}}{\partial t}+\boldsymbol{\nabla}\cdot\mathsf{P}\right)+\frac{\boldsymbol{v}}{c^2}\left(\frac{\partial E}{\partial t}+\boldsymbol{\nabla}\cdot\boldsymbol{F}\right). \tag{6.26}$$

- We can ignore $O(v/c)$ terms in converting $\boldsymbol{F}\to\boldsymbol{F}_0$; they produce $O(v^2/c^2)$ terms.
- We must keep $O(v/c)$ terms in the transformations $E = E_0 + (2v/c^2)F_0$ and $P = P_0 + (2v/c^2)F_0$.
- We can drop $(v/c^2)(\partial E/\partial t)$ which is $O(v^2/c^2)$ compared to $\nabla\mathsf{P}$.
- In spherical geometry:

$$\boldsymbol{\nabla}\cdot\mathsf{P}=\frac{\partial P}{\partial r}+\frac{(3P-E)}{r},\qquad\text{and}\qquad\boldsymbol{\nabla}\cdot\boldsymbol{F}=\frac{\partial F}{\partial r}+\frac{2F}{r}.$$

Using these results in (6.26) we obtain:

$$\varrho\frac{\mathrm{D}v}{\mathrm{D}t}=-\frac{GM_r\varrho}{r^2}-\frac{\partial p}{\partial r}-\left[\frac{1}{c^2}\frac{\partial F_0}{\partial t}+\frac{v}{c^2}\frac{\partial F_0}{\partial r}+\frac{\partial P_0}{\partial r}\right.$$
$$\left.+\frac{(3P_0-E_0)}{r}+\frac{2}{c^2}\left(\frac{\partial v}{\partial r}+\frac{v}{r}\right)F_0\right]. \tag{6.27}$$

The radiative terms on the right-hand side of (6.27) are identical to those on the left-hand side of the comoving-frame radiation momentum equation (5.28). Therefore replace them with the left-hand side of that equation. We obtain

$$\varrho\frac{\mathrm{D}v}{\mathrm{D}t}=-\frac{GM_r\varrho}{r^2}-\frac{\partial p}{\partial r}+\frac{1}{c}\int_0^\infty\chi_0(\nu_0)F_0(\nu_0)\,\mathrm{d}\nu_0\,, \tag{6.28}$$

which is the correct comoving-frame momentum equation (6.2). One can perform many other checks; all of them show consistency, so we can trust our equations.

7. Radiation Diffusion

Having derived the equations of radiation hydrodynamics, how do we solve them? For a preliminary orientation, it is instructive to count the number of variables and the number of equations available to determine them. For the material, the variables that appear in the equations are $\varrho, \boldsymbol{v}, e, p$, and T. There are a total of seven variables (the velocity has three components). To determine these quantities we have the equation of continuity, three components of the momentum equation, the gas energy equation, a mechanical equation of state $p = p(\varrho, T)$, and a caloric equation of state $e = e(\varrho, T)$. We thus have the same number of equations as we have unknown variables. The system is complete. In fact, we can ignore T and p if, as is true for, say, a polytrope, we know $p = p(\varrho, e)$.

For radiation we have to determine the scalar E, the vector $\boldsymbol{F}$, and the symmetric tensor P, which in 3D comprise one, three, and six components respectively, for a total of ten. To determine these variables we have the radiation energy equation plus three components of the radiation momentum equation. Thus *we are short by six equations*! The missing information must be supplied by a ***closure relation.*** A convenient closure relation is provided by the *Eddington tensor*, $\mathsf{f} = \mathsf{P}/E$, which, however, is not known a priori, and must be computed in an auxiliary calculation. In 1D we need only one component; in 2D we need three; and in 3D we need six. In 1D the calculations of the Eddington factor is relatively easy. In 3D it can be quite difficult, and costly.

The simplest possible case is ***diffusion.*** Here we know a priori that down the diagonal $f \equiv \frac{1}{3}$, and all off diagonal elements are zero. So let's consider this case first.

In general we will solve radiation hydro problems numerically on a discrete grid. We will discuss this procedure later; here we make only a few simple remarks. Typically the physical variables are centered in space on a ***staggered mesh*** as shown in Fig. (7.1). The centering in time can be ***leapfrog*** if we use an *explicit* method, or we can ***synchronize*** all variables at a given time level if we use an ***implicit*** method. The basic idea is that if we are given ***initial conditions*** we can construct the solution at a sequence of advanced times by means of a suitable numerical ***marching procedure.*** We do not require further information about the procedure now; we will treat it in more detail later. For now we will deal with physical issues.

7.1 Radiation Diffusion

The basic assumptions of all diffusion theories is that the test point is far from boundaries (optical depth $\tau \gg 1$), and that the particle mean free path is very small compared to a typical structural length ($\lambda_{\mathrm{p}} \ll \ell$). In the case of radiation, there are several "flavors" of diffusion.

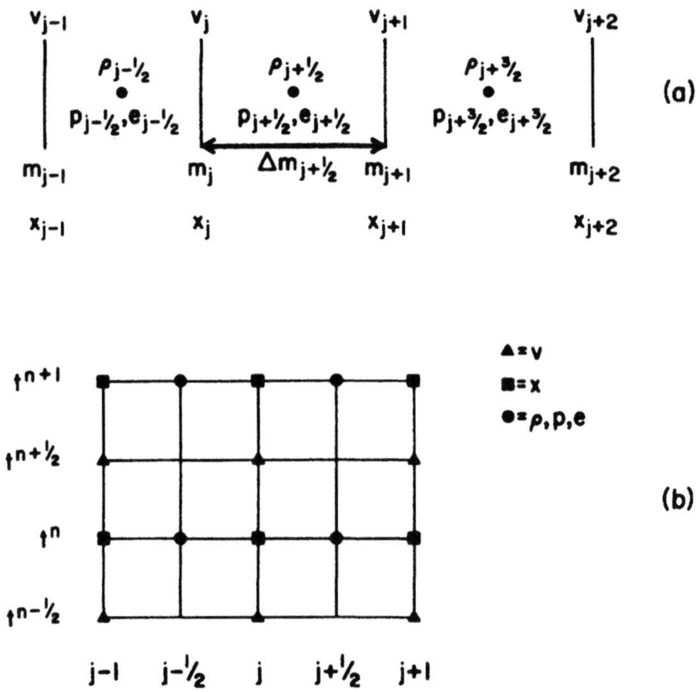

Fig. 7.1. (**a**) Centring of physical variables on Lagrangian mesh. (**b**) Spacetime centring of physical variables on Lagrangian mesh

- *Equilibrium diffusion* assumes that the radiation temperature is identical to the material temperature so that $E = 3P = a_R T^4$.
- Equilibrium diffusion applies strictly only for material very far from boundaries or regions with steep temperature gradients.
- In equilibrium diffusion it is always assumed that the radiation has a Planckian spectrum. Thus we can always replace a nongrey opacity with the Rosseland mean, and we never need multigroup (multiple frequency groups) diffusion.
- A second-order diffusion theory can be worked out for equilibrium diffusion. It allows us to develop a theory for radiation viscosity.
- In *nonequilibrium diffusion* we allow the radiation temperature $T_{\text{radiation}}$ to differ from the material temperature T_{material}. Then $E = 3P = a_R T_R^4$ where $T_R \neq T$. Thus this theory can give decent results even near a boundary surface. It is also reasonable to use for, say, the penetration of radiation from a hot source into cold material.
- In *nonequilibrium diffusion* the multigroup problem makes sense because the spectrum of radiation is not constrained to be Planckian, and, in fact, the spectral distribution of the radiation field is one of the things we want to determine.

Ordering of Terms. In *static* media we saw that in the diffusion limit all radiation properties are specified by local fluid properties and their gradients. There is only one significant parameter in the problem: (λ_p/ℓ). In *moving* media the situation is more complicated. Here we have *two* small parameters, (λ_p/ℓ) and (v/c), which are *independent.* We must now be careful in sorting terms in order to get a valid asymptotic theory:

- Zero order: drop all terms $O(\lambda_p/\ell)$, $O(v/c)$ or higher.
- First order: retain $O(\lambda_p/\ell)$, $O(v/c)$. Work in comoving frame to find that $O(v/c)$ terms are absent. (The photon horizon contracts to a point; the observer is unaware he is moving).
- "$1\frac{1}{2}$" order: retain terms $O(\frac{\lambda_p}{\ell}\frac{v}{c})$. Such terms are the next dominant terms after 1st order.
- Second order: (a) retain terms of $O(v^2/c^2)$. These terms are contained in the covariant expression for R which can be worked out to all orders in (v/c); or (b) retain terms of $O(\lambda_p^2/\ell^2)$. This expression is not used at all because it gives poor results in ordinary gas dynamics.
- In view of the above, we whall refer to the "$1\frac{1}{2}$" order expression as "2nd order".

One final remark: all of the arguments we shall use below are heuristic. A rigorous asymptotic theory is both possible and needed; this work would make a good Ph.D. thesis.

Thermalization. In static media we need require only $\tau \gg 1$, *i.e.*, $(\lambda_p/\ell) \ll 1$. In moving media we should demand in addition that a photon be destroyed in the same thermal environment as it was created. Thus we must assure that

$$t_\lambda = \frac{\lambda_p}{c} \ll t_{\mathsf{flow}} = \frac{\ell}{v} \Rightarrow \frac{\lambda_p}{\ell}\frac{v}{c} \ll 1\,. \tag{7.1}$$

This condition is almost always assured whenever diffusion is appropriate; certainly it is true in the stellar interior. For scattering, replace λ_p by Λ, the thermalization length.

Zero-Order "Diffusion". For grey matter, the radiation energy equation (5.64) becomes in the equilibrium diffusion limit

$$\begin{aligned} &4\pi\left(\frac{\eta_0}{\chi_0}\right) - cE_0 \\ &= \frac{1}{\chi_0}\left[\varrho\frac{\mathrm{D}}{\mathrm{D}t}\left(\frac{E_0}{\varrho}\right) + \frac{1}{r^2}\frac{\partial}{\partial r}\left(r^2F_0\right) + \frac{1}{3}E_0\varrho\frac{\mathrm{D}}{\mathrm{D}t}\left(\frac{1}{\varrho}\right)\right] \end{aligned} \tag{7.2}$$

and the radiation momentum equation (5.65) becomes

$$F_0 = -\frac{c}{\chi_0}\left[\frac{\varrho}{c^2}\frac{\mathrm{D}}{\mathrm{D}t}\left(\frac{F_0}{\varrho}\right) + \frac{1}{3}\frac{\partial E_0}{\partial r} + \frac{F_0}{c^2}\frac{\partial v}{\partial r}\right]\,. \tag{7.3}$$

Consider material that is so homogeneous that we can *ignore* all gradients within an interaction volume. Formally, let $(\lambda_p/\ell) \to 0$. Then the right-hand sides of both (7.2) and (7.3) vanish and we have $E_0 \to (4\pi/c)(\eta_0/\chi_0) = (4\pi/c)B_0$ and $F_0 \to 0$ Moreover, if there are no gradients, then the radiation field is isotropic, hence $P_0 = \frac{1}{3}E_0$. Thus in the comoving frame, the radiation stress–energy tensor is:

$$\mathsf{R}_0 = \begin{pmatrix} a_R T^4 & \mathbf{0} \\ \mathbf{0} & \frac{1}{3}a_R T^4 \mathsf{I} \end{pmatrix} . \tag{7.4}$$

Thus in this limit radiation behaves like an ideal fluid: it has an isotropic pressure, and there are no transport effects. Therefore there is no radiation flux.

1st Order Diffusion. Let us now look for terms that are $O(\lambda_p/\ell)$ or $O(v/c)$. Start with the radiation momentum equation

$$\underset{\mathsf{A}}{F_0} = -\frac{c}{\chi_0}\left[\underset{\mathsf{B}}{\frac{\varrho}{c^2}\frac{\mathrm{D}}{\mathrm{D}t}\left(\frac{F_0}{\varrho}\right)} + \underset{\mathsf{C}}{\frac{F_0}{c^2}\frac{\partial v}{\partial r}} + \underset{\mathsf{D}}{\frac{\partial P_0}{\partial r}}\right] . \tag{7.5}$$

From a dimensional analysis one can infer that B/A and C/A are both $O(\lambda_p v/\ell c)$. Therefore we can drop these terms for now. Thus in equilibrium diffusion, $\boldsymbol{F}_0 = -(c/\chi_0)\boldsymbol{\nabla}\cdot\mathsf{P}_0 = -(c/3\chi_0)\boldsymbol{\nabla}E_0 = -(a_R c/3\chi_0)\boldsymbol{\nabla}T^4$ or

$$\boldsymbol{F}_0 = -K_R \boldsymbol{\nabla} T , \tag{7.6}$$

where the *radiative conductivity* is

$$K_R = \frac{4a_R c T^3}{3\chi_0} . \tag{7.7}$$

Equations (7.6) and (7.7) are the same as derived earlier for *static diffusion.* Now we see that they also apply in the *comoving frame* of the *moving* material. Note the strong dependence of K_R on T. It is made even stronger by the variation of $\chi_0 \propto T^{-3.5}$ in the stellar interior. This sensitivity of K_R to T will turn out to be important when we study *thermal waves.*
For nongrey material a similar analysis yields

$$\boldsymbol{F}_0(\nu_0) = -\frac{4\pi}{3\chi_0(\nu_0)}\boldsymbol{\nabla}B(\nu_0,T) = -\frac{4\pi}{3\chi_0(\nu_0)}\frac{\partial B(\nu_0,T)}{\partial T}\boldsymbol{\nabla}T . \tag{7.8}$$

We then find that to obtain the correct integrated flux we should use the Rosseland mean χ_R in (7.6) for K_R instead of χ_0. Note in passing that F_0 is, in fact, $O(\lambda_p/\ell)\cdot E_0$. E_0 is the dominant term on the diagonal of R_0.

$$\underset{\mathsf{A}}{E_0} = \frac{4\pi}{c}B_0 - \frac{1}{\chi_0 c}\left[\underset{\mathsf{B}}{\varrho\frac{\mathrm{D}}{\mathrm{D}t}\left(\frac{E_0}{\varrho}\right)} + \underset{\mathsf{C}}{\frac{1}{r^2}\frac{\partial}{\partial r}(r^2 F_0)} + \underset{\mathsf{D}}{\frac{1}{3}E_0\varrho\frac{\mathrm{D}}{\mathrm{D}t}\left(\frac{1}{\varrho}\right)}\right] . \tag{7.9}$$

Then from dimensional analysis we can infer:

$$\mathsf{A} : \mathsf{B} : \mathsf{C} : \mathsf{D} = 1 : \left(\frac{\lambda_\mathrm{p}}{\ell}\frac{v}{c}\right) : \left(\frac{\lambda_\mathrm{p}}{\ell}\right)^2 : \left(\frac{\lambda_\mathrm{p}}{\ell}\frac{v}{c}\right) .$$

Thus the departure of E_0 from $4\pi B_0/c$ is at most $\lambda_\mathrm{p}v/\ell c$ (a result we used earlier).

Thus a consistent *first-order* expression for the radiation stress–energy tensor in the *comoving frame* of a moving fluid in the equilibrium diffusion limit is

$$\begin{aligned}\mathsf{R}_0 &= \begin{pmatrix} a_\mathrm{R}T^4 & -\dfrac{1}{3\chi_\mathrm{R}}\nabla(a_\mathrm{R}T^4) \\ -\dfrac{1}{3\chi_\mathrm{R}}\nabla(a_\mathrm{R}T^4) & \dfrac{1}{3}a_\mathrm{R}T^4\,\mathsf{I} \end{pmatrix} \\ &= \begin{pmatrix} a_\mathrm{R}T^4 & -\dfrac{K_\mathrm{R}}{c}\nabla T \\ -\dfrac{K_\mathrm{R}}{c}\nabla T & \dfrac{1}{3}a_\mathrm{R}T^4\,\mathsf{I} \end{pmatrix} . \end{aligned} \tag{7.10}$$

There are interesting contrasts between this stress–energy tensor and that of a material fluid:

- For material, Chapman–Enskog theory gives *both* viscous and heat-conducting terms to 1st order in $(\lambda_\mathrm{p}/\ell)$.
- But for radiation we get only "heat conduction".
- We will see later that radiative viscosity is $O(\frac{\lambda_\mathrm{p}}{\ell}\frac{v}{c})$. The physical reason radiation viscosity is so small is that for a material particle (momentum/energy) $\sim (mv)/(mv^2) \to O(1/v)$, while for a photon, it is $(h\nu/c)/(h\nu) \to O(1/c)$. Therefore photons are much less efficient at transferring momentum per unit energy transport.
- Note that *no* $O(v/c)$ terms appear in R_0 in the comoving frame.

Equations of Radiation Hydrodynamics in 1st Order Diffusion Limit. In the equilibrium diffusion limit R_0 can be expressed in terms of T and ∇T. No new information beyond that used in an ordinary material fluid calculation is required. Using R_0 in the general dynamical equations $R^{\alpha\beta}_{;\beta} = f^\alpha$, we find the momentum equation

$$\varrho\frac{\mathrm{D}\boldsymbol{v}}{\mathrm{D}t} = \boldsymbol{f} - \nabla\left(p + \frac{1}{3}a_\mathrm{R}T^4\right) \tag{7.11}$$

and the radiating fluid energy equation

$$\frac{\mathrm{D}}{\mathrm{D}t}\left(e + \frac{a_\mathrm{R}T^4}{\varrho}\right) + \left(p + \frac{1}{3}a_\mathrm{R}T^4\right)\frac{\mathrm{D}}{\mathrm{D}t}\left(\frac{1}{\varrho}\right) = \frac{1}{\varrho}\nabla\cdot(K_\mathrm{R}\nabla T) + \varepsilon\,. \tag{7.12}$$

These are the standard equations used in dynamical stellar structure calculations.

Let us now sketch briefly how the computation proceeds (in spherical symmetry). Suppose at t^n we know ϱ, v, e, p, and T. Then we can compute the acceleration $\dot{v}_i^n$ from the momentum equation. We use this acceleration to update the velocity: $v_i^{n+\frac{1}{2}} = v_i^{n-\frac{1}{2}} + \Delta t\, \dot{v}_i^n$. We then use the new velocity to find the cell interface positions at t^{n+1}: $r_i^{n+1} = r_i^n + \Delta t\, v_i^{n+\frac{1}{2}}$. The interface positions imply volumes, hence densities, of mass shells. We must then solve the energy equation at t^{n+1}. Assuming $e(\varrho, T)$ and $p(\varrho, T)$ are known (e.g., from tables), we have a *nonlinear equation* for T^{n+1}. The highest-order differential operator in this equation is $\nabla \cdot \nabla$, hence in discrete form the equation is *tridiagonal.* To solve it, we *linearize* it with respect to T, to derive an equation of the form:

$$-a_{i+\frac{1}{2}} \delta T_{i-\frac{1}{2}} + b_{i+\frac{1}{2}} \delta T_{i+\frac{1}{2}} - c_{i+\frac{1}{2}} \delta T_{i+\frac{3}{2}} = d_{i+\frac{1}{2}}, \qquad (i = 1, ..., N_i)\,. \tag{7.13}$$

We then solve (7.13) for the corrections δT to the current estimate of the temperature distribution. We apply the corrections, and iterate the procedure to convergence. Then we proceed with the next time-step. This algorithm is basically very simple and cheap, hence the popularity of this approach.

Second-Order Equilibrium Diffusion. Now let's go after terms that are $O(\lambda_p v/\ell c)$, which are only linear in *each* of the two small parameters, and therefore are likely the next largest terms. The new physics that emerges concerns *radiative viscosity.* This problem has been studied by Jeans, Milne, and others; the best work of that era was by L. H. Thomas [31]. Thomas started with the mixed-frame transfer equation and developed a solution for I by iteration. Given this expression for I, one can evaluate R by integration over ω and ν. This calculation is straightforward but tedious. It is impressive that Thomas was able to derive all these results and cast them into covariant form by inspection. A less cumbersome and physically more appealing approach was developed by [30]. After a considerable amount of algebra he was able to derive covariant expressions for the radiation energy density $\mathcal{E}$, the radiative energy flux $\mathcal{F}$, and the radiation pressure $\mathcal{P}$, as well as the radiative conductivity K_{R}, which is the same as given in (7.7). The *coefficient of radiative viscosity* (a world scalar) is

$$\mu_{\mathrm{R}} \equiv \frac{4}{15} \frac{a_{\mathrm{R}} T^4}{c \kappa_{\mathrm{R}}^0} = \left(\frac{T}{5\,c^2} \right) K_{\mathrm{R}}\,, \tag{7.14}$$

which can be derived to within a numerical factor of $O(1)$ by simple mean free path arguments. From these quantities we can derive a complete covariant radiation stress energy tensor. To gain further physical insight one can reduce the relativistically correct results to the comoving frame and retaining terms to $O(\lambda_p v/\ell c)$ to get

$$E_0 = a_\mathrm{R} T^4 - 5\mu_\mathrm{R}\left(3\frac{\mathrm{D}\ln T}{\mathrm{D}t} + \nabla\cdot v\right) , \tag{7.15}$$

$$(F_0)_i = -K_\mathrm{R}\left(T_{,i} + \frac{T}{c^2}a_i\right) , \tag{7.16}$$

$$P_0^{ij} = \mathcal{P}\delta^{ij} - 2\mu_\mathrm{R} D^{ij} - \zeta_\mathrm{R} v^k_{,k}\delta^{ij} , \tag{7.17}$$

where D^{ij} is the traceless rate of strain tensor,

$$D^{ij} = \frac{1}{2}(v^i_{,j} + v^j_{,i}) - (\frac{1}{3}v^k_{,k})\delta^{ij} , \tag{7.18}$$

and

$$\mathcal{P} \equiv \frac{1}{3}a_\mathrm{R} T^4 - 5\mu_\mathrm{R}\left(\frac{\mathrm{D}\ln T}{\mathrm{D}t}\right) , \tag{7.19}$$

and

$$\zeta_\mathrm{R} \equiv \frac{5}{3}\mu_\mathrm{R} . \tag{7.20}$$

- The radiant flux contains an acceleration term which is identical with Eckart's relativistic expression for the conductive energy flux in a material fluid.
- From (7.20) it appears that radiation has a substantial bulk viscosity (ζ_R). [33] showed that this is an artifact that arises because Thomas and Simon use the *material* temperature T, not the thermal-equilibrium temperature T^*.
- Using the radiative viscosity in place of the material viscosity we can define a *radiative Reynolds number*

$$\mathrm{Re}_\mathrm{R} \sim \frac{t_f}{t_\lambda} \cdot \left(\frac{\text{material kinetic energy density}}{\text{radiation energy density}}\right) . \tag{7.21}$$

  Re_R can be quite small in high-temperature flows. At first sight this result might imply that radiative viscosity might be important. However, from a dimensional analysis one can show that the radiative viscosity terms (radiative dissipation function) are only $O(v^2/c^2)$ relative to the dominant terms in the energy equation. Thus it appears that in the radiating fluid energy equation we can neglect energy dissipation by radiative viscosity.
- In the momentum equation for a radiating fluid the radiation viscosity terms are $O(\lambda_\mathrm{p} v/\ell c)$ and thus should be retained. One can show that these terms could be important in some cases, particularly when $\lambda_\mathrm{p}/\ell \gg 1$. But of course diffusion theory is not valid then, and we must make an analysis based on full (comoving frame) *transport* theory.
- Taking the full comoving-frame radiation energy equation to the diffusion limit we recover the results of 2nd-order diffusion theory result exactly. Further, one sees that the acceleration term in the comoving-frame equation comes from the inertia of the radiant flux.

- The comoving-frame radiation momentum equation in the diffusion limit reduces *exactly* to the 2nd-order diffusion theory form. One can show that the acceleration terms in the radiation momentum equation account for the inertia of radiation enthalpy, while terms like $c^{-2}\boldsymbol{F} \cdot \boldsymbol{\nabla v}$ account for the interaction of the radiation momentum density with shear flow.
- Virtually all existing (Newtonian) discussions (e.g., by [7, 13, 14, 18, 19]) of radiative viscosity in a radiating fluid are inconsistent; they omit terms of the same order as are retained.
- Analytical 2nd-order radiation diffusion theory, while giving important results of principle, isn't too useful because the new terms are negligibly small when $\lambda_\mathrm{p}/\ell \ll 1$, where the theory is valid, and become significant only in the transport regime where $\lambda_\mathrm{p}/\ell \geq 1$, where diffusion theory breaks down.
- The obvious thing to do is to use the full comoving-frame radiation energy and momentum equations [e.g., $O(v/c)$ equations from [4]] because (a) these *are* valid in the transport regime and give a correct description of the *nonlocal* energy and momentum exchange between radiation and material when $\lambda_\mathrm{p}/\ell \gg 1$, while (b) limiting to the correct form in the diffusion limit.

7.2 Nonequilibrium Diffusion

Diffusion theory as discussed thus far is very restrictive because it demands strict thermal equilibrium between the radiation and the material, which implies the radiation field is Planckian at $T_\mathrm{R} \equiv T_\mathrm{material}$. We can develop a much more powerful and useful theory by assuming $\lambda_\mathrm{p}/\ell \ll 1$ (*i.e.*, the material is opaque) but that the radiation is *not* in equilibrium with the matter. Some examples of physical situations of this type are:

- Suppose the radiation field is *imposed* from the outside with $T_\mathrm{rad} \gg T_\mathrm{material}$, as in stellar radiation falling on an interstellar nebula.
- Suppose fluid dynamics drives the material properties (e.g., T_m) faster than the matter can relax radiatively (e.g., as in a strong shock).
- Suppose the radiation field changes (for a variety of possible reasons) faster than the material can follow instantaneously.

The advantages of nonequilibrium theory are:

- The radiation energy density E can be quite different from $a_\mathrm{R} T^4_\mathrm{material}$.
- The spectrum of the radiation field can be non-Planckian.

We will again formulate the theory in the comoving frame. By exactly the same arguments used for equilibrium diffusion we again find:

$$\boldsymbol{F}_0(\nu_0) = -\frac{c}{3\chi_0(\nu_0)}\boldsymbol{\nabla} E_0(\nu_0) + O\left(\frac{\lambda_\mathrm{p}}{\ell}\frac{v}{c}\right) + O\left(\frac{\lambda_\mathrm{p}}{\ell}\right)^2 . \tag{7.22}$$

But now we do *not* replace $E_0(\nu_0)$ with $(4\pi/c)B(\nu_0, T)$. Next, examine the Lagrangian monochromatic radiation energy equation (5.25). Using (7.22) for $\boldsymbol{F}_0(\nu_0)$, and requiring isotropy (because departures from isotropy are higher than 1st order), we get:

$$\varrho\left\{\frac{\mathrm{D}}{\mathrm{D}t}\left[\frac{E_0(\nu_0)}{\varrho}\right]+\frac{1}{3}\left\{E_0(\nu_0)-\frac{\partial}{\partial\nu_0}\left[\nu_0 E_0(\nu_0)\right]\right\}\frac{\mathrm{D}}{\mathrm{D}t}\left(\frac{1}{\varrho}\right)\right\}$$
$$=\kappa_0(\nu_0)\left[4\pi B(\nu_0,T)-cE_0(\nu_0)\right]+\boldsymbol{\nabla}\cdot\left[\frac{c}{3\chi_0(\nu_0)}\boldsymbol{\nabla}E_0(\nu_0)\right]. \quad (7.23)$$

Equations (7.22) and (7.23) are the monochromatic nonequilibrium diffusion equations.

Now integrate over ν_0; note that the $(\partial/\partial\nu_0)$ term vanishes. We obtain the radiation energy equation (nonequilibrium diffusion equation)

$$\varrho\left[\frac{\mathrm{D}}{\mathrm{D}t}\left(\frac{E_0}{\varrho}\right)+\frac{1}{3}E_0\frac{\mathrm{D}}{\mathrm{D}t}\left(\frac{1}{\varrho}\right)\right]$$
$$=\boldsymbol{\nabla}\cdot\left(\frac{c}{3\bar{\chi}}\boldsymbol{\nabla}E_0\right)+c\left(\kappa_{\mathrm{P}}a_{\mathrm{R}}T^4-\kappa_{\mathrm{E}}E_0\right) \quad (7.24)$$

and the integrated flux

$$\boldsymbol{F}_0=-\frac{c}{3\bar{\chi}}\boldsymbol{\nabla}E_0\,. \quad (7.25)$$

Here κ_{P} is the *Planck mean*, and κ_{E} is the *absorption mean* defined as

$$\kappa_{\mathrm{P}}=\int_0^\infty\kappa_0(\nu_0)B(\nu_0,T)\mathrm{d}\nu_0\Big/\int_0^\infty B(\nu_0,T)\mathrm{d}\nu_0\;, \quad (7.26)$$

$$\kappa_{\mathrm{E}}\equiv\int_0^\infty\kappa_0(\nu_0)\frac{E_0(\nu_0)}{E_0}\mathrm{d}\nu_0=\int_0^\infty\kappa_0(\nu_0)\,\mathsf{e}(\nu_0)\,\mathrm{d}\nu_0\;, \quad (7.27)$$

where $\mathsf{e}(\nu_0)$ is the *spectral profile* of the radiation energy density. Likewise, $\bar{\chi}$ is defined such that

$$\frac{1}{\bar{\chi}}\boldsymbol{\nabla}E_0=\int_0^\infty\frac{1}{\chi_0(\nu_0)}\boldsymbol{\nabla}E_0(\nu_0)\mathrm{d}\nu_0 \quad (7.28)$$

(using different values along different coordinate axes if necessary). It is often convenient to introduce a *radiation temperature* T_{R}:

$$a_{\mathrm{R}}T_{\mathrm{R}}^4\equiv E_0. \quad (7.29)$$

Note that T_{R} is only a *parameter* which is related to the *total* radiation energy density E_0. In nonequilibrium diffusion, $E_0(\nu_0)$ is *not* Planckian at T_{R}!

When solving the equations of radiation hydrodynamics using nonequilibrium diffusion, we must determine *two* variables at time-level t^{n+1}, namely T and T_{R}. So now we solve the gas energy equation

$$\varrho\left[\frac{\mathrm{D}e}{\mathrm{D}t}+p\frac{\mathrm{D}}{\mathrm{D}t}\left(\frac{1}{\varrho}\right)\right]=a_{\mathrm{R}}c\left(\kappa_{\mathrm{E}}T_{\mathrm{R}}^{4}-\kappa_{\mathrm{P}}T^{4}\right)+\varrho\varepsilon \tag{7.30}$$

and the radiation energy equation

$$\varrho\left[\frac{\mathrm{D}}{\mathrm{D}t}\left(\frac{a_{\mathrm{R}}T_{\mathrm{R}}^{4}}{\varrho}\right)+\frac{1}{3}a_{\mathrm{R}}T_{\mathrm{R}}^{4}\frac{\mathrm{D}}{\mathrm{D}t}\left(\frac{1}{\varrho}\right)\right]$$
$$=\nabla\cdot\left(\frac{4a_{\mathrm{R}}cT_{\mathrm{R}}^{3}}{3\bar{\chi}}\nabla T_{\mathrm{R}}\right)+a_{\mathrm{R}}c\left(\kappa_{\mathrm{P}}T^{4}-\kappa_{\mathrm{E}}T_{\mathrm{R}}^{4}\right) \tag{7.31}$$

simultaneously. This approach is sometimes called the *two-temperature description* of a radiating fluid. As before, we linearize with respect to δT and δT_{R} at t^{n+1}. We now get a block tridiagonal system with (2×2) blocks. This system can be solved by Gaussian elimination to give the temperature corrections, and these results are used in the nonlinear system to iterate to consistency.

In the discussion above we have assumed that $\bar{\chi}$ and $\bar{\kappa_{\mathrm{E}}}$ (hence the frequency-dependent information required to compute them) are given. How should we proceed in reality?

- The simplest approximation for nongrey material is to take $\bar{\chi}=\chi_{\mathrm{R}}$ and $\kappa_{\mathrm{E}}=\kappa_{\mathrm{P}}$, evaluating both χ_{R} and κ_{P} at the material temperature T. This approach is reasonable if radiation is not too far out of equilibrium with the material. This is the usual approach.
- To do anything better we need to know the profile $e_0(\nu_0)$. An idea suggested by [10] is to take the two-temperature picture *literally* and define the *two-temperature means*

$$\kappa_{\mathrm{E}}(T,T_{\mathrm{R}})=\int_0^{\infty}\kappa_0(\nu_0,T)B(\nu_0,T_{\mathrm{R}})\mathrm{d}\nu_0\bigg/\int_0^{\infty}B(\nu_0,T_{\mathrm{R}})\mathrm{d}\nu_0 \tag{7.32}$$

and

$$\frac{1}{\bar{\chi}(T,T_{\mathrm{R}})}=\frac{\displaystyle\int_0^{\infty}\frac{1}{\chi_0(\nu_0,T)}\left[\frac{\partial B(\nu_0,T_{\mathrm{R}})}{\partial T}\right]_{T_{\mathrm{R}}}\mathrm{d}\nu_0}{\displaystyle\int_0^{\infty}\left[\frac{\partial B(\nu_0,T_{\mathrm{R}})}{\partial T}\right]_{T_{\mathrm{R}}}\mathrm{d}\nu_0}\,. \tag{7.33}$$

The basic idea here is to get something we can actually compute, while still allowing the spectrum of the radiation (characterized by T_{R}) to depart from $B_{\nu}(T_{\mathrm{material}})$, for example in the case of hot radiation penetrating cold material).

To improve the treatment further, we would have to solve the monochromatic diffusion equation. Let us consider this task briefly. The monochromatic diffusion equation is complicated by the presence of the $(\partial/\partial\nu_0)$ operator, whose presence implies that the equation is a partial differential equation in (r, t, ν_0) instead of just (r, t). How should we solve this equation? Note that the term with $(\partial/\partial\nu_0)$ is in conservative form. It will therefore vanish when integrated over all frequency. Hence it cannot affect the integrated diffusion equation. So why not just *drop* $(\partial/\partial\nu_0)$? This is a *bad idea*. A simple gedankenexperiment shows why: Consider an adiabatic enclosure of an opaque, homogeneous medium (which means that the radiant flux will be exactly zero) with only Thomson scattering ($\kappa \equiv 0$; *grey*). Then the monochromatic diffusion equation reduces to:

$$\frac{\mathrm{D}\ln E_0(\nu_0)}{\mathrm{D}t} - \frac{4}{3}\frac{\mathrm{D}\ln\varrho}{\mathrm{D}t} + \frac{1}{E_0(\nu_0)}\frac{\partial[\nu_0 E_0(\nu_0)]}{\partial\nu_0}\frac{\mathrm{D}\ln\varrho}{\mathrm{D}t} = 0\,. \tag{7.34}$$

Now *drop* the term containing $(\partial/\partial\nu_0)$. We get

$$\frac{\mathrm{D}\ln E_0(\nu_0)}{\mathrm{D}t} - \frac{4}{3}\frac{\mathrm{D}\ln\varrho}{\mathrm{D}t} \equiv 0\,, \tag{7.35}$$

which implies that $E_0(\nu_0) \propto \varrho^{4/3}$. This result is correct for the *integrated* radiation in an adiabatic enclosure, but is *wrong* for monochromatic radiation. The point is that if the volume of the enclosure decreases, then T_R rises because $T_\mathrm{R}^4 \propto V^{-4/3}$ which implies that $T_\mathrm{R} \propto V^{-1/3}$. Because the radiation temperature changes, the *spectrum* of the radiation $B(\nu_0, T_\mathrm{R})$ *necessarily* must change.

- Therefore, unless we retain $(\partial/\partial\nu_0)$ we get the *wrong* monochromatic spectral profile $\mathsf{e}(\nu_0) \equiv E_\nu/E$ because we have ignored the adiabatic red/blue shift of the radiation when the material expands/contracts.
- But the whole point of doing the monochromatic problem is to get the correct spectrum!
- So, in short, we have no choice but to solve the full equation. Direct solution may be too costly, therefore we examine an iteration scheme.

Let us discretize the frequency variable with G groups (or ODF [opacity distribution function] pickets [22]). Then we must solve G diffusion equations

$$\begin{aligned}\frac{\mathrm{D}}{\mathrm{D}t}\left(\frac{E_g}{\varrho}\right) &+ \frac{1}{3}\left[E_g - \Delta_g(\nu E_\nu)\right]\frac{\mathrm{D}}{\mathrm{D}t}\left(\frac{1}{\varrho}\right)\\ &= \frac{\partial}{\partial M_r}\left[\frac{c(4\pi r^2)^2}{3(\bar{\chi}_g/\varrho}\frac{\partial E_g}{\partial M_r}\right] + \frac{4\pi\kappa_{\mathrm{P},g}B_g - c\bar{\kappa}_g E_g}{\varrho}\end{aligned} \tag{7.36}$$

along with the material energy equation

$$\frac{\mathrm{D}e}{\mathrm{D}t} + p\frac{\mathrm{D}}{\mathrm{D}t}\left(\frac{1}{\varrho}\right) = \sum_g \frac{1}{\varrho}\left(c\bar{\kappa}_g E_g - 4\pi\kappa_{\mathrm{P},g}B_g\right)\,. \tag{7.37}$$

Here $\bar{\kappa}_g$ and $\bar{\chi}_{\mathrm{R},g}$ are appropriate direct and harmonic averages. Possible choices are: (a) ODF with $\bar{\kappa}_g = \kappa_{\mathrm{picket}}$; and $\bar{\chi}_{\mathrm{R},g} = \chi_{\mathrm{picket}}$. (b) Group means $\bar{\kappa}_g = \kappa_{\mathrm{P},g}$; and $\bar{\chi}_{\mathrm{R},g} = \chi_{\mathrm{R},g}$, computed at the material temperature T. (c) Two-temperature group means as in (7.32) and (7.33). Among these choices (a) is probably best; (b) is the most commonly used, and (c) may never have been tried.

The Δ_g in (7.36) are discrete representations of the integral of $\partial(\nu E_\nu)/\partial\nu$ over (ν_g, ν_{g+1}). In writing them, one must remember that if the material (expands | contracts), photons are (red | blue) shifted. For numerical accuracy and stability we use ***upstream differencing***. For example, for expansion, photons come *into* group g from group $g+1$. Recall that this frequency-derivative term is conservative, and its finite difference representation telescopes to zero when summed over the entire spectrum.

As before we assume the material properties are all $f(\varrho, T)$ [or $f(\varrho, T, T_{\mathrm{R}})$], and that we are using explicit hydrodynamics. Then at t^{n+1} the goal is to determine a solution vector

$$\psi_d \equiv (E_1, E_2, \ldots, E_G, T)_d, \qquad (d = 1, \ldots, D) \tag{7.38}$$

which gives the spectral distribution of the radiation field, and the local material temperature. As before, we linearize the equations and get a system of block-tridiagonal equations. Each block is of dimension $(G+1) \times (G+1)$. Therefore the computing effort for a direct solution scales as $D(G+1)^3$, and the process is iterated to consistency. If the linear system leading to (7.38) is solved by iterative methods, considerable speed-ups are possible. An alternative is to use an iteration method based on physical considerations.
A popular iteration procedure is the ***multifrequency/grey method*** (MFG) first introduced in the VERA (Variable Eddington Radiation Approximation) [11] code. The basic idea is to ***split*** the problem into steps: compute T and T_{R} at t^{n+1} using frequency-integrated equations, assuming that the spectral profiles needed to compute the necessary mean opacities are given. Then use the multigroup equations with T and T_{R} given to calculate the spectral profiles needed by the opacity weighting procedure. Thus:

- Use known profiles from t^n to evaluate $\bar{\chi}$ and κ_{E}.
- Solve the gas energy equation and integrated radiation energy equation to get $(T, T_{\mathrm{R}})^{n+1}$.
- Using these temperatures, solve the multigroup radiation equations. Move all Δ_g terms (*i.e.*, the frequency derivatives) to the right-hand side. We then have G ***uncoupled*** equations. Solve these equations in parallel. Using these results, evaluate new estimates of the Δ_g terms. Then re-solve the uncoupled system. Iterate to convergence to obtain the spectral profiles at t^{n+1}.
- Using the new spectral profiles, update the estimates of $\bar{\chi}$ and κ_{E} at t^{n+1}.
- Then do the timestep (t^n, t^{n+1}) over (possibly iterating to convergence), or kiss it goodbye and take the next timestep.

For astronomical problems MFG usual works well. But for very hard problems, *e.g.*, with material interfaces or large departures from equilibrium, it may not.

Critique of Earlier Treatments of Diffusion. The correct nonequilibrium diffusion equation was first written by John Castor [4]. Before his Lagrangian transfer equation was known, it was not possible to derive the correct equations rigorously, and most people made (incorrect) guesses from heuristic arguments.

Consider a typical approach: start from the Eulerian radiation and momentum equations, and then:

1. compute F as if it were F_0 (which it is not); and
2. substitute this expression into the energy equation, dropping all $O(v/c)$ terms on the right-hand side to get:

$$\frac{\partial E}{\partial t} = \kappa(4\pi B - cE) - \nabla \cdot \left(\frac{c}{3\chi}\nabla E\right) . \tag{7.39}$$

The problem is that this equation is neither Eulerian nor Lagrangian. Indeed, the fact is that it is just wrong:

- Although the right-hand side looks like our Lagrangian diffusion equation (7.24), the left-hand side is hopeless in that frame. It ignores the radiation work term $P\frac{\mathrm{D}}{\mathrm{D}t}(\frac{1}{\varrho})$, and fails to discriminate between $(\frac{\partial E}{\partial t})$ and $\varrho\frac{\mathrm{D}}{\mathrm{D}t}(\frac{E}{\varrho})$. So it is *not* Lagrangian.
- Likewise it isn't Eulerian because, comparing to (4.11), on the right-hand side we are missing a term $\kappa \boldsymbol{v} \cdot \boldsymbol{F}/c$ equal to the rate of work done by the radiation force, and in the flux divergence, the *dominant* term that discriminates between $\boldsymbol{F}$ and $\boldsymbol{F}_0$ (*i.e.*, $\frac{4}{3}\boldsymbol{v}E_0$) is omitted. Actually it is tricky to write down a correct Eulerian diffusion equation.

No one knows how many calculations could have been ruined by using (7.39). Moral: *be careful of frame!*

7.3 The Problem of Flux Limiting

Everyone knows that diffusion theory applies only when $\lambda_p/\ell \ll 1$. But radiation hydro with radiation ***transport*** is very hard, while radiation hydro with radiation diffusion is relatively easy. So people have tried to push radiation diffusion codes into optically thin regimes. The usual result is trouble. Why? Symptom: one finds $|\boldsymbol{F}_{\text{diffusion}}| > cE$, which implies that the signal speed is $v_E \equiv |\boldsymbol{F}|/E = \frac{1}{3}c\lambda_p \nabla E|/E \sim (\lambda_p/3\ell)c \gg c$ because $\lambda_p \gg \ell$, which is physically absurd. Root of the problem: (linear) diffusion theory ***insists*** that a photon always travels one mean-free-path per timestep, even if λ_p exceeds the free-flight length $c\Delta t$. [Remember that the linear diffusion equation has an infinite signal speed. In nonlinear diffusion the signal speed is bounded below c.]

To overcome the problem, people use an ad hoc *flux limiting* procedure invented by [2]. The idea is to control the flux with an expression of the general form:

$$\boldsymbol{F} = \frac{-c\boldsymbol{\nabla} E}{(3/\lambda_{\mathrm{p}}) + |\boldsymbol{\nabla} E|/E} \, . \tag{7.40}$$

Then when $\lambda_{\mathrm{p}} \ll \ell, \boldsymbol{F} \to -(c\lambda_{\mathrm{p}}/3)\boldsymbol{\nabla} E$, which is the usual diffusion result, whereas when $\lambda_{\mathrm{p}} \gg \ell, \boldsymbol{F} \to cE\boldsymbol{n}$, where $\boldsymbol{n}$ is chosen to point in the direction of $-\nabla E$. This kind of flux limiting procedure is certainly qualitatively reasonable; many other formulations have been suggested. But in the end, flux-limiting is a *fixup*. To really overcome the problem, one must retain the time-derivative, which becomes important on a radiation-flow timescale, in the radiation momentum equation. To gain insight, ignore all $O(v/c)$ terms:

$$\frac{\partial E}{\partial t} + \frac{\partial F}{\partial z} = \kappa(4\pi B - cE) \, , \tag{7.41}$$

$$\frac{1}{c^2}\frac{\partial F}{\partial t} + \frac{\partial (fE)}{\partial z} = -\frac{\chi}{c} F \, . \tag{7.42}$$

Replace the time derivative $(\partial F/\partial t)$ with a difference formula; for stability use a fully implicit *backward Euler* formula. Then

$$\frac{F^{n+1} - F^n}{c^2 \Delta t} + \frac{\partial (f^{n+1} E^{n+1})}{\partial z} = -\frac{\chi^{n+1}}{c} F^{n+1} \, , \tag{7.43}$$

which implies

$$F^{n+1} = -\frac{c}{\gamma + \chi^{n+1}} \frac{\partial (f^{n+1} E^{n+1})}{\partial z} + \frac{\gamma}{\gamma + \chi^{n+1}} F^n \, , \tag{7.44}$$

where $\gamma \equiv 1/c\Delta t$. To understand (7.44), define

$$\frac{1}{\lambda} \equiv \gamma + \chi = \frac{1}{c\Delta t} + \frac{1}{\lambda_{\mathrm{p}}} = \frac{1}{\lambda_t} + \frac{1}{\lambda_{\mathrm{p}}} \, , \tag{7.45}$$

which implies that the effective mean free path λ is the harmonic mean of the free-flight length λ_t and the optical mean free path λ_{p}. Then

$$F^{n+1} = -c\lambda^{n+1} \frac{\partial (f^{n+1} E^{n+1})}{\partial z} + \left(\frac{\lambda^{n+1}}{\lambda_t}\right) F^n \, . \tag{7.46}$$

Equation (7.46) gives us an algebraic expression for F^{n+1}, which we can use to eliminate the flux from the radiation energy equation.

In opaque material, $\chi \gg 1$, and $\lambda_{\mathrm{p}} \ll \lambda_t$, hence $\lambda \to \lambda_{\mathrm{p}}$ and $f \to \frac{1}{3}$. Then $F^{n+1} \to -(c\lambda_{\mathrm{p}}/3)(\partial E^{n+1}/\partial z)$ which is the correct diffusion limit. In transparent material $\lambda_{\mathrm{p}} \gg \lambda_t, \lambda \to \lambda_t, f \to 1$, and $F \to cE$. Then $F^{n+1} =$

$F^n - \lambda_t(\partial F^{n+1}/\partial z)$ which implies that a change in the local flux over a timestep reflects the flux gradient over a photon free-flight length λ_t. Now use (7.46) in the radiation energy equation to get a combined moment equation. Again replace the time derivative with an implicit time difference. We get

$$(\gamma + \kappa^{n+1})E^{n+1} - \frac{\partial}{\partial z}\left[\lambda^{n+1}\frac{\partial(f^{n+1}E^{n+1})}{\partial z}\right] = \frac{4\pi}{c}\kappa^{n+1}B^{n+1} + \gamma E^n - \frac{\gamma}{c}\frac{\partial(\lambda^{n+1}F^n)}{\partial z} . \quad (7.47)$$

In the limit of high opacity and/or long timesteps, $\kappa/\gamma \gg 1$, so

$$\frac{1}{3}\frac{\partial}{\partial z}\left(\lambda_{\mathrm{p}}^{n+1}\frac{\partial E^{n+1}}{\partial z}\right) = \kappa^{n+1}\left(E^{n+1} - \frac{4\pi}{c}B^{n+1}\right) , \quad (7.48)$$

which is just the quasistatic nonequilibrium diffusion equation. In the limit of low opacity and/or short timesteps $\kappa/\gamma \ll 1$. Setting $f = 1$ we get

$$E^{n+1} - (c\Delta t)^2\frac{\partial^2 E^{n+1}}{\partial z^2} = E^n - \frac{1}{\gamma c}\frac{\partial F^n}{\partial z} \rightarrow 2E^n - E^{n-1} . \quad (7.49)$$

To obtain the last expression we used the fully implicit representation of the radiation energy equation (not the combined moment equation) at t^n. Thus

$$c^2\frac{\partial^2 E^{n+1}}{\partial z^2} = \frac{E^{n+1} - 2E^n + E^{n-1}}{\Delta t^2} \approx \left.\frac{\partial^2 E}{\partial t^2}\right|_{t^n} , \quad (7.50)$$

which is an approximation to the wave equation (mis-centered in time because all time differences were fully implicit instead of centered).

In summary:

- The flux limiting problem is an artifact of dropping $\frac{1}{c^2}\frac{\partial F}{\partial t}$ in the radiation momentum equation.
- This problem can be overcome if we solve the time-dependent moment equations (both), with accurate Eddington factors. ($f \equiv \frac{1}{3}$ won't do in the streaming limit; this would imply a wave speed of $c/\sqrt{3}$, not c)
- While it is practical to overcome the problem in one-dimensional flow, flux limited diffusion remains the only realistic option in workhorse codes in 2D and 3D, at least at present. (An alternative is to use Monte Carlo, which will employ parallel processors.)

8. Shock Propagation: Numerical Methods

We will sketch here how to follow the propagation of a shock numerically.

References: [28, 32, 6, 12].

8.1 Acoustic Waves

Consider the propagation of 1D adiabatic acoustic waves, in planar geometry, with no external forces. The equations to be solved are:
momentum conservation

$$\frac{\mathrm{D}v}{\mathrm{D}t} = -\frac{\partial p}{\partial m} , \tag{8.1}$$

definition of velocity

$$\frac{\mathrm{D}x}{\mathrm{D}t} = v , \tag{8.2}$$

continuity

$$V = \frac{1}{\varrho} = \frac{\partial x}{\partial m} , \tag{8.3}$$

energy conservation

$$\frac{\mathrm{D}e}{\mathrm{D}t} = -p\frac{\mathrm{D}V}{\mathrm{D}t} . \tag{8.4}$$

Discretize these equations: Choose a set of mass cells m_j and a set of times t^n. The m_j are Lagrangian coordinates. As shown in Fig. (7.1), assign thermodynamic variables to cell centers: $\varrho_{j+1/2}, T_{j+1/2}, p_{j+1/2}$, and $e_{j+1/2}$. Assign position markers x_j and velocities v_j to cell surfaces. Define $\Delta m_{j+1/2} \equiv m_{j+1} - m_j$. Then discrete representations of the differential equations are:

$$(v_j^{n+1/2} - v_j^{n-1/2})/\Delta t^n = -(p_{j+1/2}^n - p_{j-1/2}^n)/\Delta m_j , \tag{8.5}$$

$$(x_j^{n+1} - x_j^n)/\Delta t^{n+1/2} = v_j^{n+1/2} , \tag{8.6}$$

$$V_{j+1/2}^{n+1} = 1/\varrho_{j+1/2}^{n+1} = (x_{j+1}^{n+1} - x_j^{n+1})/\Delta m_{j+1/2} , \tag{8.7}$$

$$e_{j+1/2}^{n+1} - e_{j+1/2}^n = -p_{j+1/2}^{()}(V_{j+1/2}^{n+1} - V_{j+1/2}^n) . \tag{8.8}$$

Here $\Delta t^{n+1/2} \equiv t^{n+1} - t^n$; $\Delta t^n \equiv \frac{1}{2}(\Delta t^{n-1/2} + \Delta t^{n+1/2})$; $\Delta m_j \equiv \frac{1}{2}(\Delta m_{j-1/2} + \Delta m_{j+1/2})$.
Given initial conditions $\{v_j\}^{n-1/2}, \{x_j\}^n, \{\varrho_{j+1/2}\}^n, \{e_{j+1/2}\}^n, \{p_{j+1/2}\}^n$, we can solve (8.5–8) in the order listed to find the solution one timestep later (t^{n+1}).

To solve Eq. (8.8), one must specify the time-level for $p_{j+1/2}$. (a) Suppose the gas is perfect. Then $p = (\gamma - 1)\varrho e$, so we can eliminate p in terms of e. We can then use the time-centered value $\frac{1}{2}(p^{n+1}_{j+1/2} + p^{n}_{j+1/2})$ (expressed in terms of e), because then the only unknown in (8.8) is $e^{n+1}_{j+1/2}$. Then the equation is trivial to solve. (b) But more generally, $p = p(\varrho, e)$. We then have two choices:

1) Use $p^{n}_{j+1/2}$, which is not time-centered. But this gives us an *explicit* formula because only $e^{n+1}_{j+1/2}$ is unknown; again the equation is trivial to solve.

2) Use the time-centered value $\frac{1}{2}(p^{n+1}_{j+1/2} + p^{n}_{j+1/2})$ and *iterate* to get consistency between $e^{n+1}_{j+1/2}$ and $p^{n+1}_{j+1/2}$. This method is *implicit.*

The explicit formula is obviously cheaper. But as we see next, we must almost always use an implicit energy equation in order to assure numerical *stability.*

8.2 Numerical Stability

Experience shows that if we solve (8.5–8) with a sufficiently small timestep Δt, we get a smooth variation of the solution in time. But if Δt rises above some critical value, the solution abruptly develops unphysical oscillations that grow and swamp the true solution. That is, the solution becomes *numerically unstable.* We must therefore do a *stability analysis* of our finite difference scheme. We will sketch here the *von Neumann local stability analysis* which provides a *necessary* (and, sometimes, but not always, sufficient) criterion for stability.

If $y^n_j = L(x, t, y)$ where L is a *linear* difference operator on a *uniform* spacetime mesh $x_j = j\Delta x, t^n = n\Delta t$, then an exact solution for y^n_j is

$$y^n_j = \sum_k A_k e^{ikj\Delta x} \xi^n_k \,. \tag{8.9}$$

The A_k's are determined by initial conditions. Each harmonic k grows or decays independently of all others. If the solution is to be *stable*, no harmonic should become unbounded. Therefore we must demand that the modulus of the *amplification factors* $\|\xi_k\| \le 1$ *for all k.*

Example: Advection. Consider the advection equation

$$\frac{\partial \varrho}{\partial t} + v_0 \frac{\partial \varrho}{\partial x} = 0 \tag{8.10}$$

where $v_0 = const \ge 0$. The solution of this equation is that the original density profile moves uniformly to the right with velocity v_0, *i.e.*, $\varrho(x,t) = f(x - v_0 t)$ where f is an arbitrary function. Suppose we represent (8.10) with the difference equation:

$$\varrho^{n+1}_j - \varrho^n_j = -v_0 \Delta t(\varrho^n_{j+1} - \varrho^n_{j-1})/2\Delta x \,. \tag{8.11}$$

This is an *explicit* difference equation. Take a trial solution of the form:

$$\varrho_j^n(k) = \xi_k^n e^{ikj\Delta x} \ . \tag{8.12}$$

Inserting (8.12) into the difference equation we find $\xi_k = 1 - i\alpha \sin k\Delta x$, where $\alpha \equiv v_0 \Delta t/\Delta x$. The amplification factor is thus $\| \xi_k \| = (1+\alpha^2 \sin^2 k\Delta x)^{1/2} \geq 1$. Therefore, the difference scheme is *unconditionally unstable*, i.e., it is unstable for *any* $\Delta t > 0$, no matter how small! Thus (8.11) is useless. So try another scheme:

$$(\varrho_j^{n+1} - \varrho_j^{n-1})/2\Delta t = -v_0(\varrho_{j+1}^n - \varrho_{j-1}^n)/2\Delta x \ . \tag{8.13}$$

This equation is known as the *leapfrog* formula. Take the same trial solution as before. Now we find $\xi_k^2 + 2i\beta\xi_k - 1 = 0$ where $\beta \equiv \alpha \sin k\Delta x$. Solving for ξ_k we find

$$\xi_k = -i\beta \pm (1-\beta^2)^{1/2} \ . \tag{8.14}$$

Thus if $\beta^2 > 1, \| \xi_k \| > 1$, and the method is *unstable*, whereas if $\beta^2 \leq 1$, $\| \xi_k \| = 1$. Thus the leapfrog scheme is *conditionally stable*, i.e., it is stable if $| \frac{v_0 \Delta t}{\Delta x} \sin k\Delta x | \leq 1$ for all k, which implies that we get stability if

$$\Delta t \leq \Delta x / v_0 \ . \tag{8.15}$$

This result is known as the *Courant condition* (or more correctly as the *Courant–Friedrichs–Lewy condition*, or *CFL condition*).

Physically the CFL condition demands that the timestep be small enough that we don't propagate the solution more than one cell-width per timestep. The Courant condition will appear again and again in what follows. The moral of this analysis is that:

- A small change in the form of a difference equation can have major (perhaps catastrophic) consequences for stability.
- One must *always* check a proposed scheme for stability.

8.3 Systems of Equations

The analysis above applies to *one* equation. Suppose we have several *coupled* equations:

$$\boldsymbol{y}^n = \mathsf{L}(x, t, \boldsymbol{y}) \tag{8.16}$$

where L is a linear operator (matrix). Represent the solution by a Fourier series with amplitudes $\boldsymbol{a}_k$. One can show that (8.16) reduces to:

$$\boldsymbol{a}_k^{n+1} = \mathsf{G}\boldsymbol{a}_k^n \tag{8.17}$$

where $\mathsf{G}(\Delta t, k)$ is called the *amplification matrix*. By analysis (see Richtmyer and Morton) one now finds that the *necessary condition for stability* is that the *spectral radius* of G be bounded by $O(\Delta t)$. Thus if $\lambda_1, \ldots, \lambda_i, \ldots, \lambda_m$ are the eigenvalues of G, then we must have

$$|\lambda_i| \leq 1 + O(\Delta t) \tag{8.18}$$

for all i, for all k, for $0 \leq \Delta t \leq \tau$.

For example, consider again acoustic waves. To simplify the algebra rewrite the momentum equation as:

$$\frac{\mathrm{D}v}{\mathrm{D}t} = -\left(\frac{\partial p}{\partial \varrho}\right)_s \left(\frac{\partial \varrho}{\partial m}\right) = \left(\frac{a^2}{V^2}\right)\left(\frac{\partial V}{\partial m}\right) . \tag{8.19}$$

Difference (8.19) as:

$$v_j^{n+1/2} - v_j^{n-1/2} = \left(\frac{\Delta t}{\Delta m}\right)\left(\frac{a^2}{V^2}\right)_j \left(V_{j+1/2}^n - V_{j-1/2}^n\right) \tag{8.20}$$

and couple to the time difference of (8.7),

$$\left(V_{j+1/2}^{n+1} - V_{j+1/2}^n\right) = \left(\frac{\Delta t}{\Delta m}\right)(v_{j+1}^{n+1/2} - v_j^{n+1/2}) . \tag{8.21}$$

Take trial solutions of the form

$$v_j^{n+1/2}(k) = (A_k)^{n+1/2} \mathrm{e}^{ikj\Delta m} \tag{8.22}$$

and

$$V_{j+1/2}^n(k) = (B_k)^n \mathrm{e}^{ik(j+1/2)\Delta m} . \tag{8.23}$$

We then find the amplification matrix:

$$\mathsf{G}(\Delta t, k) = \begin{pmatrix} 1 & i\beta^2/\alpha \\ i\alpha & 1-\beta^2 \end{pmatrix} , \tag{8.24}$$

where $\alpha \equiv 2\Delta t/\Delta m \sin k\Delta m/2$, and $\beta \equiv (a/V)\alpha$. Calculating eigenvalues we find

$$\lambda = \frac{1}{2}\left\{(2-\beta)^2 \pm \left[(2-\beta)^2 - 4\right]^{1/2}\right\} . \tag{8.25}$$

Then if $\beta^2 > 4$, $||\lambda|| > 1$ which shows that the scheme is unstable, whereas if $\beta^2 \leq 4$, then λ lies on the unit circle in the complex plane, and the scheme is stable. Therefore for stability we must have

$$-1 \leq \frac{\varrho a \Delta t}{\Delta m} \sin\left(\frac{1}{2}k\Delta m\right) \leq 1 \tag{8.26}$$

for all k, which can be guaranteed if and only if

$$\Delta t \leq \frac{\Delta m}{a\varrho} = \frac{\Delta x}{a} , \tag{8.27}$$

that is, if and only if the Courant condition is satisfied for the sound speed (which is the relevant wave speed) in the fluid.

8.4 Implications of Shock Development

- Recall that nonlinearities in adiabatic wave propagation *inevitably* produce shocks.
- If we solve *differential* equations, we can, in principle, handle a shock either as a steep front on a scale λ_p, or as a mathematical discontinuity (a kind of "internal boundary condition").
- But *difference* equations can't handle discontinuities; the solution must be smooth on a scale Δx, or the difference quotients give garbage.
- The equations written above have no dissipative terms. Therefore even if they could handle discontinuities, they would not yield an entropy increase behind the shock.
- Suppose we put in a molecular viscosity and try to resolve the shock front by brute force. This approach is doomed. We would need to reduce Δx to $O(\lambda_p)$; i.e., by *many* orders of magnitude. And if $\Delta x \to \Delta x/k$, the Courant condition would imply that $\Delta t \to \Delta t/k$, so the amount of computing required rises as k^2. This is hopeless!
- If we treat the shock as a *weak solution* (i.e., a mathematical discontinuity) we can use the *method of characteristics* with *shock finding* and *shock fitting*. This procedure can be cumbersome and hard to generalize when additional physics (e.g., radiation) is included in the calculation. Other problem with this approach are shock detection and shock dissolution.

Von Neumann and Richtmyer suggested a clever scheme that handles shocks automatically whenever and wherever they occur. The basic idea is to use an *artificial viscosity* (also called *pseudoviscosity*) which produces entropy like ordinary viscosity, but smoothes transitions in the flow over a few typical *zones* (scale $\sim \Delta x$) instead of a few mean free paths.
We can write the 1D equations for planar viscous flow exactly as:

$$\varrho \frac{\mathrm{D}v}{\mathrm{D}t} = f - \frac{\partial}{\partial z}(p+Q) \tag{8.28}$$

and

$$\frac{\mathrm{D}e}{\mathrm{D}t} + (p+Q)\frac{\mathrm{D}}{\mathrm{D}t}\left(\frac{1}{\varrho}\right) = \dot{q}\,, \tag{8.29}$$

where $\dot{q}$ is the energy input per unit mass from "external" sources (e.g., radiation) and Q is the equivalent "viscous pressure"

$$Q \equiv -\frac{4}{3}\mu' \frac{\partial v}{\partial z} \tag{8.30}$$

and

$$\mu' = \mu + \frac{3}{4}\zeta\,. \tag{8.31}$$

An analytical study of shock structure shows that viscosity *guarantees* continuous shocks. Therefore we are motivated to try to use the Von Neumann-Richtmyer scheme in a numerical scheme. A crucial question is: how do we choose the coefficient μ'? One option would be to use the formula for molecular viscosity, but with λ replaced by $\ell \sim \Delta x$. This would be a *bad idea.* There are two problems: 1) this choice of μ artificially reduces the Reynolds number *everywhere* in the flow, and thus gives a bad represention of the smooth part of the flow; 2) the thickness of a shock $\sim \lambda/(M^2-1)$; therefore weak shocks will be smeared much more than strong shocks. This gives a bad representation of the flow.

The brilliant idea of Von Neumann and Richtmyer is to use a *nonlinear* pseudoviscosity chosen such that it is as large as we like in shocks, but then drops quickly to zero outside the shock. They chose

$$Q = -\frac{4}{3}\mu_Q \left(\frac{\partial v}{\partial z}\right) \tag{8.32}$$

with

$$\mu_Q = \begin{cases} \ell^2(\mathrm{D}\varrho/\mathrm{D}t) & \text{for} \quad (\mathrm{D}\varrho/\mathrm{D}t) > 0 \quad \text{(compression)}\,, \\ 0 & \text{for} \quad (\mathrm{D}\varrho/\mathrm{D}t) \le 0 \quad \text{(expansion)}\,. \end{cases} \tag{8.33}$$

Note that $[\mu] = (\mathrm{g/cm^3})(\mathrm{cm^2}/s) \Rightarrow [\ell] = \mathrm{cm}$ (length). In practice one chooses $\ell = k\Delta z$, where Δz is an average zone size and k is a number of order unity. Thus,

$$Q = \begin{cases} \frac{4}{3}\varrho\ell^2(\partial v/\partial z)^2 & \text{for} \quad (\partial v/\partial z) < 0\,, \\ 0 & \text{for} \quad (\partial v/\partial z) > 0\,. \end{cases} \tag{8.34}$$

- The main virtue of this methods is that *it works.*
- Q comes quickly into action when the gas is compressed (e.g., in shocks).
- Q is small or zero away from shocks. Therefore it doesn't affect the background flow spuriously.
- From experience we find that we get about the same thickness for all shocks regardless of M_1. For $k \sim 1, l \sim \Delta z$.
- From an analysis of shock structure one can show that we get the correct connection between upstream and downstream flow conditions as predicted by the Rankine–Hugoniot relations provided $Q \to 0$ as $(\partial v/\partial z) \to 0$ in the upstream/downstream flow far from the shock. The Von Neumann-Richtmyer artificial viscosity meets this requirements; therefore *it preserves the Rankine–Hugoniot relations.*

8.5 Implications of Diffusive Energy Transport

In a real gas there will always be diffusive energy transport such as thermal conduction or radiation diffusion. This fact has important consequences in numerics. Consider the problem of linear heat conduction:

$$\frac{\partial T}{\partial t} = \sigma \frac{\partial^2 T}{\partial x^2} . \tag{8.35}$$

First try an *explicit formula:*

$$T_j^{n+1} - T_j^n = \left(\frac{\sigma \Delta t}{\Delta x^2}\right) (T_{j+1}^n - 2T_j^n + T_j^n) . \tag{8.36}$$

From a Von Neumann stability analysis one finds

$$\xi_k = 1 + \alpha(\cos k\Delta x - 1) , \tag{8.37}$$

where $\alpha \equiv 2\Delta t \sigma / \Delta x^2$. One thus sees that $\xi_k \leq 1$ for all α. To bound ξ_k away from -1, we must have $\alpha \leq 1$. Therefore the timestep must satisfy the condition:

$$\Delta t \leq \frac{\Delta x^2}{2\sigma} , \tag{8.38}$$

which is very restrictive. If we refine the x-grid by a factor of k, then we must reduce Δt by a factor of k^2. Therefore the computational effort rises as k^3. This result is very bad; we must find a more stable scheme. So try an *implicit scheme*:

$$T_j^{n+1} - T_j^n = \frac{1}{2}\alpha \left[\theta(\delta^2 T)_j^{n+1} + (1-\theta)(\delta^2 T)_j^n\right] , \tag{8.39}$$

where

$$(\delta^2 T)_j \equiv T_{j+1} - 2T_j + T_{j-1} \tag{8.40}$$

and

$$0 \leq \theta \leq 1 . \tag{8.41}$$

The value 0 corresponds to an explicit formula and 1 corresponds to a fully implicit formula (backward Euler). From a Von Neumann analysis we find:

$$\xi_k = \frac{1 - (1-\theta)\alpha(1 - \cos k\Delta x)}{1 + \theta\alpha(1 - \cos k\Delta x)} . \tag{8.42}$$

We see that ξ_k is always ≤ 1, and is a monotonic decreasing function of $\gamma \equiv \alpha(1 - \cos k\Delta x)$. To bound the amplification factor away from -1 we must have

$$-1 \leq \frac{1 - 2(1-\theta)\alpha}{1 + 2\alpha} \Rightarrow (1 - 2\theta)\alpha \leq 1 . \tag{8.43}$$

Thus, for $0 \leq \theta \leq \frac{1}{2}$, $\Delta t \leq \Delta x^2 / 2\sigma(1 - 2\theta)$, and for $\frac{1}{2} \leq \theta \leq 1, \Delta t$ is *unrestricted*, which implies the scheme is *unconditionally stable*. The particular choice $\theta = \frac{1}{2}$ minimizes the truncation error; this is called the *Crank–Nicholson scheme*. Having chosen an implicit scheme, we must solve a linear tridiagonal system at t^{n+1}:

$$-a_j T_{j-1}^{n+1} + b_j T_j^{n+1} - c_j T_{j+1}^{n+1} = r_j . \tag{8.44}$$

This computation is trivial, and worth the cost because there is no limit on Δt.

Moral: in any realistic problem of interest, there will always be some kind of diffusion of energy. Therefore *we will always use an implicit form for the energy equation*, even though we may choose to use an explicit form for the other equations.

8.6 Illustrative Example

Consider flow in a 1D stratified medium (*e.g.*, a star), using planar geometry with explicit hydrodynamics. As a Lagrangian variable we can use the column mass measured downward into the star

$$m(z) \equiv \int_z^\infty \varrho(z')\,\mathrm{d}z' . \tag{8.45}$$

The differential equations to be solved are:

$$\frac{\mathrm{D}v}{\mathrm{D}t} = \frac{\partial}{\partial m}(p+Q) - g , \tag{8.46}$$

$$\frac{\mathrm{D}z}{\mathrm{D}t} = v , \tag{8.47}$$

$$V = \frac{1}{\varrho} = -\frac{\partial z}{\partial m} , \tag{8.48}$$

$$\frac{\mathrm{D}e}{\mathrm{D}t} + (p+Q)\frac{\mathrm{D}}{\mathrm{D}t}\left(\frac{1}{\varrho}\right) = \dot{q} . \tag{8.49}$$

The corresponding difference equations are:

$$\begin{aligned}(v_d^{n+1/2} - v_d^{n-1/2})/\Delta t^n \\ = -g + (p_{d+1/2}^{n+\lambda} - p_{d-1/2}^{n+\lambda} + Q_{d+1/2}^{n-1/2} - Q_{d-1/2}^{n-1/2})/\Delta m_d ,\end{aligned} \tag{8.50}$$

where

$$\Delta m_d \equiv \frac{1}{2}(\Delta m_{d-1/2} + \Delta m_{d+1/2}) \tag{8.51}$$

and

$$\Delta t^n \equiv \frac{1}{2}(\Delta t^{n-1/2} + \Delta t^{n+1/2}) . \tag{8.52}$$

In general $\Delta t^{n+1/2} \neq \Delta t^{n-1/2}$ because of the requirements of stability and accuracy. Therefore to center $p^{n+\lambda}$ on the right hand side of the momentum equation define

$$\begin{aligned} t^{n+\lambda} &\equiv \frac{1}{2}[(t^n - \frac{1}{2}\Delta t^{n-1/2}) + (t^n + \frac{1}{2}\Delta t^{n+1/2})] \\ &= t^n + \frac{1}{4}(\Delta t^{n+1/2} - \Delta t^{n-1/2}) , \end{aligned} \tag{8.53}$$

which amounts to taking the average of the midpoint of the old timestep with the midpoint of the new timestep, and also choose

$$p_{d+1/2}^{n+\lambda} \equiv p_{d+1/2}^n + \frac{1}{4}(\Delta t^{n+1/2} - \Delta t^{n-1/2}) \left[\frac{(p_{d+1/2}^n - p_{d+1/2}^{n-1})}{\Delta t^{n-1/2}} \right] . \approx \left(\frac{\mathrm{d}p}{\mathrm{d}t}\right)_{d+1/2}^n \tag{8.54}$$

This is an *extrapolation* of p. For Q use

$$Q_{d+1/2}^{n-1/2} = k_Q^2 \frac{1}{2}(\varrho_{d+1/2}^{n-1} + \varrho_{d+1/2}^n)(v_{d+1}^{n-1/2} - v_d^{n-1/2})^2 \tag{8.55}$$

if $\varrho^n > \varrho^{n-1}$, and otherwise zero. Note that Q is lagged in the momentum equation (this turns out to be unimportant). *Remark:* in *implicit* schemes it is *not* necessary to extrapolate p or lag Q.
To find $v_1^{n+1/2}$ and $v_{D+1}^{n+1/2}$ we need *boundary conditions*. Some typical choices are:

Inner:
(a) $v_{D+1} \equiv 0$. Rigid core.
(b) $v_{D+1} \equiv f(t)$. Piston.
Outer:
(a) Free boundary; no force across first cell. $v_1^{n+1/2} = v_2^{n+1/2}$.
(b) Transmitting (outbound waves pass through)

$$\frac{\partial v}{\partial t} = -(a+v)\frac{\partial v}{\partial x} = \varrho(a+v)\frac{\partial v}{\partial m} .$$

(c) Zero pressure. $p_1 + Q_1 \equiv 0$ which implies

$$v_1^{n+1/2} - v_1^{n-1/2} = -g\Delta t^n + \frac{\Delta t}{\Delta m}\left(p_{3/2}^{n+\lambda} + Q_{3/2}^{n-1/2}\right) ,$$

where $\Delta m_1 \equiv \frac{1}{2}\Delta m_{3/2} + m_1$, and m_1 is the mass above the first cell.

Having updated the velocities, we can compute

$$z_d^{n+1} = z_d^n + \Delta t^{n+1/2} v_d^{n+1/2} , \tag{8.56}$$

$$V_{d+1/2}^{n+1} = \frac{1}{\varrho_{d+1/2}^{n+1}} = \frac{z_d^{n+1} - z_{d+1}^{n+1}}{\Delta m_{d+1/2}} . \tag{8.57}$$

We can then evaluate $Q^{n+1/2}_{d+1/2}$. Finally we can solve the energy equation

$$e^{n+1}_{d+1/2} - e^{n}_{d+1/2} + \left[\frac{1}{2}\left(p^{n}_{d+1/2} + p^{n+1}_{d+1/2}\right) + Q^{n+1/2}_{d+1/2}\right]\left(V^{n+1}_{d+1/2} - V^{n}_{d+1/2}\right)$$
$$= \Delta t^{n+1/2}\left[(1-\theta)\dot{q}^{n}_{d+1/2} + \theta\dot{q}^{n+1}_{d+1/2}\right] \equiv \Delta t^{n+1/2}\langle\dot{q}_{d+1/2}\rangle^{n+1/2} . \quad (8.58)$$

To complete the system, we must be given mechanical and caloric equations of state: $p = p(\varrho, T)$, $e = e(\varrho, T)$, or $e = e(\varrho, p)$. Assume the first alternative. Linearize around some guess T^*, *i.e.*, take $T^{n+1}_{d+1/2} = T^*_{d+1/2} + \delta T_{d+1/2}$. We obtain a tridiagonal system for $\delta T_{d+1/2}, (d = 1, \ldots, D)$. The solution is iterated to convergence.

The stability analysis for this system is complicated (see Richtmyer and Morton). One finds that the timestep must obey the Courant condition outside the shock. Inside the shock, one finds

$$\Delta t \leq f_Q \Delta t_{\mathrm{CFL}} , \quad (8.59)$$

where

$$f_Q = \gamma^{1/2}/2k_Q . \quad (8.60)$$

Trial calculations show that $f_Q \approx \frac{1}{2}$ works. In practice we must also limit the timestep to assure *accuracy*. Therefore we demand that Δt be chosen so that $\mid \Delta T/T \mid \leq \varepsilon_T, \mid \Delta p/p \mid \leq \varepsilon_p$, and so on. It is the accuracy requirment that restricts the timestep in an implicit calculation where there is no stability restriction on Δt.

To see how artificial viscosity works, see Fig. (8.1) which shows a shock with $p_2/p_1 = 5$ and $\gamma = 2$. Panel (a) has $k_Q = 2$, and panel (b) has $k_Q = 0$. In the latter, the shock speed is wrong, and the density shows large oscillations. This is *not* a problem of numerical instability. Rather, it is the failure of the system to convert upstream directed momentum into heat, and a failure to increase entropy, both because there is no dissipative term in the equations. The downstream cells are acting like "macro molecules", and there is no hope they will ever thermalize. This is a vivid illustration of why one *must* use artifical viscosity. (Or an equivalent mechanism.)

Total Energy Check. From the *differential* equations of the system, one can derive

$$\frac{\mathrm{D}}{\mathrm{D}t}(e + \frac{1}{2}v^2) - \frac{\partial}{\partial m}[(p+Q)v] = -gv + \dot{q} . \quad (8.61)$$

Integrating over mass, we get an exact conservation law

$$\frac{\mathrm{D}}{\mathrm{D}t}\int (e + \frac{1}{2}v^2 + gz)dm$$
$$= + \int \dot{q}\mathrm{d}m + (p+Q)_{D+1}v_{D+1} - (p+Q)_1 v_1 . \quad (8.62)$$

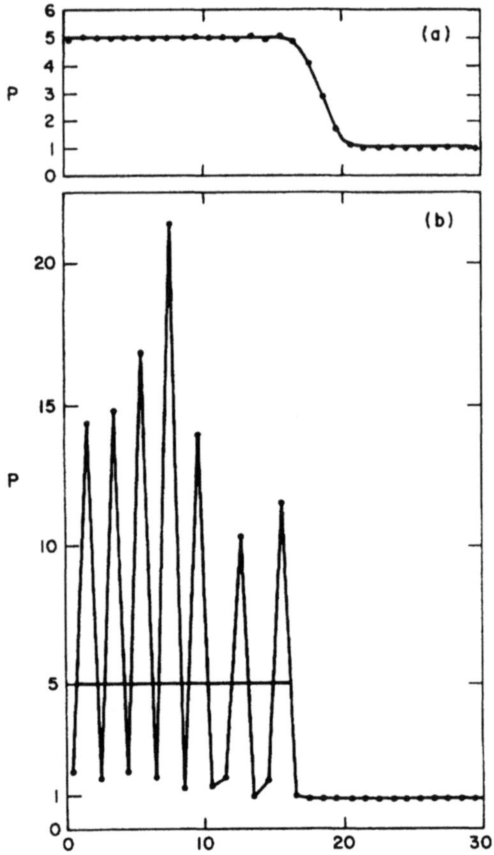

Fig. 8.1. Shock computed (**a**) with and (**b**) without artificial viscosity. From [28] by permission

The first term is the rate of change of the total energy in the system. The second is the heat deposited. And the last two terms give the surface work terms. But the *difference* equations do not guarantee the above integral exactly. So in practice we compute a discretized version of the total energy as a check:

$$\begin{aligned}\mathcal{E}^{n+1} &\equiv \sum_d \left\{ e^{n+1}_{d+1/2} \Delta m_{d+1/2} + \left[g(z^{n+1}_d - z^0_d) + \frac{1}{2}(v^{n+1}_d)^2 \right] \Delta m_d \right\} \\ &\quad + \mathcal{W}^{n+1}_1 - \mathcal{W}^{n+1}_{D+1} - \bar{q}^{n+1} \\ &= \sum_d \left[e^0_{d+1/2} \Delta m_{d+1/2} + \frac{1}{2}(v^0_d)^2 \Delta m_d \right] = \mathcal{E}^0 = \text{constant} \,. \quad (8.63)\end{aligned}$$

Here

$$\mathcal{W}^{n+1}_i \equiv \sum_{k=0}^{n} \left[\frac{1}{2}(p^k_i + p^{k+1}_i) + Q^{k+1/2}_i \right] (z^{k+1}_i - z^k_i) \qquad (8.64)$$

and

$$\tilde{q}^{n+1} \equiv \sum_{k=0}^{n} \Delta t^{k+1/2} \sum_{d} \langle \dot{q}_{d+1/2} \rangle^{k+1/2} \Delta m_{d+1/2} \,. \tag{8.65}$$

Ideally, $\mathcal{E}$ should be conserved to high accuracy. If it isn't, one should reduce the timestep.

9. Numerical Radiation Hydrodynamics

Consider now the 1D spherically symmetric flow of a single material, as might occur, for example, in a star. Use Lagrangian variables and coordinates.

9.1 Radiating Fluid Energy and Momentum Equations

All variables are measured in the comoving frame, so for simplicity drop the suffix "0". Retain all $O(v/c)$ terms in the energy equation; drop all v-dependent terms in the momentum equation except $(\mathrm{D}/\mathrm{D}t)$ (for flux limiting). Then the monochromatic energy equation (see Sect. 5.3) is

$$\begin{aligned} &\frac{\mathrm{D}}{\mathrm{D}t}\left(\frac{E_\nu}{\varrho}\right) + \left[f_\nu \frac{\mathrm{D}}{\mathrm{D}t}\left(\frac{1}{\varrho}\right) - (3f_\nu - 1)\frac{v}{\varrho r}\right] E_\nu \\ &\quad - \frac{\partial}{\partial \nu}\left\{\left[f_\nu \frac{\mathrm{D}}{\mathrm{D}t}\left(\frac{1}{\varrho}\right) - (3f_\nu - 1)\frac{v}{\varrho r}\right] \nu E_\nu\right\} \\ &= \frac{\kappa_\nu}{\varrho}\left(4\pi B_\nu - cE_\nu\right) - \frac{\partial\left(4\pi r^2 F_\nu\right)}{\partial M_r} \end{aligned} \tag{9.1}$$

and the monochromatic momentum equation is

$$\frac{1}{c^2}\frac{\mathrm{D}F_\nu}{\mathrm{D}t} + 4\pi r^2 \varrho \frac{\partial\left(f_\nu E_\nu\right)}{\partial M_r} + \frac{(3f_\nu - 1)}{r} E_\nu = -\frac{1}{c}\chi_\nu F_\nu \,. \tag{9.2}$$

Define a *sphericity factor* as

$$\ln q_\nu = \int_{r_c}^{r} \frac{3f_\nu - 1}{f_\nu r'} \mathrm{d}r' . \tag{9.3}$$

Then we can rewrite the momentum equation as

$$\frac{1}{c^2}\frac{\mathrm{D}F_\nu}{\mathrm{D}t} + \frac{4\pi r^2 \varrho}{q_\nu}\frac{\partial\left(f_\nu q_\nu E_\nu\right)}{\partial M_r} = -\frac{1}{c}\chi_\nu F_\nu \,. \tag{9.4}$$

Integrating over frequency, we get the total radiation energy equation

$$\frac{\mathrm{D}}{\mathrm{D}t}\left(\frac{E}{\varrho}\right)+\left[f\frac{\mathrm{D}}{\mathrm{D}t}\left(\frac{1}{\varrho}\right)-(3f-1)\frac{v}{\varrho r}\right]E$$
$$=\frac{1}{\varrho}\left(4\pi\kappa_{\mathrm{P}}B-c\kappa_{\mathrm{E}}E\right)-\frac{\partial\left(4\pi r^2F\right)}{\partial M_r} \tag{9.5}$$

and the total radiation momentum equation

$$\frac{1}{c^2}\frac{\mathrm{D}F}{\mathrm{D}t}+\frac{4\pi r^2\varrho}{q}\frac{\partial\left(fqE\right)}{\partial M_r}=-\frac{1}{c}\chi_{\mathrm{F}}F\,. \tag{9.6}$$

Here

$$\kappa_{\mathrm{E}}\equiv\int_0^\infty\left(\frac{E_\nu}{E}\right)\kappa_\nu\mathrm{d}\nu\,, \tag{9.7}$$

$$\chi_{\mathrm{F}}\equiv\int_0^\infty\left(\frac{F_\nu}{F}\right)\chi\nu\mathrm{d}\nu\,, \tag{9.8}$$

$$f\equiv\int_0^\infty\left(\frac{E_\nu}{E}\right)f_\nu\mathrm{d}\nu\,, \tag{9.9}$$

$$\ln q\equiv\int_{r_c}^{r}\frac{3f-1}{fr'}\mathrm{d}r'\,. \tag{9.10}$$

We can represent (9.5) and (9.6) as difference equations on a staggered grid. We locate $E, f, q, \kappa_{\mathrm{P}}$, and κ_{E} at cell centers, and F and χ_{F} at cell boundaries. For the time differencing we use an implicit scheme for stability; a common choice, if we don't care about following radiation fronts moving with $v\sim c$, is *fully implicit* (backward Euler). We then have a choice: we can use either two 1st order equations for E and F at t^{n+1}, or one 2nd order equation for E (eliminate F). The choice is made on the basis of computational convenience; the physics is unchanged once the equations are discretized.

The radiation equations above are coupled to the standard material energy and momentum equations:

$$\frac{\mathrm{D}v}{\mathrm{D}t}=-\frac{GM_r}{r^2}-4\pi r^2\frac{(p+Q)}{\partial M_r}+\frac{\chi_{\mathrm{F}}F}{\varrho c}\,, \tag{9.11}$$

or

$$\frac{\mathrm{D}v}{\mathrm{D}t}=-\frac{GM_r}{r^2}-4\pi r^2\frac{(p+Q)}{\partial M_r}+\frac{1}{q}\frac{\partial(fqE)}{\partial M_r} \tag{9.12}$$

and

$$\frac{\mathrm{D}e}{\mathrm{D}t}+(p+Q)\frac{\mathrm{D}}{\mathrm{D}t}\left(\frac{1}{\varrho}\right)=\varepsilon+\left(c\kappa_{\mathrm{E}}E-4\pi\kappa_{\mathrm{P}}B\right)/\varrho\,, \tag{9.13}$$

where Q is the pseudoviscous pressure of Sect. 8.4.

9.2 Computational Strategy

Overall Procedure. It is costly to try to solve directly the full angle-frequency dependent radiation equations coupled to the equations of hydrodynamics because the dimensionality of the problem (= 4) is too large. So we look for an approximate technique. Notice that we need only the *frequency-integrated moments* to solve the material energy and momentum equations. Therefore we are led to try a *splitting technique* in which the spacetime evolution is handled separately from the angle-frequency information. The method described below is called the *multifrequency/grey* method. For ease of exposition, assume we use *explicit hydro.*

1. Suppose we are given the solution at t^n. Update velocities to t^{n+1} (material momentum equation), hence radii and densities to t^{n+1}.
2. If we can assume that the *spectral distributions* (E_ν/E) and (F_ν/F) are known at t^{n+1} (actually they are *not*!) we can calculate $\kappa_{\rm E}$ and $\chi_{\rm F}$, hence $k_{\rm E} \equiv \kappa_{\rm E}/\kappa_{\rm P}$ and $k_{\rm F} \equiv \chi_{\rm F}/\chi_{\rm R}$. Similarly, if we can assume that the *angular distribution* of the radiation field is known at t^{n+1} (again, it isn't), then we have f_ν and q_ν, hence f and q.
3. At t^{n+1} we can now solve the gas energy equation plus the radiation energy and momentum equations. We thus obtain the run of T^{n+1}, E^{n+1}, and F^{n+1}. This system is nonlinear because the material properties depend nonlinearly on T^{n+1}; to solve it we linearize and iterate to convergence, as described earlier.

In principle, these steps give us the solution at the advanced time. But actually they don't because we *assumed* we already knew $(E_\nu/E), (F_\nu/F), f_\nu$, and q_ν at t^{n+1}, when in fact we didn't. Therefore at the advanced time level t^{n+1} we should now:

4. Solve the angle-dependent transfer equation for the geometric factors f_ν and q_ν using source-sink terms evaluated at t^{n+1} [*i.e.*, $\kappa_\nu(T^{n+1})$, $\eta_\nu(T^{n+1})$].
5. Then solve the frequency-dependent moment equations for the *spectral profiles* (E_ν/E) and (F_ν/F). Use these updated profiles to update the information in $\kappa_{\rm E}, \chi_{\rm F}, k_{\rm E}$, and $k_{\rm F}$ at t^{n+1}.
6. Go to step 3, and iterate to consistency.

- In step 4, we *ignore frequency-coupling* in order to reduce the dimensionality of the problem to (r, μ) at a given (ν, t). From this step we *derive* the geometric factors f_ν and q_ν which depend mainly on *angular* information about the radiation field.
- In step 5, we dispose of angular information by using moments, and attempt to *improve* the frequency-dependent information contained in (E_ν/E) and (F_ν/F) by accounting for the (ν, r) coupling at a given t, assuming that the geometric factors (f_ν, q_ν) are known.

- This procedure works fairly well because f_ν is only a *ratio* which is mainly *geometry* dependent, whereas (E_ν/E) and (F_ν/F) are only spectral *profiles*, which are insensitive to geometry.
- Very few calculations have been done to this level of consistency. Most either ignore the frequency-coupling in the moments, and lag the Eddington factors, or they handle the frequency-coupling but assume $f_\nu \equiv \frac{1}{3}$ (multigroup diffusion).

Spatial Differencing and Unresolved Fronts. It is straightforward to write difference formulae for the equations written above. We won't produce them here; the reader can easily work them out. But consider one very difficult problem: how can we calculate a good value for χ_F at an interface? A seemingly reasonable value giving the correct optical depth increment between cell centers is

$$\left(\frac{\chi}{\varrho}\right)_d = \frac{(\chi/\varrho)_{d-1/2}\Delta m_{d-1/2} + (\chi/\varrho)_{d+1/2}\Delta m_{d+1/2}}{\Delta m_{d-1/2} + \Delta m_{d+1/2}} . \tag{9.14}$$

But trouble arises when we fail to *resolve* a front in the flow (a shock, ionization front, ...). The problem is that the opacity can vary as T^α with $\alpha \sim 12 - 14$ in regions where H and/or He get excited and/or ionize. Therefore even small changes in T between zones imply a *huge* change in χ. This rapid variation in χ makes it hard to choose the correct value at the interface, and the errors made can clobber the energy transport at the interface. Generally the result is disasterous. How can we fix this problem? R. Christy [6] found that we can get "reasonable" results using an energy-weighted harmonic mean:

$$\left(\frac{\varrho}{\chi}\right)_d \equiv \frac{(T_{d-1/2})^4\left(\frac{\varrho}{\chi}\right)_{d-1/2} + (T_{d+1/2})^4\left(\frac{\varrho}{\chi}\right)_{d+1/2}}{(T_{d-1/2})^4 + (T_{d+1/2})^4} . \tag{9.15}$$

This recipe has been very popular, but it is only ad hoc. The *real* solution to the problem is to use an *adaptive grid.* The problem can't be solved by brute force with tiny Lagrangian zones. For example, the ionization zone in Cepheids and RR Lyrae envelopes is about 0.001 to 0.01 scale heights thick, and moves back and forth over 3–5 scale heights during a pulsation cycle. A prohibitive number of zones (and timesteps!) would be required to resolve the front. In contrast, an adaptive grid *moves with the front* and resolves it. This approach is the preferred technique.

Method of Solution. For the explicit hydro scheme sketched above, at t^{n+1} we get equations coupling $T^{n+1}, E^{n+1}, F^{n+1}$ at d = 1, ... , D. Linearize the equations around the current estimate of the solution. We get a block tridiagonal system which can be solved by Gaussian elimination, and iterated to convergence (Newton–Raphson procedure). In linearizing opacities we can write $\kappa_\mathrm{E} = k_\mathrm{E}\kappa_\mathrm{P}$ and $\chi_\mathrm{F} = k_\mathrm{F}\chi_\mathrm{R}$ and assume $(k_\mathrm{E}, k_\mathrm{F})$ remain fixed. Or (better) we could calculate $(\partial\kappa_{\mathrm{E}}/\partial T)$ and $(\partial\chi_{\mathrm{F}}/\partial T)$ from $(\partial\kappa_g/\partial T)$ and $(\partial\chi_g/\partial T)$ in groups (or an ODF) given (E_g/E) and (F_g/F).

If we do implicit hydro instead of explicit, then at t^{n+1} we must solve for r, v, ϱ, T, E, and F. We then have a block tridiagonal system with (6×6) blocks, or (7×7) with an adaptive grid. In principle one can take large timesteps with an implicit system, but in practice *accuracy* requirements severely limit Δt, *especially if there are unresolved fronts.* Again an adaptive grid is the only solution.

9.3 Energy Conservation

As for a nonradiating fluid, we can write a total energy conservation theorem:

$$\frac{\mathrm{D}}{\mathrm{D}t}\int\left(e+\frac{E}{\varrho}+\frac{1}{2}v^2-\frac{GM_r}{r}\right)\mathrm{d}M_r + \left[4\pi r^2 v(p+P+Q)\right]+[L]=\int\varepsilon\mathrm{d}M_r\,, \tag{9.16}$$

where $[X] \equiv X_{\text{outer boundary}} - X_{\text{inner boundary}}$. But as in Sect. 8.6, *difference* equations don't *guarantee* conservation, so we use a discrete representation of the conservation theorem (9.16) to *monitor* the numerical accuracy of the solution.

9.4 Formal Solution

How do we update the Eddington factors? In the past the best that has been done is to use a *static snapshot.* That is, we drop $\mathrm{D}/\mathrm{D}t$ and all velocity-dependent terms, and solve only the *static* spherical transfer equation. This procedure is probably all right in low-energy astrophysical applications where $v/c \ll 1$ and $t_\lambda \ll t_F$. But in more dynamical situations we should perhaps try to do better. Write the comoving frame transfer equation as

$$\begin{aligned}&\frac{\varrho}{c}\frac{\mathrm{D}}{\mathrm{D}t}\left(\frac{I_\nu}{\varrho}\right)+4\pi r^2\varrho\mu\frac{\partial I_\nu}{\partial M_r}+\frac{(1-\mu^2)}{r}\frac{\partial I_\nu}{\partial\mu}+\frac{1}{c}\left\{\left(\frac{3v}{r}+\frac{\mathrm{D}\ln\varrho}{\mathrm{D}t}\right)\right.\\&\quad\left.\cdot\frac{\partial}{\partial\mu}\left[\mu(1-\mu^2)I_\nu\right]-\left[(1-3\mu^2)\frac{v}{r}-\mu^2\frac{\mathrm{D}\ln\varrho}{\mathrm{D}t}\right]\frac{\partial(\nu I_\nu)}{\partial\nu}\right\}\\&=\eta_\nu-\left[\chi_\nu+(1-3\mu^2)\frac{v}{cr}-\frac{\mu^2}{c}\frac{\mathrm{D}\ln\varrho}{\mathrm{D}t}\right]I_\nu\,.\end{aligned} \tag{9.17}$$

- All terms in the curly bracket are $O(v/c)$.
- All terms in the curly bracket *vanish identically* when integrated over (μ, ν)
- The remaining terms describe transfer along straight rays in spherical geometry, and account for the velocity-dependent rate of work done by radiation pressure.

These observations suggest that when solving transfer equations to get the angular distribution of the radiation field we can simply *drop all terms in the curly bracket*. The result is called the *model Lagrangian transfer equation*, which:

- accounts for the major *geometric* effects that determine the angular distribution of the radiation, hence f_ν,
- yields the correct *integrated* radiation energy/momentum equations (including all necessary velocity-dependent terms) when integrated over (μ, ν).

The resulting problem is of the same dimensionality as a static snapshot: (r, μ) at a given (ν, t).

From the model transfer equation we can form symmetric/antisymmetric equations:

$$\begin{aligned} &\frac{\varrho}{c}\frac{\mathrm{D}}{\mathrm{D}t}\left(\frac{j_\nu}{\varrho}\right) + 4\pi r^2 \varrho\mu \frac{\partial h_\nu}{\partial M_r} + \frac{(1-\mu^2)}{r}\frac{\partial h_\nu}{\partial \mu} \\ &\quad = \eta_\nu - \left[\chi_\nu + (1-3\mu^2)\frac{v}{cr} - \frac{\mu^2}{c}\frac{\mathrm{D}\ln\varrho}{\mathrm{D}t}\right] j_\nu \end{aligned} \tag{9.18}$$

and

$$\frac{\varrho}{c}\frac{\mathrm{D}}{\mathrm{D}t}\left(\frac{h_\nu}{\varrho}\right) + 4\pi r^2 \varrho\mu \frac{\partial j_\nu}{\partial M_r} + \frac{(1-\mu^2)}{r}\frac{\partial j_\nu}{\partial \mu} = -\chi_\nu h_\nu + O(v/c)\,. \tag{9.19}$$

We can either difference these equations along tangent rays:

$$\frac{\varrho}{c}\frac{\mathrm{D}}{\mathrm{D}t}\left(\frac{j_\nu}{\varrho}\right) + \frac{\partial h_\nu}{\partial s} = \eta_\nu - \left[\chi_\nu + (1-3\mu^2)\frac{v}{cr} - \frac{\mu^2}{c}\frac{\mathrm{D}\ln\varrho}{\mathrm{D}t}\right] j_\nu \tag{9.20}$$

and

$$\frac{1}{c}\frac{\mathrm{D}h_\nu}{\mathrm{D}t} + \frac{\partial j_\nu}{\partial s} = -\chi_\nu h_\nu\,, \tag{9.21}$$

or we can write them in conservative form and difference on the (r, μ) mesh:

$$\begin{aligned} &\frac{\varrho}{c}\frac{\mathrm{D}}{\mathrm{D}t}\left(\frac{j_\nu}{\varrho}\right) + 4\pi \varrho\mu \frac{\partial (r^2 h_\nu)}{\partial M_r} + \frac{1}{r}\frac{\partial\left[(1-\mu^2)h_\nu\right]}{\partial \mu} \\ &\quad = \eta_\nu - \left[\chi_\nu + (1-3\mu^2)\frac{v}{cr} - \frac{\mu^2}{c}\frac{\mathrm{D}\ln\varrho}{\mathrm{D}t}\right] j_\nu \end{aligned} \tag{9.22}$$

and

$$\begin{aligned} &\frac{\mu}{c}\frac{\mathrm{D}h_\nu}{\mathrm{D}t} + 4\pi\varrho\mu^2 \frac{\partial (r^2 j_\nu)}{\partial M_r} + \frac{(\mu^2-1)}{r} j_\nu \\ &\quad + \frac{1}{r}\frac{\partial}{\partial\mu}\left[\mu(\mu^2-1) j_\nu\right] = -\mu\chi_\nu h_\nu\,. \end{aligned} \tag{9.23}$$

9.5 Multigroup Equations

Having determined the angular factors by the solution of the angle-dependent transfer equation, we next solve the monochromatic (multigroup) moment equations for the frequency-profile of the radiation field. We obtain appropriate difference equations by differencing the monochromatic radiation energy/momentum equations. There is one important technical point: in the term

$$-\frac{\partial}{\partial\nu}\left\{\left[f_\nu \frac{\mathrm{D}}{\mathrm{D}t}\left(\frac{1}{\varrho}\right) - \frac{(3f_\nu - 1)v}{\varrho r}\right]\nu E_\nu\right\} \tag{9.24}$$

the effective "advection velocity" of photons in *frequency space* is

$$\frac{1}{\nu}\frac{\mathrm{D}\nu}{\mathrm{D}t} = f_\nu \frac{\mathrm{D}\ln\varrho}{\mathrm{D}t} + \frac{(3f_\nu - 1)v}{r} \,. \tag{9.25}$$

As in multigroup diffusion theory, we want to use upwind differencing, and must choose the direction of the frequency difference according to the sign of $\mathrm{D}\nu/\mathrm{D}t$. Thus if $\mathrm{D}\nu/\mathrm{D}t > 0$, photons are blue-shifted during a timestep, hence we couple group g to group $g-1$; in contrast, if $\mathrm{D}\nu/\mathrm{D}t < 0$, photons are redshifted and we couple group g to group $g+1$. In diffusion theory the term in $(3f_\nu - 1) \equiv 0$, so *all* groups couple in the same direction (*i.e.*, all to $g-1$ or $g+1$). But in transport, the two terms compete, and the direction of coupling may change for different groups depending on the signs and relative sizes of the two terms.

9.6 An Astrophysical Example

Now let's look briefly at a nice example of radiation hydro in action.

9.6.1 Supernovae. *Reference*: S. Falk and D. Arnett [9].

- Type II supernova, $M \geq M_\odot$. Core exhausted to Fe.
- Endothermic reactions trigger collapse to nuclear densities.
- Infalling envelope bounces on core. Drives intense shock outward.
- Total of 10^{50}–10^{51} ergs released into envelope on a timescale $\ll$ typical dynamical timescale $t_d = R/a \sim 6$ months. Outer envelope is blown away.

The *presupernova* has the structure of a red giant:

- *core*: contains 1–2 $M_\odot, r \sim 10^7$–10^8 cm. This core collapses to a neutron star or a black hole composed entirely of Fe.
- *mantle*: contains a few $M_\odot$ composed mainly of C, O, Ne, Mg, *etc.*, with a radius of $r \sim 10^{11}$cm.
- *envelope*: composed of He (including He-burning shell). Perhaps containing H and a H-burning shell also. $r \sim 10^{13}$–10^{14} cm.
- *circumstellar shell*: 0.01 $M_\odot, r \sim 10^{15}$cm. This is the material transported outward in a stellar wind. It is important because it affects the supernova light curve.

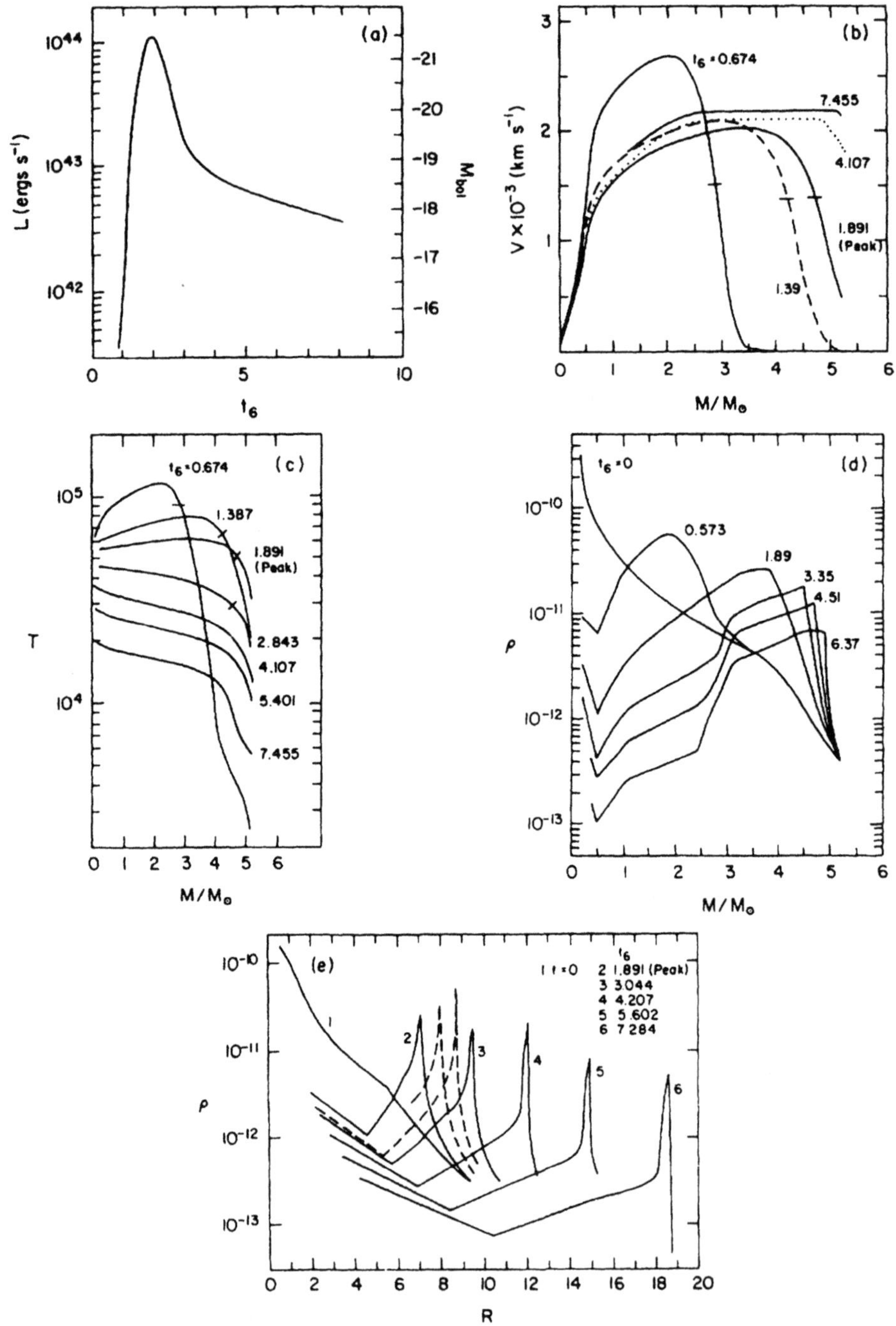

Fig. 9.1. Supernova explosion in a $5M_\odot$ supergiant with circumstellar shell. **(a)** Light curve ($t_6 = t(s)/10^6$). **(b)** Velocity as a function of mass. **(c)** Temperature as a function of mass. **(d)** Density as a function of mass. **(e)** Density as a function of radius. From [9] by permission

Basic Scenario.

- The shock emerging from the bounce is markedly *supercritical.*
- The shock drives an intense radiation front into the upstream material. In the shock precursor, the radiation pressure $\gg$ gas pressure, by about a factor of 100.
- Therefore the shock is heavily mediated by radiation, and the velocity jump is quite small. The effects of radiative diffusion are so large that the calculation can proceed many timesteps with no artificial viscosity, without having stability trouble!
- As it runs down the density gradient in the envelope, the shock strengthens, steepens, and accelerates.
- An external observer is unaware of the shock until it reaches $\tau \approx 1$ at some wavelength. The first signal to emerge is a soft X-ray burst. The luminosity maximum in the visible occurs 4 to 20 days later!
- The shock *unloads* and produces a *blowoff* moving at a few times 10^3 km/sec, *i.e.*, $v/c \sim 10^{-2}$.
- A strong viscous shock finally [penetrates the optically thin layers and heats the gas there to very high temperatures (10^6–10^7 K) and drives a hard X-ray burst. (Early calculations predicted, erroneously, a γ-ray burst!)

Falk–Arnett Calculation.

- 1D, spherical, LTE, grey.
- Assumes radiation diffusion inside, and couples to transport equations (moments) near the surface, but uses the Eddington approximation throughout.

Results are shown in Fig. (9.1) for a model with a circumstellar shell.

(a) Light curve. The computed light curve has a broad peak, like observed. This delay is the time t_d required for the radiation to *diffuse* out from the envelope (the X-ray burst is long over).
(b) Velocity profile. The shock position is shown by tick marks. At late times the envelope expands with nearly constant velocity.
(c) Temperature. The peak temperature occurs before peak luminosity. After the peak luminosity, the temperature decreases monotonically outward and in time.
(d), (e) Density profile: The striking feature is the thin, dense shell that forms after the peak luminosity and expands rapidly outward. The shell forms because photons leak out from the shock very efficiently, which implies no temperature rise behind the shock ("isothermal" shock). Therefore we get a gigantic density increase behind the shock; in fact, we get a *density inversion.* The latter may be *Rayleigh–Taylor unstable*; if so, the flow would fragment into blobs, which implies mixing and large turbulent velocities. Similar phenomena are observed in supernova remnants.

10. Adaptive-Grid Radiation Hydrodynamics

Reference: [8].

We have seen that nonlinearities in hydrodynamics and radiation hydrodynamics imply that the development of "discontinuities" in the flow is inevitable. Examples are ***shocks, contact discontinuities, dissociation/ionization fronts***, and ***chemical reaction fronts***. Physically these structures are not actually discontinuities, but are regions of very steep ***gradients*** with characteristic scales $O(\lambda)$ and $O(\lambda_{\mathrm{p}})$, where λ is a particle mean free path and λ_{p} is a photon mean free path, and ℓ, the charactersitic scale-length in the flow, is much larger than either λ or λ_{p}. Propagation of these sharp transition regions poses insurmountable difficulties for finite-difference equations unless the grid spacing is $\Delta x \sim \lambda$ or λ_{p} which is, of course, prohibitively fine.

Thus, to obtain numerical solutions for problems with fronts, we are forced to consider alternative methods:
1) We can abandon the differential equations of hydrodynamics, and replace them with integral conservation laws, and strict mathematical discontinuities imbedded in the flow. We thus obtain what are known as ***weak solutions*** of the partial differential equations. To make the method work we apply ***front fitting*** techniques using ***jump conditions*** appropriate to the front in question. 2) Or we can introduce ***artificial dissipative terms*** into the difference equations, chosen such that sharp fronts are smeared over a few typical (hydrodynamic) zone widths. An example of this technique is the von Neumann-Richtmyer ***artificial viscosity method.***

10.1 Front Fitting

There are both advantages and disadvantages to front fitting. The two most important advantages are 1) that the front has zero thickness on the grid, and 2) that the Rankine–Hugoniot relations are guaranteed exactly for a pure hydrodynamic shock. The disadvantages of this method are: 1) the front can move many front-widths in a timestep, hence we must ***find*** the position of the front; further we must ***detect*** formation of new fronts, and we must ***discard*** old fronts that have dissipated. 2) The algorithm becomes progressively more cumbersome as fronts interact, and the structure in the flow increases. 3) This approach tacitly assumes ***steady flow*** through the front. 4) We do not actually know how to write jump conditions for all kinds of fronts. In some cases they cannot exist because of ***nonlocal interactions*** (e.g., by radiation) across the front. Finally, in astrophysical flows we typically have an enormous range of characteristic time and length scales. A few of these are summarized below:

LENGTH	λ	particle mfp
	λ_p	photon mfp
	H	scale height (exponential medium)
	R	gravitational scale (power law)
	L	size of medium (flat distribution)
TIME	λ_p/c	photon flight (10^{-8} sec)
	ℓ/c	radiation flow (msec–sec)
	$\ell^2/\lambda_p c$	photon diffusion (days–years)
	ℓ/a	fluid flow (minutes)
	R/a	pulsation (hours–days)
	$(R^3/GM)^{1/2}$	gravitational collapse (years)
	(GM^3/RL)	Kelvin–Helmholtz contraction (Myears)
	Mc^2/ε	nuclear evolution (Gyears)

Clearly we would need an excessive logial structure in the code to account accurately for all possible interactions on all these time/length scales.

10.2 Artificial Dissipation

von Neumann and Richmtmeyr [32] took the point of view that all the front-handling by the code should be *automatic*. Almost 50 years ago, they showed that the method works! Dissipative terms *do* detect and track fronts automatically. In addition, one can construct dissipative terms such that jump conditions are preserved across the front. On the other hand, the method cannot handle multiple scale-length cases well. For example, using artifical viscosity fronts that are "resolved" on a hydrodynamic scale may be *under-resolved* by orders of magnitude on some other physical scale. Thus in radiation hydrodynamics shock resolution of 3 to 4 $\Delta x_{\text{typical}}$ may be perfectly all right hydrodynamically, but these zones may be *optically thick*. Hence radiation transport through these zones may be calculated completely incorrectly. Thus it is necessary to develop a fully automatic method that can resolve *all* structure in the flow. This can be done with an *implicit adaptive grid* which can move grid points into regions of steep gradients of the physical variables.

10.3 The Adaptive Grid

Definition. What is an adaptive grid? There are two ways of answering this question.

1) It is a *coordinate system.* In hydrodynamics we can have three useful coordinate systems – Eulerian: $(\boldsymbol{r}, t)$ which is fixed in the laboratory frame; Lagrangian: (m, t) or (r_0, t) which is fixed in the fluid frame; and Adaptive: (grid index k, t) which is attached to *features in the flow*, and may move with respect to both, the lab frame and fluid elements. It provides a mapping $\boldsymbol{r}(j, t)$ and $m(j, t)$.

2) To implement an adaptive grid we must provide an algorithm which determines how grid points are distributed over the domain as a function of time. This algorithm must *detect, resolve*, and *track* important flow features *automatically*. It must allow a huge *dynamic range* (10^5–10^6) in cell sizes (spacing of grid points) so that flow features are always *resolved*. It must move the gridpoints such that flow features are almost *stationary* with respect to the grid (important!) even though they may move rapidly with respect to the fluid and in real space. If we can achieve these desiderata, we can use very large timesteps, and still satisfy both the CFL condition and accuracy criteria with respect to the adaptive grid. An adaptive grid becomes powerful when we incorporate it into the set of radiation hydro equations to be solved, and solve for the positions of gridpoints *implicitly* at the advanced time along with the hydro variables.

The implicit adaptive grid just described is quite different from *rezoning* schemes, much in the same sense that implicit hydro schemes differ from explicit schemes. In a rezoning scheme we rearrange the mesh *after* a timestep, in such a way as to reduce gradients measured with respect to grid index. One problem always faced is interpolation on the new grid; this process may be noisy. If we use implicit hydro we can use small zones and still beat the CFL condition, but the problem of getting the Newton–Raphson procedure to converge will set stringent limits because a narrow front may move over several zones if Δt is too large. And the most important problem is that by rearrangement of the grid *after* a timestep, we cannot guarantee that the high-resolution regions will be located correctly after the *next* timestep. An implicit adaptive grid addresses both of these difficulties. Here *both* the fluid variables and the spatial positions of the gridpoints are regarded as unknowns at t^{n+1}. Thus we must determine both *simultaneously* at the advanced time, *i.e.*, we move the mesh *during* the timestep fully consitently with the motion of flow features. In this way we can achieve the goal of near stationarity of the flow with respect to the grid.

Adaptive-Grid Conservation Laws. In radiation hydrodynamics one always needs to distinguish between the *frame* in which the physical variables (dependent variables) are measured, and the *coordinates* chosen for the independent variables. We are always free to choose a new coordinate system if we know the transformation between the two sets of coordinates. We shall distinguish carefully between the Eulerian time-derivative $\partial/\partial t$, the Lagrangian time-derivative $\mathrm{D}/\mathrm{D}t$, and the adaptive time-derivative $\mathrm{d}/\mathrm{d}t$. Then the *fluid velocity* is $\boldsymbol{u} = \mathrm{D}\boldsymbol{r}/\mathrm{D}t$, where $\boldsymbol{r}$ is the position of a definite fluid element, the *grid velocity* is $\boldsymbol{u}_{\text{grid}} = \mathrm{d}\boldsymbol{r}/\mathrm{d}t$, where $\boldsymbol{r}$ is the position of an adaptive gridpoint, and the *relative velocity* is $\boldsymbol{u}_{\text{rel}} = \boldsymbol{u} - \boldsymbol{u}_{\text{grid}}$ measures the velocity of the fluid with respect to the moving adaptive grid.

With these definitions one can derive several basic connections between the adaptive coordinate system and the usual Eulerian and Lagranagian systems. For example, it is easy to show that the standard formula

$$\frac{\mathrm{D}f}{\mathrm{D}t} = \frac{\partial f}{\partial t} + \boldsymbol{u} \cdot \boldsymbol{\nabla} f \tag{10.1}$$

is exactly equivalent to

$$\frac{\mathrm{d}f}{\mathrm{d}t} = \frac{\partial f}{\partial t} + \boldsymbol{u}_{\text{grid}} \cdot \boldsymbol{\nabla} f \,. \tag{10.2}$$

Likewise, by following through exactly the argument that leads to the *Euler expansion formula*

$$\frac{\mathrm{D}}{\mathrm{D}t}(\ln J_f) = \boldsymbol{\nabla} \cdot \boldsymbol{u} \,, \tag{10.3}$$

when the coordinate system is anchored in the fluid so that $\boldsymbol{u}$ is the fluid velocity and

$$\mathrm{d}V_{\text{fluid}} = J_f \; \mathrm{d}V^0_{\text{fluid}} \,, \tag{10.4}$$

one can derive the *adaptive grid expansion formula*

$$\frac{\mathrm{d}}{\mathrm{d}t}(\ln J) = \boldsymbol{\nabla} \cdot \boldsymbol{u}_{\text{grid}} \,, \tag{10.5}$$

where now the coordinate system is the adaptive grid, which moves with velocity $\boldsymbol{u}_{\text{grid}}$ with respect to the laboratory frame. Similarly, the *Reynolds transport theorem*

$$\frac{\mathrm{d}}{\mathrm{d}t}\left(\int_{V_{\text{fluid}}} f \mathrm{d}V\right) = \int_{V_{\text{fluid}}} \left[\frac{\partial f}{\partial t} + \boldsymbol{\nabla} \cdot (\boldsymbol{u} f)\right] \mathrm{d}V_f \tag{10.6}$$

can be generalized to the *adaptive grid transport theorem*

$$\frac{\mathrm{d}}{\mathrm{d}t}\left(\int_V f \mathrm{d}V\right) = \int_V \left[\frac{\partial f}{\partial t} + \boldsymbol{\nabla} \cdot (\boldsymbol{u}_{\text{grid}} f)\right] \mathrm{d}V \,. \tag{10.7}$$

A moment's thought shows that all we did is change *markers* in the flow from bits of matter moving with the fluid velocity $\boldsymbol{u}$ to a logical grid moving with velocity $\boldsymbol{u}_{\text{grid}}$.

These theorems allow us to integrate over a *control volume* on the adaptive grid, and thus to obtain *integral conservation laws* on the adaptive grid. For example, the continuity equation, Euler's equation of motion, and the (adiabatic) gas energy equation become:

$$\frac{\mathrm{d}}{\mathrm{d}t}\left(\int_V \varrho \mathrm{d}V\right) + \int_{\partial V} \varrho \boldsymbol{u}_{\text{rel}} \cdot \mathbf{d}\boldsymbol{S} = 0 \,, \tag{10.8}$$

$$\frac{\mathrm{d}}{\mathrm{d}t}\left(\int_V \varrho \boldsymbol{u} \mathrm{d}V\right) + \int_{\partial V} (\varrho \boldsymbol{u}) \boldsymbol{u}_{\text{rel}} \cdot \mathbf{d}\boldsymbol{S} + \int_V p \boldsymbol{n} \cdot \mathbf{d}\boldsymbol{S} = \int \varrho \boldsymbol{f} \mathrm{d}V \,, \tag{10.9}$$

and

$$\frac{\mathrm{d}}{\mathrm{d}t}\left(\int_V \varrho e \mathrm{d}V\right) + \int_{\partial V} (\varrho e) \boldsymbol{u}_{\text{rel}} \cdot \mathbf{d}\boldsymbol{S} + \int_V p(\boldsymbol{\nabla} \cdot \boldsymbol{u}) \mathbf{d}\boldsymbol{V} = 0 \,. \tag{10.10}$$

These equations can then be differenced directly on the adaptive grid.

The Grid Equation. There are basically two adaptive grids which have been used in astrophysics. The first was developed by Karl-Heinz Winkler for his remarkable code WH80s [34]. The WH80s grid equation admits of some deep physical interpretation, and is probably the most robust and powerful (*i.e.*, highest resolution) of all schemes. On the other hand, successful management of the grid requires considerable experience and sophistication on the part of the user. A somewhat simpler and easier-to-implement scheme was suggested by Dorfi and Drury (DD) [8].

DD define a *grid concentration* giving the number of gridpoints per typical length

$$n_i \equiv \frac{X_i}{x_{i+1} - x_i} \,. \tag{10.11}$$

Here X_i is a natural length such as a scale height or the size of the domain. Next they define the *monitor function*

$$R_i = \left[1 + n_i^2 \sum_j W_j S_{ij}^2 \right]^{1/2} \,. \tag{10.12}$$

Here S_{ij} provides a measure of the *slope* of a graph of some physical variable (e.g., $m, \varrho, T, p, e, E, \chi, Q$), and W_j is a weight factor of order unity. For *logarithmic resolution*, which is useful, we can define

$$S_{ij} \equiv 2 \frac{f_{i+1,j} - f_{ij}}{f_{i+1,j} + f_{ij}} \,. \tag{10.13}$$

The basic idea of the adaptive grid is to put a lot of gridpoints, *i.e.*, make n_i large, wherever there are steep structures in the flow, *i.e.*, wherever R_i is large. Thus we could demand

$$\frac{n_i}{R_i} = \frac{n_{i-1}}{R_{i-1}} \,, \tag{10.14}$$

which, with suitable boundary conditions provides the extra equation needed to determine the grid. However (10.14) is not satisfactory in practice because we need to keep the grid *smooth*, *i.e.*, keep adjacent zones at roughly the same size. Otherwise we will get unacceptably large spatial discretization errors. So in practice we replace n_i with the *diffused* quantity

$$\tilde{n}_i \equiv n_i - \alpha(\alpha + 1)(n_{i+1} - 2n_i + n_{i-1}) \,, \tag{10.15}$$

which forces n_i to fulfill the criterion

$$\frac{\alpha}{\alpha + 1} \le \frac{n_{i+1}}{n_i} \le \frac{\alpha + 1}{\alpha} . \tag{10.16}$$

For $\alpha = 2$, $(2/3) \le (n_{i+1}/n_i) \le (3/2)$, *i.e.*, there can be about a $\pm 40\%$ change in cell size.

In addition, one can put in a *time filter* which prevents the grid from trying to readjust on unphysically short time scales. If Δt is the timestep, and if τ_g is the minimum response time desired for the grid, then we take

$$\hat{n}_i \equiv \tilde{n}_i + \frac{\tau_g}{\Delta t}(\tilde{n}_i - \tilde{n}_i^{\text{old}}). \tag{10.17}$$

This formula assures that when $\tau_g/\Delta t \to 0$ (*i.e.*, long timesteps) we permit *immediate* readjustment of the grid. Whereas when $\tau_g/\Delta t \to \infty$ (*i.e.*, short timesteps) we keep the grid *frozen* at the old value. A reasonable choice for τ_g is about 0.01 times the characteristic timescale of the problem. If τ_g is too large, the motion of the grid will be jerky.

The final form of the grid equation is

$$\frac{\hat{n}_i}{R_i} = \frac{\hat{n}_{i-1}}{R_{i-1}} . \tag{10.18}$$

This grid allows a total range in cell size of about 10^5, which is adequate to handle many (but not all) important physical problems. The final grid equation (10.18) is linearized along with the radiation hydro equations in the Newton–Raphson procedure used to solve the system. Thus the gridpoints move consistently in response to changes of the physical solution during the iteration procedure, and at convergence are situated so as to best satisfy both the fluid and grid equations. The mathematical structure of the system is a sparse matrix, consisting of a pentadiagonal set of (7×7) blocks. These can be solved in a number of ways.

10.4 The TITAN Code

The TITAN code is a one-dimensional adaptive-grid radiation hydrodynamics code intended for astrophysical problems with a broad variety of initial and boundary conditions. It employs the adaptive grid generation developed by Dorfi and Drury [8] and else follows the basic philosphy of the WH80s code. TITAN comes with a suite of test problems that cover purely hydrodynamical and radiation hydrodynamical calculations as well as time dependent radiation transport through static media. A reference manual and a user's guide as well as the code itself can be found under the URL `http://141.142.3.131/Pubs/TechReports/TechReportsSeries.html` of the National Center for Computing Applications of the University of Illinois at Urbana-Champaign.

References

1. L. H. Aller. *The Atmospheres of the Sun and Stars.* Ronald Press Co., New York, 1963.
2. M.L. Alme and J.R. Wilson. Numerical study of X-ray induced mass transfer in the HZ Herculis/Hercules X-I binary system. *ApJ*, 194:147, 1974.
3. J. R. Buchler. Radiation hydrodynamics in the fluid frame. *J. Quant. Spectrosc. Rad. Transf.*, 22:293, 1979.
4. J. I. Castor. Radiative transfer in spherically symmetric flows. *ApJ*, 178:779, 1972.
5. S. Chandrasekhar. *An Introduction to the Study of Stellar Structure.* Dover Publications, New York, 1957.
6. R. Christy. The calculation of stellar pulsation. *Rev. Mod. Phys.*, 36:555, 1964.
7. J. P. Cox and R. T. Giuli. *Principles of Stellar Structure.* Gordon and Breach, New York, 1968.
8. E.A. Dorfi and L.O'C. Drury. Simple adaptive grids for 1-D initial value problems. *Journal of Computational Physics*, 69:175, 1987.
9. S.W. Falk and W.D. Arnett. Radiation dynamics, envelope ejection, and supernova light curves. *ApJ Suppl. Series*, 33:515, 1977.
10. B. E. Freeman. The non-equilibrium diffusion equation. Technical Report 3377, Los Alamos Scientific Laboratory, Los Alamos, 1965.
11. B. E. Freeman, L.E. Hause, J.T. Palmer, S.O. Pickard, G.M. Simmsons, D.G. Williston, and J.E. Zerkle. The VERA code: A one dimensional radiation hydrodynamics code. DASA Report 2135, Systems, Science, and Software, Inc, La Jolla, 1968.
12. F.H. Harlow and A.A. Amsden. Fluid dynamics. Technical Report 4700, Los Alamos Scientific Laboratory, Los Alamos, 1971.
13. Z. Kopal. The effects of viscosity and radiative braking on stellar pulsations. *Astrophysica Norvegica*, 9:239, 1964.
14. Z. Kopal. The effects of radiative braking on free periods of stellar pulsations. *Z. für Astrophysik*, 61:156, 1965.
15. E. Landi Degl'innocenti. Polarization in spectral lines. *Sol. Phys*, 85:3, 1983.
16. E. Landi Degl'innocenti and M. Landi Degl'innocenti. Quantum theory of line formation in a magnetic field. *Sol. Phys*, 27:319, 1972.
17. E. Landi Degl'innocenti and M. Landi Degl'innocenti. Transfer equations for polarized light. *Nuovo Cimento*, 27:134, 1975.
18. P. Ledoux. Stellar stability. In L. H. Aller and D. B. McLaughlin, editors, *Stellar Structure*, page 499, Chicago, 1965. University of Chicago.
19. P. Ledoux and Th. Walraven. Variable stars. In S. Flügge, editor, *Handbuch der Physik Vol. LI*, page 353, Berlin, 1958. Springer-Verlag.
20. C.D. Levermore. A Chapman-Enskog approach to flux-limited diffusion theory. Technical Report UCID-18229, Lawrence Livermore Laboratory, University of California, Livermore, 1979.
21. R. W. Lindquist. Relativistic transport theory. *Ann. Phys.*, 37:487, 1966.
22. D. Mihalas. *Stellar Atmospheres.* W. H. Freeman, San Francisco, 1970.
23. D. Mihalas. Solution of the comoving-frame equations of transfer in spherically symmetric flows VI. Relativistic flows. *ApJ*, 237:574, 1980.
24. D. Mihalas and B.W. Mihalas. *Foundations of Radiation Hydrodynamics.* Oxford University Press, New York, 1984.
25. G. Minerbo. Maximum entropy Eddington factors. *J. Quant. Spectrosc. Rad. Transf.*, 20:541, 1978.

26. A. Munier and R. Weaver. Radiation transfer in the fluid frame Part I and II. *Computer Physics Reports*, 3:125, 1986.
27. G. Pomraning. *Equations of Radiation Hydrodynamics.* Pergamon Press, Oxford, 1973.
28. R.D. Richtmyer and K.W. Morton. *Difference Methods for Initial-Value Problems.* Interscience Publishers, New York, 1967.
29. G. B. Rybicki and A. P. Lightman. *Radiative Processes in Astrophysics.* Wiley, New York, 1979.
30. R. Simon. The conservation equations of a classical plasma in the presence of radiation. *J. Quant. Spectrosc. Rad. Transf.*, 3:1, 1963.
31. L. H. Thomas. The radiation field in a fluid in motion. *Quart. J. of Math. (Oxford)*, 1:239, 1930.
32. J. Von Neumann and R.D. Richtmyer. A method for the numerical calculation of hydrodynamic shocks. *J. Appl. Phys*, 21:232, 1950.
33. S. Weinberg. Entropy generation and the survival of proto-galaxies in an expanding universe. *ApJ*, 168:175, 1971.
34. K-.H. Winkler and M.L. Norman. WH80s: Numerical radiation hydrodynamics. In K-.H. Winkler and M.L. Norman, editors, *Astrophysical Radiation Hydrodynamics*, page 71, Dordrecht, 1986. D.Reidel.
35. R. Woolley and Stibbs D. W. N. *The Outer Layers of a Star.* Clarendon Press, Oxford, 1953.

Ernst Dorfi, contemplating photon scattering on the glacier

Radiation Hydrodynamics: Numerical Aspects and Applications

Ernst A. Dorfi

Institut für Astronomie, Universität Wien, Türkenschanzstr. 17, A-1180 Wien, Austria

1. Introduction

Since electromagnetic radiation measured at the earth is the main source of information about astronomical objects (beside lunar rocks, meteorites and neutrinos) the interaction between matter and the radiation field plays a fundamental role for our understanding of the universe. From the intensity, the spectral distribution and the shape of individual spectral lines we can deduce basic quantities like temperatures, the degree of ionization, elemental abundances, velocity fields, as well as gravitational acceleration. Clearly, a large amount of theoretical concepts as well as laboratory data are needed to extract this information on the physical states from the photons that we can detect at telescopes.

In the following sections I want to set the "scene" for some typical astrophysical applications where a different numerical strategy is necessary, *i.e.*, simulations over very large time scales covering also many spatial magnitudes.

1.1 General Remarks on the Numerical Method

Concerning the other lectures held during this course I want to emphasize some general points on the numerical technique. First, in contrast to the other speakers the method is ***implicit***, *i.e.*, the solution at the subsequent time level cannnot be calculated by some sort of extrapolation from the physical values given at the old time levels. Therefore the time step is not restricted by the Courant–Friedrichs–Levy (CFL) condition (Courant et al. 1920). Second, the method is based on a ***finite volume discretization*** ensuring the conservation properties of the hyperbolic equations. Third, the method does not involve any *Riemann-solver* since the inclusion of radiation, diffusion, nuclear or chemical reactions, opacities, equations of state of non-perfect gases, *etc.*, can lead to very complicated shock structures which destroy the nice features of analytical shock solutions. Fourth, shock fronts are treated by an ***artificial viscosity*** but the amount of viscosity can be reduced at least by a factor of 10^3 since the adaptive grid clusters around such sharp transitions. Fifthly, the individual grid points are freely moving within the computational domain and their locations are determined by the so-called ***grid equation***. This additional equation is simultaneously solved together with the physical equations. Up to now only a one-dimensional version is available and the last section discusses some applications of multi-dimensional adaptive computations.

1.2 Time Scales

It is almost typical for astrophysical problems that the time scales characterizing the physical interactions can differ by many orders of magnitude which is closely linked to the computational aspects. Usually we are interested in the evolution lasting over many short time events like travelling waves, light fronts, convective eddy turn-around times, *etc.* Nevertheless, the transport phenomena connected with these motions influence the long term behaviour and we have to be careful in computing such evolutionary sequences without missing the essential physics.

Taking a stellar object with a typical dimension R_*, a luminosity L, a mass of M_*, the mass $M_{\rm nuc}$ available for nuclear burning as well as a mean density $\bar{\rho}$, we can illustrate some of the aforementioned problems, *i.e.*,

$$\begin{aligned}
\tau_{\rm light} &= \frac{R_*}{c} \quad \left(\text{or} \simeq \frac{R_*^2}{\kappa} \text{ for diffusion } \right) , \\
\tau_{\rm dyn} &= \frac{R_*}{c_{\rm s}} \quad \left(\simeq (G\bar{\rho})^{-1/2} , \text{ for stars } \right) , \\
\tau_{\rm therm} &= \frac{E_{\rm therm}}{L} \quad \left(\simeq \frac{GM_*^2}{R_* L}, \text{ for stars } \right) , \\
\tau_{\rm nuc} &= \frac{M_{\rm nuc}\, c^2\, \varepsilon}{L} ,
\end{aligned}$$

where $c_{\rm s}$ denotes the mean sound velocity, κ the mean opacity, $E_{\rm therm}$ the total thermal energy and ε the conversion efficiency. For a $1\,M_\odot$ main sequence stars these time scales vary between 10^2 and 10^{17} s. As stated before the time scale we are interested in is much longer than the shortest one, *e.g.*, stars evolve on the nuclear time scale $\tau_{\rm nuc}$ but are disturbed by numerous "short"-time events like pulsations (typically on $\tau_{\rm dyn}$) or thermally induced secular changes on $\tau_{\rm therm}$. In many cases such different time scales are accompanied by rather stiff differential equations and every numerical scheme has to deal with these complications.

1.3 Length Scales

Beside the temporal changes of different physical processes the spatial variations of gradients, waves, *etc.*, are small compared to the typical dimensions. Again, the appropriate resolution at these regions of action is essential to get a correct answer to the problem. An accretion shock on top of a protostellar core may serve as a relevant example since for a large fraction of time during the collapse phase the luminosity (and its spectral appearance) is determined by the details of this radiating transition zone. During a protostellar collapse the matter is compressed by a factor of 10^{20}, the temperature raises by a factor of 10^6 and within such a flow we find an almost freely falling envelope enclosing a quasi-hydrostatic core separated by this strong accretion shock.

Almost all dynamical events are associated with supersonic flows leading to shock waves. Hence, taking l_p mean free path for photons and l_m mean free path for molecules we can estimate some typical length scales Δx involved in these transitions for a Mach number M,

$$\begin{aligned} \Delta x &= l_m \frac{M}{M^2 - 1}, \quad \text{in a viscous shock wave}, \\ \Delta x &\simeq l_p \ll d, \quad \text{in a supercritical shock wave}. \end{aligned}$$

This shock thickness Δx is usually very small compared to the size d of the astrophysical object. In computing such flows we have to develop a numerical technique which "recognizes" and follows the formation of characteristic structures.

Since the gravitational force is relevant in many astrophysical events the Poisson equation has to be solved for those cases. In particular, self-gravity is the driving mechanism for the formation of all complex large scale structures. This requires the solution of the Poisson equation to calculate the gravitational potential, *i.e.*, an elliptical equation defined as a typical boundary value problem. Clearly, an implicit method is required to do this task. Concerning the length scales introduced by self-gravity we can define the pressure scale height H_P within a spherical hydrostatic configuration and find for typical astrophysical situations

$$\frac{H_\mathrm{P}}{R_*} \simeq 10^{-1} \ldots 10^{-5}, \quad \text{with} \quad H_\mathrm{P} = \left(\frac{1}{P} \frac{\partial P}{\partial r} \right)^{-1},$$

i.e., very steep gradients can be encountered. Within a stellar atmosphere density and pressure vary on a scale of H_P and to get the radiative boundary correct we have to resolve these gradients. Taking a typical star the pressure ratio between the center and the atmosphere is of the order of 10^{30}, the density contrast is about 10^{20} and the opacity can vary up to a factor of 10^{12}. Since at the main sequence a normal star remains in dynamical as well as in thermal equilibrium the inner and outer values are closely linked together by the boundary conditions which have to be simultaneously fulfilled.

1.4 Interaction Between Matter and Radiation

The interaction between matter and radiation leads to a momentum and energy exchange which can be non-local, coupling regions far apart. The importance of an accurate treatment of this interaction is easily illustrated by a number of astrophysical situations. Radiation drives stellar mass loss by acting on stellar atmospheres. Radiative pressure imposes the so-called Eddington limit which defines the maximum luminosity to allow a static structure. Heating as well as cooling by photons are examples of the energy exchange between matter and radiation often accompanied with ionization fronts. The different versions as well as the derivation of the radiative transfer equations describing all these effects can be found in the lectures by Mihalas and are therefore omitted here. Stellar pulsations are caused by appropriate opacity changes, this so-called κ-mechanism uses part of the radiative flux through the star to support cyclic pulsations of the stellar envelope. Again, the details of this radiative interaction as well as the transport of photons determine the variability, *e.g.*, the shape of light curves, colour changes and the amplitudes.

A further complication inherent to radiation hydrodynamics comes from the frequency dependence of the above mentioned processes adding a new dimension of complexity to the problems. Without going into details, a certain averaging or integration over frequency is used to calculate the radiative force or heating (see lectures by Mihalas). In most applications Rosseland-means, κ_{R}, or Planck-means, κ_{P}, are used, *i.e.*,

$$\frac{1}{\kappa_{\mathrm{R}}} = \frac{\displaystyle\int_0^\infty \frac{1}{\kappa_\nu}\frac{\partial B_\nu}{\partial T}\,\mathrm{d}\nu}{\displaystyle\int_0^\infty \frac{\partial B_\nu}{\partial T}\,\mathrm{d}\nu}\,, \quad \kappa_{\mathrm{P}} = \frac{\displaystyle\int_0^\infty \kappa_\nu B_\nu\,\mathrm{d}\nu}{\displaystyle\int_0^\infty B_\nu\,\mathrm{d}\nu}\,,$$

which are nowadays available as tables in electronic form through the OPAL or OP data. OP data can be accessed interactively through:
`http://vizier.u-strasb.fr/OP.html`

1.5 Moving Fronts

Radiation fronts such as ionization zones in pulsating stars or expanding HII-regions are neither fixed in mass (Lagrangian description) nor in position (Eulerian description). Therefore, an accurate tracking of such features must relay on a numerical technique which tracks a number of grid points to such fronts. We will call such a method an adaptive grid. It redistributes a given number of grid points according to user-prescribed needs.

Shock fronts can travel over large spatial distances clearly seen from the evolution of a supernova remnant. During the so-called Sedov phase of a strong point explosion the shock radius R_{s} obeys a self-similar expansion law (with n dimensionality of the problem),

$$R_{\mathrm{s}} \propto t^{2/(2+n)}\,.$$

Depending on the chosen reference system, shock fronts can be almost stationary within a flow but their strength varies in time, *e.g.*, the accretion shocks due to different rates of infalling matter, the terminal shock of a stellar wind acting on the surrounding medium or a reverse shock propagating through the expanding clumpy SN-ejecta.

Radiative shock fronts have a more complex structure than ordinary hydrodynamical shocks which can be described by a simple discontinuity. Radiation from the heated post-shock region can leave towards the up-stream region leading to a radiative precursor. The compression at the shock front may cause variations in the equation of state by opening new degrees of freedom in a non-perfect gas or by triggering chemical reactions. Finally, the flow through the down-stream region can contain several relaxation layers where excited states, heated electrons, *etc.*, need some time (or distance) to find to their equilibrium configurations.

To summarize this brief introduction we point out that the inclusion of "realistic" physics leads to a large diversity of time and length scales, steep gradients as well as very complex structures moving through the medium. Based on these physical needs I present – at least in one spatial dimension – an implicit numerical method capable of treating such astrophysical problems.

2. Basic Equations

This section describes the general physical as well as some mathematical background of radiation hydrodynamics without a formal derivation or lengthy discussion of the underlying equations.

2.1 Radiation Hydrodynamics (RHD)

The dynamical behaviour of radiation and matter is contained in the equations of radiation hydrodynamics (RHD) which I state here in a simple form. Assuming small non-relativistic velocities, an Eulerian (*i.e.*, fixed) coordinate system I refer to the standard literature for a formal derivation or generalizations to more complex situations (*cf.*, Mihalas & Mihalas 1984). In order to demonstrate the numerical technique I can restrict the equations to their one-dimensional, spherical version. For the radiation terms I will take a LTE-source function without scattering. Denoting the gas density by ρ, the radial velocity by u, the gas pressure by P and the internal gas energy by e, as well as the mass integrated up to the radius r by m. We obtain the first part of equations describing the gas :

Equation of continuity (mass conservation):

$$\frac{\partial \rho}{\partial t} + \frac{1}{r^2}\frac{\partial r^2\, \rho u}{\partial r} = 0\,, \tag{2.1}$$

Equation of motion (momentum conservation):

$$\frac{\partial \rho u}{\partial t} + \frac{1}{r^2}\frac{\partial r^2\, \rho u\, u}{\partial r} + \frac{\partial P}{\partial r} + \frac{G\rho m}{r^2} - \frac{4\pi}{c}\rho\kappa_{\mathrm{H}} H - u_{\mathrm{Q}} = 0\,, \tag{2.2}$$

Equation of gas energy (1st law of thermodynamics):

$$
\begin{aligned}
\frac{\partial \rho e}{\partial t} &+ \frac{1}{r^2}\frac{\partial r^2 \rho e u}{\partial r} + P\frac{1}{r^2}\frac{\partial r^2 u}{\partial r} \\
&+ 4\pi\rho(\kappa_{\rm J} J - \kappa_{\rm S} S) - \rho(\epsilon_{\rm nuc} + \epsilon_{\rm Q}) = 0 \,,
\end{aligned}
\tag{2.3}
$$

where I have also introduced the terms containing the interaction with the radiation field, *i.e.*, $\kappa_{\rm H,J,S}$ defined through Eqs. (2.5) as well as the nuclear energy production rate $\epsilon_{\rm nuc}$. Details on these quantities will follow in the next section and the symbols $u_{\rm Q}$ and $\epsilon_{\rm Q}$ are the artificial viscous pressure and the energy generation, respectively (see Sect. 3.12).

The effect of self-gravity is included by the Poisson equation in spherical symmetry. Instead of computing the gravitational potential we can calculate the mass m within the radius r entering directly the equation of motion (2.2):

Poisson equation (self-gravity):

$$
m = \int_0^r 4\pi\rho r'^2 \, {\rm d}r' \,. \tag{2.4}
$$

The simplest way of taking into account the radiation field with a specific intensity $I_\nu(\boldsymbol{r}, \boldsymbol{n}, t)$ can be achieved by integrating the radiative transfer equation over frequency and solid angle ${\rm d}\Omega = \sin\theta {\rm d}\theta {\rm d}\omega$ again assuming spherical symmetry (*cf.*, lectures by Mihalas), *i.e.*,

$$
\{J, H, K\}_\nu (r, t) = \frac{1}{2}\int_{-1}^{1} I_\nu(r, \mu, t)\mu^{\{0,1,2\}} \, {\rm d}\mu \,, \tag{2.5}
$$

where $\mu = \cos\theta$ has been used.

After some algebra and together with integrating over frequency ν we end up with the so-called grey moments of the radial specific intensity $I_\nu(r, \mu, t)$

Radiation energy equation (0th moment):

$$
\begin{aligned}
\frac{1}{c}\frac{\partial J}{\partial t} &+ \frac{1}{r^2}\frac{\partial r^2 J u}{\partial r} + \frac{1}{r^2}\frac{\partial r^2 H}{\partial r} + \frac{1}{c}K\frac{1}{r^2}\frac{\partial r^2 u}{\partial r} \\
&- \frac{1}{c}\frac{3K - J}{r}u + \rho(\kappa_{\rm J} J - \kappa_{\rm S} S) = 0 \,,
\end{aligned}
\tag{2.6}
$$

Radiation flux equation (1th moment):

$$
\begin{aligned}
\frac{1}{c}\frac{\partial H}{\partial t} &+ \frac{1}{c}\frac{1}{r^2}\frac{\partial r^2 H u}{\partial r} + \frac{\partial K}{\partial r} \\
&+ \frac{3K - J}{r} + \frac{1}{c}H\frac{1}{r^2}\frac{\partial r^2 u}{\partial r} + \rho\kappa_{\rm H} H = 0 \,.
\end{aligned}
\tag{2.7}
$$

As in any moment technique we are confronted with a closure problem since we have only two equations for the three moments, the mean specific intensity $e_{\rm rad} = 4\pi/c\, J$, the radial radiative flux $F_{\rm rad} = 4\pi H$ and the radiation pressure $P_{\rm rad} = 4\pi/c\, K$. A closure condition is required to specify a further relation between the radiative moments; this condition is discussed in Sect. 2.3.

2.2 Coupling Terms

Due to the momentum and energy exchange between matter and radiation the coupling terms enter the radiation hydrodynamical equations with different weightings of the frequency-dependent opacity κ_ν, *e.g.*,

$$\kappa_{\mathrm{H}} = \int_0^\infty \kappa_\nu \frac{H_\nu}{H} \, \mathrm{d}\nu \, , \qquad H = \int_0^\infty H_\nu \, \mathrm{d}\nu \, . \tag{2.8}$$

In this example the radiative flux H_ν has to be known *a priori* to perform the integration (2.8). In many physical situations we have to iterate between the two equations (e.g., stellar atmospheres). However, there exist certain limits so that we can substitute these weightings by the Planck- or Rosseland means. For these means the integration can be performed once before any specific application. Such data are in principle available by tables (e.g., OPAL by Rogers & Iglesias 1992) or OP-project by Seaton et al. 1994). Nevertheless, in almost all RHD computations we set $\kappa_{\mathrm{H,J,S}} = \kappa_{\mathrm{R}}$, since only κ_{R} is tabulated (OPAL, OP) for different chemical mixtures. Physically it would be better if one could take the appropriate limits, *i.e.*, $\kappa_{\mathrm{H}} = \kappa_{\mathrm{R}}$ and $\kappa_{\mathrm{J,S}} = \kappa_{\mathrm{P}}$.

2.3 Closure Condition

As seen before it is still necessary to close the radiative moment equations. This can be achieved by either calculating the full radiation field or much easier by applying the much simpler Eddington approximation. First, the Eddington factor (see lectures by Mihalas) is defined through

$$f_\nu = \frac{K_\nu}{J_\nu} \tag{2.9}$$

as the ratio between the zeroth and second moment (*cf.*, 2.5). Knowing this ratio obviously closes the set of radiative moments.

Assuming a nearly isotropic radiation field at a point r, the specific intensity $I_\nu(r,\mu)$ is expanded up to the first power of the azimuthal angle θ with $\mu = \cos\theta$,

$$I_\nu(r,\mu) = a_\nu(r) + b_\nu(r)\mu \, . \tag{2.10}$$

According to the expressions (2.5) we calculate the corresponding moments

$$\begin{aligned} \{J,H,K\}_\nu(r) &= \tfrac{1}{2}\textstyle\int_{-1}^{1} [a_\nu(r) + b_\nu(r)\mu] \, \mu^{\{0,1,2\}} \, \mathrm{d}\mu \\ &= \left\{a_\nu(r), \tfrac{1}{3}b_\nu(r), \tfrac{1}{3}a_\nu(r)\right\} \end{aligned} \tag{2.11}$$

and get therefore simply

$$f_\nu = \frac{K_\nu(r)}{J_\nu(r)} = \frac{1}{3} \, . \tag{2.12}$$

In the case of linear anisotropy in μ we can use $f = 1/3$ for our computations.

Another useful description of the Eddington factor is based on the work of Lucy (1971, 1976) where the radiation field of an extended static atmosphere in spherical symmetry is treated by a geometrical approximation. Denoting the stellar radius by R_*, the radiation field at radius r is approximated by the two intensities I_ν^+ and I_ν^- which are discontinuous at the attenuation angle

$$\mu_* = \sqrt{(1 - y^2)} \ , \quad \text{and} \quad y = \frac{R}{r} \ . \tag{2.13}$$

Hence, adopting the Heaviside function $H(x) = 1$ for $x \leq 1$ and $H(x) = 0$ for $x < 0$, the intensity is written as

$$I_\nu(r, \mu) = H(\mu - \mu_*)\, I_\nu^+ + H(\mu_* - \mu)\, I_\nu^- \ . \tag{2.14}$$

Calculating moments (2.5) for such an intensity one can derive

$$J_\nu - 3\, K_\nu + \frac{1}{2}\mu_*\, H_\nu = 0 \ , \tag{2.15}$$

which reduces to $J_\nu = 3\, K_\nu$ at the stellar surface $r = R_*$ when $\mu_* = 0$. Following Lucy (1971), the moment equations can be solved for a static atmosphere yielding

$$J_\nu = H_\nu(1)\left[W + \frac{3}{4}\bar{\tau} \right] \ , \tag{2.16}$$

$$H_\nu = H_\nu(1)\, y^2 \ , \tag{2.17}$$

where a dilution factor W and a modified optical depth $\bar{\tau}$ are introduced through

$$W = \frac{1}{2}\left[1 - \sqrt{1 - y^2} \right] , \qquad \bar{\tau} = \int_0^y \kappa_\nu \rho \, \mathrm{d}y \ . \tag{2.18}$$

Combining Eqs. (2.16), (2.15) the Eddington factor (2.9) is obtained

$$f_\nu = \frac{K_\nu}{J_\nu} = \frac{1}{3} + \frac{1}{6}\sqrt{1 + y^2}\frac{y^2}{W + \frac{3}{4}\bar{\tau}} \ . \tag{2.19}$$

Applying the definition of μ_* in Eq. (2.13) the last expression reads

$$f_\nu = \frac{1}{3} + \frac{1}{3}\mu_* \frac{1 - \mu_*^2}{1 - \mu_* + \frac{3}{2}\bar{\tau}} \approx \frac{1}{3} + \frac{1}{3}\mu_*(1 + \mu_*) \quad \text{for} \quad \mu_* \to 1 \tag{2.20}$$

having the limit of $f_\nu = 1$ for large distances $y \to 0$, *i.e.*, $\mu_* \to 1$ and $\bar{\tau} \to 0$. Note that the derivative of f_ν is discontinuous at the stellar photosphere, *i.e.*, $r = R_*$ and from the last expressions one finds $\mathrm{d}f_\nu/\mathrm{d}r \to -\infty$ for $y \to 1$.

The shortcomings of such a simplification are discussed in more detail in Sect. 3.14, where also the technique is outlined how the static radiative transfer equation can be solved by a method based on characteristics. Nevertheless, for a first attempt to do radiation hydrodynamical computations it is strongly suggested to start with the $f = 1/3$ which avoids additional complications and reduces the CPU-time by about a factor of two compared to a full solution including the Eddington factor.

2.4 Opacity

The stellar opacities are of fundamental importance for radiation hydrodynamical computations. Only recently the quality of these data has been improved and several long standing problems related to pulsational instabilities have been solved (see, *e.g.*, reviews by Gautschy & Saio 1995, 1996). Based on very time-consuming quantum mechanical computations the Rosseland means for different chemical compositions (X, Y, Z) (where X, Y, and Z are the mass fraction of Hydrogen, Helium, and heavier elements, respectively) are available at prescribed points of density and temperature, usually for $R = \rho/T_6^3$ and $T_6 = T/10^6$ K. Hence, we have to interpolate $\kappa_{\rm R} = \kappa_{\rm R}(\rho, T)$ or $\kappa_{\rm R} = \kappa_{\rm R}(R, T)$ for our purposes and we need also smooth derivatives with respect to density and temperature in order to obtain rapid convergence in the Newton–Raphson iteration. The later is needed as a consequence of the imlicit formulation of our scheme (*cf.*, Sect. 3.4).

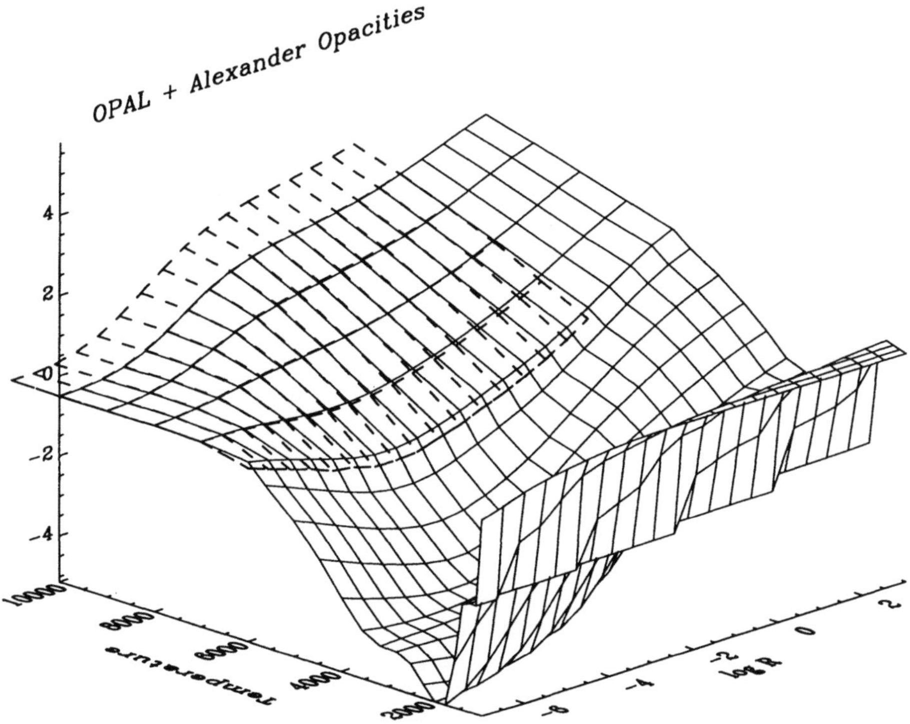

Fig. 2.1. Part of a typical Rosseland opacity $\kappa_{\rm R}$ used in radiation hydrodynamics as a function of temperature T and $R = \rho/T_6^3$ with $T_6 = T/10^6$ K. This figure is a combination of two different sources. The full lines represent low temperature opacities from Alexander et al., while the dashed lines show OPAL data (see text for details)

Figure 2.1 shows the low-temperature part of the Rosseland mean κ_R of the opacity as a function of temperature $10\,000 \geq T/[\mathrm{K}] \geq 2\,000$ and $R = \rho/T_6^3$. The chemical composition corresponds to $X = 0.7$, $Y = 0.28$ and $Z = 0.02$. This figure demonstrates a typical problem encountered in RHD computations, namely, that different sources have to be combined to one opacity table. The OPAL opacities are available for temperatures above 6000 K. The low temperature part is due to Alexander et al. (1995) with a upper limit at 10 000 K but down to 1000 K including also molecular opacities. In the overlapping region both tables are tabulated on different T, ρ and R values and have therefore to be interpolated. Although the differences are small one has to be carefully in order to avoid further artificial edges when joining such tables. In particular the low-temperature region is important to determine the atmospheric structure. In the case of extended or dynamical atmospheres, opacities influence, e.g., the shape of the light curves, the formation of dust particles, the strength of the radiative force, and the density and temperature stratification in this transition region to the interstellar medium. Note, that these combined tables are then used for further smooth interpolations within the given data points as outlined in Sect. 5.1. Since these tables depend strongly on the chemical composition in stellar evolution calculations a set of different tables is needed to take into account the chemical evolution of the matter.

2.5 Equation of State

The thermodynamical properties of the fluid are determined by the equation of state (EOS). In most astrophysical applications we cannot assume a perfect gas since ionization of atoms, dissociation of molecules, as well as degeneracy of electrons are usually encountered in radiating stellar plasmas. Hence, we have either to calculate the thermodynamic variables for every set of input parameters or we can utilize an EOS table which is less costly than direct computations. In Fig. 2.2 one of the thermodynamic adiabatic indices, namely

$$\Gamma_1 = \frac{\rho}{P}\frac{\partial P}{\partial \rho} = \frac{\partial \ln P}{\partial \ln \rho}\,, \tag{2.21}$$

is plotted against the temperature for a constant value of $\rho = 10^{-8}\ \mathrm{g\,cm^{-3}}$. This particular EOS was calculated by Wuchterl (1990) containing H, H_2 and He with their thermodynamic properties and solar metal abundances. The graph of Γ_1 reveals a number of prominent thermodynamic features and can serve as a test for the numerical evaluation of the derivatives used in relation (2.21). As discussed in Sect. 5.1 the interpolation is performed via rational splines which show some small oscillations in the vicinity of steep gradients and which can reduce the convergence rate during the Newton-iteration (*cf.*, Sect. 5.3).

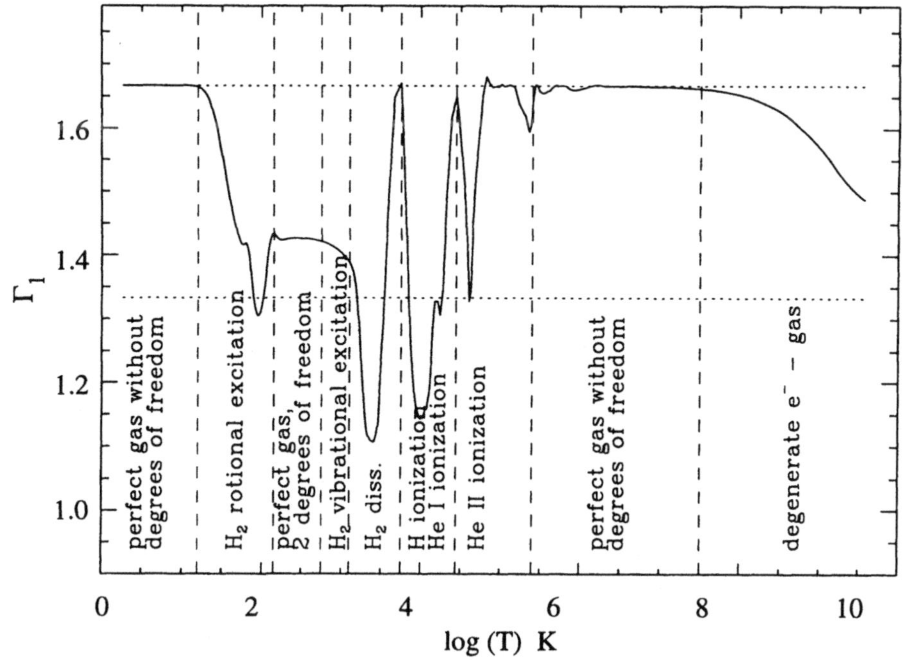

Fig. 2.2. The adiabatic index Γ_1 plotted against temperature at a constant density of $\rho = 10^{-8}\,\mathrm{g\,cm^{-3}}$ depicts the various thermodynamical effects entering a typical equation of state (see text for more details)

In Fig. 2.2 the two horizontal dashed lines indicate the values of $\Gamma_1 = 5/3$ and $\Gamma_1 = 4/3$, respectively. The vertical lines are plotted to separate the various thermodynamical effects entering this equation of state. In particular, at lower temperatures the properties of H_2 are important for stellar collapse calculations since the dissociation of H_2 is responsible for the so-called second collapse after which a hydrostatic core is formed (see also Sect. 6.3 for more details). At temperatures up to a few 100 K we find a perfect gas without internal degrees of freedom, *i.e.*, $\Gamma_1 = 5/3$. Moving up to higher temperatures, the rotational and subsequently the vibrational excitation of H_2 reduces Γ_1 significantly which is followed by the minimum of Γ_1 caused by the dissociation of molecular hydrogen. Around 10 000 K the ionization of hydrogen starts leading again to a strong decrease of Γ_1. Within this ionization feature of hydrogen also the ionization zone of He I is located. Increasing the temperature helium becomes fully ionized as it is clearly seen in the figure. The ionization of heavier elements around 10^5 K produces a small decrease from $\Gamma_1 = 5/3$ but up to a few million Γ_1 remains constant until the degeneracy of electrons begins to push the adiabatic index against the ultra-relativistic limit of $\Gamma_1 = 4/3$.

It is clear that most thermodynamical features are accompanied by changes in the opacity tables. Notice that the numerical constants and the details of the thermodynamics used for the opacities need not be identical to those applied for the equation of state if they come from different sources. Hence, special care is necessary to check whether the energies at which these changes occur are exactly at the same values in both set of tables. Such a discrepancy between the equation of state and the opacity would lead to a number of unphysical effects.

If nuclear reactions induce a change of the chemical composition a direct computation of the EOS might, however be unavoidable. Due to the conservative form of our RHD equations the EOS is needed at least in the form of two tables, namely $P = P(\rho, e)$ and $T = T(\rho, e)$. The density ρ by Eq. (2.1) and the specific internal energy e by Eq. (2.3) specify the thermodynamic basis in a compositionally homogeneous medium.

2.6 Transport Theorem

In order to formulate the physical equations on an adaptive grid we encounter the technical problem to restate the usual conservation relations of the radiating fluid on the moving coordinate system in such a way that they are still valid for an arbitrarily moving mesh and that they are computationally well-posed. To systematize notation we recognize three different time derivatives in an adaptive system. First, the Eulerian derivative $\partial/\partial t$ taken with respect to the fixed coordinates in space, secondly the Lagrangian derivative $\mathrm{D}/\mathrm{D}t$ with respect a the moving fluid element, and thirdly the adaptive derivative $\mathrm{d}/\mathrm{d}t$ taken with respect to the fixed values of the adaptive mesh. Note however that the physical quantities are still the Eulerian variables.

This change of coordinate systems can formally be understood by the Euler expansion formula. Taking a fluid element with an initial volume $\mathrm{d}V_0$ and moving at velocity $\boldsymbol{u}$ we get the the volume at later time by

$$\mathrm{d}V = J_f\,\mathrm{d}V_0\,, \qquad \text{and} \qquad \frac{\mathrm{D}\ln J_f}{\mathrm{D}t} = \nabla\cdot\boldsymbol{u}\,. \tag{2.22}$$

This expansion formula can be generalized to arbitrary motions by

$$\mathrm{d}V = J_g\,\mathrm{d}V_0\,, \qquad \text{with} \qquad \frac{\mathrm{d}\ln J_g}{\mathrm{d}t} = \nabla\cdot\boldsymbol{u}^{\mathrm{grid}}\,, \tag{2.23}$$

when $\mathrm{d}/\mathrm{d}t$ is now the derivative with respect to a motion defined through $\boldsymbol{u}^{\mathrm{grid}}$. Note, that the velocity $\boldsymbol{u}^{\mathrm{grid}}$ is not defined through this equation (*cf.*, Eq. (3.7) and Sect. 7.6).

In accordance with the standard Reynolds transport theorem of fluid mechanics we can obtain the so-called ***adaptive transport theorem*** (Winkler & Mihalas 1986) which is the formal basis of our adaptive conservative formulation of the RHD equations

$$\begin{aligned}\frac{\mathrm{d}}{\mathrm{d}t}\left(\int_V f\,\mathrm{d}V\right) &= \int_V \left[\frac{\partial f}{\partial t} + \nabla\cdot\boldsymbol{u}^{\mathrm{grid}} f\right]\mathrm{d}V \\ &= \int_V \frac{\partial f}{\partial t}\,\mathrm{d}V + \int_{\partial V} f\boldsymbol{u}^{\mathrm{grid}}\cdot\mathrm{d}\boldsymbol{S}\,. \end{aligned} \tag{2.24}$$

In Eq. (2.24) V denotes a definite volume corresponding to the fixed values of the adaptive mesh, $\mathrm{d}\boldsymbol{S}$ is the outwards-pointing surface element of the surface ∂V. Note that setting $\boldsymbol{u}^{\mathrm{grid}} = \boldsymbol{0}$ or $\boldsymbol{u}^{\mathrm{grid}} = \boldsymbol{u}$ reveals standard Eulerian (fixed in space) or Lagrangian (fixed in flow) versions of the expansion formula.

3. Solution Strategy

The following sections describe the solution strategy of the RHD equations based on an implicit conservative discretization on an adaptive grid. The set of the discrete algebraic equations is then used throughout the computations presented in the subsequent sections. Further improvements and new developments are discussed in Sect. 7.6.

3.1 Integral Form of the RHD Equations

From the work of Tscharnuter and Winkler (1979) it has become clear that a so-called compact form of the different terms is desirable to calculate radiating flows under the influence of self-gravity. At that time a fully conservative formulation has been unknown and it has been difficult to find a stable difference scheme. Following the results of Winkler and Norman (1986) and in the light of the other contributions presented in this volume it becomes clear that a *finite volume* discretization is essential to ensure physical solutions on a discrete space, in particular to obtain the right jump conditions across shock fronts and to get the correct wave speeds. Therefore, we have to reformulate the RHD equations to make them suitable for an adaptive grid based on this finite volume approach.

To begin with we take a volume V contained within the surface ∂V and we integrate the equations over this time-dependent volume using the Gauss theorem to manipulate the divergence terms into flows across the spherical surfaces.

We end up with the following set of integrated RHD equations:

$$\frac{\partial}{\partial t}\int_V \rho\,\mathrm{d}V \quad + \quad \int_{\partial V} \rho\, u\,\mathrm{d}A = 0\,, \tag{3.1}$$

$$\begin{aligned}\frac{\partial}{\partial t}\int_V \rho u\,\mathrm{d}V \quad &+ \quad \int_{\partial V} \rho u\, u\,\mathrm{d}A + \int_V \nabla P\,\mathrm{d}V \\ &+ \quad \int_V \frac{Gm\rho}{r^2}\,\mathrm{d}V - \frac{4\pi}{c}\int_V \kappa_{\mathrm{H}}\rho H\,\mathrm{d}V = 0\,,\end{aligned} \tag{3.2}$$

$$\begin{aligned}\frac{\partial}{\partial t}\int_V \rho e\,\mathrm{d}V \quad &+ \quad \int_{\partial V} \rho e\, u\,\mathrm{d}A + \int_V P\,\nabla\cdot u\,\mathrm{d}V \\ &- \quad 4\pi\int_V \rho(\kappa_{\mathrm{J}} J - \kappa_{\mathrm{S}} S)\,\mathrm{d}V = 0\,,\end{aligned} \tag{3.3}$$

$$m \quad - \quad \int_0^r 4\pi r'^2 \rho\,\mathrm{d}r' = 0\,, \tag{3.4}$$

$$\begin{aligned}\frac{1}{c}\frac{\partial}{\partial t}\int_V J\,\mathrm{d}V \quad &+ \quad \frac{1}{c}\int_{\partial V} J u\,\mathrm{d}A + \int_V \nabla\cdot H\,\mathrm{d}V \\ &+ \quad \frac{1}{c}\int_V K\,\nabla\cdot u\,\mathrm{d}V - \frac{1}{c}\int_V \frac{3K - J}{r} u\,\mathrm{d}V \\ &+ \quad \int_V \rho(\kappa_{\mathrm{J}} J - \kappa_{\mathrm{S}} S)\,\mathrm{d}V = 0\,,\end{aligned} \tag{3.5}$$

$$\begin{aligned}\frac{1}{c}\frac{\partial}{\partial t}\int_V H\,\mathrm{d}V \quad &+ \quad \frac{1}{c}\int_{\partial V} H u\,\mathrm{d}A + \int_V \nabla K\,\mathrm{d}V \\ &+ \quad \int_V \frac{3K - J}{r}\,\mathrm{d}V + \frac{1}{c}\int_V H\nabla u\,\mathrm{d}V \\ &+ \quad \int_V \kappa_{\mathrm{H}}\rho H\,\mathrm{d}V = 0\,.\end{aligned} \tag{3.6}$$

As mentioned in the previous section, additional relations between the physical quantities are required to close this system. We need the various frequency integrated and differently weighted opacities $\kappa_{\mathrm{J}}(\rho, T)$, $\kappa_{\mathrm{S}}(\rho, T)$ and $\kappa_{\mathrm{H}}(\rho, T)$, an equation of state $P = P(\rho, e)$ and $T = T(\rho, e)$ as well as the source function $S = S(T)$ and the Eddington factor $f = K/J$.

Due to the purely analytical manipulations the set of equations is equivalent to the standard notation as presented in Sect. 2.1. This integral formulation is used for the subsequent discretization and we can proceed by taking now the volume V to be between adjacent computational cells, *i.e.*, the numerical method is based on so-called *finite volumes*.

3.2 Symbolic Notation

We get the discrete version of the RHD equations by applying the rules which transform the differential operators into finite volume operators. Therefore, we have to define the various differences which will be used throughout the following sections for any physical quantity X defined at the grid point r_l, *i.e.*, $X_l = X(r_l)$. Note that also the position r_l is determined implicitly according to the grid equation (*cf.*, Sect. 4.).

symbol	operator	description
δX_l	$= X_l^{(n)} - X_l^{(o)}$	temporal difference
ΔX_l	$= X_l - X_{l+1}$	spatial difference
$\overline{X_l}$	$= 0.5\,(X_l + X_{l+1})$	spatial mean

3.3 Moving Coordinates

Since the coordinates are not fixed in space we have to specify the *relative velocity* u^{rel} between the gas velocity and the grid motion. We use an arbitrary coordinate system but do not transform the variables into this system, *i.e.*, use an adaptive coordinate system but still Eulerian (or Lagrangian if the equations are formulated in Lagrangian coordinates) variables. We can now define the relative velocity by changes of the grid point position

$$u^{\mathrm{rel}} = u - u^{\mathrm{grid}} = u - \frac{\delta r}{\delta t} \,. \tag{3.7}$$

According to the implicit nature of the grid equation the actual grid velocity u^{grid} is given only after convergence of the iteration at the new time level since the grid equation is solved simultaneously with the physical equations. Another definition of the grid velocity is presented in Sect. 7.6 but not implemented up to now in the current RHD-codes.

3.4 Implicit Discretization

In the following sections we introduce a difference scheme which allows a stable computation of radiating and self-gravitating flows, at least in one dimension. There exists no rigorous mathematical proof for the stability but in a large variety of examples we never encountered a numerical instability, in particular the adaptive mesh is very sensitive to such situations because a rapid oscillation of any variable will increase the clustering of grid points making the numerical difficulties even more visible.

The unknown variables are defined on a so-called *staggered mesh* where scalars like density, pressure, energy densities are defined inside a computational cell and vectors like velocity and fluxes are defined at the cell boundaries (Fig. 3.1). Hence, we have to provide a description of how to transform the variables from the cell boundary to the cell's interior and vice versa. Up

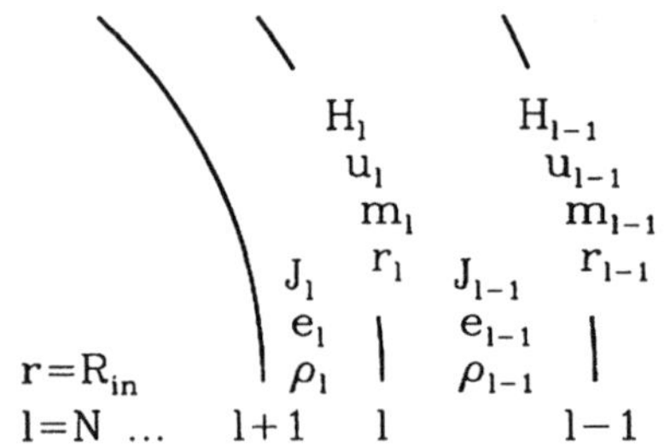

Fig. 3.1. Location of variables on a staggered mesh. Scalars are defined within a cell, vectors at the cell boundaries. The grid index l is increased inwards

to now this is done by a simple mean between the left and right values and in the final version of the discrete RHD equations of Sect. 3.13 we will use this spatial mean for the gas momentum flux as well as for the radiation flux. A better but much more sophisticated way is presented in Sect. 7.6 which up to now has not been incorporated in any RHD code. Note, that a more complicated description rapidly increases the need for CPU-time for evaluating the corresponding Jacobi-matrix (*cf.*, Sect. 5.4). The whole computational domain is divided by N grid points and the grid index is running inwards, *i.e.*, $l = 1$ defines the outer boundary at $r_1 = R_{\text{out}}$ whereas $i = N$ gives the inner boundary located at $r_N = R_{\text{in}}$. Note also that r_l appears as a dependent variable since the location of an individual radius point is determined by the grid equation (Sect. 4.).

Compared to the work of Tscharnuter and Winkler (1979), the key difference to find the following conservative formulation comes from the fact that the implicit nature does not require a form of the equations like

$$\frac{\partial}{\partial t}\rho_l + \nabla\cdot\rho_l \boldsymbol{u}_l = 0 \tag{3.8}$$

but any combination of ρ_l with other variables is suited as long as the corresponding Jacobi-Matrix remains non-singular.

An advantage of implicit schemes comes from the large time steps that can be used during the computations (see Sect. 5.3 for more details). All *explicit* numerical schemes like the ones described in the lectures by LeVeque and Müller have to obey for stability reasons the restrictive *Courant–Friedrichs–Lewy (CFL)* condition restated here in its simplest form for a grid with spacing of Δx,

$$t_{\text{CFL}} = \min_{\text{all cells}} \frac{\Delta x}{|u| + c_{\text{s}}} . \tag{3.9}$$

The Courant number can be defined through

$$N_{\text{CFL}} = \frac{\delta t}{t_{\text{CFL}}} \tag{3.10}$$

and any explicit scheme requires $N_{\text{CFL}} \leq 1$.

With an implicit scheme, we have to deal with large nonlinear algebraic systems for the unknowns $X^{(n)}$ at the new time step. If $\mathcal{G}$ denotes the discrete version of the differential equations one gets a system like

$$\frac{X^{(n)} - X^{(o)}}{\delta t} + \mathcal{G}(X^{(n)}, X^{(o)}) = 0 \tag{3.11}$$

and this system is typically ***nonlinear*** and has therefore to be solved iteratively, *e.g.*, by a ***Newton–Raphson iteration*** for the variables $X^{(n)}$ at the new time level. Usually the time step is not restricted but the convergence radius of the Newton iteration places some limits. Another condition on the time steps comes from accuracy considerations, *e.g.*, the time step can be restricted to limit the allowed maximum variation of variables as time advances.

3.5 Time-centering

The computation of oscillatory solutions requires special care for the temporal evolution, in particular the numerical damping in time has to be reduced. By separating the time derivative from the radial terms the system of equations is written symbolically as

$$\frac{\partial h}{\partial t} = \mathcal{H}(t) \tag{3.12}$$

and a Taylor expansion is performed to calculate the finite difference approximation. If Eq. (3.12) is substituted,

$$\begin{aligned} \frac{h(t+\delta t) - h(t)}{\delta t} &= \frac{\partial h}{\partial t} + \frac{1}{2}\frac{\partial^2 h}{\partial t^2}\delta t + \ldots \\ &= \mathcal{H}(t) + \frac{1}{2}\frac{\partial \mathcal{H}}{\partial t}\delta t + \ldots \end{aligned} \tag{3.13}$$

it becomes clear that the temporal error scales with δt. Repeating this procedure for $\mathcal{H}$ one gets

$$\frac{\partial}{\partial t}\mathcal{H} = \frac{\mathcal{H}(t+\delta t) - \mathcal{H}(t)}{\delta t} + \frac{1}{2}\frac{\partial^2 \mathcal{H}}{\partial t^2}\delta t + \ldots \; . \tag{3.14}$$

Now, substituting the last expression into Eq. (3.13) and collecting terms we end up with

$$\frac{h(t+\delta t) - h(t)}{\delta t} = \frac{1}{2}\left[\mathcal{H}(t+\delta t) + \mathcal{H}(t)\right] + \frac{1}{4}\frac{\partial^2 \mathcal{H}}{\partial t^2}\delta t^2 + \ldots \tag{3.15}$$

and this expression goes with the second order of δt. Hence, the accuracy order of the temporal discretization can be increased if the radial part of the equations is time-centered with $\theta = 1/2$, *i.e.*,

$$\langle \mathcal{H}(t) \rangle = \theta \mathcal{H}(t+\delta t) + (1-\theta)\mathcal{H}(t) \; . \tag{3.16}$$

If $\mathcal{H}$ is a linear operator this is equivalent to center the physical variables like

$$\begin{aligned} \langle X(t) \rangle &= \theta\, X(t+\delta t) + (1-\theta)\, X(t) \\ &= \theta\, X^{(n)} + (1-\theta)\, X^{(o)}, \qquad 0 \le \theta \le 1 \, , \end{aligned} \tag{3.17}$$

where $\theta = 1$ gives a fully implicit scheme. As demonstrated before the mean value of $\theta = 0.5$ leads to second order accuracy in time. However, in the case of nonlinear equations both procedures are not equivalent. Nevertheless, from numerical experience I prefer the time-centered variables and thus the code uses differential operator on time-centered variables rather than time-centered operators on physical variables. For stability reasons a typical value of $\theta = 0.51$ may be necessary (see Feuchtinger & Dorfi 1995).

3.6 Adaptive RHD Equations

Based on the finite volume conservative discretization on an adaptive grid the basic form of all physical equations is given by

$$\frac{\mathrm{d}}{\mathrm{d}t}\int_V X\,\mathrm{d}V + \int_{\partial V} X u^{\mathrm{rel}}\,\mathrm{d}A - \int_V (X_{\mathrm{source}} - X_{\mathrm{sink}})\,\mathrm{d}V = 0\,. \tag{3.18}$$

Depending on the physical nature of the source and sink terms their transformation onto the moving coordinates can be easily achieved by applying the adaptive transport theorem (*cf.*, Sect. 2.6). The following sections explain in more detail how the different terms can be discretized in a stable manner. Since we are carrying the integrated mass m as an independent variable to obtain the gravitational acceleration we can use the temporal change of the integrated mass (divided by 4π) to calculate the fluxes across the cell boundaries,

$$-\frac{\delta m_l}{\delta t} = \langle r_l^2\rangle\langle\widetilde{\rho^{\mathrm{ad}}}\rangle\left[\langle u_l\rangle - \frac{\delta r_l}{\delta t}\right]\,, \tag{3.19}$$

where we have expanded all terms explicitly according to our time-centered description of Eq. (3.17) as well as our advection procedure as stated in Eq. (3.25). The derivation of the Jacobi-Matrix simplifies when taking Eq. (3.19) (instead of the direct use of $r^2 u \rho^{\mathrm{ad}} u$) and it has in general no influence on the numerical solution as long as the flow structure is well represented on the adaptive grid. An advantage of using $\delta m/\delta t$ comes from keeping the discretization stencil 5-point for the momentum equation. The direct use of $r^2 u \widetilde{\rho^{\mathrm{ad}}} u$ would result in a 6-point scheme due to the interpolation necessary on the staggered mesh. Note also that for clarity the explicit notation of the time-centered variables will be omitted in the following sections.

3.7 Discretization of Gradients and Divergence Terms

Starting with a gradient of the physical quantity X we can write the volume integrated values between the cell boundaries $(l+1, l)$ after a division with 4π

$$\left.\int_V \nabla X\,\mathrm{d}V\right|_l = r_l^2\,\Delta X_l\,, \tag{3.20}$$

where the staggered mesh allows an intuitive physical interpretation since, e.g., a pressure difference ΔP between adjacent cells acts directly on the surface area r_l^2.

The volume integrated divergence is straight forward due to the Gauss theorem and therefore it is discretized as

$$\int_V \nabla \cdot X \, \mathrm{d}V \bigg|_l = \Delta(r_l^2 X_l) \,. \tag{3.21}$$

3.8 Diffusion

Diffusion terms are not included in the basic set of RHD equations presented in Sect. 3.1 but clearly a number of astrophysical phenomena can be characterized by a diffusion process. Although modifying the nature of the differential equations from a hyperbolic type to a second order derivative in space and by adding parabolic terms to the equations the implicit method can deal with such terms without major changes. In the case of diffusive terms, *e.g.*, an explicit method suffers severely from the limits imposed on the time step by

$$\delta t \leq \min_{\text{all cells}} \frac{1}{6} \frac{(\Delta x)^2}{D} \tag{3.22}$$

denoting the diffusion coefficient by D, *i.e.*, the time step has to be reduced by the minimum spatial resolution squared. Note that the exact factor in front of condition (3.22) depends on the discretization scheme used. Hence, a number of numerical methods employs an operator splitting technique (see also lectures by LeVeque) where the parabolic parts are solved by an implicit method still allowing the CFL time step of Eq. (3.9) for the hyperbolic parts of the equations.

A particular example including diffusive terms is given in Sect. 6.1, where a few models are plotted describing the SNR-evolution with particle acceleration. In this case a diffusion term in the cosmic ray energy equation has to be added to the usual term of the energy equation. The following (volume integrated) diffusion term is discretized in spherical symmetry

$$\int_V \nabla \cdot D \nabla X \, \mathrm{d}V \bigg|_l = \Delta \left[r_l^2 D_l \frac{\Delta X_l}{\Delta r_l} \right] \tag{3.23}$$

and in more detail

$$\frac{\Delta X_l}{\Delta r_l} = \frac{X_l - X_{l+1}}{r_l - r_{l+1}} \,, \tag{3.24}$$

since the scalar quantity X_l is located inside a cell whereas the radii r_l and r_{l+1} are defined at the cell boundary. I emphasise this point because taking, *e.g.*, the mean value $\langle X_l \rangle$ defined at the "correct" position of r_l can lead to some numerical instabilities as described already by Tscharnuter and Winkler (1979).

3.9 Advection

Any numerical scheme for hydrodynamics different from a Lagrangian has to deal with the transport of quantities from one computational cell to the neighbouring ones. The corresponding physical terms are called *advection* and their numerical properties play an important role also for adaptive conservative computations where this transport is done with a relative velocity instead of the usual gas velocity. The conservation form ensures that any loss or gain in one cell due to this transport is compensated by the same amount of gain or loss in the neighbouring cell. As demonstrated in the lectures by LeVeque and Müller these advection terms can be discretized, *e.g.*, by employing a Riemann solver or some approximations to the physical nonlinear flow structure. The resulting flux over a boundary is reconstructed from the different fluxes carried by the various waves. The complexity of this procedure depends critically on the EOS and on the wave speeds. In the case of radiating fluids the velocity of light can place rather strict conditions on the time step. Furthermore, the nature of the various waves becomes more complex. Hence, up to now only the simplest "wave" is taken into account to calculate the advection terms in radiation hydrodynamics: The direction of the flow is used to construct the flux. Therefore, the advection terms for the quantity X will be discretized in spherical geometry by

$$\int_{\partial V} X u^{\mathrm{rel}} \,\mathrm{d}A \simeq \Delta\left(r^2\, \widetilde{X^{\mathrm{ad}}}\, u^{\mathrm{rel}}\right) \tag{3.25}$$

for stability reasons $\widetilde{X^{\mathrm{ad}}}$ has always to be taken from *up-stream*. Depending on the accuracy the advected quantity $\widetilde{X^{\mathrm{ad}}}$ can be obtained from

- constant slopes within a cell, *donor-cell*
- linear monotonized slopes, *monotonic advection*
- piecewise parabolic slopes, *PPM*
- or some other high order method.

Without reviewing the vast mathematical literature on these points I want to state that almost all examples presented here rely on an *upwind monotonic advection* since together with the adaptive grid such an advection scheme provides in most cases the necessary accuracy. Therefore the advection can be written as

$$\widetilde{X_l^{\mathrm{ad}}} = \begin{cases} X_{l-1}^{\mathrm{ad}}\,, & \text{if } u_l^{\mathrm{rel}} < 0 \\ X_l^{\mathrm{ad}}\,, & \text{otherwise} \end{cases} \tag{3.26}$$

together with the monotonized slopes

$$X_l^{\mathrm{ad}} = X_l + \begin{cases} \dfrac{\Delta X_l\, \Delta X_{l-1}}{0.5\,(X_{l-1} - X_{l+1})}\,, & \text{if } \Delta X_l\, \Delta X_{l-1} > 0 \\ 0\,, & \text{otherwise}\,. \end{cases} \tag{3.27}$$

This version of monotonized slopes is based on the work of van Leer (1977), where a slope within a cell is only introduced when there is a clear "trend" in the same direction within the neighbouring cells. In the case of developing numerical oscillations one encounters a typical wave length of the instability of one cell size and the above monotonicity criterion will cut these oscillations through replacing the slope by a constant, *i.e.*, the advection is truncated to the stable first order donor-cell scheme. This procedure guarantees also non-oscillatory solutions in the vicinity of shock fronts as long as the grid provides enough resolution to evaluate the shock transition correctly. An example of a small post-shock oscillation is presented in Fig. 4.6 which was caused by insufficient spatial resolution.

From counting the indices it is evident that X_l^{ad} depends on the three points $l-1, l, l+1$. The flux at location r_l is calculated from X_l^{ad} and X_{l-1}^{ad} depending on the direction of the relative flow and finally, the changes of X_l within a cell are given by the difference between the left hand boundary situated at r_{l+1} and the right hand boundary at r_l. Hence, collecting this information the whole advection term based on such a monotonic description depends on quantities with indices $l-2, l-1, l, l+1, l+2$. This so-called 5-point stencil enters then directly into the structure of the corresponding Jacobi matrix (Sect. 5.4).

From the statements above it is also clear that up to now the non-uniformity of the grid as well as the geometric properties have not been taken into account for calculating the monotonized slopes but further remarks on this subject can be found in Sect. 7.6.

3.10 Initial Conditions

Starting the implicit computation requires the *full* solution of the system of RHD equations together with an appropriate ***grid distribution***. This initial model must lie within the convergence radius of the Newton iteration otherwise the computation cannot be started. Since a number of problems attacked by implicit methods are mixed boundary and initial-value problems the demands on initial models can be very severe and special care must be taken to obtain suitable initial models as discussed in more detail in Sect. 4.10, where the chosen examples illustrate very different initial conditions.

In the case of a protostellar collapse the initial model is simple because constant values for density, pressure and zero velocity in a Jeans-unstable fragment can be assumed. The grid is distributed according to a constant point concentration $n = \mathrm{const.}$ As the collapse proceeds the structures develop together with the adaptive grid as discussed in more detail in Sect. 6.3. The creation of hydrostatic initial models requires more advanced methods involving in most cases the numerical integration of the stationary stellar structure equations. As demonstrated in Sect. 4.10 the grid adaption can be achieved by a pseudo-time evolution. Such a dust-free hydrostatic initial model is used to start a stellar wind calculation where after a while a dust driven wind develops from a carbon-rich asymptotic giant branch (AGB)-star. The outer boundary is kept as a Lagrangian boundary for some time, then the whole structure expands and when the outer gas reaches its final escape velocity the boundary is changed to an outflow boundary condition. The details of these models are described in Sect. 6.4.

To summarize: it can be complicated to create initial models appropriate for implicit computations. RHD-problems are in general mixed boundary/initial value problems and it can be difficult to advance the solution in time with time steps longer than the CFL time step (3.9). Small fluctuations in the initial conditions can generate various waves which must then be followed on such small time steps.

3.11 Boundary Conditions

The boundary conditions for the RHD equations have to be specified together with the corresponding boundary conditions for the adaptive grid which are presented in Sect. 4.5. Usually, the boundary conditions can easily be incorporated into the staggered mesh as seen in the next examples. I want to emphasize at this point that at both ends of the computational domain boundary conditions are supplied in a way as to keep the structure of the corresponding Jacobi-matrix as regular as possible. For the computations a 5-point stencil is adopted and therefore any boundary condition relating two points either to the left for the outer conditions or to right for the inner condition is included without changing the overall pattern of the Jacobi pentadiagonal block structure (see Sect. 5.4). A periodic boundary condition joins the innermost to the outermost grid points leading to a different structure in the Jacobi matrix which modifies the standard inversion procedure of the pentadiagonal block matrix.

Starting with a simple case of fixed values X_{bound} for the variable X the boundary conditions read as follows

$$X_{1,N} - X_{\text{bound}} = 0 \,, \tag{3.28}$$

where the index 1 denotes the outer boundary and N the inner boundary, respectively. Clearly, this equation leads to an entry of 1 in the corresponding Jacobi-matrix and no problems are expected for the subsequent inversion during the iteration. At the innermost point a regularity condition is often encountered yielding equations like

$$\left.\frac{\partial X}{\partial r}\right|_{\text{inner}} = 0 \,, \quad \text{or} \quad X_N - X_{N-1} = 0 \,, \tag{3.29}$$

where the second expression represents the discrete version of the regularity condition. The case of a divergence free outer boundary at r_N can be realized by

$$X_N r_N^2 - X_{N-1} r_{N-1}^2 = 0 \,. \tag{3.30}$$

An outer Lagrangian boundary requires the solution of the momentum balance and an adaptive grid moving at the same speed u_N. The full discrete equation of motion is given in Sect. 3.13 and the boundary conditions for the grid are described in Sect. 4.5. Taking an external pressure of P_{ext}, a total mass M, no mass flow over the boundary and expanding all terms, the velocity u_N at the boundary is given through

$$\begin{aligned}\delta(u_N \overline{\rho_N \Delta V_N}) \quad &+ \quad \delta(\overline{m_{N-1}})\, \widetilde{u_{N-1}^{\mathrm{ad}}} + \delta t r_N^2 (P_{\mathrm{ext}} - P_{N-1}) \\ &+ \quad \frac{4\pi\delta t G M}{r_N^2}\, \overline{\rho_N \Delta V_N} - \frac{4\pi\delta t}{c}\, \overline{\kappa_{\mathrm{R},N}} H_N\, \overline{\rho_N \Delta V_N}\,. \end{aligned} \tag{3.31}$$

This equation corresponds to the finite volume discretization of the momentum equation containing also self-gravity and radiation pressure. Note that the radiation flux H_N has to be determined from the radiative boundary conditions as described in Sect. 3.15.

From the examples above it should be clear how the various boundary conditions can be included in the discrete version of the RHD equations. If the boundary conditions are not correctly formulated, then in most cases the Jacobi matrix will become singular during the Newton iteration, or at least, the time step will remain rather small.

3.12 Artificial Viscosity

As outlined in the previous sections the present numerical scheme for RHD computations does not make use of any kind of Riemann solver. Therefore, in order to treat shock fronts a so-called *artificial viscosity* is used in order to broaden shock waves over a few computational cells. This idea has been proposed by von Neumann & Richtmyer (1950) by adding an additional pressure Q in the vicinity of a shock so that one gets the correct jump conditions when integrating over the discontinuity. Their original work is restricted to plane geometry in the case of other metrics we have to use the geometry-independent formulation developed by Tscharnuter and Winkler (1979). Instead of introducing a viscous pressure Q they generalized the concept by the viscous pressure tensor which is formulated in analogy to a physical viscous pressure. Following their derivation with a traceless tensor – allowing homologous contractions without pressure generation – the pressure tensor $\mathbf{Q}$, the viscous momentum transfer u_{Q}, and the viscous energy dissipation ϵ_{Q} are given by

$$\mathbf{Q} \;=\; \mu_{\mathrm{Q}} \left[(\nabla \mathbf{u}) - \frac{1}{3} \nabla \cdot u\, \mathbf{I} \right], \tag{3.32}$$

$$u_{\mathrm{Q}} \;=\; -\nabla \cdot \mathbf{Q}\,, \tag{3.33}$$

$$\epsilon_{\mathrm{Q}} \;=\; -\frac{1}{\rho}\, \mathbf{Q} \cdot (\nabla \mathbf{u})\,, \tag{3.34}$$

where $(\nabla \mathbf{u})$ is the symmetrized velocity gradient and $\mathbf{I}$ the unity tensor. The artificial viscosity coefficient μ_{Q} consists of a linear and a quadratic term weighted by q_1 and q_2, *i.e.*,

$$\mu_Q = -q_1 l_{\rm visc} c_s + q_2^2 l_{\rm visc}^2 \min(\nabla\cdot u, 0) \ . \tag{3.35}$$

The definition of μ_Q include a typical viscous length scale $l_{\rm visc}$. Inspecting the last expression one sees that the linear term in $l_{\rm visc}$ scales with the sound velocity c_s and is always present when $q_1 \neq 0$. Such a term has been included in a number of explicit computations to damp out small scale oscillations near contact discontinuities. For all computations I have done so far, such oscillations never appeared and therefore I set $q_1 = 0$. The second term is quadratic in $l_{\rm visc}$ and it is evident that *compressive and non-homologous* motions, *i.e.*, $\nabla\cdot u < 0$ and $tr(\mathbf{Q}) \neq 0$ produce a viscous pressure. The amount of artificial viscosity is then determined by these length scales $q_1 l_{\rm visc}$ and $q_2 l_{\rm visc}$ and the following table summarizes the main difference between adaptive and non-adaptive artificial viscosity formulations, namely that the adaptive grid clusters around discontinuities which enables much lower values of these length scales. In non-adaptive computations the viscosity has to be of the order of the typical cell size Δx and q_2 is set to the number of cells over which the shock front is smeared out. In adaptive and implicit computations a length scale $l_{\rm visc}$ can be defined over which a shock should be broadened and then the grid shrinks around the shock front to resolve the structure having this typical thickness of $l_{\rm visc}$. Hence, the thickness of the shock front can be prescribed *a priori* and does not relate to the resolution Δx available in non-adaptive computations. Note, that in explicit computations such a fine grid around shock fronts reduces the time step through the CFL-condition (Eq. (3.9)).

grid type	viscosity scale	shock scale
non-adaptive	$q_{1,2} l_{\rm visc} \simeq \Delta x$	grid size
adaptive	$q_{1,2} l_{\rm visc} \leq 10^{-3}\, r$	"physical scale"

According to the discretization scheme the artificial viscosity term are written like

$$\mu_{Q,l} = q_1 \overline{r_l}\, c_{s,l} - (q_2 \overline{r_l})^2 \min\left(\frac{\Delta(r_l^2 u_l)}{\Delta V_l}, 0\right) , \tag{3.36}$$

including the linear part with $c_{s,l}$, the adiabatic sound velocity at r_l. The corresponding discrete and volume integrated versions of u_Q and ϵ_Q are given in the next section. Clearly, the viscous momentum transfer u_Q enters the equation of motion and the viscous energy generation adds to the gas energy equation. The action and smallness of this artificial viscosity can be inferred from a shock tube calculation depicted in Fig. 4.2 or from the SNR evolution where the three discontinuities moving with the ejected material are presented at one year after the explosion in Fig. 6.1.

3.13 Discrete RHD Equations

Adopting the previous notation and including the artificial viscosity terms the full set of discrete RHD equations is given in symbolic notation as a set of nonlinear algebraic equations for all basic variables like ρ, u, e, J, H, m and r at the grid index l. The time-centering as stated in Sect. 3.5 has been omitted for clarity.

$$\frac{\delta(\rho_l \Delta V_l)}{\delta t} + \Delta(r_l^2 \widetilde{\rho_l^{\mathrm{ad}}} u_l^{\mathrm{rel}}) = 0\,, \tag{3.37}$$

$$\begin{aligned} \frac{\delta(u_l\, \overline{\rho_l \Delta V_l})}{\delta t} &+ \Delta(-\frac{\overline{\delta m_l}}{\delta t} \widetilde{u_l^{\mathrm{ad}}}) + r_l^2 \Delta P_l \\ &+ \frac{4\pi G m_l}{r_l^2} \overline{\rho_l \Delta V_l} - \frac{4\pi}{c} \overline{\kappa_{\mathrm{R},l}} H_l\, \overline{\rho_l \Delta V_l} \\ &+ \frac{2}{3 r_l} \Delta \left\{ \mu_{\mathrm{Q},l}\, \rho_l \left(\frac{\Delta u_l}{\Delta r_l} - \frac{\overline{u_l}}{\overline{r_l}} \right) \right\} = 0\,, \end{aligned} \tag{3.38}$$

$$\begin{aligned} \frac{\delta(\rho_l e_l \Delta V_l)}{\delta t} &+ \Delta(-\frac{\delta m_l}{\delta t} \widetilde{e_l^{\mathrm{ad}}}) \\ &+ P_l \Delta(r_l^2 u_l) + 4\pi \kappa_{\mathrm{R},l}\, \rho_l\, (J_l - S_l)\, \Delta V_l \\ &+ \frac{2}{3} \mu_{\mathrm{Q},l}\, \rho_l\, \Delta V_l \left(\frac{\Delta u_l}{\Delta r_l} - \frac{\overline{u_l}}{\overline{r_l}} \right)^2 = 0\,, \end{aligned} \tag{3.39}$$

$$\Delta m_l - \rho_l\, \Delta V_l = 0\,, \tag{3.40}$$

$$\begin{aligned} \frac{1}{c} \frac{\delta(J_l \Delta V_l)}{\delta t} &+ \frac{1}{c} \Delta \left(\widetilde{ -\frac{\delta m_l}{\delta t} \left[\frac{J}{\rho} \right]_l^{\mathrm{ad}} } \right) + \Delta(r_l^2 H_l) \\ &+ \frac{1}{c} K_l \Delta(r_l^2 u_l) - \frac{1}{c} (3K_l - J_l) \frac{\overline{u_l}}{\overline{r_l}} \Delta V_l \\ &+ \kappa_{\mathrm{R},l} \rho_l\, (J_l - S_l)\, \Delta V_l = 0\,, \end{aligned} \tag{3.41}$$

$$\begin{aligned} \frac{1}{c} \frac{\delta(H_l\, \overline{\rho_l \Delta V_l})}{\delta t} &+ \frac{1}{c} \Delta(\overline{r_l^2} \widetilde{H_l^{\mathrm{ad}}} \overline{u_l^{\mathrm{rel}}}) + r_l^2 \Delta K_l + \frac{3\overline{K_l} - \overline{J_l}}{r_l} \\ &+ \overline{\kappa_{\mathrm{R},l}}\, H_l\, \overline{\rho_l \Delta V_l} + \frac{1}{c} H_l r_l^2 \Delta u_l = 0\,. \end{aligned} \tag{3.42}$$

An appropriate volume definition of

$$\Delta V_l = \frac{1}{3} \left(r_l^3 - r_{l+1}^3 \right) \tag{3.43}$$

as well as the integrated continuity equation has been used, *i.e.*,

$$-\frac{\delta m_l}{\delta t} = r_l^2 \widetilde{\rho_l^{\mathrm{ad}}} u_l^{\mathrm{rel}}\,. \tag{3.44}$$

Note that due to the staggered mesh, *e.g.*, the mean mass between two adjacent cell $\overline{\rho \Delta V} = \frac{1}{2}(\rho_l V_l + \rho_{l+1} V_{l+1})$ as well as other mean values have been used for the gas momentum equations (3.38) and the radiative flux equation (3.42). This system has to be solved numerically together with the other physical relations such as $P = P(\rho, e)$ or $\kappa_{\mathrm{R}} = \kappa_{\mathrm{R}}(\rho, T)$ and the adaptive grid derived in Sect. 4.

3.14 Radiative Closure Condition

The numerical treatment of radiative transfer requires some additional remarks. The radiation field is described by the first moments as presented in Sect. 2.1 and the representation necessitates an additional relation, e.g., between J and K to close the system. This so-called radiative closure condition via the Eddington factor has been discussed in Sect. 2.3. The following sections are devoted to the calculation of this closure condition from a solution of the time-independent radiative transfer equation (see lectures by Mihalas). As pointed out in the introduction, for a number of astrophysical problems the time scale to adjust the geometry of the radiation field is short compared with the outer physical time scales and therefore a time-independent version of the radiative transfer equation can be used to calculate the ratio between the zeroth and second moment. However, the method outlined in the next sections can be easily generalized to handle also the time-dependence of the Eddington factor. The specific intensity $I_\nu(r,\mu)$ obeys then the static radiation transfer equation in spherical geometry with θ angle between radius and line of sight ($\mu = \cos\theta$)

$$\mu \frac{\partial I_\nu}{\partial r} + \frac{1-\mu^2}{r}\frac{\partial I_\nu}{\partial \mu} = \kappa_\nu (S_\nu - I_\nu) \tag{3.45}$$

introducing also the source function S_ν. Clearly, the full solution of this equation increases the dimensionality of the problem because the intensity $I_\nu(r,\mu)$ depends on the radius r *and* the angle $\cos\theta = \mu$. Although the next formulae look rather complicated the numerical treatment is simple and mainly a book-keeping of the different variables.

Since Eq. (3.45) is a linear partial differential equation for $I_\nu(r,\mu)$, the theory of characteristics can be utilized to reduce and solve such an equation by integrating along their characteristics (e.g., Yorke 1977, Balluch 1988). The characteristics are light rays along a constant *impact* parameter given by

$$p = r\sqrt{1-\mu^2} = \text{const.}\,, \tag{3.46}$$

also plotted in Fig. 3.2. The ray with impact parameter p_j cuts different radii r_l at varying angles $\mu_{l,j}$ (Fig. 3.2). If the inner boundary of the computation, $R_{\rm in}$, is not located at the stellar center a number of additional impact parameters have to be introduced inside the innermost sphere defined by $R_{\rm in}$. Hence, the total number of impact parameters is $K \geq N$, where N is the total number of radial grid points. Clearly, the values of the impact parameters p_j are defined through the radial values r_l to avoid further interpolations on the p-grid.

Transforming from (r,μ)-grid to (r,p)-grid the intensity can be calculated by integrating along these characteristics

$$I_\nu = I_\nu(0)\mathrm{e}^{-\tau_\nu} + \int_0^{\tau_\nu} \mathrm{e}^{-(\tau_\nu - \tau'_\nu)} S_\nu(\tau'_\nu)\,\mathrm{d}\tau'_\nu\,, \tag{3.47}$$

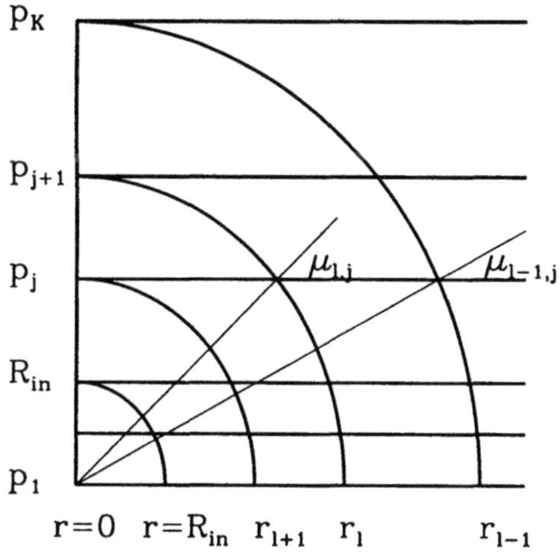

Fig. 3.2. The impact parameters p_j are chosen such that their corresponding rays are tangent to the radial shells. They intersect the shells r_l, at angles $\theta_{l,j}$, whose direction cosines are $\mu_{l,j} = \cos\theta_{l,j}$.

where the optical depth τ_ν has been introduced through

$$\tau_\nu = \int \kappa_\nu \, \mathrm{d}(r\mu) \,. \tag{3.48}$$

Since the parameter p cannot distinguish between $\mu < 0$ and $\mu > 0$ we have to define the outward and inward intensities

$$I_\nu^\pm(r,\mu) = I_\nu(r, \pm\mu) \,, \qquad \mu \geq 0 \,. \tag{3.49}$$

To arrive at a discrete version of the above integral we assume a linear dependence of the source function S_ν within a computational cell and drop ν for the sake of simplicity in the following. The inward $I_{j,l}^+$ and outward $I_{j,l}^-$ at the grid point r_l read as follows

$$\begin{aligned} I_{j,l}^+ &= \mathrm{e}^{-\Delta\tau_j^+} I_{l+1}^+ + \mathrm{e}^{-\Delta\tau_j^+} \int_0^{\Delta\tau_j^+} \mathrm{e}^{\tau'} S(\tau' + \tau_{l+1}) \, \mathrm{d}\tau' \,, \\ I_{j,l}^- &= \mathrm{e}^{-\Delta\tau_j^-} I_{l-1}^- + \mathrm{e}^{-\Delta\tau_j^-} \int_0^{\Delta\tau_j^-} \mathrm{e}^{\tau'} S(\tau_{l-1} - \tau') \, \mathrm{d}\tau' \,, \end{aligned} \tag{3.50}$$

where the different indices $l+1$ or $l-1$ are caused by the left or right starting point of the integration. The optical depths are defined along each p_j-path determining μ and given by

$$\begin{aligned} \Delta\tau_j^+ &= \kappa_l \Delta(\mu r)_l \,, \\ \Delta\tau_j^- &= \kappa_{l-1} \Delta(\mu r)_{l-1} \,. \end{aligned} \tag{3.51}$$

For the second part of expressions (3.50) the discrete version based on the linear interpolation of $S(\tau)$ leads to

$$\frac{1}{2}\left[S_{l-1}\frac{\Delta\tau_j^+ - 1 + \mathrm{e}^{-\Delta\tau_j^+}}{\Delta\tau_j^+} + S_l\left(1-\mathrm{e}^{-\Delta\tau_j^+}\right) + S_{l+1}\frac{1-\mathrm{e}^{-\Delta\tau_j^+}(1+\Delta\tau_j^+)}{\Delta\tau_j^+}\right], \tag{3.52}$$

$$\frac{1}{2}\left[S_{l}\frac{\Delta\tau_j^- - 1 + \mathrm{e}^{-\Delta\tau_j^-}}{\Delta\tau_j^-} + S_{l-1}\left(1-\mathrm{e}^{-\Delta\tau_j^-}\right) + S_{l-2}\frac{1-\mathrm{e}^{-\Delta\tau_j^-}(1+\Delta\tau_j^-)}{\Delta\tau_j^-}\right]. \tag{3.53}$$

In the case of small optical depths $\Delta\tau^\pm \ll 1$ the weighting factors for the source function appearing in the last expressions should be expanded to increase the numerical accuracy.

3.15 Radiative Boundary Conditions

Specifying the outer radiation field by $I_{\nu,\mathrm{ext}}$ as well as the central source h_ν both specific intensities $I_\nu^\pm$ can be obtained by integrating Eqs. (3.50). Therefore the following boundary conditions are needed

$$\begin{aligned} I_\nu^-(r_1,p) &= I_{\nu,\mathrm{ext}}(p)\,, & 0 \le p \le r_1\,,\\ I_\nu^+(p,p) &= I_\nu^-(p,p)\,, & r_N \le p \le r_1\,,\\ I_\nu^+(p,p) &= I_\nu^-(p,p) + h_\nu(p)\,, & 0 \le p \le r_N\,. \end{aligned} \tag{3.54}$$

The first row of (3.54) allows for every impact parameter p_j to start the inwards integration of I_ν^- until the inner boundary is reached. This can happen at $r=0$ for all impact parameter $p_j \ge r_1$. In this case the regularity condition on the axis (Fig. 3.2 and second line of 3.54) demands that the in-going intensity I_ν^- must equal the outgoing intensity I_ν^+. If the inner boundary is not placed at the center $r=0$ the innermost impact parameters cut through the inner sphere, *i.e.*, $p_j < r_1$ and a central source h_ν is added to the incoming radiative field (third row of 3.54). The case of an optically thick central source added into a non-grey inner boundary is discussed in Sect. 6.2, where a non-grey radiative transfer calculation is performed on top of grey RHD models to obtain U,B,V colours of RR Lyrae pulsations.

3.16 Eddington Factor

After having obtained all intensities $I_\nu^\pm(r_l,p_j)$ the angle-dependence of the radiation field can be eliminated by an integration over all impact parameters. As inferred from Fig. 3.2 a fixed radius r_l cuts various impact parameter under different angles $\mu_{l,j}$ and therefore the integration over the angle μ can be replaced by an integration over the impact parameter p. Recalling the definitions of the radiative moments (2.5) the discrete versions reads as a summation over all p_j with $1 \leq j \leq K$:

$$J_\nu(r_l) = \frac{1}{2}\int_0^1 \left(I_\nu^+(\mu) + I_\nu^-(\mu)\right) \mathrm{d}\mu = \frac{1}{2}\sum_{j=1}^{K} a_j \left(I_{jl}^+ + I_{jl}^-\right), \quad (3.55)$$

$$H_\nu(r_l) = \frac{1}{2}\int_0^1 \left(I_\nu^+(\mu) - I_\nu^-(\mu)\right) \mu \,\mathrm{d}\mu = \frac{1}{2}\sum_{j=1}^{K} b_j \left(I_{jl}^+ - I_{jl}^-\right), \quad (3.56)$$

$$K_\nu(r_l) = \frac{1}{2}\int_0^1 \left(I_\nu^-(\mu) + I_\nu^-(\mu)\right) \mu^2 \,\mathrm{d}\mu = \frac{1}{2}\sum_{j=1}^{K} c_j \left(I_{jl}^+ + I_{jl}^-\right), \quad (3.57)$$

where I_{jl}^+ and I_{jl}^- denote the specific intensities at (r_l, p_j). The weighting factors of the integrals a_k, b_k and c_k are calculated from geometrical arguments and regularity conditions on $I_{jl}^{\pm}$ and I refer to the original work of Yorke (1980) for their evaluation. Finally, the discrete version of the Eddington factor $f\nu$ at the grid point r_l is obtained,

$$f_\nu(r_l) = \frac{K_\nu(r_l)}{J_\nu(r_l)} . \quad (3.58)$$

For the solution of the radiative transfer equation the outer boundary condition for the radiation intensity is given at $r = r_1$ through

$$J(r_1) = J_{\mathrm{ext}} + \frac{1}{\bar{\mu}}\left[H(r_1) + H_{\mathrm{ext}}\right] , \quad (3.59)$$

with

$$\bar{\mu} = \frac{H^+(r_1)}{J^+(r_1)} , \quad (3.60)$$

the ratio of the outgoing radiative flux H^+ to the outgoing energy density J^+.

4. Adaptive Grids

As stated in the introduction a number of astrophysical problems require the capability of dealing with physical variables which vary over many orders of magnitude. The occurrence of steep and propagating fronts like shock waves or ionization fronts led to the idea to redistribute a given number of grid points according to these numerical needs (Winkler 1975). In particular, the protostellar collapse calculations have triggered this development since the evolution of the accretion shock determines the optical appearance of the collapsing cloud (see also Sect. 6.3). In contrast to the large number of adaptive grids which are used in engineering or aerodynamics and which are so-called boundary fitted, the adaptive grids we have in mind are variable in time and depend on the physical solution. They are not generated in advance but they evolve in time as the physical situation changes its nature. Hence, the location of an individual grid point depends on the solution and on the resolution available. However, we have to ensure that the number of grid points is sufficient to resolve all structures.

4.1 Basic Grid Properties

Although the grid points are freely moving several requirements have to be fulfilled throughout the entire computation:

1. monotonicity, *i.e.*, $r_{l+1}(t) < r_l(t)$ for all t, $1 \leq l \leq N$,
2. smooth variations in space and time,
3. concentration of points at "important" places, automatic relaxation of the grid to a regular spacing in the case of smooth structures.

The derivation of the following grid equation is based on these three points and they are automatically incorporated. Next we define the *point concentration* at index l by

$$n_l = \frac{\mathcal{X}_l}{x_l - x_{l+1}} , \tag{4.1}$$

where $\mathcal{X}_l$ denotes a typical (local) scale. In the simplemost cases we use a grid equation like

$$n_l = \text{const.} \tag{4.2}$$

and take, e.g., $\mathcal{X}_l = b - a$ if we want an equidistant grid on the interval $[a, b]$. To be more flexible we can write, e.g., $\mathcal{X}_l = \overline{x_l}$ and create a logarithmic equidistant grid. Summarizing these two example we see that specifying $\mathcal{X}_l$ tailors the grid in the case where no structures are present. Hence, we can use $\mathcal{X}_l$ to prescribe the grid locations, e.g., more zones at a boundary, in the center, or wherever.

4.2 Desired Resolution

The so-called desired resolution $\mathcal{R}$ contains all information *what* should be resolved during the computation (*e.g.*, density, opacity, shock waves, ionization zones) and clearly depends on the problem to be solved. In what follows it is essential that $\mathcal{R}$ is a *positive* quantity. Taking these two building blocks we formulate the basic form of the grid equation

$$n \propto \mathcal{R} . \tag{4.3}$$

The constant of proportionality will be determined by the total number of points and the separation of the boundaries. In praxi we use the path-length of a graph of $f(x)$, *i.e.*,

$$\mathcal{R} \propto \sqrt{1 + \left(\frac{\mathrm{d}f}{\mathrm{d}x}\right)^2} \tag{4.4}$$

because this gives a natural concentration at locations of steep gradients and the points are distributed uniformly in arc-length along the graph of f. It is then obvious that the discrete case of M functions can be generalized to

$$\begin{aligned} \mathcal{R}_l &\propto \sqrt{1 + \sum_{j=1}^{M} \left(\frac{\mathcal{X}_l}{F_{l,j}} \frac{f_{l,j} - f_{l+1,j}}{x_l - x_{l+1}} \right)^2} \\ &= \sqrt{1 + n_l^2 \sum_{j=1}^{M} g_j \left(\frac{f_{l,j} - f_{l+1,j}}{F_{l,j}} \right)^2} , \end{aligned} \tag{4.5}$$

where the scaling factors $F_{l,j}$ as well as different weighting factors g_j for the M physical quantities have been introduced. As an initial guess for the weighting factors g_j it is recommended to take a "democratic" approach by setting $g_j = 1$ for all j. The spatial scaling factor $\mathcal{X}_l$ is defined through Eq. (4.1). This leads to our basic form of the discrete adaptive grid which fulfills already conditions (1) and (3)

$$n_l \propto \mathcal{R}_l , \tag{4.6}$$

i.e., the points remain monotonic for single valued functions $f(x)$ and are concentrated at steep gradients due to the large variations of $\mathrm{d}f/\mathrm{d}x$. From the numerical solution it becomes evident that the grid should not vary too rapidly in space and time, *e.g.*, neighbouring cells should have comparable dimensions or move at almost the same speed.

4.3 Spatial and Temporal Smoothing

The grid equation as stated in Eq. (4.6) reacts immediately to changes in the physical variables and allows very abrupt changes of the cell size. From numerical stability and smoothness we want to avoid such a rapid increase or decrease of the cell sizes, *e.g.*, the mass contained in two neighbouring cell should be comparable. Hence, we will limit variation of the point concentration through the ratio between two zones

$$\frac{\alpha_g}{\alpha_g + 1} \le \frac{n_l}{n_{l+1}} \le \frac{\alpha_g + 1}{\alpha_g} , \tag{4.7}$$

where α_g is a free parameter controlling the maximum permitted variation in cell sizes. The values for this spatial smoothing are typically around $\alpha_g = 2$ which allows a shrinking or expansion of the cell size by about 30%. The condition (4.7) can be transformed by replacing $\mathcal{R}_l$ through a spatially smooth quantity like

$$n_l \propto \sum_j \mathcal{R}_j \left(\frac{\alpha_g}{\alpha_g + 1} \right)^{|l-j|} . \tag{4.8}$$

Inspecting the above equation we notice that any perturbation occurring at $\mathcal{R}_j$ will be "felt" at a grid point r_l by a decreasing influence, namely by a reduction factor of $[(\alpha_g/(\alpha_g+1)]^{|l-j|}$. Hence, the further away a point concentration occurs, the less this concentration influences the local point concentration. To proceed we can replace the sum on the rhs by a difference on the lhs

$$\hat{n}_l = n_l - \alpha_g(\alpha_g+1)\,(n_{l-1} - 2n_l + n_{l+1}) \propto \mathcal{R}_l \;. \tag{4.9}$$

This differential version is preferable since also the discrete physical equations are written in such a local way that only the most neighbouring grid points enter the finite differences. A formulation like Eq. (4.8) couples all grid points together and fills up our Jacobi-Matrix which has to be inverted at every time step (see Sect. 5.3). In all computation presented in the following sections we have used $\alpha_g = 2$ allowing typical grid variations around 30% when going from one cell to their neighbouring ones.

The temporal smoothing can be done in a similar way by damping the grid motions over an exponential time scale τ_g,

$$\mathcal{R}(t) = \int_0^\infty \mathcal{R}(t-t')\exp\left(\frac{-t'}{\tau_g}\right)\frac{\mathrm{d}t'}{\tau_g} \;, \tag{4.10}$$

or in a discrete differential version

$$\tilde{n}_l^{(n)} = \hat{n}_l^{(n)} + \frac{\tau_g}{\delta t}\left(\hat{n}_l^{(n)} - \hat{n}_l^{(o)}\right) \;, \tag{4.11}$$

where the integral operator in Eq. (4.10) has be replaced by the corresponding discrete differential operator. Note that the difference operator used in Eq. (4.11) is the first order approximation to the differential operator belonging to the Greens function defined through Eq. (4.10). In many cases the computation runs smoothly with an instantaneous grid adaption, *i.e.*, $\tau_g = 0$.

4.4 Grid Equation

Now we can collect the previous terms and eliminate the constant of proportionality by equating at two neighbouring grid points the ratio between the spatially as well as temporally smoothed point concentration of Eq. (4.10) and the desired resolution Eq. (4.5) yielding the final version of the grid equation, *i.e.*,

$$\frac{\hat{n}_l}{\mathcal{R}_l} = \frac{\hat{n}_{l-1}}{\mathcal{R}_{l-1}} \tag{4.12}$$

Several remarks should be added concerning this grid equation. The constant of proportionality can be eliminated in numerous ways, e.g., by the inverse of Eq. (4.12) or by the product $n\mathcal{R}$, *etc.* However, according to numerical experiments the version presented in Eq. (4.12) leads to the smallest number of iterations although the other elimination procedures work as well. Remembering the previous derivation of the grid equation we see that the spatial as well as the temporal smoothness of the grid variation requires most of the manipulation on the point concentration and from the usage of a discrete diffusion operator acting in the space of indices (4.9) and the exponential damping (4.11) the necessary property of monotonicity is also ensured. It is also clear that there may be other possibilities of get rid of the constant of proportionality by writing, *e.g.*,

$$\frac{\mathrm{d}}{\mathrm{d}t}\left(\frac{\tilde{n}}{\mathcal{R}}\right) = 0, \quad \text{or} \quad \nabla\left(\frac{\tilde{n}}{\mathcal{R}}\right) = 0\,. \tag{4.13}$$

Again, from numerical experience such formulations may work but difficulties arise in the case of shock reflections at rigid boundaries, or in more general situations where the *number* of features to be resolved by the grid equation is changing rapidly with time.

Another remarks concerns the grid points involved in Eq. (4.12) because at every index l the two left and right neighbouring grid indices enter, *i.e.*, we arrive at a 5-point stencil, $(i-2, i-1, i, i+1, i+2)$ yielding at least for the grid equation a pentadiagonal Jacobi matrix (*cf.*, Sect. 5.4). Hence, we have to specify boundary conditions for two grid points at both ends of our computational domain.

4.5 Grid Boundary Conditions

Taking N grid points $(x_1 > x_2 > \ldots > x_N)$ the boundary conditions can be stated by demanding a constant point concentration at the boundary

$$n_1 = n_2, \quad \text{and} \quad n_{N-1} = n_{N-2}\,. \tag{4.14}$$

The innermost and outermost mesh points x_N and x_1 can still move according to physical requirements, *e.g.*, fixed in space by specifying the old position equal to the new position

$$x_1^{(n)} = x_1^{(o)}\,. \tag{4.15}$$

A Lagrangian boundary is constructed by calculating the flow velocity, *e.g.*, u_1 at the outer boundary

$$u_1^{\mathrm{rel}} = 0, \quad \textit{i.e.}, \quad x_1^{(n)} = x_1^{(o)} + \delta t \left\langle u_1 \right\rangle, \tag{4.16}$$

where we have used the time-centered velocity $\langle u_1 \rangle$. In a fully implicit version we simply take $u_1^{(n)}$. Note that at a Lagrangian boundary the velocity u has to be determined from the equation of motion. A moving piston with velocity w_N at the inner boundary leads to another typical example,

$$x_N^{(n)} = x_N^{(o)} + \delta t\, w_N\,. \tag{4.17}$$

All such kinds of boundary conditions can be combined together so that we can have, *e.g.*, a outer boundary fixed in space and a moving piston at the inner boundary over vice versa. However, we have to take care of the physical situation: We cannot specify for example an outflow boundary condition for the physical equations and force the grid to move in a Lagrangian way (see also Sect. 3.11).

4.6 Grid Motion

As seen in the adaptive RHD-equations (Sect. 3.6) the grid motion enters through the advection terms and the transport of all physical quantities across the cell boundaries scales with $u^{\rm rel}$. Hence, the overall motion of individual grid points should be limited to avoid advection errors and in some cases the grid adaption can be optimized by a temporal smoothing. The temporal grid time scale τ_g is then related to a physical time scale typical for the problem, e.g., $\tau_g \simeq \tau_{\rm dyn}$. However, to allow a readjustment of grid points the grid scale τ_g must be larger than the smallest physical time scale relevant for the physical system.

We can restrict the temporal variation by introducing a local description of the damping, e.g., like $\tau_g = \tau_g(r)$ if many orders of magnitude have to be covered during the simulation. Due to the diffusive nature of the spatial smoothing in Eq. (4.9) we have formally an infinite propagation speed. Any new feature developing somewhere in the flow causes all grid points to move towards this phenomenon and increase there the local resolution. It is therefore possible that small motions near the center can result in rather large motions at the outer boundary, in particular for the case of a logarithmic grid as defined through $\mathcal{X}_l \propto x$. Accordingly, the variation of $u^{\rm rel}$ scales in the same manner, yielding large advection flows without any transport of physical material just by the displacement of grid points on the physical background. Such situations can partially be cured through the aforementioned scaling with $\tau_g = \tau_g(r)$. Nevertheless, a better handling of these rapid grid motions remains to be understood.

4.7 Remarks on the Grid Equation

The grid equation is always solved simultaneously with the physical equations, so the location of an individual grid point is not known a priori but only after a successful solution of the coupled system. At each time step a new mapping between the physical space and the space of indices is constructed.

We have outlined the basic properties of an adaptive grid but each specific problem requires the appropriate definitions of the point concentration n, the resolution $\mathcal{R}$ together with the physical quantities to be traced (physical variable, scaling and weighting). In addition we have to specify the spatial tolerance by $\alpha_g \simeq 2$ and the temporal variation through τ_g related to typical time scales. Although the number of the parameters entering the grid equation seems numerous their action is well defined and will be demonstrated in the subsequent simple problems for which also analytical solutions exist.

4.8 First Example: Simple Test Function

To begin with, it is quite useful to see how a single test function like

$$f(x) = \frac{1}{2}\left[1 + \tanh(a(x - b))\right] \exp\left[-((x - b)/c)^2\right] \tag{4.18}$$

is resolved on the adaptive grid. The function contains a steep gradient defined through the hyperbolic tangent with a scale length $a^{-1} \simeq 10^{-3}$ as well as a smooth variation contributed by a Gaussian term $c = 0.2$ both centered around $b = 0.4$ on the unit interval $[0, 1]$. Such structures are frequently encountered in RHD computations since the thickness of shock waves is usually set by the artificial viscosity length scale (3.35) to 10^{-3} of the local radius. For the purpose of demonstration $f(x)$ is normalized to unity as seen in Fig. 4.1a. Therefore, further scaling is not necessary and the adaptive grid with $N = 100$ points clusters near the steep gradients. In Fig. 4.1b the function (4.18) is plotted against the grid index i to illustrate how the spatial smoothing spreads the gradients within the grid. The ratio of the point concentration (4.1) has to be bound between $\alpha_g/(\alpha_g + 1) = 2/3$ and $(\alpha_g + 1)/\alpha_g = 3/2$ as shown by the two dotted lines in Fig. 4.1c. The resolution achievable can be inferred from Fig. 4.1d where the dashed line marks the resolution on a equidistant grid with $n_i = 10^2$. The peak resolution corresponds to an equidistant grid with about 15 000 points.

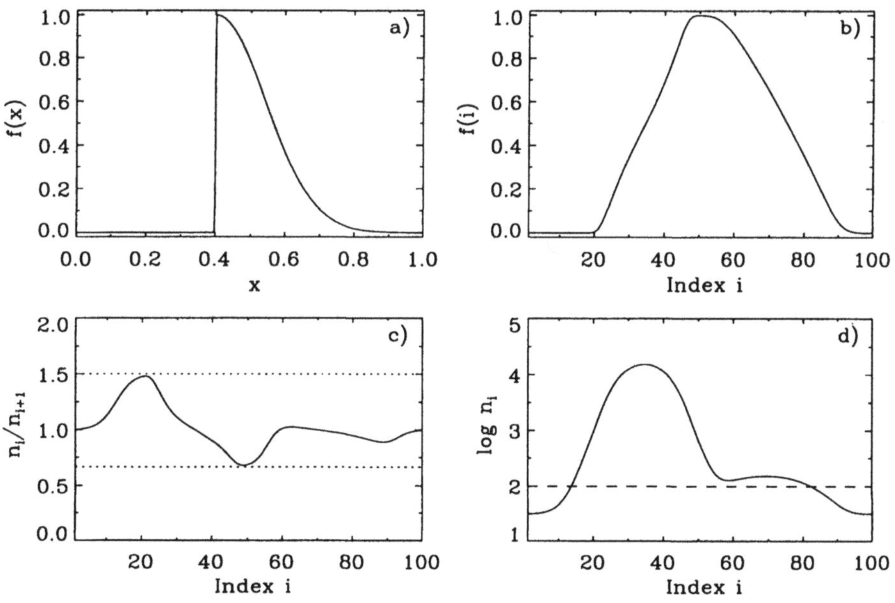

Fig. 4.1. A simple test function $f(x)$ in the physical space x in panel a) and on the adaptive grid as a function of the index i in panel b); the ratio of the point concentration n_i/n_{i+1} as well as n_i illustrate the grid distribution used to resolve the steep gradients

At the boundaries the conditions avoid gradients in the point concentration and the values of n_i agree on both sides indicating that the total number of grid points is sufficient to resolve all structures. A rough estimate on the number of grid points necessary to cover m orders of magnitude is given through

$$m\frac{\ln(10)}{\ln(1+1/\alpha_g)} \approx 2.3\, m\, \alpha_g \,, \tag{4.19}$$

since the grid can concentrate at most by the factor $1 + 1/\alpha_g$ per computational cell.

The adaptive grid presented in Fig. 4.1 is obtained by relaxation of the initially equidistant points to their final positions and when $f(x)$ is given by an analytical expression like (4.18) only the solution of a pentadiagonal matrix is involved during the iteration. More complicated grid distributions for starting RHD computations are discussed in Sect. 4.10.

4.9 Second Example: Shock Tube Problem

Since the review by Sod (1978) on numerical methods for pure hydrodynamics the so-called shock tube problem is used to illustrate the properties of new numerical techniques. The initial conditions are simple but discontinuous and for early phase analytical solutions are obtainable from Riemann's theory. All variables are normalized and defined on the unit interval $[0, 1]$ where at $x = 0.5$ an initial discontinuity separates a hot, high density gas ($\rho = 1$, $P = 1$) from a cold low-pressure medium ($\rho = 0.125$, $P = 0.1$). The gas is initially at rest and satisfies an equation of state of a perfect gas with $\gamma = 7/5$. At both ends a reflecting boundary is imposed yielding a number of nonlinear waves such as contact discontinuities, shock reflections, shock merging, reflection of a rarefaction wave, as well as the interaction of a rarefaction with a shock wave.

Since all variables are already normalized and the problem is stated in plane geometry we can take a very simple choice. However, the performance of the grid depends on the value of the temporal smoothing τ_g since we have to deal with reflections of waves at the boundaries. Taking $N = 100$ grid points we expect a mean spacing of $\Delta x \simeq 10^{-2}$ and peak velocities around $u \simeq 1$. From this two numbers we estimate a typical flow time across one cells of $\tau \simeq 10^{-2}$ and therefore as a first guess we can take a $\tau_g = 10^{-3}$, *i.e.*, ten times smaller than our physical time scale τ to allow an almost freely moving grid. As demonstrated in Figs. 4.4 and 4.5 the choice of τ_g is not so critical as long as τ_g is shorter than the physical time scales. For clarity the "standard" grid parameter for the shock tube problem are repeated as follows

$$\begin{aligned} f_1 &= \rho\,, & F_1 &= 1\,, & w_1 &= 1\,,\\ f_2 &= e\,, & F_2 &= 1\,, & w_2 &= 1\,,\\ \mathcal{X}_l &= 1\,, & \alpha_g &= 2\,, & \tau_g &= 10^{-3}\,. \end{aligned}$$

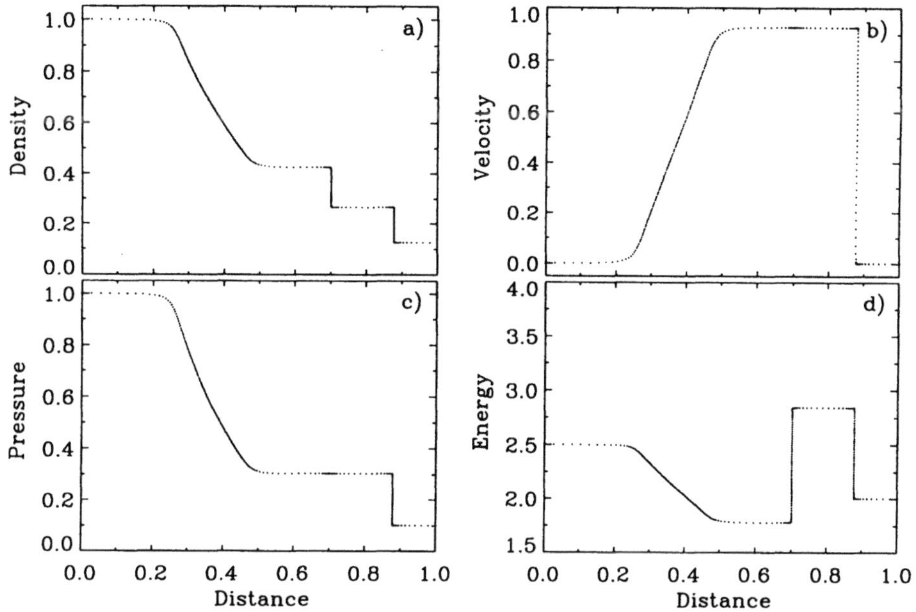

Fig. 4.2. Shock tube problem at $t = 0.23$ calculated with $N = 300$, donor-cell advection and $\tau_g = 10^{-3}$

The physical solution consists of two nonlinear waves travelling to the right and a rarefaction wave propagating to the left away from the initial discontinuity at $x = 0.5$. Figure 4.2 plots the variables [a) density, b) velocity, c) pressure and d) internal specific energy] at $t = 0.23$ before the first interaction at the reflecting boundary and the three different waves can be clearly distinguished. Going from left to right we see the rarefaction wave, followed by a contact discontinuity and the shock wave at the rightmost position. Although this solution has been calculated with a very simple first-order donor-cell scheme (*cf.*, Sect. 3.9) the adaptive grid is almost "sitting" on the structures keeping all advection errors small. The length scale of the artificial viscosity is 10^{-3} and we find a shock thickness of this order without any oscillations. Due to the crude nature of our difference scheme the only deviations from the analytical solution occur at the edges of the rarefaction wave. However, despite of this nice-looking solution any high order advection scheme suitable for an adaptive grid is preferable since the numerical shortcomings of the discretization scheme cannot be compensated by an adaptive grid, in particular for higher dimensions where the number of cells is more limited than in these simple problems.

In Fig. 4.3 the paths of individual grid points are plotted and the development of the three waves can be followed by the increased density of the lines where the expansion wave is less pronounced than the remaining features. At $t = 0.29$ we see the shock reflection occurring at the right boundary and

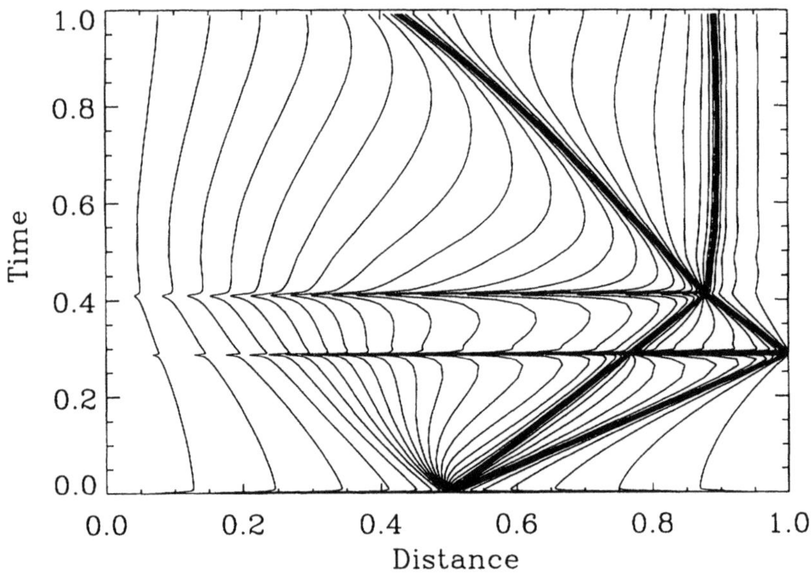

Fig. 4.3. The adaptive grid for $N = 100$ points in the shock tube problem with $\alpha_g = 2$ and $\tau_g = 10^{-3}$. Every second grid point is plotted and the different waves appear as contour line concentrations. The horizontal features are caused by the shock reflection at the right boundary ($t = 0.29$) and the interaction of the shock wave with the contact discontinuity around $t = 0.41$ (see text for more details)

during this time the shock becomes invisible for the grid so all points move away to resolve the remaining features. As soon as the shock leaves the wall the grid rushes back to pick up the variations in density and energy as forced by the desired resolution. The same situation happens at $t = 0.41$ where the shock wave meets the contact discontinuity and suddenly two structures can be resolved on the same fine grid. Hence, the points start to spread out and are pulled in again as the two waves separate later in time. This grid motion is clearly noticeable as the horizontal features in the (x, t)-plot and their time scale is controlled by τ_g. As stated before at such points the quality of the advection scheme determines the overall errors. In principle, it should be possible to avoid such rapid grid motions around the interaction of discontinuities. Up to now only the paper of Winkler et al. (1985) dealt with a so-called asymmetric time-filtering in the context of their complex formulation of another adaptive grid technique where the grid motion posses some "memory". Although the grid motion is disturbed by these interaction regions we have not encountered any problems during these phases. The whole evolution requires about 100 time steps and due to the implicit nature of the numerical technique the typical Courant-numbers (Eq. (3.10)) are in the range of $N_{\mathrm{CFL}} \simeq 10^2$–$10^3$.

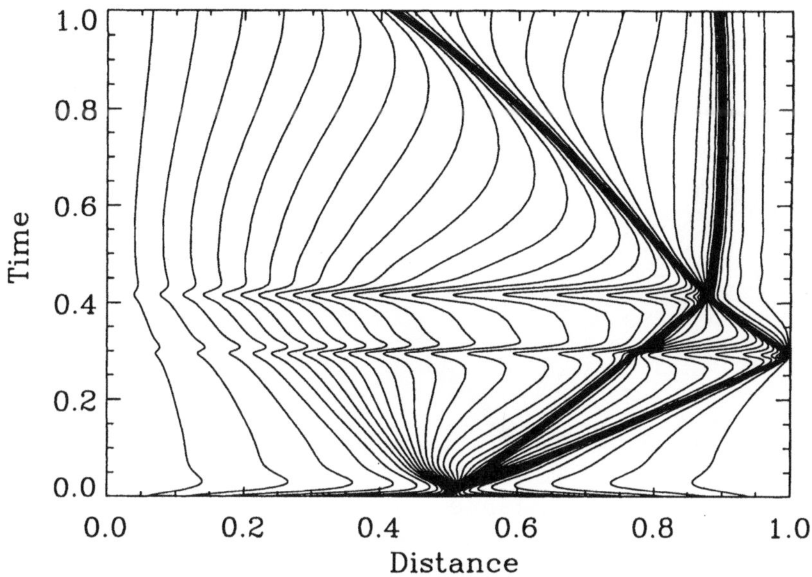

Fig. 4.4. Same (x, t)-plot as Fig. 4.3 but $\tau_g = 10^{-2}$ to demonstrate the robustness of the grid equation (see text for more details)

To study the effect of the temporal smoothing the grid parameter is increased to $\tau_g = 10^{-2}$, the order of our typical interaction time scale. The corresponding (x, t)-plot of the moving grid points is presented in Fig. 4.4, where we see how τ_g controls the grid point motion during these interaction time scales around $t = 0.29$ and $t = 0.41$. The grid points can still trace the wave propagation but looking closely to the wave interaction we observe a small deviation in the propagation speed before and after the interaction, in particular for the contact discontinuity. This is caused by the grid points which cannot come back fast enough to catch the steep density and energy gradients. The larger value of τ_g mainly affects the long term evolution of the contact discontinuity since the shock wave can steepen itself depending on the length scaled defined by the artificial viscosity.

Figure 4.5 illustrates even more dramatically the action of temporal damping with $\tau_g = 10^{-1}$. This value is clearly too large for the shock tube problem and the space-time diagram reveals that the grid point motion can only partly follow the waves. Nevertheless, the numerical solution as plotted in Fig. 4.6 shows some basic properties of the physical solution but the discontinuity as well as the shock wave are strongly smoothed and most of the grid points are still around the initial discontinuity at $x = 0.5$. At time $t = 0.29$ we can barely see the shock reflection at the right boundary.

As seen from the solution in Fig. 4.6 the coarseness of the grid leads to some small post-shock oscillation because the length scale of the artificial viscosity of 10^{-3} is too small to broaden the shock sufficiently for typical

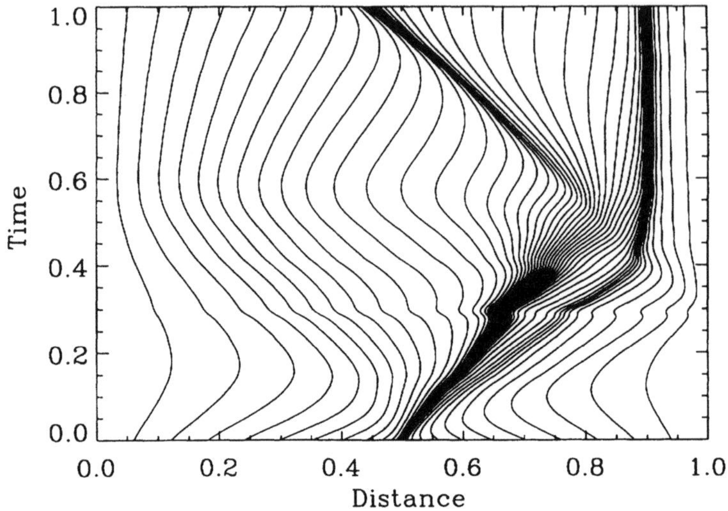

Fig. 4.5. Same (x,t)-plot as Fig. 4.3 and Fig. 4.4 but $\tau_g = 10^{-1}$ being too large for the problem. The points are not tracing the features although the numerical solution plotted in Fig. 4.6 still resembles the physical solution

cell sizes. In cases with smaller temporal parameters τ_g, the thickness of the shock is determined by the viscosity length scale and the grid points are free to cluster as dense as needed to resolve this jump.

From the previous figures and the discussion we can conclude that τ_g must be as small as possible to allow an optimal grid distribution and grid motion. Plotting the numerical solution obtained on an adaptive grid, it becomes evident whether the grid points have been able to follow the structures. In most cases the computations run smoothly without any time constant, *i.e.*, $\tau_g = 0$, in particular for all kinds of expanding flows. The choice for the grid parameter is usually rather easy and values of $g_i = 1$ are in almost all cases sufficient to resolve the structures. In the case of large differences or linear versus logarithmic variations of the physical variables the smaller limits can be more weighted than the broad variations. Usually, all weightings are between $1 \leq g_i \leq 5$ for normalized variables. Although the adaptive grid is quite flexible one should avoid to load too many distribution functions. From numerical experience we can state that in general two or three functions such as density, energy and/or opacity are enough to obtain a spatially well resolved computation. However, collapse calculations like the formation of protostars are more complicated to perform. This will be discussed in some detail in Sect. 6.3.

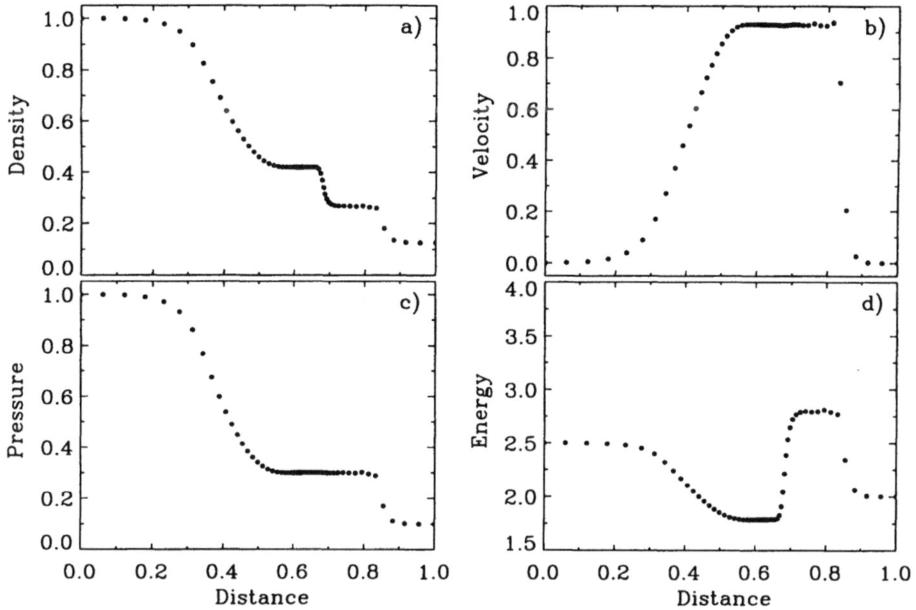

Fig. 4.6. Shock tube problem at $t = 0.20$ calculated with $N = 100$ grid points, donor-cell advection and an adaptive grid with a time constant $\tau_g = 10^{-1}$ yielding a non-optimal mesh distribution

4.10 Initial Grid Distributions

Using an adaptive grid, we need an initial grid distribution which is already a solution of the grid equation (4.12). Hence the physical variable have to be known at the locations of the grid points which are determined by an implicit description and whose positions depend on the physical variables themselves. In particular, the construction of appropriate hydrostatic initial models suited for our adaptive grid has caused problems (*cf.*, Dorfi & Gautschy 1989) and requires some care. The basic numerical strategy consists of three steps and the initial grid distribution is created by a pseudo-time evolution. First, an initial physical model is constructed either by taking an analytical solution or by integration of a simplified system of time-independent ordinary differential equations. Second, the positions at which the initial model is evaluated are, *e.g.*, equidistant or come from a self-adjusting integration step size of the integrator used for the ordinary differential equations. In general these point locations are not in agreement with the demands imposed by the desired resolution of our adaptive grid. Therefore we can keep the initial physical structure fixed in space and apply a linear interpolation to estimate the physical values at the positions in between the initial points if the points are sufficiently dense spaced. To summarize this step we need all M functions $f_j(x)$, $1 \leq j \leq M$ entering our desired resolution (Eq. (4.5)) to be given at all locations x of our computational domain. Third, we start the pseudo-time evolution and let grid points move to their final positions by solving a penta-diagonal matrix. Due to the fact that $f_l = f(x_l)$ are fixed at all points x_l during the pseudo-time evolution, the block-pentadiagonal matrix is reduced during this relaxation to a much simpler penta-diagonal matrix necessary to iterate the nonlinear grid equation. Note that the time constant τ_g applied in this pseudo-time evolution is not related to any physical scale or to a value of τ_g used in the subsequent dynamical calculations.

As an example let us consider the initialization of a stellar model. Starting with a hydrostatic initial model the full set of RHD equations is reduced to the static case. For this purpose all time derivatives as well as the gas velocity are set to zero leading to the following system of ordinary differential equations (ODEs)

$$\frac{\mathrm{d}m}{\mathrm{d}r} = 4\pi r^2 \rho \,, \tag{4.20}$$

$$\frac{\mathrm{d}P}{\mathrm{d}r} = -\frac{Gm\rho}{r^2} + \frac{4\pi}{c}\kappa_{\mathrm{H}}\rho H \,, \tag{4.21}$$

$$\frac{\mathrm{d}T}{\mathrm{d}r} = -\frac{\pi}{4\sigma}\frac{\kappa_{\mathrm{H}}\rho H}{fT^3} - \left(\frac{\mathrm{d}f}{\mathrm{d}r} + 3f - 1\right)T \,, \tag{4.22}$$

$$\frac{\mathrm{d}H}{\mathrm{d}r} = -\frac{2}{r}H \,. \tag{4.23}$$

Augmented by the appropriate constitutive relations the static RHD system is integrated numerically using a standard integrator for ODEs. Such an integration has to begin at the photosphere where total mass, effective temperature and luminosity are initially specified for the mass equation, the radiation momentum equation and the radiation energy equation. An estimate of photospheric gas pressure provides the initial value for the hydrostatic equilibrium and is given by the conditions of a grey atmosphere. This estimate has to be determined iteratively until a given outer radius is reached. At this point the boundary condition for the radiation field yields an external temperature. A comparison between the computed and the desired external temperature then improves the new value of photospheric pressure and so on. Normally this iteration converges within less than 20 steps. If a variable Eddington factor f has to be used an additional iteration becomes necessary where $f = 1/3$ can be taken for an initial guess. The resulting structure is improved by a new f-profile obtained by solving the static transfer equation (*cf.*, Sect. 3.16). The spatial derivative of f in Eq. (4.22) is computed by means of a one-dimensional rational spline interpolation. After 3–4 iterations the structure of the atmosphere remains unchanged within an relative accuracy of 10^{-5}. The solution of the static radiation transfer equation is also needed to specify the outer boundary conditions for the radiation field through $J = \bar{\mu}^{-1} H$ where $\bar{\mu}$ reflects the angular intensity distribution of the radiation field. Applying the Eddington approximation one obtains $\bar{\mu} = 0.5$

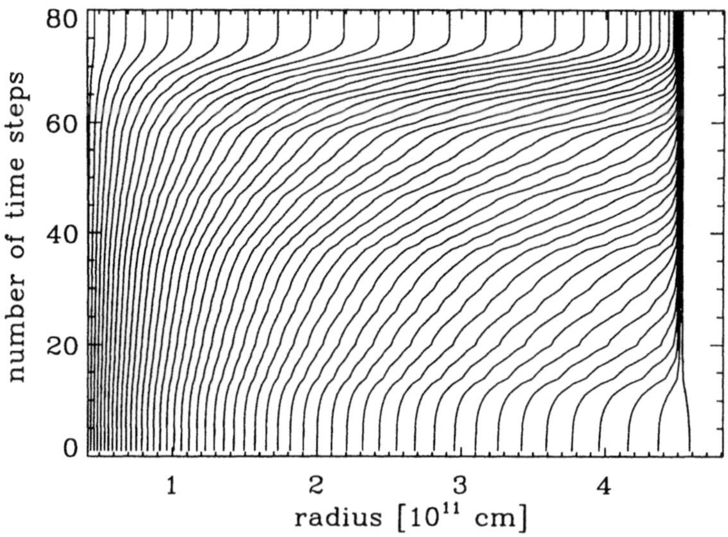

Fig. 4.7. The grid relaxation towards an initial RR Lyrae model. Every 4th grid point is plotted as a function of the number of time steps (see text for more details)

when neglecting a stellar atmosphere. In the case of extended atmospheres $\bar{\mu} = 1$ is the limit for free streaming of photons.

The subsequent relaxation of the grid points on the computed initial conditions is clarified by inspecting Fig. 4.8, where the position of every forth grid points is plotted during such a relaxation procedure. After integration of a hydrostatic RR Lyrae model we determine the physical variables to be tracked during the subsequent computations which in this case are the density by $\log(\rho)$, the internal energy by $\log(e)$, the opacity by $\log(\kappa)$ as well as the linear function of the thermodynamic index ∇_{ad}. According to our notation we have $M = 4$ and we must also specify the weighting for these quantities. We take $g_{1,2,3} = 1$ and since ∇_{ad} has a small absolute variation (however, these variations are very important to calculate the driving zones), we take $g_4 = 5$. The relaxation procedure is initiated by a small time step, *i.e.*, $\delta t \ll \tau_g$ and we use $\tau_g = 1$ which determines the pseudo-time. At the beginning, the time step adjusts to a value where the nonlinear iteration is still converging and will increase to values $\delta t \gg \tau_g$, when all grid points have reached their final position. In Fig. 4.7 this grid motion is illustrated by plotting the radial position of every 4th point out of $N = 200$ as a function of the number of time steps and we see how the points cluster around the steep gradients occurring around the ionization zone.

The typical point concentration reaches at the ionization zone values of $n = 10^5$ meaning that the relative resolution is 10^{-5} of the local radius. This grid feature is depicted in Fig. 4.8, where only the temporal evolution of the radial section is plotted around the ionization zone. According to the

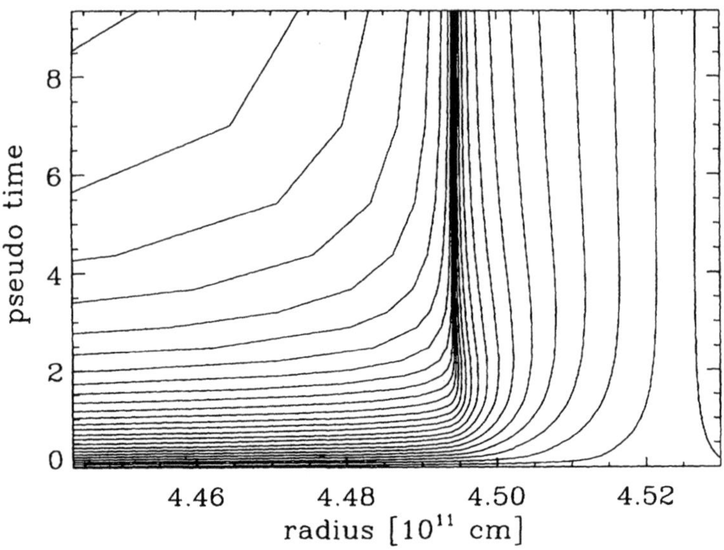

Fig. 4.8. The grid relaxation towards an initial RR Lyrae model around the ionization zone. Note the change in scale at the x-axis. Every 4th grid point is plotted as a function of the pseudo-time (see text for more details)

variation of the pseudo-time on the y-axis for $t \leq 1$ the rapid grid motion towards the "important" structures becomes clearly visible whereas at $t \geq 1$ the grid point paths begin to steepen indicating that the grid points are approaching their final locations. This division must happen around $t = \tau_g$ since for time steps $\delta t > \tau_g$ the grid becomes less and less influenced by the old time level. This plot nicely illustrates the grid motion without any crossing of grid lines as enforced by the derivation of the grid equation. Applying such a relaxation procedure we can always create initial grid distributions from equidistant mesh points.

The construction of an initial grid leads to some small interpolation errors in the physical variables as explained by the following example. If we take the mass conservation in spherical symmetry we recall the simple differential equation (4.20)

$$\frac{\mathrm{d}m}{\mathrm{d}r} = 4\pi\rho r^2 , \tag{4.24}$$

which can be integrated with any integrator for ordinary differential equations. However, the discrete integrated version of this equation reads in our staggered mesh notation at the grid point r_l

$$m_l - m_{l+1} = \frac{4\pi}{3}\rho_l \left(r_l^3 - r_{l+1}^3\right) . \tag{4.25}$$

Both versions agree to first order but the integrator solving Eq. (4.20) usually adopts some internal representation of the differential operator like, e.g., a

Runge–Kutta integration. It is evident that the mass integration based on Eq. (4.20) and our discrete finite volume version (4.25) can disagree by a small fraction which depends on the number of grid points, on the accuracy of the integrator as well as on the location of our grid points. For stellar initial models this total mass difference between the two versions is typically less than 10^{-3} and must decrease if N the total number of grid points used for the dynamical computations is increased. This discrepancy between the system of ordinary differential equations and the time-independent limit of the discrete RHD-equations is unavoidable.

To summarize the above discussion which focused on the mass equation: We have to be careful when constructing hydrostatic initial models since the hydrostatic equilibrium is determined by a particular numerical representation of the ODE-solver. In order to start with an initial model being strictly in hydrostatic equilibrium (as defined through our discrete set of equations) we therefore begin our computations by applying the full set of time-dependent RHD-equations on the initial models but take a large time step. This procedure can be used to check the quality of the initial model because if the model is stable and close enough to equilibrium the time step increases immediately and all remaining small velocities are damped out. On the other hand, if the initial model created is not well represented on the adaptive grid the time step remains small without reaching the time-asymptotic solution.

5. Further Computational Needs

5.1 Rational Spline Interpolation

Since our numerical method uses an implicit discretization (*cf.*, Sect. 3.4) we obtain a nonlinear system of algebraic equations that must be solved by some iteration. This is discussed in Sect. 5.3. During such an iteration not only the actual physical variables but also all functions like the EOS, opacity, Eddington factor have to be known, *i.e.*, $P = P(\rho, e)$, $T = T(\rho, e)$, *etc.* In particular, for any iteration scheme based on a Newton–Raphson procedure also the derivatives of these quantities with respect to the unknowns must be provided and have to be smooth to allow for a rapid convergence. In the case of analytic expressions this task can be simple but for most "real" astrophysical problems the needed relations are given as tables which have then to be interpolated together with smooth derivatives. Due to the usual spareness of the data points special care is necessary to avoid spurious oscillations around steep gradients.

From experiences with a large number of numerical computations with tables, I can strongly recommend the so-called rational splines as discussed in Späth (1983). In the case of 2-dimensional interpolation of a function $\mathcal{F}$ given at locations x_i and y_j the values in between are approximated by the cartesian product of two rational base functions $g(x)$ and $g(y)$, *i.e.*,

$$
\begin{aligned}
\mathcal{F}_{ij}(x,y) &= \sum_{k=1}^{4}\sum_{l=1}^{4} a_{ijkl}\, g_k(x)\, g_l(y)\ , \qquad (5.1)\\
g_k(x) &= a_{1k}\, t + a_{2k}\, s + a_{3k}\,\frac{t^3}{1+qs} + a_{4k}\,\frac{s^3}{1+qt}\ ,\\
\text{and}\quad t &= \frac{x-x_i}{x_{i+1}-x_i}\ , \quad s = 1-t\ , \quad x_i \le x \le x_{i+1}\ .
\end{aligned}
$$

In principle the weighting q can be different at every interval and there is no general adopted value for this parameter. Usually, one has to try several values of q, normally around $q = 1$ and one has to look at the variations of $\mathcal{F}$ but also at the smoothness of the derivatives $\partial\mathcal{F}/\partial x$ and $\partial\mathcal{F}\partial y$, which are needed during the iterations. At each grid point (x_i, y_j), 16 values of a_{ijkl} have to be calculated in advance and stored for each function to be used by the computation. Some hints for optimizing this interpolation are given in Sect. 5.2. Note, that $q = 0$ yields the standard cubic splines. A typical example of 1-dimensional interpolation through rational splines is illustrated by the cooling curve (Fig. 6.4) used in the SNR computations.

5.2 CPU-Time Requirements

The CPU-time necessary clearly depends on the problem and so I want to make only some general remarks for the present implicit method. The main CPU-time consuming operations are:

- the *opacity and EOS* and their derivatives with respect to the physical variables X either by interpolation or through direct calculations,
- the *Matrix inversion* which scales like $M^3\, N$ when M denotes the number of unknowns per grid point and N is the total number of grid points,
- the overall *Newton–Raphson iteration* which takes typically a few iterations to converge,
- the total *number of time steps* to cover the long term evolution of the astrophysical system but the time steps are usually much larger than the CFL-time steps of explicit computations,
- and in cases where the *Eddington factor* has to be obtained from the radiative transfer equation the total CPU-time is usually doubled.

Nevertheless, most 1-dimensional problems can be solved these days on fast workstations with CPU times ranging from minutes in the case of SNRs (Sect. 6.1), up to hours or days in the case of long-term pulsational calculations (Sect. 6.2) or dust driven winds (Sect. 6.4) and weeks for protostellar collapse calculations (Sect. 6.3), where also a convective energy transport has to be included.

5.3 Iteration Procedure and Matrix Inversion

By denoting all unknown variables at grid index l by

$$\boldsymbol{X} = X_l^{(n)} = (x_{1,l}^{(n)}, \dots, x_{M,l}^{(n)}) = (\rho_l^{(n)}, u_l^{(n)}, J_l^{(n)}, r_l^{(n)}, \dots)\,, \tag{5.2}$$

the system of the discretized nonlinear equations (Sect. 3.13) together with the boundary conditions (Sect. 3.11) can formally be written at all grid points r_l

$$\mathcal{G}_m(\langle X_l \rangle) = 0\,, \qquad 1 \leq m \leq M\,, \tag{5.3}$$

which includes time-centering (Sect. 3.5) between the new time level $t^{(n)}$ and the old one $t^{(o)}$. Clearly, we have $\delta t = t^{(n)} - t^{(o)}$. This nonlinear algebraic system (5.3) contains

$$\left.\begin{array}{ll} M & \text{equations (\# of unknowns per point)} \\ N & \text{grid points} \end{array}\right\} = M \times N \text{ unknowns}$$

and they are given *implicitly*, *i.e.*, they cannot be calculated by a more or less complicated extrapolation from the old time level.

Such a nonlinear system is solved iteratively with a Newton–Raphson iteration, *i.e.*, the system is expanded up to first order in a Taylor-series around the new estimate $\boldsymbol{X}^{(i+1)}$

$$\mathcal{G}(\boldsymbol{X}^{(i+1)}) = \mathcal{G}(\boldsymbol{X}^{(i)}) + \frac{\partial \mathcal{G}}{\partial \boldsymbol{X}^{(i)}}(\boldsymbol{X}^{(i+1)} - \boldsymbol{X}^{(i)}) = 0\,, \tag{5.4}$$

where the Newton-iteration assumes that $\boldsymbol{X}^{(i+1)}$ is already a solution of the nonlinear system, *i.e.*, $\mathcal{G}(\boldsymbol{X}^{(i+1)}) = 0$. The right part of this expansion (5.4) is used to calculate the correction

$$\delta \boldsymbol{X} = \boldsymbol{X}^{(i+1)} - \boldsymbol{X}^{(i)} \tag{5.5}$$

by inverting the linear system

$$\delta \boldsymbol{X} = -\left(\frac{\partial \mathcal{G}}{\partial \boldsymbol{X}}\right)^{-1} \mathcal{G}(\boldsymbol{X}^{(i)})\,, \tag{5.6}$$

where $\dfrac{\partial \mathcal{G}}{\partial \boldsymbol{X}}$ is the $(NM \times NM)$-Jacobi matrix of the algebraic system (5.3).

This procedure is repeated until

$$\min_{l,m} \frac{|\delta X_{m,l}|}{|X_{m,l}| + \eta_{m,l}} \leq \varepsilon \tag{5.7}$$

is reached for all variables and indices. We introduced $\eta_{m,l}$ to specify a lower boundary for variables which can become zero, *e.g.*, for the velocity a fraction of the sound velocity can be used to set an upper iteration accuracy, *i.e.*, $\eta_{\text{vel},l} = 10^{-2}\, c_{\text{s},l}$. In most cases the Newton–Raphson iteration converges in less than 5 cycles to an accuracy of $\varepsilon = 10^{-6}$. In principle, the Jacobi matrix is a $(NM \times NM)$-matrix but most entries are zero due to the 5-point stencil adopted for the discretization scheme.

Since system (5.6) must be solved iteratively it can happen that no convergence is found. Assuming no coding error in the vast number of derivatives to be calculated, either the inversion procedure can become ill-conditioned or the time step is too large yielding starting values for the iteration too far away from the solution at the old lime level, a case being discussed in Sect. 3.5.

5.4 Structure of the Jacobi Matrix

Since the nonlinear algebraic system (5.3) is generated by the discretization rules the corresponding $(NM{\times}NM)$-Jacobi matrix must reveal the discretization scheme having a very specific structure. The matrix is mostly empty since only the variables at the neighbouring grid points $(l-2, l-1, l, l+1, l+2)$ enter the discrete equations. In particular, the matrix has a ***block penta-diagonal*** form since a each grid index $1 \le l \le N$ one gets rows of 5 $(M{\times}M)$-submatrices

$$\begin{pmatrix} \ddots & \ddots & \ddots & \ddots & \ddots & \ddots & & \\ \cdots\ 0 & \dfrac{\partial \mathcal{G}_{m,l}}{\partial x_{m,l-2}} & \dfrac{\partial \mathcal{G}_{m,l}}{\partial x_{m,l-1}} & \dfrac{\partial \mathcal{G}_{m,l}}{\partial x_{m,l}} & \dfrac{\partial \mathcal{G}_{m,l}}{\partial x_{m,l+1}} & \dfrac{\partial \mathcal{G}_{m,l}}{\partial x_{m,l+2}} & 0\ \cdots \\ & \ddots & \ddots & \ddots & \ddots & \ddots & \ddots \end{pmatrix} \tag{5.8}$$

and, e.g., one to of these $(M \times M)$-submatrices looks like,

$$\frac{\partial \mathcal{G}_{m,l}}{\partial x_{m,l-1}} = \begin{pmatrix} \dfrac{\partial \mathcal{G}_{1,l}}{\partial x_{1,l-1}} & \dfrac{\partial \mathcal{G}_{1,l}}{\partial x_{2,l-1}} & \dfrac{\partial \mathcal{G}_{1,l}}{\partial x_{3,l-1}} & \cdots \\ \dfrac{\partial \mathcal{G}_{2,l}}{\partial x_{1,l-1}} & \dfrac{\partial \mathcal{G}_{2,l}}{\partial x_{2,l-1}} & \dfrac{\partial \mathcal{G}_{2,l}}{\partial x_{3,l-1}} & \cdots \\ \vdots & \vdots & \vdots & \ddots \end{pmatrix} = A_{l,l-1} \,.$$

Adopting this last abbreviation, $A_{l,l-1}$, and recalling the notation (5.5) for the corrections δX_l to be added to the values X_l

$$\delta X_l = (X_{1,l}^{(i+1)} - X_{1,l}^{(i)}, X_{2,l}^{(i+1)} - X_{2,l}^{(i)}, \cdots) \,, \tag{5.9}$$

the system (5.4) can be written like

$$\begin{aligned} A_{l,l-2}\delta X_{l-2} \ &+\ A_{l,l-1}\delta X_{l-1} \ +\ A_{l,l}\delta X_l \ + \\ &+\ A_{l,l+1}\delta X_{l+1} \ +\ A_{l,l+2}\delta X_{l+2} \ =\ -\mathcal{G}(X_l^{(i)}) \,, \end{aligned} \tag{5.10}$$

revealing the penta-diagonal structure. Note that all A's are $(M{\times}M)$-matrices and all X's are M-vectors. The inner boundary conditions $l = N$ imply

$$A_{N,N+1} = A_{N,N+2} = 0 \,, \tag{5.11}$$

whereas the outer boundary $l = 1$ yields

$$A_{1,-1} = A_{1,0} = 0 \, . \tag{5.12}$$

This special structure allows an easy inversion of the Jacobi matrix (5.6) by means of a recursion which leads to an upper triangular block system of the form

$$\delta X_l = U_l \, \delta X_{l+1} + V_l \, \delta X_{l+2} + w_l \, , \tag{5.13}$$

where after the forward substitution in (5.10) the following expressions appear:

$$Y_l = \left([A_{l,l-2}U_{l-2} + A_{l,l-1}]\, U_{l-1} + A_{l,l-2}V_{l-2} + A_{l,l} \right)^{-1} , \tag{5.14}$$

$$U_l = Y_l \left[[A_{l,l-2}U_{l-2} + A_{l,l-1}]\, V_{l-1} + A_{l,l+1} \right] , \tag{5.15}$$

$$V_l = Y_l \, A_{l,l+2} \, , \tag{5.16}$$

$$w_l = Y_l \left[-\mathcal{G}(X_l^{(i)}) - [A_{l,l-2}U_{l-2} + A_{l,l-1}]\, w_{l-1} - A_{l,l-2}w_{l-2} \right] . \tag{5.17}$$

From the last expressions it becomes clear that such a triangular block form is obtained by inversion of the $(M \times M)$-matrix (5.14) in each row of the matrix (5.8). Starting with $l = 1$ due to the aforementioned boundary conditions (5.12) one gets $Y_1 = A_{1,1}^{-1}$ and U_1, V_1 and w_1 can be calculated. Proceeding until $l = N$ is reached the boundary conditions (5.11) leads to $U_N =$, $V_N = 0$ and $V_{N-1} = 0$ and $\delta X_N = w_N$ follows immediately. Now the back substitution in recursion (5.13) is performed to compute all necessary corrections $\delta \boldsymbol{X}$ to the physical variables $\boldsymbol{X}$. These iterations are repeated until the desired accuracy (5.7) is achieved, or until it becomes evident that the iteration will diverge because the new solution is too far from the first estimate.

The numerical inversion of large matrices can cause problems, in particular when wide ranges in the variables are expected. As stated in the introduction almost all astrophysical objects cover many orders of magnitude and it is therefore necessary to normalize the equations before the inversion of the Jacobi matrix is performed. From numerical experience I can recommend the following normalization factors $G_{m,l}$ for the discrete set of the six RHD equations (see Sect. 3.13.):

$$G_{m,l} = \left(y_{m,l}^{(o)} \Delta V_l^{(o)} \right)^{-1} , \tag{5.18}$$

with

$$y_{m,l} = \left(\rho_l ,\, \rho_l \max(|u_l|, c_{s,l}),\, \rho_l e_l ,\, m_l ,\, J_l ,\, \max(|H_l|, J_l) \right) . \tag{5.19}$$

Hence, at each grid point all RHD equations are divided by the "conserved" quantity at the old time level or when the velocity u_l or the radiative flux H_l can become zero or negative by some physical limit. After such a normalization the inversion behaves more stable and the temporal variation of these "conserved" quantities (5.19) will be around one. The grid equation is already normalized through the physical scaling factors.

5.5 Time-Step Control

After a successful iteration the relative variation between the two time levels is calculated at each grid point r_l for all M variables $(x_{1,l}, x_{2,l}, \ldots, x_{M,l})$

$$s_m = \max_l s_{m,l} = \frac{|\delta X_{m,l}|}{|X^{(n)}_{m,l}| + \eta_{m,l}} . \tag{5.20}$$

These values s_m are used to determine the next time step δt from the old time step $\delta t^{(o)}$ by the simple condition

$$\delta t = \begin{cases} \frac{1}{2}\delta t^{(o)} , & \text{if} \qquad\qquad s_m > s_m^{\max} \\ \delta t^{(o)} , & \text{if} \quad \frac{1}{2}s_m^{\max} \leq s_m \leq s_m^{\max} \\ \frac{3}{2}\delta t^{(o)} , & \text{if} \qquad\qquad s_m < \frac{1}{2}s_m^{\max} , \end{cases} \tag{5.21}$$

where $s_m^{\max}$ is the maximum allowed variation in the variable x_m. Typical values are $s_m^{\max} = 10^{-1}$. As inferred from conditions (5.21) the next time step is reduced by a factor of 2 in case of larger variations but only increased by a factor of 1.5 in case of smaller variations than $s_m^{\max}/2$. Such a hysteresis in the time step control avoids oscillations in time step sizes. Of course, in case that the Newton–Raphson iteration diverges, the time step has to be decreased until convergence is achieved.

The total number of iteration can be reduced usually by one if the starting value $\boldsymbol{X}^{(1)}$ of the Newton-iteration (5.6) is obtained through an extrapolation from the last two time levels, *i.e.*,

$$\boldsymbol{X}^{(1)} = \boldsymbol{X}^{(o)} + \frac{\delta t}{\delta t^{(o)}} \left(\boldsymbol{X}^{(o)} - \boldsymbol{X}^{(oo)}\right) , \tag{5.22}$$

where the ratio of the new estimated time step (5.21) to the old time step $\delta t^{(o)}$ enters and $\boldsymbol{X}^{(oo)}$ denotes the solution before last.

6. Computational Examples

6.1 Evolution of Supernova Remnants (SNRs)

The hydrodynamical evolution of Supernova Remnants (SNRs) may serve as an astrophysical example where a number of tests can be performed on the quality of the numerical computations. We have to deal with explosive motions with high Mach numbers and for later times the self-similarity of the pressure driven expansion allows a direct comparison with analytical solutions (Taylor 1950, Sedov 1954). At the late stages of the SNR evolution radiative cooling of the expanding shell becomes important and the high compression ratios occurring then in the dense shell demand locally a high resolution to follow their development. Finally, since SNRs are considered to be the most likely sources of galactic cosmic rays (*e.g.*, Axford 1981) the contribution of these high energy particles to the overall dynamics of the interstellar medium can only be determined by numerical computations.

The computations presented are performed with $N = 300$ grid points distributed between $r_N = 10^{11}$ cm and $r_1 = 10^{21}$ cm. The adaptive grid contains the density as well as the internal gas energy, the time smoothing τ_g can be set to zero but the spatial scaling has to be logarithmic and the physical scales are defined through the harmonic mean to prevent the grid from "tearing" at the discontinuities, *i.e.*,

$$\begin{aligned} f_{1,l} &= \rho_l\,, & F_{1,l}^{-1} &= \left(f_{1,l+1}^{-1} + f_{1,l}^{-1}\right)/2\,, & w_1 &= 1\,, \\ f_{2,l} &= e_l\,, & F_{2,l}^{-1} &= \left(f_{2,l+1}^{-1} + f_{2,l}^{-1}\right)/2\,, & w_2 &= 1\,, \\ \mathcal{X}_l &= \overline{r_l}\,, & \alpha_g &= 2\,, & \tau_g &= 0\,. \end{aligned}$$

The energy of the supernova of $E_{\mathrm{SN}} = 8 \times 10^{50}$ erg is deposited purely as thermal energy within a radius of 10^{12} cm. The density structure of the progenitor star is very simple; a constant density up to 10^{14} cm and then an exponential decrease towards the external density of $n_{\mathrm{ext}} = 5\ \mathrm{cm}^{-3}$. The external material is at rest and the pressure is given by the parameters of a rather warm interstellar medium of $T_{\mathrm{ext}} = 20\,000$ K. The ejected mass $M_{\mathrm{ej}} = 3\,M_\odot$ fixes the mean density of this simple progenitor star but the detailed structure of the progenitor star is not important for the further evolution of the SNR. This particular example includes also the acceleration of energetic particles at shock waves as discussed later and the external cosmic ray pressure is set equal to the external gas pressure, *i.e.*, $P_{\mathrm{c,ext}} = P_{\mathrm{g,ext}}$. Note, that the models presented in the next figures are calculated including the acceleration of high energy particles at the shock waves. As outlined in Sect. 4.10, the initial grid distribution can be obtained from the pseudo-time evolution.

At the beginning the evolution of the remnant is described by a simple ballistic expansion where the external medium has almost no influence on the motion of the ejected material. Nevertheless, the flow configuration contains several interesting features. First, the ejected material is separated from the interstellar medium by a contact discontinuity. In front of this discontinuity we expect a strong shock wave propagating through the external medium. Since the expansion works against the interstellar pressure the ejected material slows down and the so-called *reverse shock* decelerates the SN ejecta although the whole structure is moving outwards with more than $10^4\ \mathrm{km\,s^{-1}}$.

Figure 6.1 shows such a flow configuration at one year after the explosion. In order to show the resolution of the adaptive grid every grid point is plotted in range between $0.5 \leq r/[10^{17}\ \mathrm{cm}] \leq 1.4$. The contact discontinuity is located almost at 10^{17} cm seen by the drop in the density (Fig. 6.1a). In front of it the external shock wave becomes visible a the velocity jump of $15\,000\ \mathrm{km\,s^{-1}}$ (Fig. 6.1b) with an increase of the gas pressure by more than 6 orders of magnitude (Fig. 6.1c), the corresponding Mach number is around

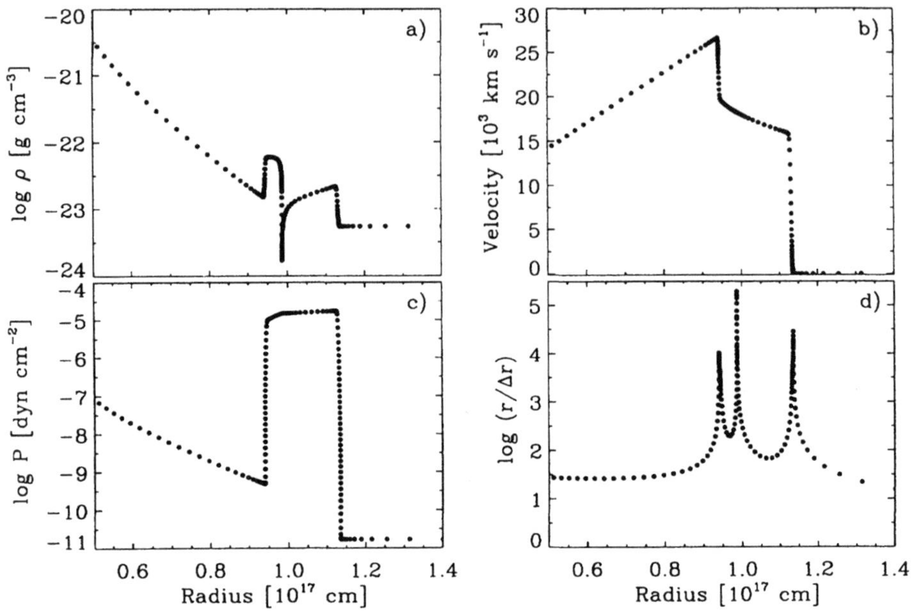

Fig. 6.1. A typical SNR shock configuration after one year, running into a constant interstellar medium

$M = 1000$. Behind this forward shock the spherical geometry causes the sharp decrease of the density towards the contact discontinuity. Inside the exploded star we have a linearly increasing velocity field up to the point where the reverse shock decelerates the expanding material as depicted in Fig. 6.1b and the Mach number of the reverse shock is $M = 140$. In Fig. 6.1d the relative spacing $\log[r/\Delta r]$ demonstrates the clustering of the grid points in the vicinity of steep gradients.

This evolutionary stage lasts until the ejected material becomes comparable to the swept-up interstellar material, *i.e.*,

$$R_{\text{sw}} = \left(\frac{3M_{\text{ej}}}{4\pi\rho_{\text{ext}}}\right)^{1/3} . \tag{6.1}$$

During the ballistic phase the supernova energy is purely kinetic energy and assuming a linear velocity field inside the ejecta we can define a typical ejection velocity through

$$v_{\text{ej}} = \sqrt{\frac{10E_{\text{SN}}}{3M_{\text{ej}}}} . \tag{6.2}$$

A typical sweep-up time scale is then given by

$$t_{\text{sw}} = \frac{R_{\text{sw}}}{v_{\text{ej}}} , \tag{6.3}$$

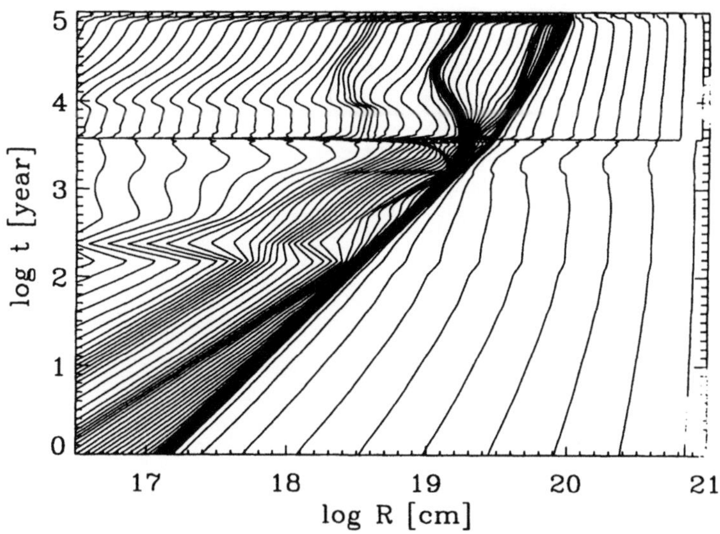

Fig. 6.2. Space-time diagram of the grid points during a SNR-evolution with $E_{\mathrm{SN}} = 8 \times 10^{50}$ erg exploding into a medium at rest with $n_{\mathrm{ext}} = 0.3\,\mathrm{cm}^{-3}$

which corresponds to about 3750 years for an external density of $n_{\mathrm{ext}} = 0.3\,\mathrm{cm}^{-3}$ and a SN-energy of $E_{\mathrm{SN}} = 8 \times 10^{50}$ erg. After this age the reverse shock starts to propagate inwards and re-heats the interior to high temperatures followed by the pressure-driven expansion which is known as the Sedov–Taylor phase. In Fig. 6.2 the space-time evolution of the grid points is plotted between 1 and 10^5 years in the range 10^{17} cm and 10^{21} cm. Up to the sweep-up time t_{sw} we see the ballistic expansion where all discontinuities are moving outwards together. The innermost line-concentration begins to move to the left and the horizontal line at $\log t = 3.57$ marks the reverse shock running towards the interior. Since the temporal smoothing τ_g is set to zero all grid points have to react to this motion immediately, in particular visible through the paths of grid points located in the external medium. Inside the forward shock the remaining density structures are resolved by the grid although they have no dynamical influence on the overall expansion. During the expansion phase as well as during the Sedov phase the adaptive grid allows large time steps. Since only every third time step is plotted in Fig. 6.2 some grid lines appear to have sharp corners. As also shown in Fig. 6.5 the evolution for times $t > t_{\mathrm{cool}}$ is strongly influenced by radiative cooling which causes a further rearrangement of the grid points after $\log t > 4.7$.

Figure 6.3 displays the location of the reverse and forward shock in a logarithmic diagram during the transition from the expansion phase to the pressure-driven Sedov phase. The estimate of t_{sw} from Eq. (6.3) is drawn as a vertical dashed line yielding almost an asymptotic line for the locations of the inward propagating reverse shock. During the following Sedov phase

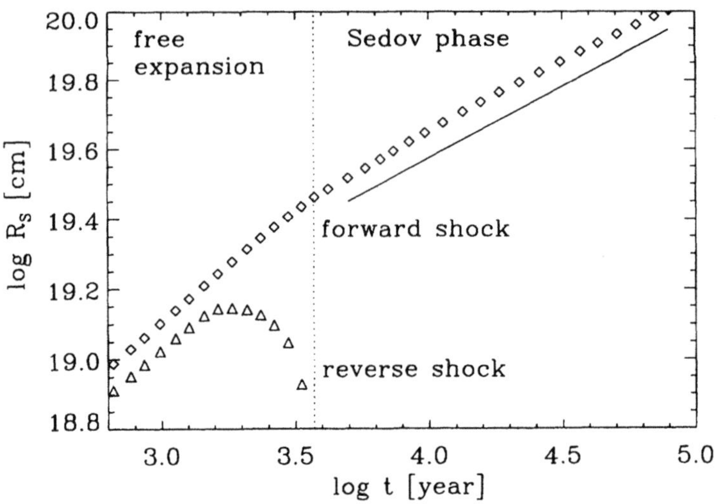

Fig. 6.3. Space-time diagram of the two SNR shock waves compared to a self-similar Sedov-solution (shifted straight line)

the motion obeys a self-similar solution with $R_s \propto t^{2/5}$ which is plotted for comparison as a shifted straight line. During such self-similar expansions the adaptive grid allows large time steps because the structures remain almost stationary on the moving mesh.

As mentioned before, the late SNR phases are dominated by radiative cooling and a typical *cooling curve* is plotted in Fig. 6.4. The symbols above the curve denote the chemical elements mainly contributing to the features seen in the curve. In particular, below 10^4 K the radiative cooling becomes very ineffective and the huge drop in cooling efficiency near 10^4 K works like a thermostat for H II-regions. Around 3×10^7 K another minimum is found because all elements are fully ionized and the plasma can therefore radiate only through free-free transitions (*i.e.*, bremsstrahlung) leading to a cooling law of $\Lambda(T) \propto T^{1/2}$. For many applications and analytical considerations Kahn's approximation of the cooling function in the range between $2 \times 10^5 \leq T/[\mathrm{K}] \leq 10^7$ (Kahn 1976) has been adopted which can be stated as a simple power law of the temperature

$$\Lambda(T) = 1.3 \times 10^{-19} T^{-1/2} \,\mathrm{erg\,cm^3 s^{-1}} \tag{6.4}$$

shown in Fig. 6.4 by the dashed straight line. Such a cooling curve is based on equilibrium conditions which are not fulfilled in many cases for a shock-heated gas. Nevertheless, as a first approximations this energy loss term can be used during the late stages of a SNR evolution. I depict this curve also to urge the use of a smooth interpolation scheme (see Sect. 5.1) since any further (numerical) oscillations will significantly reduce the convergence rate in the Newton–Raphson iteration.

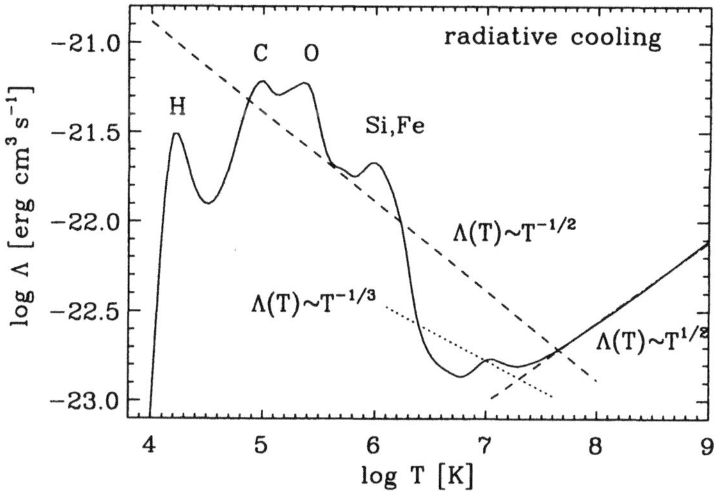

Fig. 6.4. The radiative cooling curve $\Lambda(T)$ plotted with three analytical approximations (dashed and dotted lines) used in the literature

The cooling phase in a SNR is characterized by the formation of a dense shell behind the compressed interstellar medium. Assuming, *e.g.*, Kahn's cooling law (6.4) the time and the radius where such a shell is formed can be estimated by (see, *e.g.*, McKee 1982)

$$R_{\rm shell} \simeq 20 \left(\frac{E_{\rm SN}}{10^{51}\,{\rm erg}}\right)^{0.29} n_0^{-0.41}\ {\rm pc} \tag{6.5}$$

and n_0 denotes the number density of the external medium in cm^{-3}. Due to the adiabatic expansion the temperature of the SNR decreases during the Sedov phase. Below 10^7 K the radiative cooling sets in triggered by the recombination of the heaviest elements. In Fig. 6.5 four snapshots of the SNR with $E_{\rm SN} = 8 \times 10^{50}$ erg and $n_0 = 5\,\mathrm{cm}^{-3}$ leading to $R_{\rm shell} = 9.7\,\mathrm{pc}$ are plotted at at 1.4×10^4, 3.6×10^4, 6.1×10^4 and 9.7×10^4 years. The development of such a dense shell (Fig. 6.5a) is exhibited and in good agreement with the analytical estimate of Eq. (6.5). The shock structure reveals an important reduction of the shock speed seen through the drop in the velocity (Fig. 6.5b). The leftmost curves correspond still to a Sedov-like structure as inferred from the small pressure variations inside the remnant (Fig. 6.5c). The second curve clearly exhibits the cooling region behind the radiative shock, where the post shock temperature can even drop below the up-stream temperature since the adiabatic expansion in the post shock region adds the radiative cooling (Fig. 6.5f). The very tenuous SNR interior is at least two orders of magnitude less dense than the post shock region and is therefore unable to cool since the radiative cooling scales with the particle density squared. As plotted in Fig. 6.5d the shock wave is accelerating energetic particles and this acceleration process depends on the overall compression ratio of the shock wave.

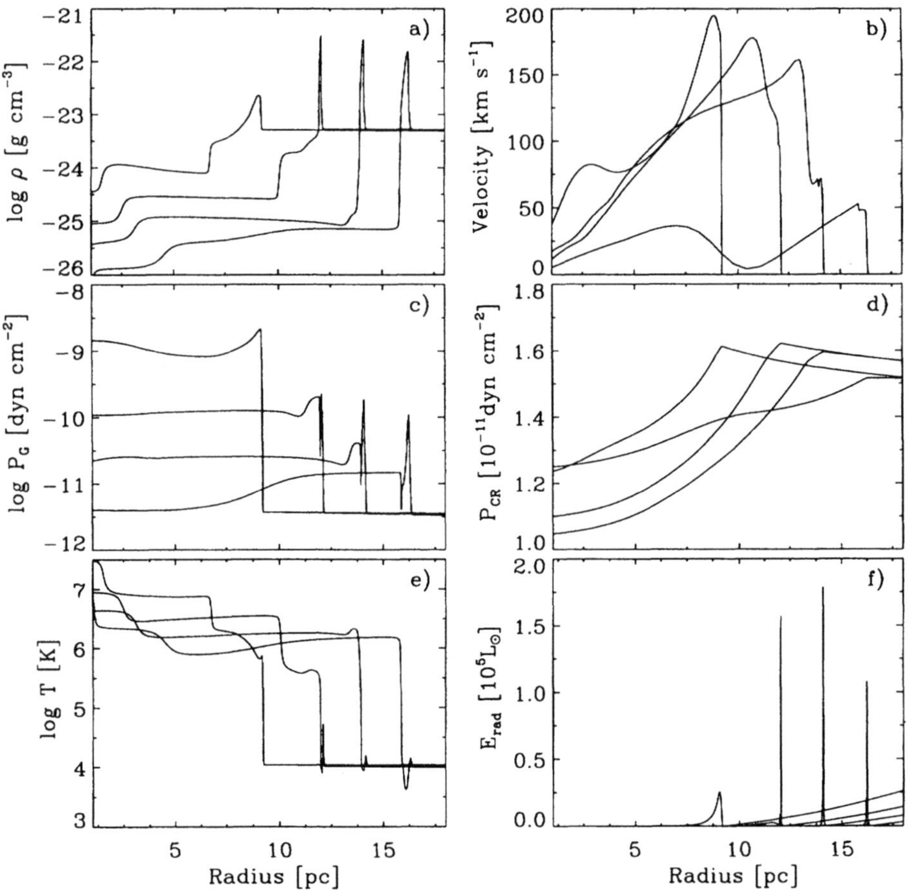

Fig. 6.5. Late SNR evolution dominated by radiative cooling (see text for more details)

The accelerated particles increase the cosmic-ray pressure P_{CR} at the shock wave and this gyro-resonant acceleration process can be characterized in the hydrodynamical limit by a diffusive process (*e.g.*, Drury 1983 for a comprehensive review on particle acceleration). Mainly two effects modify the SNR evolution at these late phases. First, these high-energy particles can also diffuse into the up-stream region of the shock front generating a precursor as visible in Fig. 6.5d where the incoming plasma is already decelerated. Second, the additional particle pressure prevents the cooling region from further collapsing. Note the linear scale on the particle pressure plot 6.5d. At later times, the temperature of the post shock region decreases further and the cooling becomes less effective as can be seen in the thickening of the dense post-shock shell. The thickness of the cooling zone is determined by the efficiency of the cooling and the maximum of the cooling curve $\Lambda(T)$ is around

10^5 K as inferred from Fig. 6.4. Since this radiative cooling is mainly observable in X-rays the details of this cooling region are important for comparison with observations. Any resolution error will change the fluxes because the energy radiated away scales like $\rho^2 \Lambda(T)$ and a poor resolution of the density structures influences the cooling and this post-shock pressure drop reacts immediately on the shock propagation. Hence, these later SNR phases depend strongly on the quality of the numerical scheme, in particular on the ability to trace very fine spatial structures. The radiative energy losses are depicted in Fig. 6.5, where the luminosity is plotted in units of $10^5\,L_\odot$. The emission is strongly peaked at the dense shell demanding a very fine resolution.

6.2 Nonlinear Stellar Pulsations

Since the first attempts to calculate nonlinear pulsation models it has become clear that the numerical representation of the driving zones is crucial for obtaining accurate results. The stellar body is divided into discrete shells and hence almost all computations have been performed on a Lagrangian grid which keeps the mass fixed to a computational cell (or grid index). Within the ionization zones the resolution varies during a pulsation cycle because the location of the ionizing and driving regions do not remain constant with respect to the Lagrangian mass coordinate (see, *e.g.*, the proceedings edited by Buchler (1990) for a detailed discussion of these issues). Due to the vast number of observational tests a number of long standing problems based on the Lagrangian numerical tradition have been emerged, *e.g.*, the numerical light curves are very "noisy", the amplitudes do not agree with observed ones, the damping due to large values of the artificial viscosity influences the results, the shape of light curves differs in the Fourier coefficients (see Eq. (6.6)), *etc.* The majority of these effects can be solved with adaptive methods although better tables for the opacity had an enormous impact on stellar pulsation theory (*e.g.*, reviews by Gautschy & Saio 1995,1996).

The inner boundary of the pulsation models is located above the nuclear active region, usually around a gas pressure of $\log P \simeq 12$ where the amplitude of the oscillations is practically zero. The outer boundary we specify within the stellar atmosphere, where the gas pressure is several orders of magnitude lower than the photospheric value. This procedure avoids artificial driving of the stellar pulsation. We start from an initially hydrostatic configuration and disturb the model by a very small random velocity field.

RR Lyrae stars are low mass ($M \simeq 0.5\,M_\odot$) variable stars with a relatively low metal content as they belong to Population II stars, *i.e.*, for the models discussed we used (X,Y,Z) = (0.7, 0.299, 0.001). These stars exhibit a narrow range of visual magnitudes and their periods range typically from about 0.2 days to 1.2 days. Depending on the amplitude, period, and the shape of the light curve, several different classes of pulsations have been introduced. The RRab subclass is characterized by a steep ascending branch with generally asymmetric light curves belonging to the fundamental mode

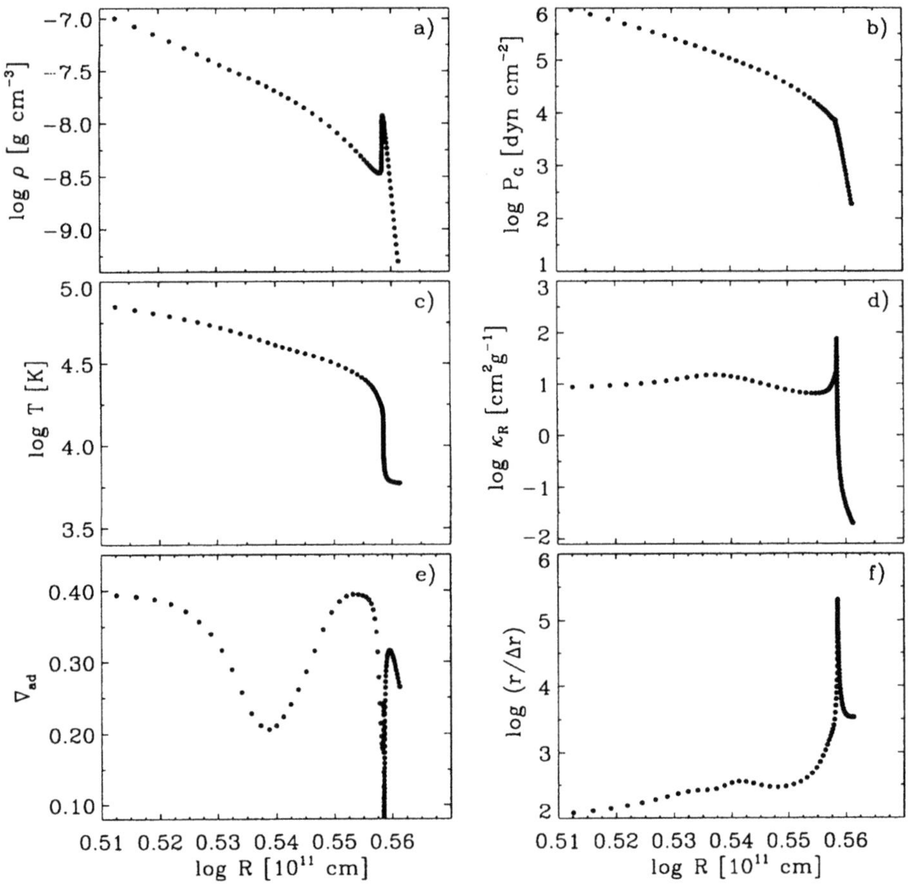

Fig. 6.6. RR Lyrae hydrostatic initial model

pulsation, whereas the RRc subclass exhibits almost sinusoidal light curves generated by a first overtone pulsation. The typical range of luminosities peaks at about $60\,L_\odot$ which makes these stars less suitable to estimate the distances of external galaxies. Hence, observational studies are restricted to our galaxy. Furthermore, it is well known that these stars undergo nonlinear pulsations in the fundamental mode, in the first overtone, or in both modes simultaneously. A large number of numerical computations have been carried out by various authors (*e.g.*, Stellingwerf 1975, Kovács & Buchler 1993, Bono & Stellingwerf 1995, Feuchtinger & Dorfi 1994, 1996, 1997) to explore the stability of RR Lyrae models. Since RR Lyrae stars are extensively studied objects from observational grounds as well as from theoretical modelling, any result based on a new development can be compared to a number of results available it the literature.

The outer radial structure of the hydrostatic initial model is plotted in Fig. 6.6 between 3.26×10^{11} cm and 3.72×10^{11} cm. Every grid point is shown individually to illustrate the good resolution of the physical quantities. In this case, the stellar parameters are given by $M = 0.65\,M_{\odot}$, $L = 52.5\,L_{\odot}$, and $T_{\rm eff} = 6800\,{\rm K}$. Focusing on the outer parts of the model, we clearly can identify the stellar atmosphere by the rapid spatial decrease of the density (Fig. 6.6a) and gas pressure (Fig. 6.6b). The temperature (Fig. 6.6c) remains constant within the atmosphere. Within this structure we see also the density inversion due to the opacity (Fig. 6.6d) which varies over 5 orders of magnitude. Note, that these are purely radiative models where no convective energy transport is included. Below the stellar photosphere are located the driving zones as demonstrated through the variations of the thermodynamic quantity $\nabla_{\rm ad} = \partial \ln T / \partial \ln P$, which clearly mirrors the stellar driving zones (Fig. 6.6e) and varies within the ionization zones (*e.g.*, Kippenhahn & Weigert 1990). It is therefore essential to distribute enough grid points in these regions to obtain an accurate description of the stellar pulsation. At a radius of about 3.45×10^{11} cm ($\log R = 0.54$ [10^{11} cm]) we find the He II-ionization zone. At 3.618×10^{11} cm ($\log R = 0.56$ [10^{11} cm]) the H-ionization zone is superimposed on the broader He I-ionization zone. The quantity in Fig. 6.6f displays the point concentration and reflects the clustering of the grid points near prominent physical features. Since the stellar pulsation is driven by these variations it is important to resolve the gradients of the physical quantities during the pulsational cycle.

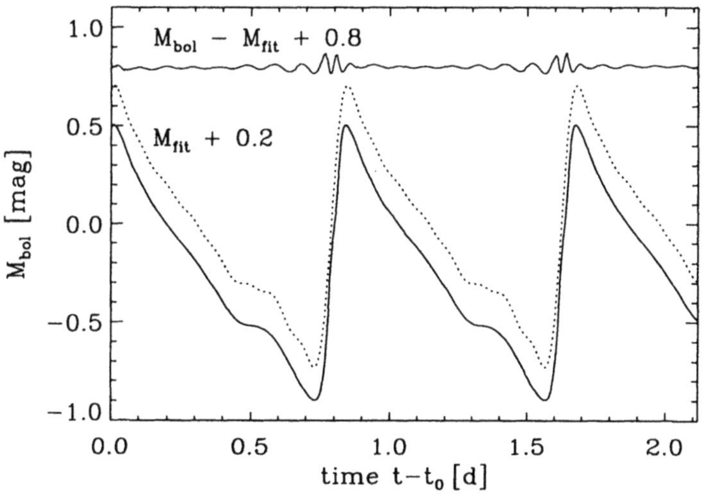

Fig. 6.7. Calculated RR Lyrae light curve in bolometric magnitudes $M_{\rm bol}$ for $M = 0.578\,M_{\odot}$, $L = 64.3\,L_{\odot}$ and $T_{\rm eff} = 6500\,{\rm K}$ (solid line) together with the Fourier fitted light curve shifted by 0.2 mag (dashed line) and the difference $M_{\rm bol} - M_{\rm fit}$ (thin solid line), shifted by 0.8 mag

I want to emphasize that inside the minimum of the HeII-ionization zone the computations use a Lagrangian grid to avoid the small advection errors always present in case of transport over cell boundaries. A more detailed discussion on this Lagrangian switching point is found in Sect 7.1. The following computations are done with $N = 200$ grid points resolving the density, the internal energy as well as ∇_{ad}, *i.e.*,

$$\begin{aligned}
f_{1,l} &= \rho_l \,, & F_{1,l}^{-1} &= \left(f_{1,l+1}^{-1} + f_{1,l}^{-1}\right)/2 \,, & w_1 &= 1 \,, \\
f_{2,l} &= e_l \,, & F_{2,l}^{-1} &= \left(f_{2,l+1}^{-1} + f_{2,l}^{-1}\right)/2 \,, & w_2 &= 1 \,, \\
f_{3,l} &= \nabla_{\mathrm{ad},l} \,, & F_{3,l} &= 1 \,, & w_3 &= 5 \,, \\
\mathcal{X}_l &= \overline{r_l} \,, & \alpha_g &= 2 \,, & \tau_g &= 0 \,.
\end{aligned}$$

The Fourier decomposition of RR Lyrae light curves has been carried out by several authors to compare the light and radial velocity curves of nonlinear pulsation models with observational results (*e.g.*, Simon & Teays 1982, Simon 1985). The observed or theoretical light $L(t)$ curves are least square fitted as a function of time t by

$$L(t) = A_0 + \sum_{k=1}^{K} A_k \cos\left(k\omega(t - t_0) + \Phi_k\right) \,. \tag{6.6}$$

K is usually set to 8 and A_k, Φ_k, ω and t_0 denote the Fourier amplitudes and phases, the frequency and an arbitrary zero point, respectively. The parameters involved in Eq. (6.6) are determined from a nonlinear least square fit. From the Fourier amplitudes and the phases the following low order combinations are derived

$$R_{21} = A_2/A_1 \,, \quad \Phi_{k1} = \Phi_k - k\Phi_1 \,, \tag{6.7}$$

and then employed for the comparison between theory and observations. Since the values of R_{21} as well as the Fourier phases are very sensitive to details of the shape and amplitude of the light curve, the theoretical models must be compared to observations in a more quantitative way (see also Sect. 7.1).

Figure 6.8 plots a typical RR Lyrae light curve ($M = 0.65\,M_\odot$, $L = 52.5\,L_\odot$, $T_{\mathrm{eff}} = 7000\,\mathrm{K}$) resulting from the adaptive RHD-computations. Since the grid points located within the driving zones provide sufficient resolution the light curves exhibit a very smooth character usually not obtained with Lagrangian codes. As pointed out earlier, the ionization zones remain neither constant in mass nor in space and a varying number of grid points within such zones yields a varying driving force becoming visible in fluctuations of the light curves. However, the period of the oscillation is quite robust and therefore the difference in the oscillation frequency between adaptive and Lagrangian computation is usually small.

On top of these RHD grey computations, a frequency dependent radiation transfer can be done to obtain the spectral information. The methods are described in Sect. 3.14. Besides the frequency-dependent opacities κ_ν, no

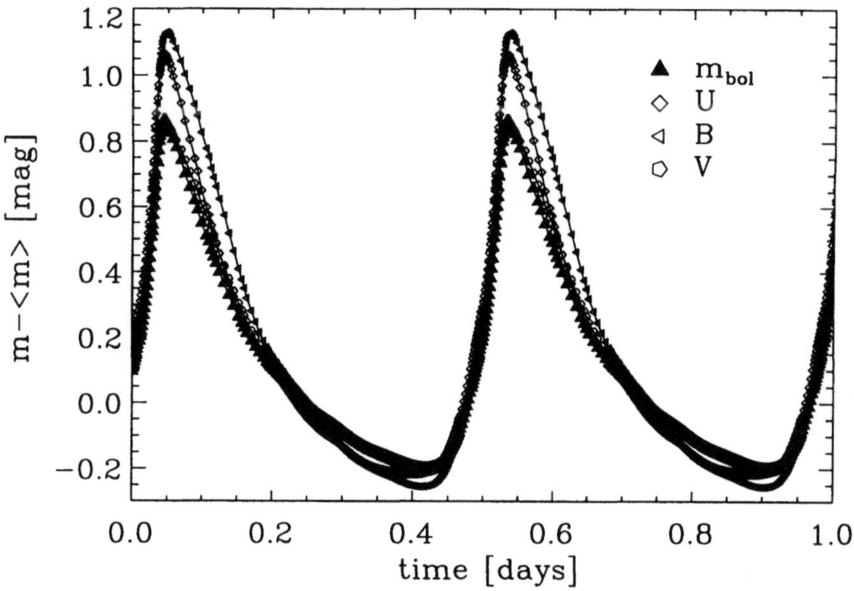

Fig. 6.8. Calculated RR Lyrae light curve in bolometric magnitudes for $M = 0.65\,M_\odot$, $L = 52.5\,L_\odot$ and $T_{\text{eff}} = 7000\,\text{K}$ together with the U,B,V colours. All curves are plotted around their mean value $\langle m \rangle$ to show the difference in the colours over a pulsational cycle

new ingredients are needed to perform such calculations. It is clear that the frequency-integrated means like, *e.g.*, the Rosseland mean κ_{R} used for the RHD computations, have to be obtained from the same frequency-dependent opacities κ_ν. The non-grey computations have to begin deep enough in the stellar interior for all frequencies of the radiation field to be still isotropic. Applying the theory of radiative transfer, the boundary condition of a black body with a temperature $T = T(r_N)$ and a Rosseland mean $\kappa_{\text{R}} = \kappa_{\text{R}}(r_N)$ can be specified at the inner boundary by

$$h_\nu(\mu) = \frac{\mu}{\kappa_\nu}\frac{\partial B_\nu}{\partial T}\frac{3\kappa_{\text{R}}\,H}{2a_{\text{R}}\,T^3}\,, \qquad H = \frac{L}{16\pi^2\,r_N^2}\,, \tag{6.8}$$

where μ is the angle between the outgoing ray and the radial direction (see Sects. 3.14 and 3.15).

After the non-grey radiative transfer is computed, the theoretical spectra are folded with the spectral filter curves to obtain the standard U, B, V photometric colours as plotted in Fig. 6.8. All curves are normalized to their mean values over a period so that the differences during a pulsational cycle become evident.

Hydrogen deficient carbon stars (HdC) are characterized by a peculiar chemical composition and they consist mainly of helium ($Y \simeq 0.9$) and carbon ($Z_{\text{C}} \simeq 0.07$) with very low hydrogen contributions ($X \simeq 10^{-4}$). Some

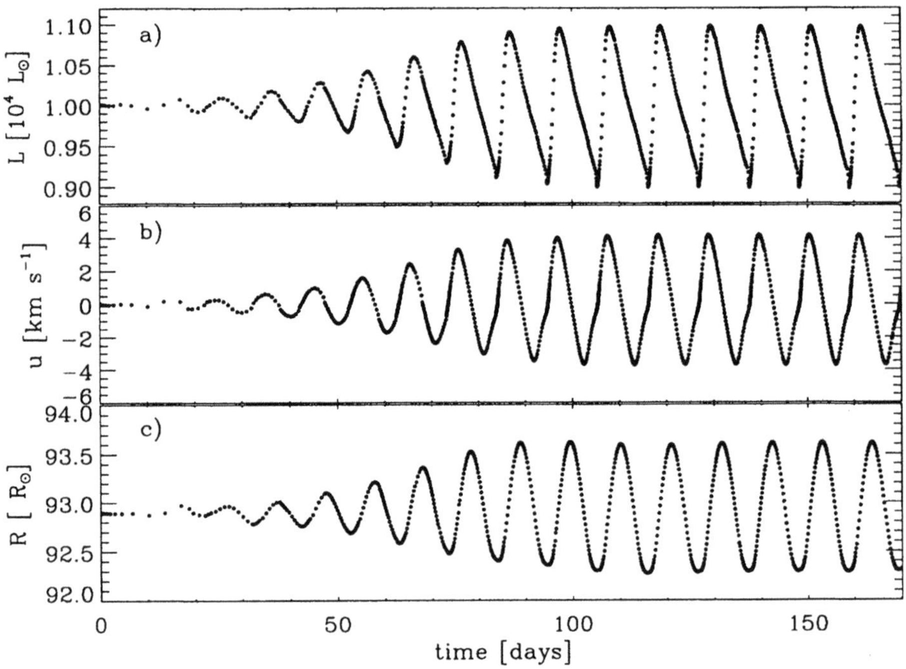

Fig. 6.9. A nonlinear stellar pulsation of a hydrogen deficient carbon star initiated by a small random velocity fluctuations with $\overline{u} = .01c_s$. a) The stellar luminosity in units of $10^4\,L_\odot$. b) The photospheric velocity in $\mathrm{km\,s^{-1}}$. c) The radius of the photosphere in units of $R_\odot$

of these HdC are also known as R CrB stars, where the large luminosity variations can be explained by obscuring of newly formed dust particles in the expanding carbon-rich circumstellar envelope. Most of these HdC stars exhibit radial pulsations with large amplitudes in the atmospheric velocities but small changes in the luminosities. The origin of the HdC stars is still a controversial issue since the apparent absence of hydrogen is difficult to explain. Theories of last helium flash scenarios and white dwarf merging are discussed in the literature (e.g., Paczynski 1971, Schönberner 1986, Iben et al. 1995).

HdC-stars can exhibit a rapid growth of nonlinear pulsations due to so-called strange modes. Figure 6.9 shows an example with $M = 1\,M_\odot$, $L = 10^4\,L_\odot$ and $T_{\rm eff} = 6000\,\mathrm{K}$. The nonlinear pulsation manifests itself quickly as a high-overtone mode with small luminosity and velocity variations with a period of 10.5 days. The photospheric velocity oscillates between $\pm 4\,\mathrm{km\,s^{-1}}$, the relative radius variations are $\Delta R/R = 0.014$ and the scale of the bolometric luminosity is given by $\Delta M_{\rm bol} = 0.2\,\mathrm{mag}$.

Due to the increased resolution available by the adaptive-grid method the nonlinear pulsations can be followed in much greater detail over a long time

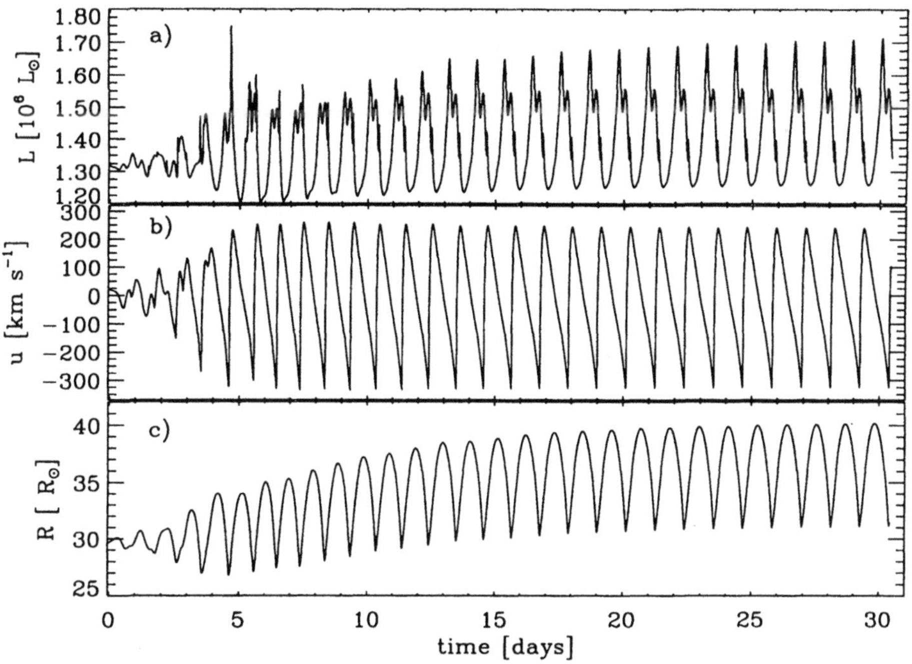

Fig. 6.10. LBV Pulsation showing a transition into another mode

interval than hitherto. Since the structures remain almost stationary on the adaptive grid the numerical time steps increase and the computations can be performed on workstations. In order to compare these results with earlier investigations we have only presented pulsation calculations with the *old* Los Alamos opacities. However, it has become evident that the details of the light variations during the pulsation depend not only on the opacities used but at the same level on the exact treatment of the radiative transfer in the outer layers. All models presented here are calculated with the inclusion of a stellar atmosphere to obtain the right boundary conditions in the case of outwards propagating waves. The pulsations of HdC stars are highly nonadiabatic leading typically to large velocity and small luminosity variations.

Luminous blue variables (LBV) are among the most luminous stars shining with luminosities up to $10^6 L_\odot$ and having masses up to around $100\,M_\odot$. Due to their high luminosities and variability they are of great interest for theoretical and observational studies, in particular as possible steps in the extragalactic distant ladder. As encountered in many stars with a high luminosity to mass ratio, strong non-adiabatic pulsations are excited (see, *e.g.*, Gautschy & Saio 1995, 1996). Figure 6.10 depicts an example with $M = 90\,M_\odot$, $L = 1.32 \times 10^6 L_\odot$, $T_{\rm eff} = 35\,900\,{\rm K}$ and $R_* = 29.9\,R_\odot$. According to a linear stability analysis this star is unstable in the $2^{\rm nd}$-overtone with a period of $P_2 = 0.88$ days. Our radiation-hydrodynamic simulations are exited with random fluctuations in the velocity field with $0.1\,c_s$. The pulsation evolves into an oscillation pattern which differs from the linear prediction as can be inferred from the long-term behaviour plotted in Fig. 6.10. Part of the kinetic energy is transferred into the stellar atmosphere which thereupon expands. As a consequence, the pulsation build up about a different equilibrium radius at $R^{\mathrm{new}} = 1.18R_*$ and $T_{\mathrm{eff}}^{\mathrm{new}} = 33\,000$ which leads to an increased pulsation period of $P_2^{\mathrm{new}} = 1.13$ days. This scaling agrees with $P \propto \bar{\rho}^{\,-1/2}$. Note, that such behaviour can only be investigated by nonlinear computations where a linearization around the initial equilibrium model is not assumed. In the case of more unstable modes, a more complicated light curve is expected which possibly leads to an explanation of the observed LBV eruptions.

6.3 Protostellar Collapse

Protostellar collapse calculations are a long standing problem for computational RHD. Beginning with the work of Larson (1968) it became evident that due to an accretion shock around a hydrostatic core very complex flow structures are expected and that the energy generation in the shock fronts is the main source of the emerging radiation. The initial conditions are taken to be a Jeans-unstable interstellar cloud embedded in an external isotropic radiation field, typically with $T_{\mathrm{ext}} = 100$ K. The collapse proceeds in several phases where a number of different physical effects modify the flow structure. First, at the beginning the collapsing gas remains isothermal since the optical depth is so small that all compressive energy gains can be radiated away. As the material becomes denser the energy gets trapped, the gas is heated up and around 2 000 K molecular hydrogen starts to dissociate (see, e.g., the decrease of Γ_1 in Fig. 2.2) and reduces the gas pressure which triggers the so-called second collapse. Second, a hydrostatic core is formed which will contract on its thermal time scale during the further evolution. This core is now surrounded by a strong radiating shock wave which decelerates the infalling material and allows a gravitational settling on the growing central core. Third, the large temperature and density variations yield large opacity changes within the flow which necessitate a calculation of the Eddington factor. Having a very isotropic radiation field near the center the radiation is dominated further out by the accretion shock and finally becomes again isotropized very far out by the dust particles embedded in the gas. Note that this accretion of material is time-dependent and the shock front will vary in strength and position. The overall evolution occurs in an extremely nonhomologous way. All these complications discussed in the previous sections impose very strict conditions on the numerical treatment since many orders of magnitude in space and time have to be covered. Actually, these problems led to the first developments of adaptive grids (Winkler 1975).

The protostellar collapse calculation (G. Wuchterl, private communication) presented is performed with $N = 1000$ grid points and the following set of grid parameters

$$\begin{aligned} f_{1,l} &= \rho_l\,, & F_{1,l}^{-1} &= \left(f_{1,l+1}^{-1} + f_{1,l}^{-1}\right)/2\,, & w_1 &= 1\,, \\ f_{2,l} &= e_l\,, & F_{2,l}^{-1} &= \left(f_{2,l+1}^{-1} + f_{2,l}^{-1}\right)/2\,, & w_2 &= 1\,, \\ \mathcal{X}_l &= \overline{r_l}\,, & \alpha_g &= 2\,, & \tau_g &= 0. \end{aligned}$$

To provide a smoother variation in time it can be advantageous to set, e.g., $\tau_g = 10^{-3} t_{\mathrm{ff},l}$, where the local free-fall time up to the radius r_l has been introduced:

$$t_{\mathrm{ff},l} = \sqrt{\frac{3\pi}{32G\bar{\rho}_l}} \quad \text{with} \quad \bar{\rho}_l = \frac{3m_l}{4\pi r_l^3}\,. \tag{6.9}$$

The grid is determined by the variations in the density and the internal gas energy. Due to the large dynamical range all variables are scaled by the harmonic mean. The temporal smoothing is done locally by allowing the grid to move at maximum only at a small fraction of the local free-fall time scale.

The first full hydrodynamical calculation of a radiating collapse flow including realistic opacities and an appropriate EOS have been performed by Winkler and Newman (1980ab). A typical snapshot of the flow structure at 3.8×10^5 years is depicted in Fig. 6.11 between 10^9 cm and 10^{17} cm. The density (Fig. 6.11a) reveals a free falling envelope with $\rho \propto r^{-3/2}$, an accretion shock at $\log r = 11.85$ and a hydrostatic core inside. The scale in the density at this stage is about 20 orders of magnitude. The velocity structure of Fig. 6.11b clearly shows how the infalling material is decelerated at the shock front from $25\,\mathrm{km\,s^{-1}}$ to a small fraction of the sound velocity. Together with the luminosity in units of $L_\odot$ plotted in Fig. 6.11e it becomes clear that almost all radiation is generated in the accretion shock and only slightly modulated by the infalling plasma. Figure 6.11c shows the variation of the gas pressure ranging over almost 30 orders of magnitude. The temperature profile exhibits several interesting features seen in Fig. 6.11d. The modulations in the collapsing material are causes by changes in the opacity (Fig. 6.11f). Around 300 K the dust particles loose their icy mantles and at 1800 K the remaining core of the dust particles get evaporated leading to the bumps in the opacity. The radiating shock front appears as a sharp peak in the temperature which is also accompanied by large variations of the opacity. Going further inside the accretion shock is accompanied by an increase of the temperature due to a slow contraction of the protostellar core. Note that the temperature inversion within the central parts of the hydrostatic core is still revealing the variations in the accretion history. At earlier times the material has been falling with lower velocities on the smaller and less compact central mass and therefore the compressive heating has been less effective. Hence, during the contraction towards the main sequence this temperature inversion has to vanish by heating from the outer hotter layers. Since the primordial

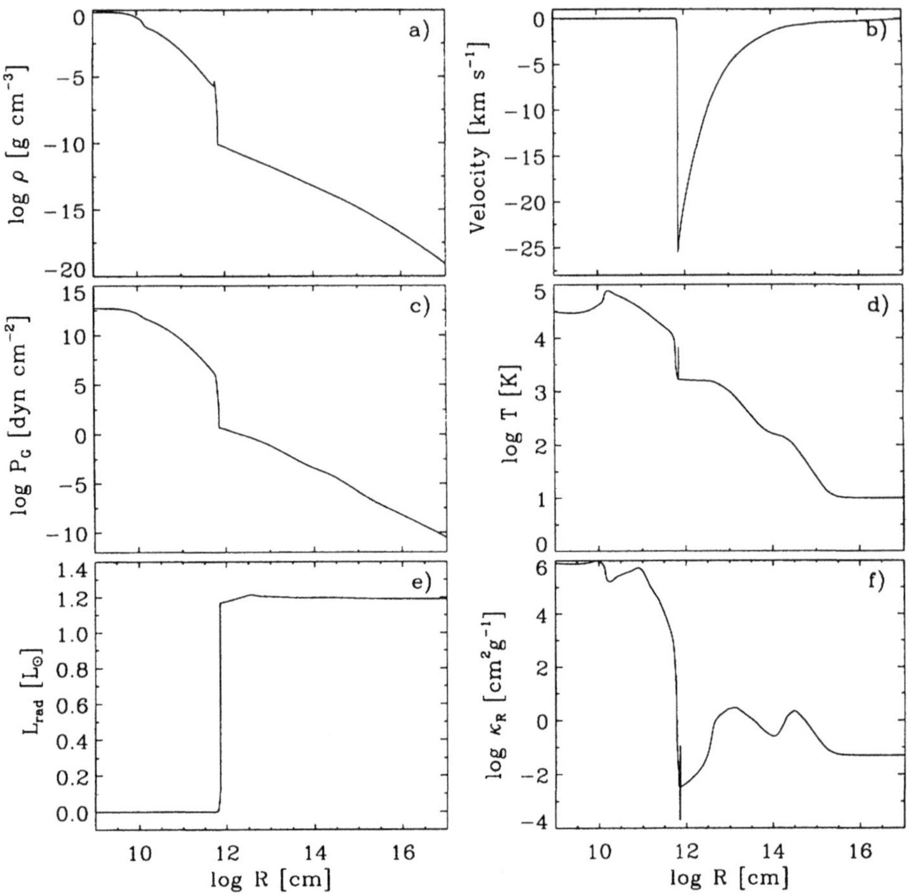

Fig. 6.11. Protostellar collapse of a $1\,M_\odot$ fragment at $t = 3.8 \times 10^5$ years (see text for more details). Courtesy of G. Wuchterl

deuterium is the first material to start a nuclear energy generation the onset of nuclear burning will not occur in the center.

To illustrate the resolving power of the adaptive grid, the vicinity of the accretion shock is depicted in more detail in Fig. 6.12, where except for the luminosity the same variables are plotted as in Fig. 6.11. The radius is given in units of the shock radius $R_s = 10.17\,R_\odot$ and one tick mark on the x-axis corresponds to $(\Delta r)/r = 10^{-4}$. However, the grid clustering leads near the shock wave to almost $(\Delta r)/r = 10^{-7}$ as can be inferred from Fig. 6.12f. The incoming gas with a temperature of $T_1 = 1620\,\mathrm{K}$ is heated up in the radiative shock wave to $T_2 = 6700\,\mathrm{K}$ and cools down again in the post region to a minimum temperature of $T_3 = 1\,680\,\mathrm{K}$ at $R/R_s = 0.9899$ which cannot be shown on the scale of the plot. Hence, this structure is described as a supercritical shock wave with $T_1 \simeq T_3 \ll T_2$ (Zeldovich & Raizer 1969). As

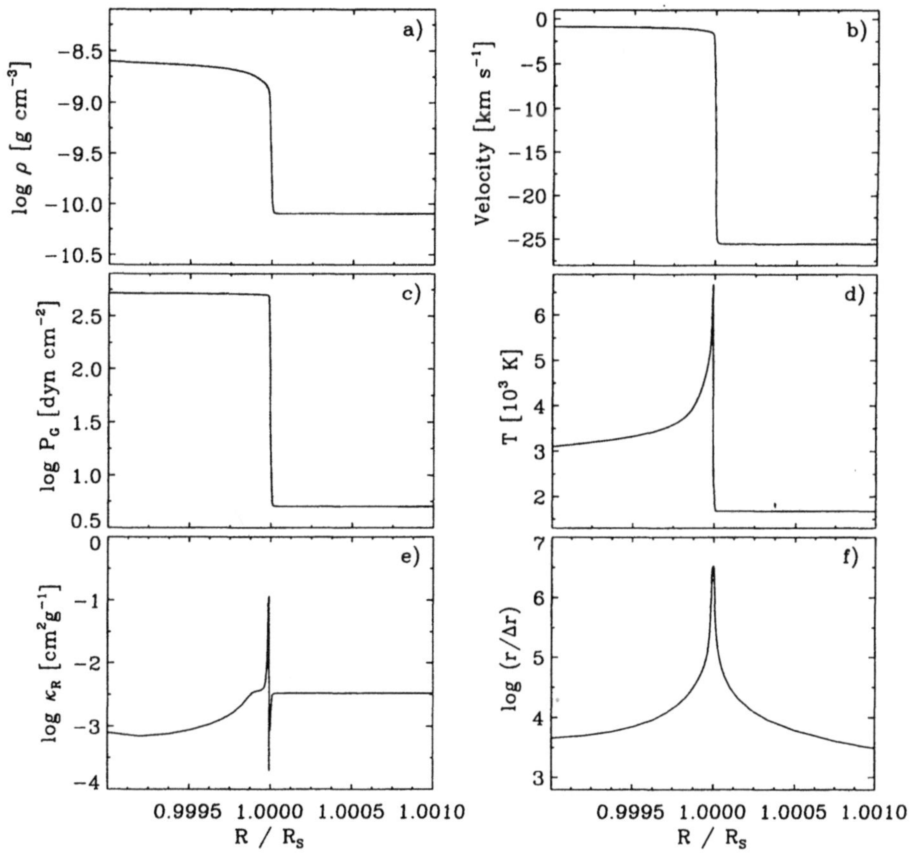

Fig. 6.12. Structure of the accretion shock front at $R_s = 7.08 \times 10^{11}$ cm (see text for more details). Courtesy of G. Wuchterl

mentioned in the introduction (Sect. 1.3) the length scale of such a shock is of the order of the mean free path l_p of the photons. The stability of this accretion fronts has been investigated by Balluch (1988). He demonstrated that using an adaptive grid permits the artificial viscosity length scale to be reduced to values smaller than l_p.

6.4 Dust-Driven Winds

The next example deals with a stellar outflow produced by the radiation pressure on newly formed dust grains. The outer layers of extended, luminous old star with typically $M \simeq 1\,M_\odot$ and $L \simeq 10^4\,L_\odot$ and low effective temperatures around $T_{\rm eff} \simeq 3000\,$K get enriched by heavier elements, mostly carbon and oxygen produced through nuclear burning in the stellar interior and mixed upwards by convective motions (see, *e.g.*, Iben & Renzini 1983 or Lattanzio 1995 for reviews on the stellar evolution aspects). In the atmospheric layers dust particles can condense out of the gas and these freshly formed particles provide an important source of opacity, the radiation pressure acts on them and by collisions with the dust particles a massive but slow outflow is driven. Since a large fraction of these old red giants are pulsationally unstable, they have therefore large-amplitude waves propagating through the atmosphere. The adiabatic expansion in the post-shock regions leads to a temperature reduction and the material lifted up by the pulsations is dense enough to form a large number of small solid particles which can than grow further in the expanding and accelerating flow. Hence, the set of RHD equations has to be supplemented by the equations of time-dependent dust formation (Gail & Sedlmayr 1988, Gauger et al. 1990). A discrete version based on the same finite volume discretization as presented in Sect. 3.13 can be found in Dorfi and Höfner (1991).

The equations of RHD including the time-dependent dust formation give a set of $M = 13$ nonlinear coupled equations and the solution strategy outlined in the previous sections can be used to calculate the complex interaction between dust, radiation and matter. The solutions illustrating the behaviour of the adaptive grid with $N = 500$ grid points are dominated by multiple shock waves propagating through the circumstellar envelope. The radiation pressure on the dust grain provides the driving force to accelerate the flow. The stellar pulsation is simulated at the inner boundary by a moving piston,

$$\begin{aligned} R_N(t) &= R_N(0) + \Delta U \frac{P}{2\pi} \sin\left(\frac{2\pi t}{P}\right) , \\ L_N(t) &= 4\pi R_N(t)\, H_N , \end{aligned} \tag{6.10}$$

where the pulsation period P and the velocity amplitude ΔU are free parameters. The luminosity L_N at the inner boundary undergoes the corresponding sinusoidal variation to account for the periodic modulation of the luminosity caused by a κ-mechanism. The initial models can be obtained from a dust-free stellar atmosphere and as the pulsation starts the gas expands and cools. Dust particles can condense which leads to a further expansion. The outer boundary is taken to move according to the Lagrange conditions (3.31) and the whole structure grows in time until the outermost layers reach the escape velocity. Now the outer boundary can be replaced by a simple outflow condition, *i.e.*, $\partial u/\partial r = 0$ which can easily be implemented (see, *e.g.*, Eq. (3.29)). The boundary conditions of the radiation field are given by Eqs. (3.59) and (3.60).

The three time levels of a dust driven wind plotted in Fig. 6.13 are based on the following set of parameters, $M_* = 1\,M_\odot$, $L_* = 10^4\,L_\odot$ and $T_{\rm eff} = 2600\,$K with a carbon abundance of $\varepsilon_{\rm C}/\varepsilon_{\rm O} = 1.8$, a piston velocity of $\Delta U = 2\,{\rm km\,s^{-1}}$ and a period of $P = 650$ days. The inner boundary is situated at $R_N = 0.91\,R_*$ with $L_* = 4\pi\sigma R_*^2 T_{\rm eff}^4$. The density structure (Fig. 6.13a) exhibits the extended outflow plotted up to 20 stellar radii $R_* = 445\,R_\odot$. The velocity (Fig. 6.13c) shows the shock waves propagating outwards and the typical outflow velocities are around $25\,{\rm km\,s^{-1}}$. Note that within two stel-

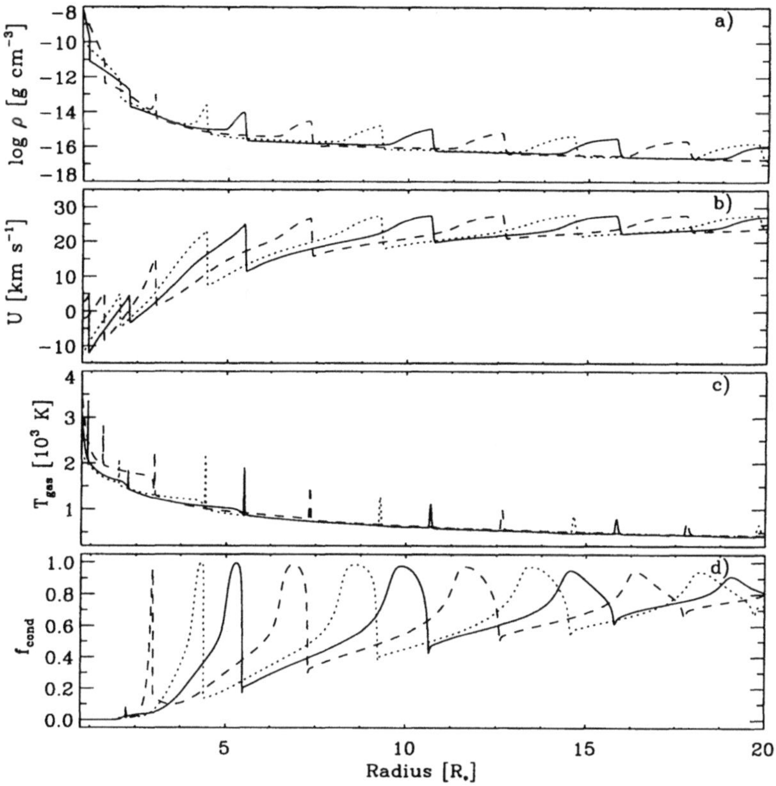

Fig. 6.13. A typical flow pattern with a dust-driven wind (see text for more details)

lar radii the material is partly falling towards the star and the shock waves have to travel against this incoming flow. The gas temperature of Fig. 6.13c reveals the radiating shock waves by the temperature increases followed by a very thin relaxation layer. Since a number of molecules are excited by these shock waves the peak temperatures can be measured and the adaptive grid provides the necessary resolution to calculate this non-adiabatic shock transitions. The innermost shock waves are located at large optical depths and the absorption due to dust particles can block the radiation field. The radiation field heats the dust particles which are emitting radiation themselves also back into the extended atmosphere. This so-called backwarming can supress the formation of new particles, the expansion is slowed down, the gas becomes denser again which increases again the production rate of dust particles. Due to this time-dependence a number of these dusty outflows are unstable with respect to the so-called dust-induced κ-mechanism (Fleischer et al. 1995, Höfner et al. 1995). The lowest panel (Fig. 6.13d) plots the degree of condensation $f_{\rm cond}$ of the carbon-rich dust indicating that a large fraction of the carbon available for the formation of dust particles has been transformed into solid grains. $f_{\rm cond} = 1$ means that all available carbon is condensed into dust particles.

6.5 Radiative Transfer

The system of RHD equations can be closed by calculating the Eddington factor from the time-independent radiative transfer equation (Sect. 3.16). In Fig. 6.14a the behaviour of the Eddington factor as well as the temperatures (Fig. 6.14b) are plotted in more detail where the solid and dashed lines are depicting two different time levels. For the earlier model the dotted line shows the radiative temperature $T_{\rm rad}$ shifted by 1500 K since it runs together with the gas temperature $T_{\rm gas}$ except within the relaxation layers of the radiating shocks. From $T_{\rm rad}$ the aforementioned effect of backwarming is clearly visible at the inner shock wave.

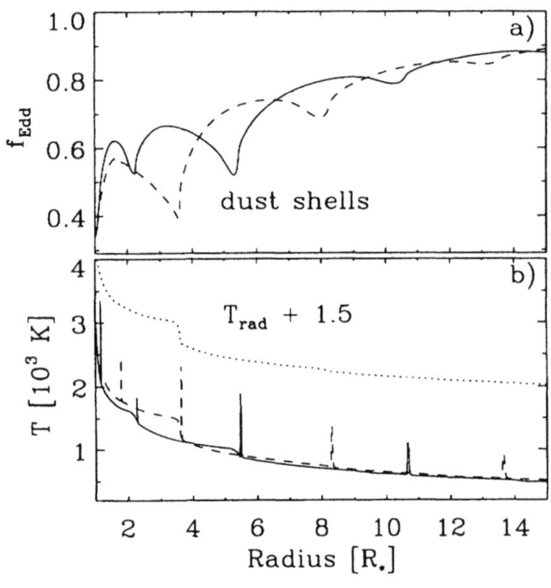

Fig. 6.14. Behaviour of the Eddington factor, the gas temperature (solid and dashed lines) and radiation temperature (dotted line shifted by 1500 K) in a dust driven wind. The solid and dashed curves in the top panel refer to two different time levels

Within the dust driven winds presented in Sect. 6.4 the condensation of solid particles leads to an increase of the opacity and therefore to a isotropisation of the radiation field in these zones of dust particles. As demonstrated in Fig. 6.14a the Eddington factor varies according to these changes and the two time levels illustrate clearly how the dust particles produced in discrete shells reduce f towards the isotropic value of $1/3$. Since the shells get thinner and the central source receds, the influence of the shells is reduced and the Eddington factor increases towards the asymptotic value of $f = 1$ at large distances.

Adopting the Lucy-approximation (2.16) for an extended static atmosphere an Eddington factor $f_{\rm L}$ can be estimated. For the two models shown

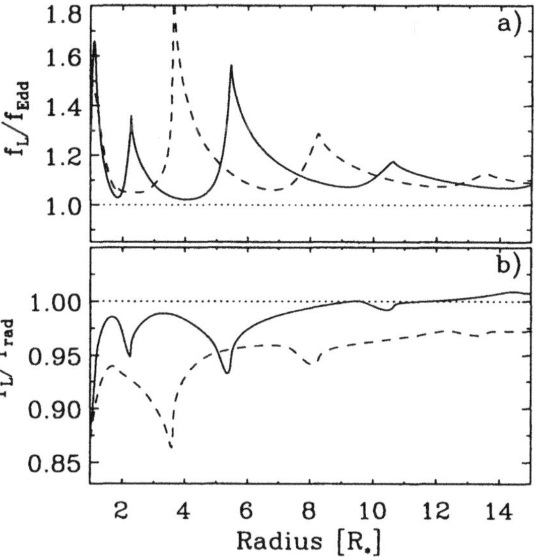

Fig. 6.15. a) The ratio of the Eddington factor f to the Eddington factor calculated from the Lucy-approximation f_{L} and b) the radiation temperature T_{rad} to the temperature T_{L} obtained by the Lucy-approximation. The same two time levels as in the previous Figure are plotted

in the previous figure the ratio of f_{L}/f is plotted in Fig. 6.15 to illustrate the differences between the full radiative transfer treatment and such approximations. Although the overall structure is maintained the deviations are pronounced in the vicinity of dust shells and in the innermost parts where the dust formation process is very sensitive to temperature changes. The corresponding radiation temperature based on the Lucy-approximation T_{L} is plotted against T_{rad} obtained from the full set of RHD equations. The errors depend strongly on the particular structure of the RHD model and almost everywhere the radiation temperature is underestimated. However, this comparison is based on values of T_{L} and f_{L} calculated on top of a fully solved RHD system and does not include the back reaction of the different temperature and radiation pressure structure on the stellar outflow. Hence, taking the differences into account already seen in this figure I expect much larger deviations when the wind structure is obtained without computing the Eddington factor. First, the temperature stratification differs in the dust forming regions and secondly, the growth time of dust particles in the outflow is determined by the radiative forces which accelerate the particles.

7. Discussion

This section deals in some detail with further extensions as well as problems concerning the numerical method presented.

7.1 Internal Accuracy

The internal accuracy of the computations depend on various factors. First, the number of grid points determines the spatial resolution possible and hence the local discretization errors which are unavoidable in numerical computations. The CPU-time required scales linearly with the radial zones and therefore I recommend larger number of radial zones than usually taken for one-dimensional calculations. Most applications presented in the previous sections use at least 300 grid points. To check the results it can be quite useful to repeat some test cases with different numbers of radial points. Second, the way how the differential operators are represented on the discrete cells, in particular how well material can be transported through the numerical grid. These so-called advection terms have been discussed in detail within the mathematical part of the lecture notes by LeVeque. However, in many situations the advection schemes are developed for an equidistant radial spacing and/or plane geometries. Without taking into account such special properties of higher order schemes the gained accuracy can be lost by computing the local fluxes across the zones. Nevertheless, I want to emphasize again that the adaptive grid technique is not restricted to a particular discretization scheme as long as it is suited to handle a time-variable grid position.

In Fig. 7.1 the long-term evolution of two RR Lyrae pulsation calculations are plotted versus the pulsational cycles. The different curves show computations with different Lagrangian switch points (see Sect. 6.2), *i.e.*, the grid point inside which the grid is moving according to a Lagrangian motions without advection terms. The full symbols correspond to a switch point at $\log T_{\mathrm{Lag}} = 4.6$, the open symbols to $\log T_{\mathrm{Lag}} = 4.9$. The pulsation characteristics are almost unchanged as long as this switch point is not located too deep $\log T_{\mathrm{Lag}} \leq 4.7$ to reduce the advection errors. The pulsational amplitude Fig. 7.1a becomes constant at different levels also depicted by the Fourier ratio R_{21} in Fig. 7.1c. The period is also almost constant to a level of 10^{-3} as inferred from Fig. 7.1. Since the equation of internal energy (2.3) is solved the conservation of the total energy can be used to check the internal energy and Fig. 7.1d depicts its evolution. Unfortunately, the total energy errors increases linearly but this discretization error is unavoidable because the same number of grid points are used to resolve a more compact structure during the contracted phase and a more extended stellar structure during the expanded pulsational phase. Nevertheless, this error seems not to influence the pulsational properties as seen in the previous figures.

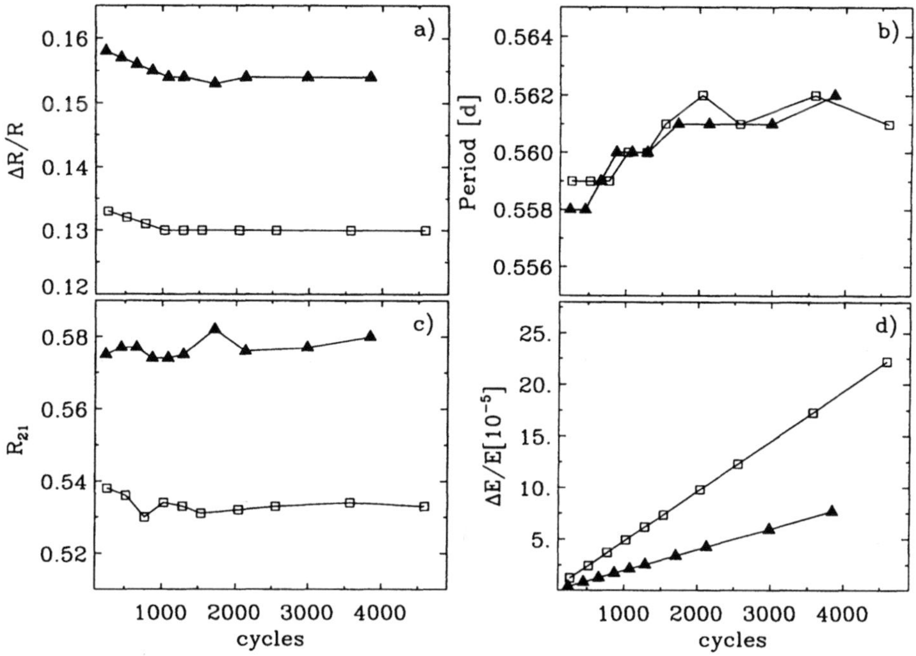

Fig. 7.1. Long term evolution of pulsational calculations

7.2 Problems

Associated with the use of the physical equations like the internal gas energy equation one has to monitor the conservation of total energy during the computational sequence and the errors have to be smaller than any other energy content influencing the physical problem. The better the physical quantities are represented on the grid the smaller the errors occur in conserving the total energy, both in space and time. In principle it is possible to use the total energy equation to calculate the internal gas energy or temperature, but the strict enforcement of total energy conservation will load all computational errors onto the internal gas energy leading in some cases to negative values. Such situations can occur when, *e.g.*, a very cold plasma is streaming at high velocities where the kinetic energy of the gas totally dominates the gas internal energy. Hence, in such cases small errors in iterating the gas velocity immediately show up in the gas internal energy, possibly as negative values.

As seen in the previous examples the artificial viscosity can be reduced by large orders of magnitude through an adaptive grid. Nevertheless, some small amount of viscosity is still left and will influence the overall properties, *e.g.*, of pulsational calculations. In Fig. 7.2 the dependence of the pulsational properties is plotted against the artificial viscosity coefficient (3.35) in units of 10^{-3}. The value of q_2 is reduced to a value around 10^{-3} which defines the local thickness of a shock front. Taking smaller values of the viscosity leads

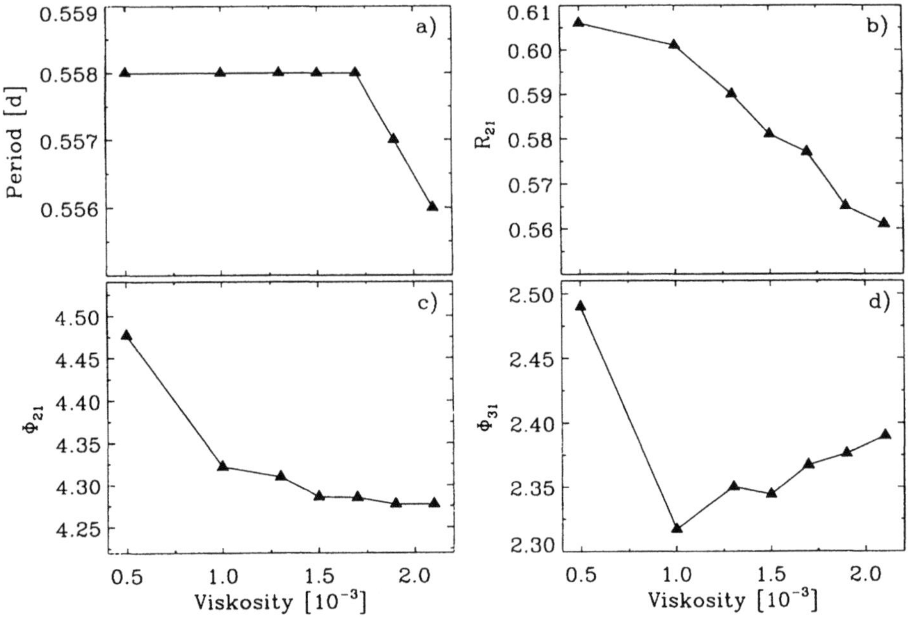

Fig. 7.2. Pulsational properties as a function of the artificial viscosity parameter

to a constant pulsation period (Fig. 7.2a). The dissipation of the code is further reduced as seen through the variation of the Fourier parameter (6.6) in Figs. 7.2b,c,d. The results presented in the Sect. 6.2 have been obtained with $q_2 = 1.5 \times 10^{-3}$ because the observed RR light curves impose a limit on these variations. In real stars also nonlinear dissipation mechanism will operate to limit such pulsations although their physical nature may not be described adequately by such a viscosity term.

As pointed out in the previous section it is necessary to check the development of conservation of total energy during the computational run. It is clear that this errors must be smaller than the energies involved in the physical processes. As already mentioned the errors result from the resolution and the representation of the quantities on a finite cell size. On the other hand the advection scheme plays a crucial role when either large velocities or rapid grid motions occur and higher order advection schemes might be invoked to limit these errors.

7.3 Advantages and Disadvantages of the implicit formulation

As mentioned earlier, for the iterative solution of our finite volume discretization a well-behaved convergence is necessary for which the corresponding Jacobi matrix needs to be determined. The matrix can be calculated either analytically or numerically but for most astrophysical applications with large variations in the physical variables an analytic evaluation yields faster convergence rates. Hence, any small change of the discretization scheme enters the Jacobi-matrix by numerous changes in the derivatives which makes the testing and debugging of such an implicit code rather lengthy. Due to the large matrices which have to be inverted at every time step, this method has been applied only to one-dimensional problems but applications to higher dimensions are discussed in Sect. 7.5. In any case the derivatives of opacity and the equation of state are needed.

The generation of suitable initial conditions requires some work as outlined in Sect. 4.10. First, a full solution of the RHD system is necessary and second, the grid points must be moved to their positions determined implicitly by the grid equation.

According to the implicit formulation it is possible to calculate stationary as well as dynamic situations with one and the same code allowing, *e.g.*, a direct comparison with solutions obtained by integrating the time-independent ordinary differential equations. We can study the convergence towards stationary solutions and the adaptive grid provides a high resolution at locations of steep gradients.

7.4 Nuclear and Chemical Networks and Convection

The method presented is not restricted to a particular set of RHD equations. In Sect. 6.1 the evolution of a SNR has been followed by the equations of radiation hydrodynamics including an equation for the cosmic ray particle pressure and radiative losses for the thermal plasma. The dust driven winds (Sect. 6.4 adopt the full set of RHD equations together with 5 equations describing the time-dependent evolution of dust particles. Hence, it is clearly possible to extend the method to multi-fluid versions which include nuclear or chemical networks. In principle, the implicit nature will allow such a change but as discussed in Sect. 5. the CPU-time scales with the number of equations to the third power. However, as briefly discussed in the next section a so-called hybrid version of an implicit code has been developed by Tscharnuter (1987) where due to a Legendre expansion at every grid point a large number of unknowns $M \simeq 50$ have to be determined. Such computations are still possible on present-day's workstations.

Another extension of the method involves a better coupling between gas and radiation by introducing the frequency dependence of the opacity and the radiative transfer. Hence, instead of having only two gray radiative moments the number is increased by the amount of frequency points used in the computations which clearly exceeds the CPU capacity when using "realistic" numbers ($M \simeq 1000$) of frequency points. However, since the radiation properties are usually varying on different time scales it may not be necessary to solve all variables with the same implicit scheme.

In a number of astrophysical objects energy transport by convective motions plays an important role. In the case of hydrostatic equilibrium the

Schwarzschild criterion (1906) limits the maximum of the temperature gradient for stability against such convective motions. The lack of a full theory of convection makes it in particular difficult to construct models including this additional energy transport and approximations at various levels of complexity to describe convective motions are available in the literature and critically reviewed, e.g., by Baker (1987). A particular model, based on averaging fluctuating physical quantities over spheres by Kuhfuß (1986) has been reformulated for adaptive grids and implemented in RHD computations by Wuchterl (1995). The work of Gehmeyr and Winkler (1992a,b) deals with another convective scheme on adaptive grids which has been used by Gehmeyr (1993) to calculate convective RR Lyrae pulsations.

7.5 Multidimensional Versions

Due to the large Jacobi matrices involved in any implicit method, up to now no 2D fully implicit codes have been developed. The numbering of grid points in higher dimensions yields a large number of non-zero, non-diagonal elements in the Jacobi matrix which makes the inversion rather time consuming. Nevertheless, a so-called hybrid version in axial symmetry has been developed by Tscharnuter (1987) which utilizes an expansion into Legendre polynomials to calculate axial symmetric self-gravitating RHD flows. Instead of having the physical variables on grid points the appropriate Legendre coefficients are advanced in time increasing the unknowns per grid point. Due to such an expansion the discretization is done only in radial direction which leads to a diagonal block structure of the Jacobi matrix. The inversion procedure outlined in Sect. 5.4 can directly be used to solve the system of discrete equations.

A different approach has been used by Balluch (1995) who generalized the adaptive grid to two dimensions in order to minimize the advection errors. He solves the hydrodynamical equations explicitly on the moving coordinates but advances the grid according to an implicit equation. The basic idea of this coordinate transformation method is to perform the coordinate transformation before discretization and to construct a grid on which the physical variables are more smoothly distributed than in physical space. This transformation leads to generally curvilinear meshes with a rather complicated metric. Again, we refer to Balluch (1995) for a detailed discussion.

7.6 Improvements and Further Recommendations

This section deals with some new insights on the discretization on adaptive grids. Up to now and due to complex programming these new developments are not included in any adaptive RHD-code.

We can use the adaptive transport theorem (Eq. (2.24)) (G. Wuchterl, private communication) to define the grid velocity (3.7) in an alternative way by setting $f = 1$, *i.e.*,

$$\frac{\mathrm{d}}{\mathrm{d}t}\int_V \mathrm{d}V = \int_{\partial V} u^{\mathrm{grid}}\,\mathrm{d}A\,, \tag{7.1}$$

or in a discrete version in our symbolic notation,

$$\frac{4\pi}{3}\frac{\delta(\Delta r^3)}{\delta t} = \Delta(4\pi r^2 u^{\mathrm{grid}})\,. \tag{7.2}$$

Rearranging the terms and specifying the above formula at each grid point we end up at a radius r with

$$u^{\mathrm{grid}} = \frac{1}{3}\frac{\delta r^3}{r^2\delta t} = \frac{1}{3}\frac{\delta r}{\delta t}\left[1 + \frac{r}{r^{\mathrm{o}}} + \left(\frac{r}{r^{\mathrm{o}}}\right)^2\right]\,, \tag{7.3}$$

which is a more geometrical definition than the original one of Eq. (3.7). First, we see that in plane symmetry we get the old definition of u^{grid}. Secondly, from $\delta r = r - r^{\mathrm{o}}$ we calculate

$$u^{\mathrm{grid}} = \frac{\delta r}{\delta t}\left[1 - \frac{\delta r}{r} + \frac{1}{3}\left(\frac{\delta r}{r}\right)^2\right]\,,$$

where for large radii $r \to \infty$ also the old u^{grid} definition is revealed. According to the derivation from the adaptive transport theorem this velocity is now also based on a finite volume formulation.

As seen in the finite difference formulations of Sect. 3.7 the variables defined within a computational cell are located in the center of this cell. As demonstrated by Mönchmeier and Müller (1989) in the context of explicit methods the centering of variables on a non-equidistant, non-planar grid becomes of particular importance in multidimensional computations. Hence, future versions of RHD codes should discretize all geometrical terms according to their description. Clearly, such an explicit dependence of the radial centering increases the derivatives to be calculated for the Jacobi matrix.

Acknowledgement. After having survived the initial shock to deliver the lectures at the Saas-Fee course on very short notice and lately the growing force to finish the manuscript I would like to thank both of the organizers for inviting me to give these lecteres. It has been pleasant and very interesting to be on the "other side" of a Saas-Fee course and I must admit that skiing as well as lecturing can be a dangerous job in Switzerland, in particular on the icy slopes down the glacier of Les Diablerets or together on the skilift with Alfred. Clearly, this work would have been imposible without the help of my colleagues in Vienna: I am greatly indebted to Drs. Michael Feuchtinger, Susanne Höfner and Günther Wuchterl for providing me results obtained with the same numerical method. The work on pulsating stars has been supported by the Austrian Fonds zur Förderung der wissenschaftlichen Forschung (FWF) under project number S7305-AST.

References

1. Alexander, D.R., Ferguson, J.W.: 1994, Astrophys. J. **437,** 879
2. Axford, I.W.: 1981, in *IAU Symp. 94, Origin of Cosmic Rays,* Eds. G. Setti, G. Spada, A.W. Wolfendale, D. Reidel, Dordrecht, p. 339
3. Baker, N.: 1987, in *Physical Processes in Comets, Stars and Acitve Galaxies,* Eds. E. Meyer-Hoffmeister, H.C. Thomas and W. Hillebrand, Springer, Berlin, p. 105
4. Balluch, M.: 1988, Astron. Astrophys. **200**, 58
5. Balluch, M.: 1991, Astron. Astrophys. **243**, 205
6. Balluch, M.: 1995, Comp. Phys. Comm. **89,** 91
7. Bono G., Stellingwerf R.F.: 1994, Astrophys. J. Suppl. **93,** 233
8. Buchler, J.R. (Ed.): 1989, *The Numerical Modelling of nonlinear Stellar Pulsations,* NATO Workshop, Les Arcs, Kluwer, Dordrecht
9. Cox, J.P., Giuli, R.T: 1968, *Principles of Stellar structure,* Gordon and Breach Sci. Publ., New York
10. Courant, R., Friedrichs, K., Lewy, H.: 1928, Math. Ann. **100,** 32
11. Dorfi, E.A., Drury, L.O'C.: 1987, J. Comp. Phys. **69,** 175
12. Dorfi, E.A., Feuchtinger, M.U.: 1991, Astron. Astrophys. **249,** 417
13. Dorfi, E.A., Feuchtinger, M.U.: 1995, Comp. Phys. Comm. **89,** 69
14. Dorfi, E.A., Gautschy, A.: 1989, in: 'The Numerical Modelling of Nonlinear Stellar Pulsations', eds. J.R. Buchler, NATO Workshop, Les Arcs, Kluwer, Dordrecht, p. 289
15. Dorfi, E.A., Höfner, S.: 1991, Astron. Astrophys. **248,** 105
16. Drury, L.O'C.: 1983, Rep. Prog. Phys. **46**, 973
17. Feast, M.W., Walker A.R.: 1987, Ann. Rev. Astron. Astrophys. **25,** 345
18. Feuchtinger, M.U., Dorfi, E.A.: 1994, Astron. Astrophys. **291,** 209
19. Feuchtinger, M.U., Dorfi, E.A.: 1996, Astron. Astrophys. **306,** 837
20. Feuchtinger, M.U., Dorfi, E.A.: 1997, Astron. Astrophys. **322,** 817
21. Feuchtinger, M.U., Dorfi, E.A., Höfner, S.: 1993, Astron. Astrophys. **273,** 513
22. Fleischer, A.J., Gauger, A., Sedlmayr, E.: 1995, Astron. Astrophys. **297,** 543
23. Frost, C.A., Lattanzio, J.: 1995, in: *Stellar Evolution: What should be done?,* Proc. of the 32nd Liege Coll., p. 307
24. Gail, H.P., Sedlmayr, E.: 1988, Astron. Astrophys. **206,** 153
25. Gautschy, A., Saio, H.: 1995, Ann. Rev. Astron. Astrophys. **33,** 75
26. Gautschy, A., Saio, H.: 1996, Ann. Rev. Astron. Astrophys. **34,** 551
27. Gauger, A., Gail, H.P., Sedlmayr, E.: 1990, Astron. Astrophys. **235,** 345
28. Gehmeyer, M.: 1993, Astrophys. J. **412,** 341
29. Gehmeyer, M., Winkler, K.-H.A.: 1992a, Astron. Astrophys. **253,** 92
30. Gehmeyer, M., Winkler, K.-H.A.: 1992b, Astron. Astrophys. **253,** 101
31. Höfner, S., Feuchtinger, M.U., Dorfi, E.A.: 1994, Astron. Astrophys. **297,** 815
32. Iben, I.Jr., Renzini, A.: 1983, Ann. Rev. Astron. Astrophys. **21,** 271
33. Iben, I.Jr., Tutukov, A.V., Yungelson, L.R.: 1996, Astrophys. J. **456,** 738
34. Kahn, F.D.: 1976, Astron. Astrophys. **50,** 145
35. Kippenhahn, R., Weigert, A.: 1990, *Stellar Structure and Evolution,* Springer, Berlin
36. Kovaćs, G., Buchler J.R.: 1993, Astrophys. J. **404,** 765
37. Kuhfuß, R.: 1986, Astron. Astrophys. **160,** 116
38. Kürschner, R.: 1994, Astron. Astrophys. **285,** 897
39. Larson, R.: 1969, Mon. Not. Roy. Astron. Soc. **145,** 271
40. Lucy, L.: 1971, Astrophys. J. **163,** 95
41. Lucy, L.: 1976, Astrophys. J. **205,** 482

42. McKee, C.F.: 1982, in: *Supernovae: A Survey of Current Research,* Eds. M.J.Rees and R.J. Stoneham, D. Reidel, Dordrecht, p. 433
43. Mihalas, D., Mihalas, B.W.: 1984, *Foundatios of Radiation Hydrodynamics,* Oxford University Press, New York
44. Mönchmeier, R., Müller, E.: 1989, Astron. Astrophys. **317,** 351
45. Nemec, J.M., Matthews, J.M.: 1993, *New Perspectives on Stellar Pulsation and Pulsating Variable Stars,* eds. J.M. Nemec and J.M Matthews, IAU Colloquium 139, Victoria (Canada), Cambridge Univ. Press
46. Paczynski, B.: 1971, Acta Astronomica **21,** 1
47. Richtmyer, R.D., Morton, K.W.: 1967, *Difference Methods for Initial Value Problems,* Interscience, New York
48. Rogers F.J., & Iglesias, C.A.: 1992, Astrophys. J. Suppl. **79,** 507
49. Schönberner, D.: 1986, in: *Hydrogen Deficient Stars and Related Objects,* eds. K. Hunger, D. Schönberner and N.K. Rao, Reidel, Dordrecht, p. 471
50. Schwarzschild, K.: 1906, Gött. Nachr. **1,** 41
51. Seaton, M.J., Yan Yu, Mihalas, D., Prandhan, A.K.: 1994, Mon. Not. Roy. Astron. Soc. **266,** 805
52. Sedov, L.I.: 1959, *Similarity and Dimensional Methods in Mechanics,* Academic Press, New York
53. Simon N.R.: 1985, Astrophs. J. **299,** 723
54. Simon N.R., Teays T.J.: 1982, Astrophs. J. **261,** 586
55. Simon N.R.: 1987, in: *Stellar Pulsation,* Eds. A.N. Cox et al., Lecture Notes in Physics **274,** Springer, Berlin, p. 148
56. Sod, G.R.: 1978, J. Comp. Phys. **27,** 1
57. Späth, H.: 1983, *Spline-Algorithmen,* Oldenburg–Verlag, München
58. Stellingwerf, R.F.: 1975, Astrophys. J. **199,** 705
59. Taylor, G.I.: 1950, Proc. R. Soc. London **A201** 159
60. Tscharnuter, W.M.: 1987, Astron. Astrophys. **188,** 55
61. Tscharnuter, W.M., Winkler, K.-H.,A.: 1979, Comp. Phys. Comm. **18,** 171
62. van Leer, B.: 1977, J. Comp. Phys. **23,** 276
63. von Neumann, J., Richtmyer, R.D., 1950, J. Appl. Phys. 21, 232
64. Winkler, K.-H.A.: 1975, PhD thesis, Univ. Göttingen, also preprint Max-Planck-Institut für Astrophysik, MPA90, Garching, engl. translation: Lawrence Livermore Lab. Rep. UCLR-Trans 11206, Univ. of. Calif., 1977
65. Winkler, K.-H.A., Newman, M.J.: 1980a, Astrophys. J. **236,** 201
66. Winkler, K.-H.A., Newman, M.J.: 1980b, Astrophys. J. **238,** 311
67. Winkler, K.-H.A., Norman, M.L., Mihalas, D.: 1984, Quant. Spect. Radiat. Trans. **31,** 473
68. Winkler, K.-H.A., Mihalas, D., Norman, M.L., 1985, Comp. Phys. Comm. **36,** 121
69. Winkler, K.-H.A., Norman, M.L., 1986, in: *Astrophysical Radiation Hydrodynamics,* eds. K.-H. Winkler and M.L. Norman, NATO-ASI Series C, Vol. 188, D. Reidel, Dordrecht, p. 223
70. Woodward, P.R., Colella, P.: 1984, J. Comp. Phys. **54,** 115
71. Wuchterl, G.: 1990, Astron. Astrophys. **238,** 83
72. Wuchterl, G.: 1995, Comp. Phys. Comm. **89,** 119
73. Yorke, H.W.: 1980, Astron. Astrophys. **86,** 286
74. Zel'dovich, Ya.B., Raizer, Yu.P.: 1967, *Physics of Shock Waves and High-Temperature Hydrodynamic Phenomena,* Vol. I,II, Academic Press, New York

Ewald Müller on the way down to Aigle, relaxing from 20 hours of lecturing

Simulation of Astrophysical Fluid Flow

Ewald Müller

[1] Max-Planck-Institut für Astrophysik, Karl-Schwarzschild-Str. 1, D-85748 Garching, Germany
[2] Institute for Theoretical Physics, University of California, Santa Barbara, California 93106-4030, USA

1. Introduction

The Oxford English dictionary defines simulation as "an attempt to deceive". It is one of the aims of this lecture to demonstrate that this definition is not appropriate in astrophysics. Instead, I hope to show that simulation is an indispensable and expedient tool for understanding many astrophysical phenomena. In particular, I try to illustrate that simulation is sometimes crucial in providing the link between a multitude of bewildering observations and a variety of theoretical ideas and models existing for an astrophysical phenomenon.

Another aim of my lecture is to discuss the techniques, the results, and the implications of simulations of astrophysical fluid flow. Obviously, a complete coverage of this vast and rapidly growing field of research is far beyond the scope of this lecture. Thus, I have restricted myself to the discussion of a personally biased set of a few specific astrophysical flow problems, the main bias being my past and/or present personal involvement into the simulation of these flow problems.

In Sect. 2, I will discuss why and how simulations are done and what computational resources they require. Then I briefly address some basic numerical issues, like what are Eulerian and Lagrangian schemes, what is an explicit and implicit method, and how are accuracy and efficiency of a method related to each other. Concerning the finite volume methods used for integrating the hydrodynamic equations the reader is referred to the lecture by Le Veque.

Section 3 deals with the simulation of core collapse supernovae. After a discussion of the observational facts, the physics of spherical core collapse is reviewed. Then I discuss observations of supernova SN 1987A, which forced supernova model builders to seriously consider nonspherical models. Triggered by these observations an ever growing set of simulations has been performed, which convincingly showed that various kinds of generic instabilities occur in core collapse supernovae giving rise to mixing and non-radial mass motion in the stellar envelope and in the star's iron core during the actual supernova explosion. These simulations are reviewed in some detail, because they provide a very instructive example demonstrating the necessity for and the potential of hydrodynamic simulations. Next, I discuss simulations of axisymmetric and non-axisymmetric rotational core collapse. The section ends with an overview of the gravitational radiation emitted during core collapse because of convective mass motion, anisotropic neutrino emission and rotation.

In Sect. 4, I discuss the interaction of hydrodynamics and thermonuclear burning. Here I have not intended to give a review of the respective simulations of astrophysical phenomena, like *e.g.*, thermonuclear supernovae or novae, but I have rather tried to point out the underlying physical processes and technical aspects, which have to be considered for these kind of simulations. After a discussion of relevant time scales and various types of nuclear burning, solution methods for (nuclear) reaction networks are analyzed in some detail. Next, I discuss various aspects, which have to be taken into account when coupling nuclear reaction networks with hydrodynamic codes. Some specific astrophysical applications of thermonuclear hydrodynamics are briefly mentioned, too. At the end of the section I present some instructive numerical experiments, which show the traps and difficulties encountered when simulating thermonuclear detonations.

In Sect. 5, I review simulations of astrophysical jets the focus being on relativistic extragalactic jets. In order to provide the necessary background, I first discuss some observational facts and summarize the well-known properties of classical jets. Next, special relativistic hydrodynamics is briefly introduced (see also Sect. 8 of the lecture by Le Veque). The remainder of the section deals with the morphology and dynamics of relativistic jets including two short subsections on two recent developments, namely the long term evolution of relativistic jets and the simulation of parsec-scale jets.

Finally, in Sect. 6, the method of Smoothed Particle Hydrodynamics (SPH) is reviewed. SPH is a free-Lagrange method solving the equations of hydrodynamics without a computational grid. This latter property and the simplicity of the basic SPH algorithm are the two main reasons, why SPH is so popular, especially among astrophysicist. Besides presenting the basic SPH algorithm, I also review extensions of SPH, which allow one to include self-gravity and to use a variable smoothing length for improving the spatial resolution. After addressing some computational aspects of SPH, I conclude with a discussion of two test calculations, which demonstrate the capabilities and limits of SPH.

2. Simulations: A Link Between Observation and Theory

In astrophysics one is in the extraordinary situation, that the objects of interest (*e.g.*, stars or galaxies) are not accessible to any kind of manipulation. No experiments can be performed, and observations of a particular one-time event (*e.g.*, of a specific supernova) cannot be repeated. Hence, astrophysicists have to rely solely on the information they can receive from astrophysical phenomena via electromagnetic radiation, particle radiation, like *e.g.*, neutrinos or cosmic rays, and gravitational radiation.

The situation is further complicated by three facts. Firstly, the physical processes giving rise to the astrophysical phenomenon may occur deep inside the observed object, *i.e.*, we only have indirect evidence of these processes. This is the case, for example, in a core collapse supernova explosion where the explosion of the star is triggered by the gravitational collapse of its iron core in a fraction of a second while the observable (electromagnetic) supernova explosion only begins hours later (see Sect. 3). Secondly, the processes may involve extreme conditions which are experimentally inaccessible in the laboratory. This implies that one has to rely on extrapolation of known physics into regimes of density, temperature, magnetic field strengths, gravity, *etc.*, which are many orders of magnitude beyond our well-tested laboratory knowledge. Thirdly, the processes one tries to understand often occur on time scales long compared to the human life span, *i.e.*, one is stuck with a snapshot of the phenomena. Such a case we will encounter in the discussion of extragalactic jets, which are collimated flow structures emanating from active galactic nuclei and extending up to several hundred thousands of light years into intergalactic space (see Sect. 5). The inferred time scale for the formation of these jets is several million years.

The snapshot restriction is partially alleviated, because observations show similar objects (*e.g.*, stars or galaxies) in a large variety of states suggesting these states may be connected or even form an evolutionary sequence. Hence, snapshots of different objects belonging to the same class provide information on the evolution of this class of objects. However, one first has to prove that the evolutionary hypothesis is indeed correct and that a specific object belongs to the considered class.

How does one proceed under the circumstances just described? Well, one constructs a theoretical model which incorporates all the physical processes thought to be of importance for the astrophysical phenomenon. Such a theoretical model in almost all cases involves some degrees of freedom which manifest themselves as parameters of the model. The model, if of any value at all, should make some definite predictions about the properties or behaviour of the object or phenomenon. In order to compute these predictions one (except for very simple models) requires simulation, because elaborated models consist of a set of nonlinear structure equations or evolutionary equations. Comparing the model predictions with observations allows one to constrain the parameters of the model or to verify or falsify certain model assumptions. In the latter case, one can try to improve or modify the model to achieve a better agreement with the observation, and start the procedure all over again. Thus, simulation provides the link between theory and observation in the iterative process to understand observed astrophysical phenomena. This is especially true for processes which are only indirectly observable, because they occur either inside the object or over time scales long compared to the human life span. In both cases simulation allows one to make phenomena "visible", which cannot be seen otherwise, and to determine from a variety of possible physical processes those which explain the observations.

2.1 Procedure and Resources

2.1.1 Hydrodynamic Approximation. The matter of astrophysical objects can often be approximated as a gas or fluid. Two requirements must be fulfilled in order for this continuum approximation to hold. Firstly, one has to assume that the microscopic behaviour of single particles can be neglected. This requires that

$$\lambda \ll L\,, \tag{2.1}$$

where λ is the collisional mean free path of the particles and L is a characteristic macroscopic linear dimension of the system, or a scale over which the distribution function varies significantly. If Eq. (2.1) holds, the concept of a fluid element can be introduced, the linear size of the fluid element being small compared to L but large compared to λ. This implies that the number of particles in the fluid element is large, and hence mean physical quantities, like *e.g.*, , the fluid density ϱ or the fluid velocity $\boldsymbol{v}$ can be defined for the fluid element. The velocity of the individual particles is then given by $\boldsymbol{u} = \boldsymbol{v} + \boldsymbol{w}$, *i.e.*, it possesses a random component $\boldsymbol{w}$ above the mean. Because $\lambda \ll L$, however, the particles only perform a random walk about the mean motion $\boldsymbol{v}$, and consequently the fluid element remains well defined during the evolution except for some small loss of particles near its surface, which, if necessary, can be well described by an additional diffusion term in the hydrodynamic equations.

Secondly, the forces between particles must be saturating or of "short range", because otherwise collective effects must be taken into account. Formally this can be expressed as

$$\lim_{N\to\infty}\left(\frac{E}{N}\right) = \text{const.}\,, \tag{2.2}$$

where E/N is the energy per particle. For a fluid element with volume V one can then define an energy density $\varepsilon \equiv E/V = n(E/N)$ where $n \equiv N/V$ is the number density, and a pressure $p = n\,(\partial\varepsilon/\partial n) - \varepsilon$ exerted on the "walls" confining the fluid element. Examples of non-saturating forces are gravity and electromagnetic forces which both scale as $\left(\sim r^{-1}\right)$. In the case of gravity the energy per particle $E/N \sim N^2$ for bosons and $E/N \sim N^{4/3}$ for fermions, respectively. Consequently, gravity must be included in the hydrodynamic equations as a macroscopic external force. Although electromagnetic forces are long range too, they saturate in electrically neutral systems due to screening.

If the hydrodynamic approximation holds, in the simplest case, the evolution of the matter in the astrophysical object is governed by a set of conservation laws for mass, momentum and energy (see Sect. 2 of the lecture by

Le Veque for the actual formulation of the hydrodynamic equations). In the general case, additional equations and/or additional source terms have to be considered, which *e.g.*, describe the effects of viscosity, nuclear burning, the diffusive transport of matter or radiation, the evolution of magnetic fields and their coupling to matter (magnetohydrodynamics), the evolution of radiation fields and their interaction with matter (radiation hydrodynamics; see lectures by Dorfi and Mihalas), and self-gravity. Moreover, in certain applications the conservation equations have to be formulated consistently with special or general relativity (see Sect. 5, and Sect. 8 of the lecture of Le Veque and Sects. 5 and 6 of Mihalas).

In astrophysical applications one usually has to solve the hyperbolic set of the inviscous and compressible Euler equations instead of the parabolic set of the viscous and compressible Navier–Stokes equations, because (except in shock waves) viscosity and heat conduction due to physical processes are often negligibly small. However, because of truncation errors and because of finite spatial resolution, all numerical methods proposed for the solution of the Euler equations posses some finite intrinsic numerical viscosity, whose form and size depends on the numerical method, on the grid resolution, and to some extent of the flow problem itself. The effect of the numerical viscosity on the solution can only be quantified by resolution studies, *i.e.*, by performing a series of simulations of the same flow problem using different grid resolution. Thus, when simulating astrophysical flow one is in the strange situation that one tries to solve the inviscous Euler equations, but instead solves some viscous variant, which is also different from the Navier–Stokes equations.

2.1.2 Discretization. The evolution equations are a set of coupled nonlinear first-order (in time) partial differential equations. One way of solving this set of equations is to discretize them in time and space. By this discretization the partial differential equations are transformed into a set of coupled nonlinear algebraic equations, which can be solved on a computer with appropriate numerical techniques. As discretization introduces unavoidable errors, it is crucial to use discretization schemes, which minimize the errors. This is at the heart of the "art of computing".

Guided by the conservation properties underlying the hydrodynamic equations applied mathematicians and physicists have been able to develop accurate and stable high-resolution finite volume schemes (see Sect. 3 of the lecture by Le Veque). In finite volume schemes the (finite) computational domain is discretized into a finite number of zones or cells (typically 10^2 to 10^3 per spatial dimension). The hydrodynamic scalar fields (*e.g.*, density) and the components of the vector fields (*e.g.*, momentum) are approximated by a set of discrete values, which are the (approximate) cell averages of the respective variable. Time is discretized too, the evolution being approximated by a series of finite time steps (typically 10^3 to 10^5 per simulation).

Besides finite volume schemes, another completely different method is widely used in astrophysics for integrating the hydrodynamic equations. This method is Smoothed Particle Hydrodynamics, or SPH for short. In SPH the fluid or gas is described by a finite set of "particles" (typically 10^3 to 10^5), which move with the flow (see Sect. 6 for more details).

2.1.3 Required Computer Resources. In order to derive an estimate of the computer resources needed for a hydrodynamic simulation let us first estimate the number of floating point operations (flops). Per computational zone one has to evolve at minimum three variables (or equivalently three equations) in a one-dimensional simulation (density, one component of the momentum vector, energy) and at maximum about 10 to 20 variables in a three-dimensional simulation involving a small reaction network with a dozen or so nuclear species. The computational grid will have from 10^3 (in 1D) up to 10^7 (in 3D) zones. The number of time steps to be performed ranges from about 10^3 for simple dynamic flows to more than 10^5 in cases where part of the flow is subsonic, or where some other physical process (like, *e.g.*, neutrino transport) limits the size of the time step to a value much smaller than the dynamic time scale. Finally, per zone, per variable and per time step the modern algorithms (incorporating Riemann solvers) require about 10^2 to 10^3 flops. Significantly more floating point operations will be necessary, when one uses a complicated equation of state and/or has to evaluate reaction rates and transport coefficients. Multiplying all these factors one finds that one needs to perform of the order of 10^{10} (in 1D) to 10^{16} (in 3D) floating point operations per simulation.

Presently, a cheap desktop workstation is able to perform about 30 Mflops per second and a state-of-the-art parallel computer (e.g., a CRAY T3E with 512 processor elements) about 30 Gflops per second when simulating hydrodynamic flows. Consequently, a simple 1D simulation requires only a couple of minutes on a desktop workstation, while large 3D simulations will need of the order of several 100 hrs on the fastest computers available today.

From the above numbers one can also estimate that a large 3D simulation requires computers with up to 10 GBytes of main memory, and will produce output, which can exceed 1 GByte per model or 100 GBytes per simulation. This shows that data handling and data analysis of large 3D hydrodynamic simulations is a nontrivial problem. From my personal experience I can say that the analysis of a simulation typically requires about twice to three times the effort than does the preparation and performance of the simulation itself. In this respect, as well as in many other aspects, large-scale computing is quite similar to performing large experiments.

2.2 Some Basic Issues

2.2.1 Lagrangian and Eulerian Formulation. The hydrodynamic equations can be formulated with respect to two distinct classes of coordinate systems called Lagrangian and Eulerian coordinates, respectively (*e.g.*, Potter 1973, Chap. IX/1).

Eulerian or *spatially fixed* coordinates, *i.e.*, "purely" geometric coordinates, specify time-independent points in space. At each point the motion of the fluid is given by $\boldsymbol{v}(x_1, x_2, x_3, t)$. In Lagrangian formulation the coordinates (x_1, x_2, x_3) are the components of the radius vector $\boldsymbol{r}$ of a *fixed* fluid element, *i.e.*, the coordinates are a function of time *and* of three parameters which identify the element.

The Lagrangian formulation guarantees that no numerical, *i.e.*, unphysical diffusion of momentum, heat or composition occurs during a simulation, because the nonlinear advection terms responsible for the occurrence of numerical diffusion are not present in the Lagrangian formulation of the hydrodynamic equations. Numerical diffusion causes especially difficult problems when trying to model (thermonuclear) burning fronts (see Sect. 4). The advantage of the Lagrangian formulation breaks down, however, and severe numerical difficulties arise when multi-dimensional problems are to be treated. The comoving grid in general will become very distorted leading *e.g.*, to grid tangling in the case of shear or vortex flow. Then one is forced to *rezone* the grid. Even when the rezoning is done carefully, which is not as easy and straightforward as one might think on first glance (note *e.g.*, that the conservation of mass, linear and angular momentum, and of energy should not be violated by the rezoning algorithm), a significant amount of numerical diffusion of the various quantities is introduced. Thus the major advantage of the Lagrangian approach is lost.

For multi-dimensional problems, therefore, Eulerian hydro-codes using time-independent spatial coordinates are to be preferred, since the grid remains regular. Obviously then special efforts have to be used to minimize the inevitable numerical diffusion. This can be achieved using more accurate, higher-order schemes (see Sect. 3 of the lecture by Le Veque).

Within the last twenty years, a new type of Lagrangian hydrodynamics method has been developed, the so-called *Smoothed Particle Hydrodynamics* or SPH method (Lucy 1977; Gingold & Monaghan 1977; Monaghan 1982). SPH is a free-Lagrange method, in which spatial gradients are evaluated without the use of a grid. Thus the method does not suffer from the problems caused by grid tangling and subsequent rezoning. The SPH method will be discussed in Sect. 6.

2.2.2 Explicit and Implicit Methods. Following Potter (1973, Chaper II/4) let us consider a system defined by the state vector $\boldsymbol{u}(\boldsymbol{r}, t)$, in the space domain $R = R(\boldsymbol{r})$. If $\boldsymbol{u} = \boldsymbol{u}^0$ is defined at time $t = 0$, and if $\boldsymbol{u}$ is defined on the surface S of R for all time t, we wish to determine $\boldsymbol{u}$ for all time t in R. The state of the system may be obtained for all time t as solutions to the initial value equation

$$\frac{\partial \boldsymbol{u}}{\partial t} = L\boldsymbol{u} \; . \tag{2.3}$$

In general, L is a nonlinear operator which is algebraic for ordinary differential equations, and is a spatial differential operator for Eq. (2.3) being a partial differential equation.

Neglecting higher than second-order terms the most general discretization of Eq. (2.3) with respect to time is given by (*e.g.*, Potter 1973)

$$\boldsymbol{u}^{n+1} = \boldsymbol{u}^n + L\boldsymbol{u}^n(1-\varepsilon)\Delta t + L\boldsymbol{u}^{n+1}\varepsilon\Delta t\,, \tag{2.4}$$

where $\boldsymbol{u}^n$ and $\boldsymbol{u}^{n+1}$ are the state vectors of the system at adjacent time points t^n and $t^{n+1} = t^n + \Delta t$. Here ε is an interpolation parameter, $0 \le \varepsilon \le 1$, and second-order accuracy is only maintained when $\varepsilon = 1/2$. In the special case when $\varepsilon = 0$, the new state $\boldsymbol{u}^{n+1}$ is defined explicitly by the known state $\boldsymbol{u}^n$ at the previous time step. In this event the method is called *explicit*, while otherwise if $\varepsilon \neq 0$ the method is called *implicit.*

Explicit schemes are only stable if the size of the time step is restricted by the famous *Courant–Friedrichs–Lewy condition* (see Sect. 3.1.10 of the lecture by Le Veque), which says that no information is allowed to travel more than one zone within a single time step. For the special case of a two-dimensional Cartesian grid the CFL condition is given by

$$\Delta t \le \Delta t_{\mathrm{CFL}} = \mathrm{Min}_{ij}\left\{\frac{|u_i|}{\Delta x_i} + \frac{|v_j|}{\Delta y_j} + c_{ij}\sqrt{\left(\frac{1}{\Delta x_i}\right)^2 + \left(\frac{1}{\Delta y_j}\right)^2}\right\}^{-1}. \tag{2.5}$$

The minimum in Eq. (2.5) is computed with respect to all zones (i,j), and $\Delta x_i, \Delta y_j, u_i, v_j$ and c_{ij} are the grid spacing and the flow velocities in x- and y-direction, and the local sound speed, respectively.

In general implicit schemes allow for larger time steps but at the expense of solving a system of nonlinear algebraic equations. This is achieved by linearizing the system and iterating the solution (*e.g.*, with the Newton–Raphson method). Thus implicit schemes require in each time step several (typically 3 to 5) times the solution of a linear system, *i.e.*, several matrix inversions. Obviously implicit schemes become prohibitively time and storage consuming when directly used for two or even three-dimensional problems, because the order of the matrix to be inverted is given by the product of the number of variables times the number of zones in each spatial direction, *i.e.*, $NV \cdot N1 \cdot N2 \cdot N3$.

Although operator splitting techniques can help to significantly reduce this problem (see Sect. 4 of the lecture by Le Veque) implicit schemes have not been widely used in astrophysics because of two further difficulties. Firstly the evaluation of the Jacobian required for the iterative solution of the nonlinear system is a problematic task since some variables may only be given in tabular form (equation of state, opacities *etc.*). In addition, experience shows that errors made in the evaluation and in the programming of the numerous derivatives are difficult to locate, *i.e.*, their removal may take a significant amount of the code development time. Secondly, when features are present in the flow across which one or several variables vary sharply and which are neither stationary in the Lagrangian nor in the Eulerian reference frame (*e.g.*, shocks and in particular accretion shocks) the advantage of the implicit approach can be greatly reduced. As convergence is only obtained if in all zones none of the iterated variables varies by more than about 10 to 30% from time step to time step, sharp non-stationary features in the flow will lead to a more or less severe time step restriction. This restriction can only be overcome with an adaptive grid, which however poses other sometimes even more challenging problems concerning the stability of the numerical scheme and the stiffness of the algebraic system (see lecture by Dorfi).

2.2.3 Accuracy and Efficiency. Because multi-dimensional calculations place such a demand on resources and time, one should expend considerable effort in finding the best procedure. Accuracy – determined by testing versus known solutions, and efficiency – getting maximum performance from existing hardware, should be the main concerns.

The computational expense of a given calculation is proportional to the number of zones calculated and the number of time steps taken. Thus, in 3D,

$$\text{Load} \sim (l/\Delta x)^3\,(\tau/\delta t)\;, \tag{2.6}$$

where l is a characteristic length, Δx is the (equidistant) zone size, τ is the time the calculation is to be run, and δt is the average size of a time step. For an explicit method, the size of the time step is limited by the CFL condition $\delta t \leq \Delta x / c_s$ (see previous subsection), so that

$$\text{Load} \sim l^3\,\tau\,c_s\,/\,(\Delta x)^4\;. \tag{2.7}$$

If one method (A) can produce equally accurate results, but uses n times the number of zones needed by another method (B), then the ratio of expense is

$$\text{Load(A)}\,/\,\text{Load(B)} \;=\; n^4\;. \tag{2.8}$$

The memory requirement scales as n^3; in practice this is often the inflexible limit. One may argue for more computer time but the memory size is more difficult to change.

Estimating n for different pairs of codes is not trivial, but depends upon the problem to some extent. An important problem of interest is the preservation of temperature or heat differences (when modeling burning fronts) and of compositional differences (when modeling mixing processes). A useful test problem is one which propagates a compositional step-function through the mesh, with no pressure gradients (alternatively one may use compensating steps in density and temperature). The result should simply be a translation of the step to new positions. Numerical errors will tend to broaden the step into a "ramp" which spreads over several zones. Plotting the width, in zones, of the ramp as a function of the number of zones propagated through is a useful way to compare quality of numerical methods.

Such a comparison is shown in Fig. 2.1, which is taken from Fryxell, Müller & Arnett (1989), who have tested the performance of a number of

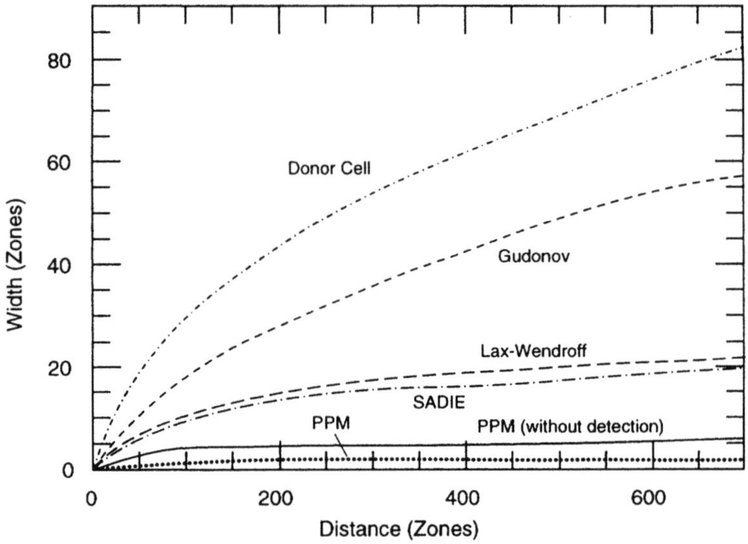

Fig. 2.1. Comparison of several numerical methods for the translational motion of a contact discontinuity. The various curves give the width of the contact discontinuity (in grid zones) as a function of the number of zones it has been propagated through the grid. (From Fryxell *et al.* 1989)

popular methods of multi-dimensional numerical hydrodynamics on the step test problem. These methods included Lax–Wendroff, Donor Cell or first-order upwind, Godunov's method, a flux-vector splitting method (SADIE; Arnold 1985), and the piecewise parabolic method (PPM) of Colella & Woodward (1984). Most of these methods are discussed in Sect. 3 of the lecture by Le Veque.

Fryxell, Müller & Arnett (1989) considered an equidistant one-dimensional grid, where a contact discontinuity has been propagated through 700 zones. Fig. 2.1 then shows that a version of PPM, without special software for detection of such contact discontinuities, has a ramp width of about 6 zones. Lax–Wendroff gives a ramp about 20 zones wide. PPM, with contact detection, gives a ramp only two zones wide. SADIE, which does shocks very well, is as bad as Lax–Wendroff for this test. For both these methods, $n \simeq 10$. First order methods, such as donor cell and Godunov, are much worse ($n \simeq 40$ and $n \simeq 30$).

The computational load for a given accuracy of SADIE or Lax–Wendroff ($n \simeq 10$, which is second best after PPM on the step test problem) relative to PPM (with contact detection switched on) is 10^3 (in 2D) and 10^4 (in 3D). This assumes that each grid point requires the same effort, which is not the case, but this is not a major error for problems which have significant microphysics. Note that the ratio of speed for a large parallel computer ($\simeq 100$ Gflops) and a fast workstation ($\simeq 100$ Mflops) is approximately 10^3, which is comparable to the difference in 3D performance for a resolution ratio of $n \simeq 6$.

3. Simulations of Core Collapse Supernovae

Supernovae (SNe) are one of the most exciting astrophysical phenomena, in particular if they occur in our own galaxy as the historical supernovae did or if they explode in our immediate galactic neighborhood as the spectacular SN 1987A in the Large Magellanic Cloud (LMC). Besides being observationally exceptional events supernovae also pose one of the most challenging theoretical problems in astrophysics involving multi-dimensional hydrodynamics, plasma physics, weak interactions, neutrino transport, nuclear physics, general relativity, thermonuclear burning fronts and turbulent combustion.

Supernova explosions are either powered by the gravitational binding energy released during the collapse of a stellar core to a neutron star (Baade & Zwicky 1934) or a black hole, or by the energy released during explosive burning of degenerate thermonuclear fuel (Hoyle & Fowler 1960).

After briefly discussing observational facts the (prompt and delayed) explosion mechanism of core collapse supernovae is reviewed. Subsequently, multi-dimensional simulations of core collapse supernovae are addressed including Rayleigh–Taylor instabilities in supernova envelopes, neutrino driven convection, and the effects of rotation. Finally, the gravitational wave signature of core collapse supernovae is discussed. Obviously, many other aspects of core collapse supernovae had to be omitted in my lecture (*e.g.*, progenitor evolution, influence of magnetic fields, high density equation of state, nucleosynthesis, light curve modeling, spectral synthesis, radio signature, remnant evolution, SN 1987A), and some of the addressed topics are discussed only relatively briefly. For additional information, the reader is kindly referred to other reviews on the subject (*e.g.*, Weiler & Sramek 1988; Arnett *et al.* 1989a; Hillebrandt & Höflich 1989; Arnett 1996; and articles in Petschek 1990; Ray & Velusamy 1991; Bludman *et al.* 1994, 95; McCray & Wang 1996; Ruiz-Lapuente *et al.* 1997).

3.1 Supernova Observations

3.1.1 Classification, Light Curves and Spectra. Observationally, supernovae are divided into two distinct major spectral classes, which are distinguished by the ***absence*** (Type I) or ***presence*** (Type II) of hydrogen lines in their spectra at any time (Minkowski 1941; Fig. 3.1). This immediately implies that if no spectrum is obtained for a supernova, its type cannot be determined. Due to a late discovery, incomplete coverage or a too small apparent brightness there is a lack of spectral data for many SNe.

During the past two decades observations have further revealed that both types do not form a homogeneous set, but can be classified into distinct subclasses which are called SNe Ia, Ib, Ic, II-L and II-P, respectively (Fig. 3.1; for a review see, *e.g.*, Leibundgut 1995; Filippenko 1997).

SNe II are divided on the basis of their light curves into the photometric "plateau" (SNe II-P) and "linear" (SNe II-L) subclasses (Barbon *et al.* 1979). The rare class of SN IIn (*e.g.*, SN 1988Z) and the possible existence of a separate class of faint SNe II similar to SN 1987A are not discussed here (*e.g.*, Filippenko 1991). The light curves of SNe II-L decline linearly past the epoch of maximum brightness, while those of SNe II-P exhibit a distinct phase of almost constant luminosity, the plateau, past maximum light between days ≈ 40 to ≈ 80 (Fig. 3.2).

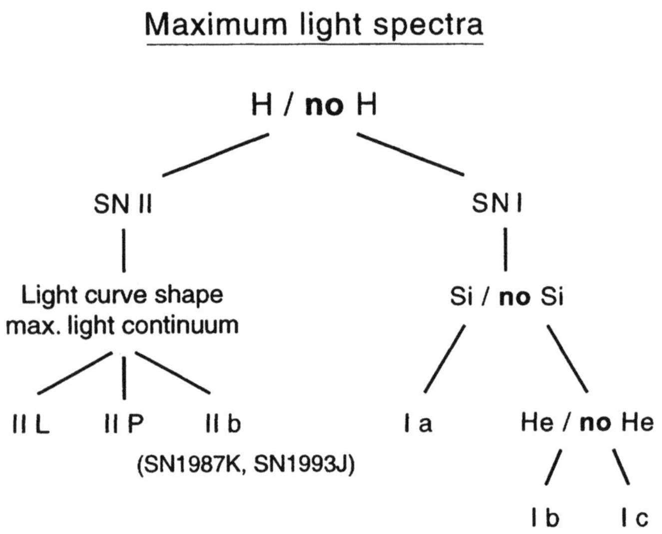

Fig. 3.1. Classification scheme of supernovae (from Harkness & Wheeler 1990)

Because of its distinctive plateau it seems to be safe to classify a supernova as type II-P on its light curve shape alone. However, disregarding the absolute luminosities the similarity of the light curve shapes of SNe I and SNe II-L (Fig. 3.2) implies that it may be dangerous to distinguish between these two classes on the basis of their light curves, in particular if one keeps the scatter of the individual supernova light curves around the mean light curve in mind. In case of incomplete light curve coverage a classification without spectral information will be almost certainly misleading.

Compared to SNe II the light curve shapes of SNe I are broadly similar (Doggett & Branch 1985) except that the light curves of SNe Ib/c seem to be somewhat broader than that of SNe Ia on average (Fig. 3.2). All SNe I light curves show a pronounced smooth maximum followed by an exponential decay, which is observed for SNe II a late epochs, too (Fig. 3.2). SNe Ia are intrinsically the brightest of all supernova types (Fig. 3.2). SNe Ib/c are about

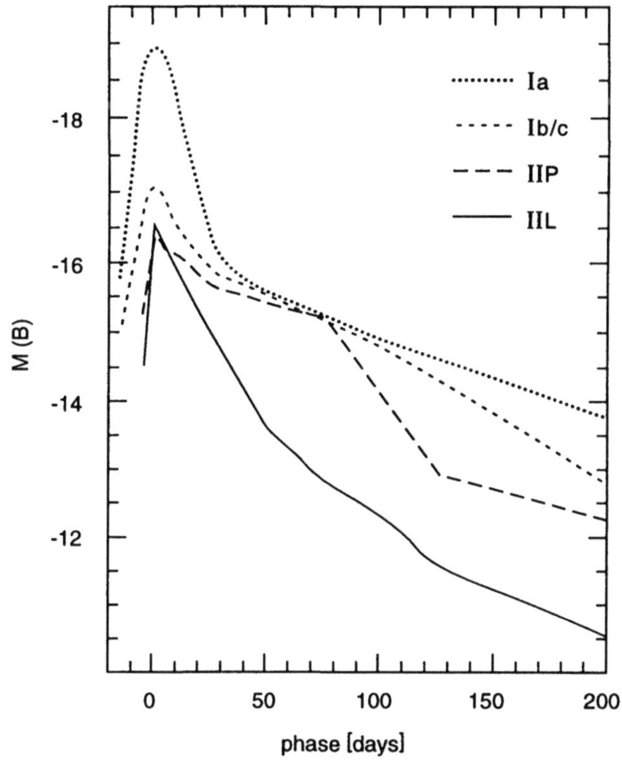

Fig. 3.2. B template light curves for Type Ia, Ib/Ic, II-P and II-L supernovae (from Cappellaro *et al.* 1997)

factor of 4 dimmer than SNe Ia (Uomoto & Kirshner 1985) and seem to be on average slightly brighter than SNe II (Leibundgut 1996).

The three subclasses of the hydrogen-deficient (Type I) SNe are defined according to features in their optical spectra within about one month past maximum brightness (Figs. 3.1 and 3.3).

Contrary to Type Ia supernovae (*e.g.*, SN 1981B in Fig. 3.3), SNe Ib/c (*e.g.*, SN 1984L in Fig. 3.3) do not show the Si II λ 6355 Å blue-shifted absorption feature at λ 6150 Å (Bertola 1964; Wheeler & Levreault 1985), and their infrared spectra do not exhibit the absorption feature at $\sim$ 20 days, which characterizes the J, H and K light curves of SNe Ia (Elias *et al.* 1985). In addition, at early times the spectra of SNe Ib contain strong optical He I lines, most prominently a distinctive absorption dip at $\approx \lambda\,5700$ Å resulting from a Doppler-shifted He I line at $\lambda\,5876$ Å, which are not present in the spectra of SNe Ia at those epochs (Wheeler & Harkness 1990).

SNe Ic lack obvious lines of both Si II and He I in their optical spectra. This subclass was proposed by Harkness & Wheeler (1990) and Wheeler & Harkness (1990) substituting the original categories "helium-rich" and "helium-

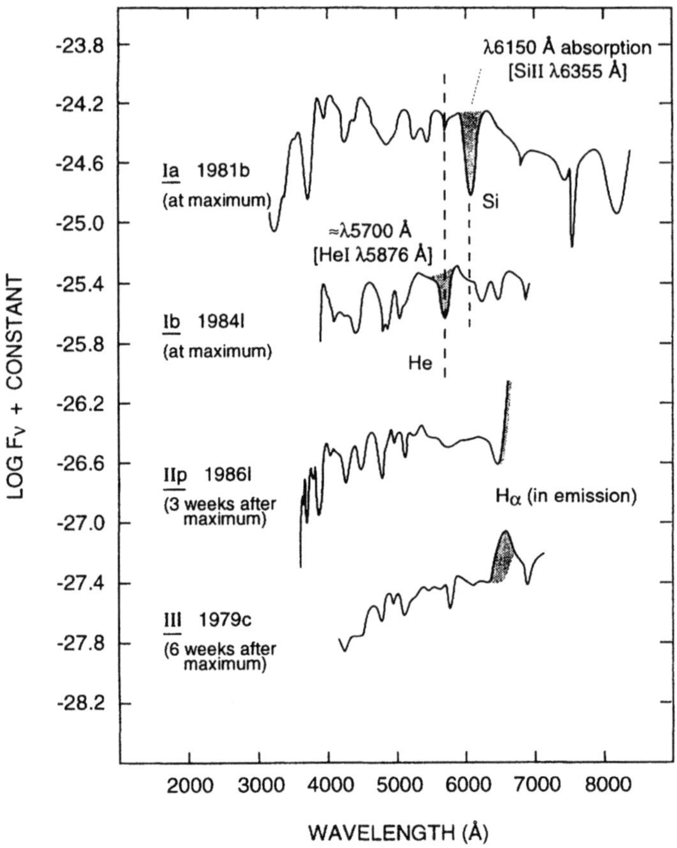

Fig. 3.3. Spectra of a Type Ia, Ib, II P and II L supernovae (from Wheeler *et al.* 1987)

poor" (Wheeler *et al.* 1987) for SNe Ib. However, whether SNe Ic are really a distinct additional subclass requiring substantially different evolution and progenitors or whether they are just extremely helium deficient SNe Ib is still unclear (*e.g.*, Leibundgut 1995; Filippenko 1997; Clocchiatti & Wheeler 1997). Support for the latter "continuity hypothesis" comes from observations, which show the presence of at least *some* helium in a subset of SNe Ic, *e.g.*, such as SN 1987M (Clocchiatti *et al.* 1996) and SN 1994I (Filippenko *et al.* 1995).

Late time spectra of SNe I are very different from spectra near maximum light. At late epochs, spectra of SNe Ia are dominated by strong blends of emission lines of [Fe II], [Fe III] and [Co III] with Ca H and K in absorption but no evidence for O (Axelrod 1980), while those of SNe Ib/c show strong lines of O I, Mg I and Ca II in emission (Gaskell *et al.* 1986; Axelrod 1988; Fransson 1988).

The spectra of SNe II are dominated by strong P-Cygni Balmer lines of hydrogen near maximum light and in the nebular phase (Zwicki 1965). The spectra further show [O I] lines, which are strong in some events but only weak in others, and permitted and forbidden Ca II lines (*e.g.*, Hillebrandt & Höflich 1989). Whether SNe II-L and II-P have distinctively different spectra and whether the observed variations in Type II spectra reflect some intrinsic differences in their physical properties is unclear, because the sample of well-studied SNe II-P is still too small (Harkness & Wheeler 1990).

Finally, peculiar supernovae have been observed, like SN 1987K and 1993J, which changed from canonically looking SNe II at early times to SNe Ib like objects with a little hydrogen at late epochs (Filippenko 1988). For this reason these supernovae are classified as SNe IIb (Woosley *et al.* 1987; Filippenko 1988). The metamorphosis of type shows that the current supernova classification scheme predominantly reflects differences in the pre-supernova evolution and the structure and composition of the progenitor, but not the underlying explosion mechanism (see Sect. 3.1.3).

3.1.2 Frequency and Spatial Distribution of Supernovae. A first attempt to determine the frequency of SNe has been made by Zwicky (1938) more than fifty years ago. His study and various later investigations (for a review see, *e.g.*, van den Bergh & Tammann 1991; Tammann 1994) have qualitatively shown that the frequency of SNe is strongly dependent on the intrinsic properties of the parent galaxy, in particular on its morphological type and luminosity. Furthermore, SNe rates depend on the supernova type. Quantitative uncertainties in the determination of supernova rates arise, because (i) a significant fraction of all SNe are discovered by chance, rather than during systematic search programs; (ii) the spectroscopic and photometric coverage of SNe is often incomplete even the type being unknown for a significant fraction of all discoveries; and (iii) inclination corrections to the SN discovery frequency in spirals are both large and uncertain as first pointed out by Tammann (1974).

One of the latest and most comprehensive determination of the frequency of SNe is due to Cappellaro *et al.* (1997). Their data sample joins the logs of five SN searches (Asiago, Crimea, Calán-Tololo and OCA photographic surveys and the visual search by Evans) and consists of 110 SNe. References to earlier rate determinations can be found in Cappellaro *et al.* (1997) and in the review by van den Bergh & Tammann (1991).

The SN frequency determined by Cappellaro *et al.* (1997) in different types of galaxies for different types of SNe is summarized in Table 1. The errors given in Table 1 are total errors estimated by adding quadratically the errors due to statistics, parameters (absolute and limited magnitudes, light curve shapes, *etc.*) and both nuclear and inclination biases. The rates are given in supernova frequency units or SNu, where 1 SNu = 1 supernova per 100 yr per $10^{10}\,L_{\odot}$. Because of normalization to galaxy luminosity and because Cappellaro *et al.* (1997) assumed a value of the Hubble constant H of 75 km/s/Mpc the rates scale as $(H/75)^2$.

In accordance with previous investigations Cappellaro *et al.* (1997) find that SNe Ib/c and SNe II do only occur in spiral and irregular galaxies. They further find that the rate of SNe Ia increases by about a factor of 2 moving from ellipticals (0.13 SNu) to late spirals (0.24 SNu). This confirms the work of Cappellaro *et al.* 1993, but is in contradiction to the rates obtained by Tammann *et al.* (1994), who determined a 4 times larger rate for the occurrence of SNe Ia in ellipticals but a similar rate in spirals. According to Turatto *et al.* (1994) this disagreement is probably due to a bias of SN discoveries in the "fiducial sample" of galaxies used by Tammann *et al.* (1994). Further results are that the most prolific galaxies are late spirals where 2/3 of the SNe are SNe II (0.88 SNu), and that SNe Ib/c are relatively rare events which amount to less than 40% of all SNe I (0.16 SNu in late spirals).

Table 3.1. Supernova rates according to Cappellaro *et al.* (1997) in supernova frequency units or SNu, where 1 SNu = 1 supernova per 100 yr per $10^{10}\,L_\odot$. A Hubble constant H of 75 km/s/Mpc was assumed, *i.e.*, the rates scale as $(H/75)^2$

galaxy type	SN rate [SNu] Ia	Ib/c	II
E - S0	0.15 ± 0.06		
S0a - Sb	0.20 ± 0.07	0.11 ± 0.06	0.40 ± 0.19
Sc - Sd	0.24 ± 0.09	0.16 ± 0.08	0.88 ± 0.37

From Table 1 one can estimate the expected rate of SNe in the Galaxy assuming its type to be Sb and its luminosity to be $2 \times 10^{10}\,L_\odot$. In a millennium one expects 4 ± 1 SN Ia, 2 ± 1 SN Ib/c and 12 ± 6 SN II, or a total of 18 ± 8 supernovae. Hence, the average interval between two supernova explosions is 40 – 100 years for our Galaxy. Note that the frequency at which galactic supernovae are observed on earth can be quite different from the supernova birth rate due to light-travel and extinction effects (Clark *et al.* 1981).

3.1.3 Progenitors. According to present knowledge SNe II are caused by the core collapse of a massive ($M \gtrsim 9M_\odot$) evolved star. This view is strongly supported by observations (in particular of neutrinos) of SN 1987A (Arnett *et al.* 1989a). SNe Ia are most likely thermonuclear explosions of degenerate white dwarfs. SNe Ib/Ic are thought to be driven by the collapse of the evolved cores of massive stars, which have lost their hydrogen or even part of their helium envelope by either a stellar wind (Wolf–Rayet progenitors with $M \gtrsim 30M_\odot$) or, if being a member of a close binary, by mass transfer (helium star progenitors with $3M_\odot \lesssim M \lesssim 5M_\odot$ and initial masses $12M_\odot \lesssim M_{\rm i} \lesssim 18M_\odot$). Some observational facts and theoretical considerations which support this general picture are outlined subsequently. A more detailed discussion of the progenitor issue can be found in Wheeler & Harkness (1990), Branch *et al.* (1991), Woosley & Weaver (1994), Nomoto *et al.* (1995), Nomoto *et al.* (1997) and Woosley & Eastman (1997), and in the references therein.

SNe Ia occur in all kind of galaxies and, in particular, also in elliptical galaxies (see Sect. 3.1.2), show no preference for spiral arms when they occur in spiral galaxies (Maza & van den Bergh 1976), and further show no evidence for hydrogen in their spectra (see Sect. 3.1.1). This suggests that their progenitors are low mass stars and belong to an old population of stars, because massive (young) stars ($M \gtrsim 10 M_\odot$) are associated with spiral arms. As single stars are unlikely candidates (because it is difficult to understand, how they can get rid of their H-envelope and still explode; see *e.g.*, Woosley 1990) the favourite model of Type Ia SNe involves an accreting white dwarf in a binary system, which when approaching the Chandrasekhar mass ignites its thermonuclear fuel (He or a mixture of C/O) under degenerate conditions. The resulting explosion completely disrupts the star leaving no compact remnant. The light curve is powered by the decay of the radioactive ^{56}Ni, which was synthesized during the explosion, into ^{56}Co and eventually into ^{56}Fe (Hoyle & Fowler 1960; Arnett 1969; Nomoto & Sugimoto 1977).

SNe Ib/c occur in spiral arms and close to or in H II regions (Porter & Filippenko 1987), which suggests massive progenitors ($\approx 20 - 30 M_\odot$, where $M_\odot$ is the solar mass) that have been stripped of their hydrogen (SNe Ib) and possibly helium (SNe Ic) envelopes. Massive progenitors are further supported by the fact that SNe Ib/c, contrary to SNe Ia, are strong radio emitters (Weiler & Sramek 1988). This is most easily explained, if the progenitor star lost its envelope by a stellar wind: the interaction of the supernova blast wave with the wind then can produce the observed radio emission. The small amount of ^{56}Ni synthesized in the explosion of a massive star is also consistent with the observation that SNe Ib/c are intrinsically dimmer than SNe Ia (Porter & Filippenko 1987; see Sect. 3.1.1). Still controversial is the mechanism, which leads to the (partial or complete) loss of the H/He-envelope prior to the explosion (*e.g.*, Branch *et al.* 1991; Leibundgut 1996). The envelope might be lost either by a stellar wind, *i.e.*, the progenitors are Wolf–Rayet stars (Wheeler & Levreault 1985; Begelman & Sarazin 1986; Gaskell *et al.* 1986; Schaeffer *et al.* 1987; Woosley *et al.* 1993; Woosley & Eastman 1997) or, if the progenitor is a member of a close binary, due to mass exchange with its companion (Ensman & Woosley 1988; Shigeyama *et al.* 1990; Nomoto *et al.* 1990; Nomoto *et al.* 1994; Woosley *et al.* 1994a; Nomoto *et al.* 1997; Woosley & Eastman 1997).

Observational data strongly indicate that SNe II originate from massive hydrogen rich Population I stars, because they only occur in the arms of spiral galaxies (Tammann 1982; Maza & van den Bergh 1976), show strong hydrogen lines and are (at least the Type II-L) radio emitters (Weiler & Sramek 1988). Before the discovery of SN 1987A it was generally believed that the progenitors of Type II SNe are red super-giants (*e.g.*, Woosley & Weaver 1986) and not blue super-giants as in the case of SN 1987A. Stellar evolution calculations show that only (single) stars with a mass $M \gtrsim 8M_\odot$ eventually form an iron core and explode (if their H-envelope has not been lost during the evolution in which case they may produce a SN Ib/c) as a SN II (*e.g.*, Woosley & Weaver 1986).

The progenitors of SNe II-L and SNe II-P probably only differ in the amount of mass of the H-envelope (Chevalier 1984; Woosley & Weaver 1994). While the plateau in the light curve of SNe II-P, which is caused by the recombination of hydrogen, is indicative of a large hydrogen mass ($\gtrsim 10M_\odot$), its absence in SNe II-L implies, that their progenitors have lost most but not all of their H-envelope (Litvinova & Nadyozhin 1983). In this picture both kinds of SNe II are caused by the same explosion mechanism (*i.e.*, by core collapse). A different picture was suggested by Dogget & Branch (1985) on the basis of the similarity of the light curves of SNe Ia and SNe II-L. They argued that SNe II-L might be caused by a thermonuclear explosion (see also Wheeler *et al.* 1987).

3.2 Physics of Spherical Core Collapse

Theoretical and numerical models of core collapse SNe (*i.e.*, SNe II and SNe Ib/c, see Sect. 3.1) are based on the idea, that the implosion of the iron core of a massive main sequence star ($M \gtrsim 9\,M_\odot$) at the end of its thermonuclear evolution causes a supernova explosion with violent mass ejection ($v_{\rm eject} \approx 10^4\,$km/sec) and the formation of a compact remnant, *i.e.*, a neutron star or a black hole (for a review, see *e.g.*, Bethe 1990; Hillebrandt 1994). The energy source for the explosion is the gravitational binding energy of the forming neutron star, which is of the order

$$E_{\rm b} \approx 3 \times 10^{53} \left(\frac{M}{M_\odot}\right)^2 \left(\frac{R}{10\,{\rm km}}\right)^{-1} \,{\rm erg}\,, \tag{3.1}$$

where M and R are the mass and the radius of the neutron star, respectively. As predicted by theoretical models and as proven through observations of SN 1987A, the overwhelming fraction of the binding energy (about 99%) is carried away by neutrinos. Roughly 1% of the binding energy is transfered into kinetic energy ($\approx 10^{51}$ erg) of the ejecta, and only about 10^{-4} of the liberated energy is emitted in form of electromagnetic radiation ($\approx 10^{49}$ erg).

3.2.1 Conditions at the Onset of Core Collapse. At the onset of collapse the central density and temperature of the degenerate core of a massive star are $\varrho_{\rm c} \approx 10^{10}\,{\rm g\,cm^{-3}}$ and $T_{\rm c} \approx (8-10) \times 10^9\,$K, respectively (*e.g.*, Woosley & Weaver 1994). The core is mainly made out of iron group nuclei (hence ***iron core***) and collapses on a time scale of $\tau_{\rm coll} \approx 0.1$ sec, its dynamical (or free fall) time scale $\tau_{\rm dyn} \propto \bar{\varrho}^{-1/2}$ ($\bar{\varrho}$ is the average core density) being of the order of 1 ms (Fig. 3.4).

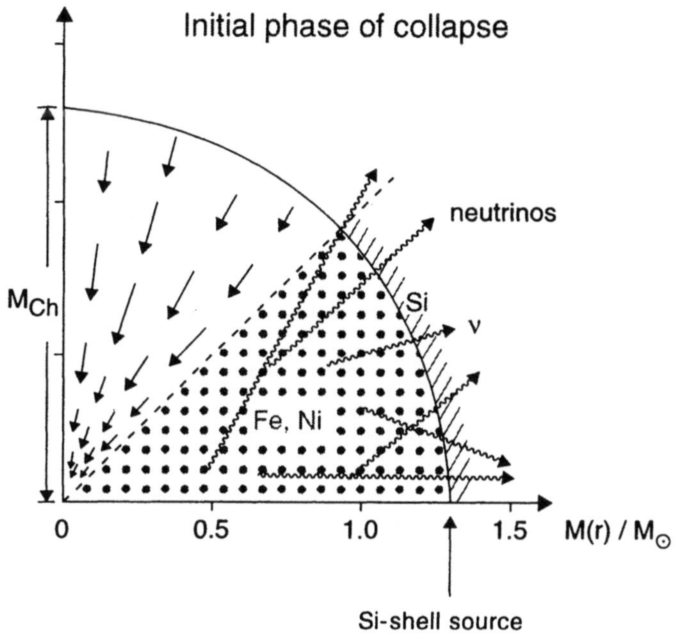

Fig. 3.4. Shortly after the onset of core collapse: The iron core, *i.e.*, the part of the star inside the Si-shell source, has a mass of about the Chandrasekhar mass and has begun to collapse homologously. Neutrinos, which are produced by electron captures, are freely escaping from the iron core at this stage

Because of the high densities and temperatures in the iron core strong and electromagnetic interactions are in equilibrium. In this (local) nuclear statistical equilibrium (NSE) the composition of matter is determined by the Saha equation and is, therefore, independent of reaction rates. Furthermore, the time scale of strong and electromagnetic reactions $\tau_{\rm rate} \ll \tau_{\rm dyn}$, *i.e.*, hydrodynamic processes can be considered as adiabatic changes for these reactions.

In the iron core energy transport by diffusion of photons and electrons can be neglected, because the corresponding diffusion length $\lambda_{\rm diff}$ is much smaller than the typical density-scale height $(d\ln \varrho/dr)^{-1}$. Energy transport by heat conduction of degenerate electrons or ions and the viscosity of core matter can be neglected, too (Flowers & Itoh 1976). Consequently, except for shock waves and energy transport by neutrinos *no* dissipative process or transport phenomenon is of importance for the dynamics of the core, *i.e.*, the iron core matter can be treated as an ideal fluid.

3.2.2 Weak Interactions and Neutrino Transport. For densities $\varrho \lesssim 3 \times 10^{12}\,{\rm g\,cm^{-3}}$ weak interactions are not in equilibrium, because $\tau_{\rm weak} \gtrsim \tau_{\rm coll}$. Important weak processes, which have to be considered in core collapse simulations, are electron captures (EC) on *free* protons and neutrino absorption on neutrons, neutrino-electron scattering and coherent scattering of neutrinos off nuclei (*e.g.*, Bruenn 1985). Electron captures on nuclei are blocked due to shell effects, because in a neutron-rich NSE composition the most abundant nuclei possess closed neutron shells (Fuller 1982).

In principle neutrinos and anti-neutrinos of all flavours, *i.e.*, ν_e, $\bar{\nu}_e$, ν_μ, $\bar{\nu}_\mu$, ν_τ and $\bar{\nu}_\tau$ can participate in the weak processes. However, due to the strong electron degeneracy ($\eta_e \gtrsim 20$) for $\varrho \gtrsim 10^{13}\,\mathrm{g\,cm^{-3}}$ and thus fairly low temperatures ($kT \propto 1\,\mathrm{MeV}$), the number density of electron neutrinos is much larger than the number densities of all other kind of neutrinos, because these neutrinos can only be created via pair and plasmon processes (*e.g.*, Bruenn 1985).

Initially neutrinos can leak out of the core, because the density (and hence the opacity of core matter) is relatively low (Fig. 3.4). The leakage of neutrinos results in a deleptonization of the core, *i.e.*, the electron fraction (per nucleon) Y_e decreases. The neutrino opacity during collapse is dominated by coherent scattering off nuclei mediated by neutral currents (Freedman 1974). The mean free path of neutrinos, therefore, is given by (Hillebrandt 1987)

$$\lambda_\nu \approx 10^7\mathrm{cm}\left(\frac{\varrho}{10^{12}\,\mathrm{g\,cm^{-3}}}\right)^{-1}\frac{\langle \mathrm{A}\rangle}{\langle \mathrm{N}\rangle^2}\left(\frac{\mathrm{E}_\nu}{10\,\mathrm{MeV}}\right)^{-2}, \tag{3.2}$$

where $\langle A\rangle$ and $\langle N\rangle$ are the average mass number and neutron number of the core matter, and where E_ν is the neutrino energy. In shocked matter the main contributions to the neutrino opacity are due to scattering off free nucleons and absorption on neutrons (Bruenn 1985).

For densities $\varrho \gtrsim 3\times10^{12}\,\mathrm{g\,cm^{-3}}$ the neutrino diffusion time scale $\tau_{\mathrm{diff}} > \tau_{\mathrm{coll}}$, which implies that the total lepton number $Y_\mathrm{L} = Y_e + Y_\nu$ is constant during further collapse (Mazurek 1974; Sato 1975a,b). This epoch of the collapse is called ***neutrino trapping*** (see Fig. 3.5). In analogy to photons a neutrinosphere can be defined as the surface, where the optical depth is in the range

$$2/3 \lesssim \tau(r) = \int_r^\infty \frac{\mathrm{d}r}{\lambda_\nu} \leq 1\,. \tag{3.3}$$

Note that the location of the neutrinosphere depends on the energy of the neutrinos, *i.e.*, the neutrino sphere has a variable spatial extent. The typical extension is of the order $\Delta r_\nu \approx (10-30)\,\mathrm{km}$ (Janka & Hillebrandt 1989a,b). Whereas inside and near the neutrinosphere transport processes are of importance, the neutrinos can freely stream further outside.

Inside the neutrinosphere the trapped neutrinos thermalize by neutrino-electron scattering. The chemical potential μ_ν of the Fermi distribution of the trapped neutrinos grows with increasing density until ***β-equilibrium*** is established at densities $\varrho \gtrsim 3\times10^{12}\,\mathrm{g\,cm^{-3}}$, *i.e.*, the weak reactions come into equilibrium and the chemical potentials (including the rest mass) of the electrons, protons, neutrons and neutrinos fulfill the relation $\mu_e + \mu_p =$

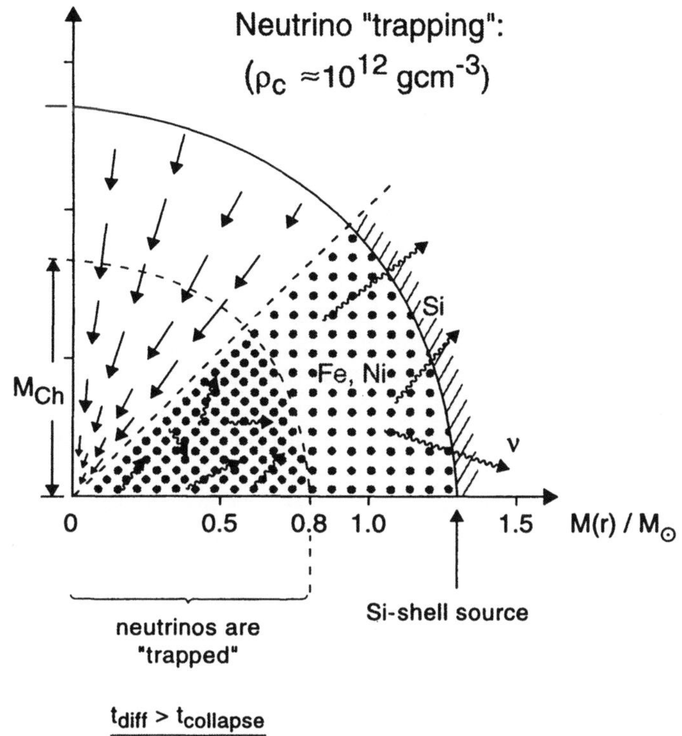

Fig. 3.5. Core collapse after neutrino trapping: The iron core has separated into a homologously and subsonically collapsing inner core and a supersonically falling outer core. The mass of the inner core is about equal to the Chandrasekhar mass at this epoch. The Chandrasekhar mass has decreased from its initial value, because of electron captures ($M_{\rm Ch} \propto Y_{\rm e}^2$). At densities larger than about $10^{12}\,{\rm g\,cm^{-3}}$ neutrinos are trapped, *i.e.*, their diffusion time scale is longer than the collapse time scale

$\mu_{\rm n} + \mu_{\nu_{\rm e}}$. An important consequence is that the entropy remains constant after β-equilibrium is reached, *i.e.*, the further collapse proceeds adiabatically unless shocks are created.

3.2.3 Equation of State (EOS). For a density $\varrho = 10^{10}\,{\rm g\,cm^{-3}}$, a temperature $T = 10^{10}\,{\rm K}$ and matter mainly composed of iron one has

$$P_{\rm e} \gg P_{\rm ion} \gg P_{\rm rad} \ , \tag{3.4}$$

where $P_{\rm e}, P_{\rm ion}$ and $P_{\rm rad}$ are the pressure of the relativistic and degenerate electron gas ($\propto (Y_{\rm e}\varrho)^{4/3}$), the ion pressure and the radiation pressure, respectively. This inequality shows that the iron core is stabilized against gravity by relativistic electrons. It further implies that the adiabatic index γ of the matter in the iron core at the onset of collapse is close to the critical value $\gamma_{\rm crit}$, where the fundamental radial mode of the core becomes unstable (against collapse) (Ledoux 1945):

$$\gamma = \left.\frac{\partial \ln P}{\partial \ln \varrho}\right|_{Y_e,S} \approx \gamma_{\rm crit} = \frac{4}{3} \,. \tag{3.5}$$

A further small but not negligible contribution to the total pressure is the (negative!) pressure due to the Coulomb interaction of the ions (*e.g.*, Hillebrandt 1994).

At "low" densities ($\varrho \lesssim 10^{12}\,\mathrm{g\,cm^{-3}}$) free nucleons are rare and nuclei are well separated. Therefore, both can be treated as Boltzmann particles. The EOS is then determined, provided the ground state binding energies and the nuclear partition functions are known (*e.g.*, El Eid & Hillebrandt 1980). At "high" but sub-nuclear densities ($10^{12}\,\mathrm{g\,cm^{-3}} \lesssim \varrho \lesssim 2\times 10^{14}\,\mathrm{g\,cm^{-3}}$) the free nucleon density increases and starts to influence the properties of the nuclei. The nuclei become more closely packed and begin to interact. In this "high" density regime a microscopic model is required to compute the EOS, which incorporates nucleon-nucleon and nucleus-nucleus interactions, effects due to the finite size of the nuclei and lattice effects due to Coulomb correlations (for a review see, *e.g.*, Hillebrandt 1990; Vautherin 1994). At nuclear saturation density a phase transition occurs and nuclei and free nucleons merge into homogeneous, asymmetric ($Y_e \approx 0.3$) nuclear matter (Lattimer *et al.* 1985). The EOS gets very complicated, as many-body forces, relativistic effects, and extra hadronic and mesonic degrees of freedom must be taken into account (*e.g.*, Vautherin 1994).

Published EOSs show that the adiabatic index lies in the range $1.27 \lesssim \gamma \lesssim 1.325$ for densities $\varrho \lesssim 2 \times 10^{14}\,\mathrm{g\,cm^{-3}}$ and entropies $S \approx 1\,k_\mathrm{B}/\mathrm{nucleon}$ (El Eid & Hillebrandt 1980; Hillebrandt & Wolff 1985; Lattimer *et al.* 1985), where k_B is Boltzmann's constant. If the density exceeds nuclear saturation density the EOS stiffens considerably due to the repulsive part of the nuclear forces ($\gamma > 2.5$).

For a qualitative discussion of the influence of the EOS on the collapse dynamics it has been proven useful to approximate the EOS by a simple polytropic law $P = K(Y_e, S)\varrho^\gamma$, where the electron fraction Y_e, the entropy S and the adiabatic index are in general density-dependent (van Riper 1978; Tohline 1984; Cooperstein & Baron 1990; Janka *et al.* 1993).

3.2.4 Collapse and Bounce. According to Chandrasekhar (1939) a star, which is supported against gravity by the (zero temperature) pressure of its relativistic electrons, can only be stabilized, if its mass is less than a critical mass

$$M_\mathrm{Ch} = 1.457\,(2Y_e)^2\,M_\odot \,, \tag{3.6}$$

which is the famous Chandrasekhar mass. At finite temperature the critical mass of a stable iron core should increase, but this increase is almost perfectly canceled by a decrease of the critical core mass due to Coulomb lattice correlations (Baron & Cooperstein 1990).

At the end of Si-core burning the electron fraction in the iron core of a massive star has a value $0.42 \lesssim Y_e \lesssim 0.44$, because of electron captures which occurred during oxygen and (core) silicon burning. The central entropy of the core is of the order 1 k_B/nucleon (Nomoto & Hashimoto 1988; Woosley & Weaver 1986, Woosley & Weaver 1988). The mass of the iron core eventually exceeds the Chandrasekhar mass, because (i) the mass of the core grows as a result of Si-shell burning, (ii) electron captures further reduce the lepton number in the core as neutrinos are lost from the star, and (iii) the pressure in the core is reduced due to (endothermic) photo-disintegration of nuclei by energetic photons (Fig 3.4). The relative importance of the last two processes depends on the stellar mass. For very massive stars ($M \gtrsim 20\,M_\odot$) photo-disintegrations are more important than EC's, but the opposite is true for less massive stars, because of their lower central entropies

The crucial quantity for the dynamical behavior of the core is the "effective" adiabatic index (van Riper & Lattimer 1982)

$$\Gamma \equiv \left.\frac{\partial \ln P}{\partial \ln \varrho}\right|_M , \qquad (3.7)$$

which describes the change of the pressure along a collapse trajectory of a given mass element. If the EOS is approximated by a polytropic form, the "effective" adiabatic index before neutrino trapping ($Y_l = Y_e$) can be written for an adiabatic collapse ($\delta S = 0$) as

$$\Gamma = \gamma + \frac{4}{3}\frac{1}{Y_e}\frac{\delta Y_e}{\delta \ln \varrho} , \qquad (3.8)$$

i.e., $\Gamma < \gamma$ if $\delta Y_e < 0$. This demonstrates that the density dependent electron captures reduce the adiabatic index, and therefore destabilize the core.

The initial entropy of the core is small ($S \approx 1\,k_B$/nucleon) and remains small during collapse ($\Delta S \lesssim 0.5$), a crucial fact first noticed by Bethe *et al.* (1979). This near adiabaticity of the collapse has important consequences. Firstly most of the nucleons remain bound in nuclei, because the entropy of a nucleus of mass number A is much less (about a factor A) than the entropy of A free nucleons. Therefore, the pressure is dominated by relativistic electrons (and neutrinos after trapping) until nuclear matter density is reached, *i.e.*, $\gamma \lesssim 4/3$. This in turn implies that the collapse of (non-rotating) iron cores cannot be stopped at densities less than nuclear matter density.

Analytical calculations (Goldreich & Weber 1980; Yahil & Lattimer 1982; Yahil 1983) and numerical simulations show that the dynamics of core collapse can be understood qualitatively by means of self-similar solutions, *i.e.*, solutions which have the same structure at all times. According to Goldreich & Weber (1980) there exists an exact stable homologous (*i.e.*, $v(r) \propto r$) solution for marginally stable polytropes with $P = \kappa_0 \varrho^\gamma$ and $\gamma = 4/3$. Obviously, the size of the homologous region cannot extend out to arbitrarily large radii, because the infall velocity cannot exceed the local free-fall velocity. Applying the Goldreich & Weber solution to realistic iron cores shows that the mass of the homologous part of the core shrinks (in mass) during collapse, because $\kappa_0 \propto Y_e^{4/3}$, which decreases through electron captures.

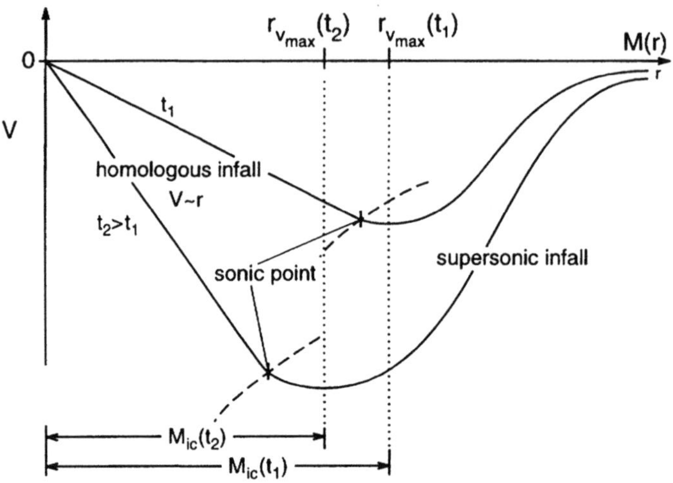

Fig. 3.6. Schematic profiles of the infall velocity versus radius at time t_1 and $t_2 > t_1$ showing the decrease (in radius and mass) of both the homologous core and the inner core (IC) due to electron captures occurring during core collapse. The edge of the homologous core coincides with the position of the sonic point, and the edge of the IC is defined by the minimum of the infall velocity. The dashed lines give the (negative of the) local sound speed

Yahil & Lattimer (1982) and Yahil (1983) have extended the work of Goldreich & Weber (1980) to polytropes, where $\gamma \equiv \Gamma$ is constant in space and time and $6/5 \leq \gamma \leq 4/3$. They found self-similar solutions for the whole core, which contain the Goldreich & Weber solution in the limit $\gamma = 4/3$, but which are not exactly homologous for smaller values of the adiabatic index. In the Yahil & Lattimer solutions the core splits into a subsonically and approximately homologously collapsing inner part and into an outer core, which is falling supersonically.

Near the edge of the homologous core exists a sonic point, where the infall velocity exceeds the local sound speed (Fig. 3.6). Somewhat further out in radius, at $r = r_{v_{max}}$, the (absolute value of the) infall velocity reaches a maximum. The mass inside the (time-dependent) radius $r_{v_{max}}(t)$ is called the mass of the *inner core* (henceforth IC). According to numerical simulations the mass of the inner core lies in the range

$$0.6\,M_\odot \lesssim M_{\rm IC} \lesssim 0.9\,M_\odot\,. \tag{3.9}$$

When the central density of the core exceeds nuclear matter density the EOS stiffens, *i.e.*, γ increases from a value somewhat below 4/3 to a value 2.5 − 3.0. According to the virial theorem the core can reach a new equilibrium position, if the (pressure weighted) average $\bar{\gamma} > 4/3$ in the core. Starting at the center of the core successive mass shells are stopped. Pressure waves move outwards in radius and accumulate within a millisecond near the sonic point, where they steepen into a shock wave (Fig. 3.7). Due to its inertia the

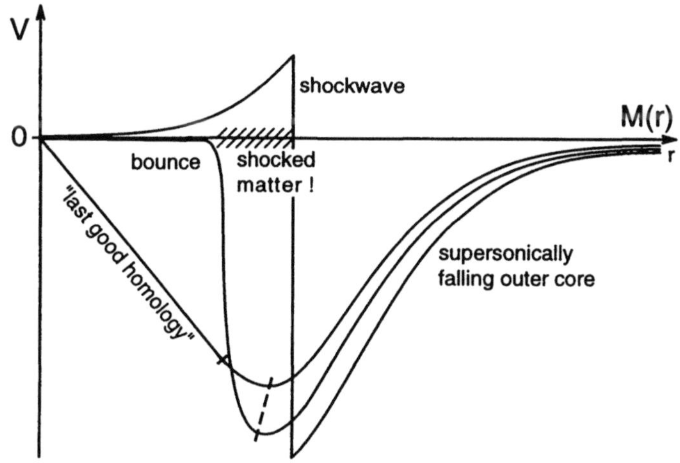

Fig. 3.7. Schematic velocity profiles versus radius at the point of "last good homology", at bounce and at the time, when the shock wave has just detached from the inner core

homologous core overshoots its equilibrium position and rebounds behind the shock wave, which has already started propagating outward. The shock detaches from the rebounding core at $r = r_{v_{\max}}$, *i.e.*, at the outer edge of the IC (Fig. 3.7).

This so-called *bounce* of the core occurs on the time scale of a radial pulsation, which is of the order 1 millisecond. For a stiff nuclear EOS the maximum density reached during bounce is less than two times nuclear saturation density, whereas for a soft nuclear EOS the collapse can continue up to 3 – 4 times, and for a very soft nuclear EOS (with $\gamma = 2$) even up to 10 times nuclear saturation density, if effects of general relativity are included in the calculation (Baron *et al.* 1985a,b).

3.2.5 Prompt Explosions and Shock Propagation. The initial energy of the shock wave is approximately equal to the kinetic energy of the IC at the moment $t = t_{\mathrm{h}}$ of last good homology (Brown *et al.* 1982), *i.e.*,

$$E^{(\mathrm{i})}_{\mathrm{shock}} \approx E^{\mathrm{IC}}_{\mathrm{kin}}\ (t = t_{\mathrm{h}}) \ . \tag{3.10}$$

During bounce the kinetic energy of the inner core is quickly transferred to the shock wave, which is an efficient absorber of sound waves. Note that the kinetic energy of matter between the sonic point (*i.e.*, the edge of the homologous core) and $r_{v_{\max}}$ (*i.e.*, the edge of the IC) is dissipated before the shock detaches from the IC and, therefore, has to be included into the initial energy of the shock wave (Fig. 3.7). The moment of last good homology is defined as the time, when the center of the core begins to be decelerated by the stiffening nuclear EOS.

Numerical simulations and analytical estimates (Mönchmeyer 1993) show that the initial energy of the shock wave is of the order of

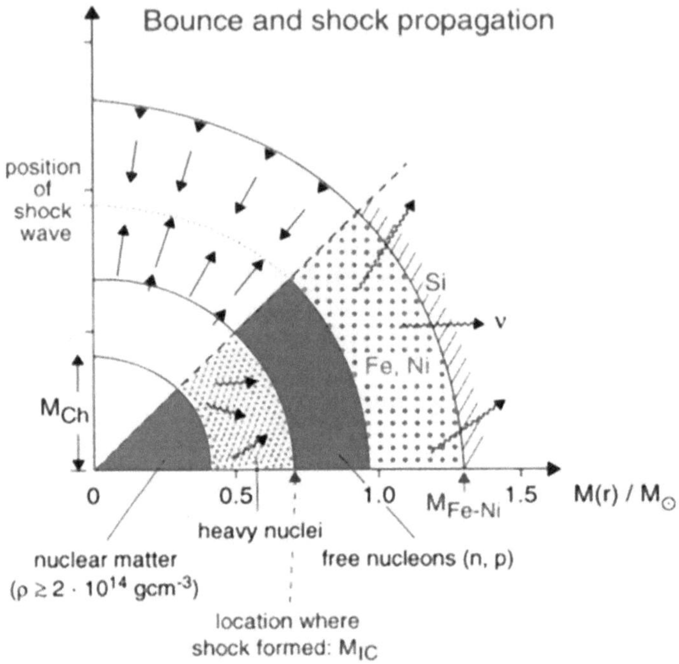

Fig. 3.8. Core collapse after shock formation: Due to the stiffness of nuclear matter the collapse has been stopped and a shock wave has formed at the edge of the inner core, which has approached approximate hydrostatic equilibrium. The central part of the inner core consists of homogeneous nuclear matter, while matter in its outer layers is made up predominantly of heavy neutron-rich nuclei. Matter from the outer core, which falls through the outward propagating shock wave is photo-disintegrated into free nucleons. This is a strongly endothermic process, which weakens the shock and eventually turns it into a standing accretion shock

$$E^{(i)}_{\text{shock}} \approx (4 - 10) \times 10^{51}\,\text{erg}\;, \tag{3.11}$$

much more than is required to explain the observed (kinetic and electromagnetic) energy output of a supernova. However, while propagating outward the shock wave is severely damped by photo-disintegration of heavy nuclei into free nucleons (if $S \gtrsim 5\,k_B$/nucleon) or alpha particles (if $3\,k_B/\text{nucleon} \lesssim S \lesssim 5\,k_B$/nucleon), the energy loss being of the order (Hillebrandt, W. & Müller 1981; Mazurek 1982; Fig. 3.8)

$$E_{\text{loss}} = (1.6 - 1.8) \times 10^{51}\,\text{erg} \,/\, 0.1\text{M}_\odot\;. \tag{3.12}$$

After the shock wave has crossed the neutrino sphere, it is further weakened by neutrino losses.

The shock wave also gains energy from the dissipated kinetic energy of matter passing through it. However, for spherically symmetric models it can be shown (Mönchmeyer 1993) that the kinetic energy is not sufficient to guarantee an expansion of the shock heated matter if $\gamma \lesssim 1.4$. Even in the layers, where nuclei are completely disintegrated ($\gamma \gtrsim 1.5$), most of the dissipated kinetic infall energy is needed to bring this matter into hydrostatic equilibrium while it is added to the quasi-static proto-neutron star. When most of the initial shock energy is lost due to photo-disintegration, the shock is no longer strong enough to lead to complete distintegration of the nuclei in the post-shock region and the value of the post-shock γ drops below 1.4. The shock must turn into an accretion shock, if no further energy source is available. Such an additional source of energy is the heating of the shocked matter immediately behind the shock by energetic neutrinos diffusing out of the opaque central regions of the collapsed core (see below).

Numerical simulations had to be performed to clarify, whether the energy losses of the shock are so severe that it stalls and turns into an accretion shock, or whether the shock wave is able to reach the outer layers of the star and cause a supernova explosion. As the (crucial) energy loss due to photo-disintegration is proportional to the difference between the mass of the iron core and that of the IC at bounce (Fig. 3.8), it is obvious that the best candidates for Type II SNe are stars with small iron cores ($M \lesssim 1.2\,M_{\odot}$) and low initial entropy, the latter resulting in a large mass of the IC (Hillebrandt 1987). In particular, the energy loss given in Eq. (3.12) implies a limiting mass, which can be photo-disintegrated by the shock, of the order

$$M_{\rm core} - M_{\rm IC} \lesssim 0.45\,M_{\odot}\ . \tag{3.13}$$

For an IC with a mass in the range $0.8\,M_{\odot} \lesssim M_{\rm IC} \lesssim 0.9\,M_{\odot}$ as found in numerical simulations a so-called *prompt explosion* can only occur, if $M_{\rm core} < 1.35\,M_{\odot}$.

Among researchers in the field there is now general agreement that when the detailed effects of neutrino transport are taken into account currently available initial models and in particular models with iron cores more massive than about 1.35 $M_{\odot}$ cannot explode by the prompt mechanism, no matter what the right equation of state is (for more details, see *e.g.*, Müller 1990).

3.2.6 Delayed Explosions. A more viable Type II supernova explosion mechanism which relies on neutrino heating was found by Wilson (1985) (Bethe & Wilson 1985; Wilson *et al.* 1986; Colgate 1989; Janka 1993). In this so-called *delayed explosion mechanism* energy is deposited in the layers between the nascent neutron star and the stalled prompt supernova shock during a period of a few 100 ms by absorption of a small fraction (between 1% and 10%) of the neutrinos emitted from the collapsed core (Fig. 3.9). The neutrino energy deposition increases the pressure behind the shock front and the respective layers of the star begin to expand. Because of this expansion an extended region of low density, but quite high temperature is created above the surface of the proto-neutron star. The persistent energy input by neutrinos maintains high pressure in the radiation dominated "hot bubble" region, thus driving the shock outward, which eventually causes the supernova explosion.

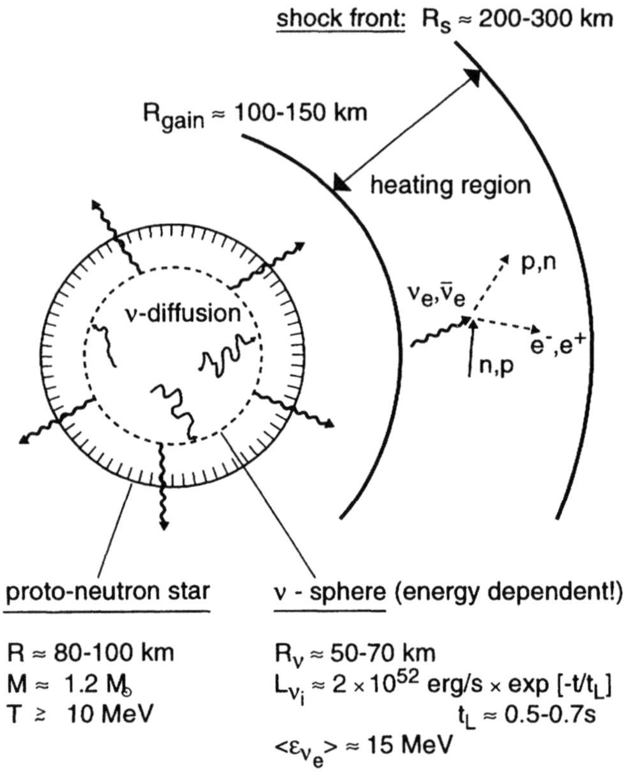

Fig. 3.9. Sketch of the proto-neutron star and the "hot bubble" region several ten milliseconds after core bounce (see text for details)

During the phase of matter accretion, lasting for up to more than hundred milliseconds after core bounce, material is falling through the standing accretion shock. Initially cooling processes dominate in this material, which radiates away its gravitational binding energy by neutrinos, and settles onto the proto-neutron star. As the neutron star becomes more and more compact and the accretion rate shrinks with the decreasing density of the infalling layers of the progenitor star, there is a moment, when net heating by neutrino energy deposition *must* win against neutrino cooling.

To see this, let us compare the most important neutrino processes in the hot, dissociated post shock material during the accretion phase, namely electron/positron absorption on protons/neutrons and the inverse reactions:

$$e^- + p \longleftrightarrow \nu_e + n \ , \qquad e^+ + n \longleftrightarrow \bar{\nu}_e + p \ . \tag{3.14}$$

The cooling rates per baryon due to the emission of ν_e and $\bar{\nu}_e$, respectively, in these capture reactions can approximately be written as (see Janka 1993 for a discussion and justification of the involved approximations)

$$Q^{-}_{\nu_e,\bar{\nu}_e} \cong 2.4 \cdot \left(\frac{T}{1\,\text{MeV}}\right)^6 \cdot \begin{Bmatrix} Y_\text{p} \\ Y_\text{n} \end{Bmatrix} \quad \left[\frac{\text{MeV}}{\text{s} \cdot \text{nucleon}}\right] . \tag{3.15}$$

The corresponding heating rates by the inverse processes are

$$Q^{+}_{\nu_e,\bar{\nu}_e} \cong 110 \cdot \frac{L_{52}}{r_7^2} \frac{\langle \varepsilon^2 \rangle}{(15\,\text{MeV})^2} \cdot \begin{Bmatrix} Y_\text{n} \\ Y_\text{p} \end{Bmatrix} \quad \left[\frac{\text{MeV}}{\text{s} \cdot \text{nucleon}}\right] . \tag{3.16}$$

Here, T denotes the gas temperature. Y_n and Y_p are the mean concentrations (*i.e.*, numbers per baryon) of free neutrons and protons, respectively. L_{52} is the ν_e- or $\bar{\nu}_\text{e}$-luminosity in 10^{52} erg/s, r_7 gives the distance from the compact neutrino source in 10^7 cm, and $\langle \varepsilon^2 \rangle$ defines the average squared energy of the neutrinos.

In the thin, hydrostatic, isothermal atmosphere of a neutron star with the pressure being dominated by a Boltzmann gas of baryons, the density declines exponentially with height. For the typical temperatures $T \gtrsim 1\,\text{MeV}$ this implies that at sufficiently small densities relativistic electrons, positrons, and photons will start to yield the main contribution to the pressure, *i.e.*, $P \propto T^4$. Hence, the matter in the gravitational field of the neutron star must be stabilized by a temperature gradient instead of a density gradient. In case of an adiabatic atmosphere, *i.e.*, $\varrho \propto T^3$, this requires a temperature decline $T \propto r^{-1}$. However, for neutrino heating being balanced by neutrino cooling, $Q^+ = Q^-$, Eqs. (3.15) and (3.16) imply $T \propto r^{-1/3}$. Hence, net energy deposition in a region close to the newly formed neutron star is unavoidable in the described situation.

Simulations showed that neutrino heating indeed works as predicted by the analytical considerations discussed above (Wilson 1985; Bethe & Wilson 1985; Wilson *et al.* 1986; Wilson & Mayle 1988; Colgate 1989; Bruenn 1993; Janka 1993). However, as pointed out by Janka (1993; see also Wilson & Mayle 1988), who performed one-dimensional simulations of the post bounce evolution until more than 10 seconds after bounce, observed explosion energies cannot be obtained through energy deposition by the slowly decaying neutrino emission during the Kelvin–Helmholtz cooling of the proto-neutron star. Instead, he found that energetic explosions require the occurrence of an early phase of high neutrino luminosities (*e.g.*, boosted by convection; see next section) with efficient heating in the vicinity of the neutron star. The main reason for this is that a sufficiently large mass must be heated and must store the energy deposited by neutrinos. Since processes like neutrino-electron scattering and $\nu\bar{\nu}$–annihilation are much less efficient in transferring neutrino energy to the stellar matter, the supernova explosion does not gain much energy any more at the late epochs when the densities around the proto-neutron star are already rather low.

3.3 Observations Demanding Nonspherical Models

Various independent observational evidence exists that substantial non-radial motion and mixing has occurred during the explosion of SN 1987A (for a review see, *e.g.*, Arnett *et al.* 1989a). This evidence includes the early detection of X-rays (*e.g.*, Dotani *et al.* 1987; Sunyaev *et al.* 1987; Wilson *et al.* 1988; Pendleton *et al.* 1995) and γ-rays (Matz *et al.* 1988; Mahoney *et al.* 1988; Sandie *et al.* 1988; Cook *et al.* 1988; Gehrels *et al.* 1988; Teegarden *et al.* 1989), both of which can most easily be explained if radioactive ^{56}Co had been mixed from the interior regions of the star into the envelope where the optical depth was much smaller. Assuming that the radioactive ^{56}Co is homogeneously mixed over a certain interior region of the star Arnett (1988) found that the radius of that region has to be of the order of 20 to 30% of the stellar radius in order to explain the time behaviour of the observed X- and γ-ray flux from SN 1987A.

This interpretation is supported by observations of the line widths of infrared emission lines of Fe II (Erickson *et al.* 1988; Haas *et al.* 1990; Colgan *et al.* 1994) and of Ni II, Ar II, and Co II (Rank *et al.* 1988; Barthelmy *et al.* 1989; Witteborn *et al.* 1989; Colgan *et al.* 1994). According to Haas *et al.* (1990) and Colgan *et al.* (1994) the inferred expansion velocity of the bulk of the Fe II is about 2000 km/s, and up to 4000 km/s for a significant fraction (8 to 30%) of the iron mass. However, this measured bulk velocity of Fe II is significantly higher than the $\approx$ 800 km/s predicted by spherically symmetric theoretical models at this epoch for those stellar layers, which should contain the iron and nickel synthesized during the explosion. Moreover, these layers were still inside the photosphere at the time of the infrared observations.

Analyzing the light curves of several Fe, Co and Ni infrared lines Li *et al.* (1993) further found that the observations can be best explained, if these elements were mixed outward macroscopically and inhomogeneously in the form of about 60 to 100 identical clumps. Evidence for mixing and clumping also comes from asymmetries of the spectral line profiles (Spyromilio & Meikle 1990; Stathakis *et al.* 1991; Spyromilio & Pinto 1991; Spyromilio *et al.* 1993; Spyromilio 1994). A similar analysis showed evidence for clumping in SN 1987F (Filippenko & Sargent 1989) and SN 1988A (Spyromilio 1991).

Another observational hint stems from the sudden appearance of fine-structure in the Balmer lines H_α–H_δ between day 20 and day 30 (so-called Bochum event; Hanuschik & Thimm 1990; Hanuschik *et al.* 1993; see also Chugai 1991), which most likely is due to a local change in the excitation and ionization conditions, probably caused by radioactive decay of mixed-up ^{56}Co in the layers just above the photosphere. Similar observations in SN 1985L (Filippenko & Sargent 1986) and SN 1985P (Chalabaev & Cristiani 1991) indicate that such a kind of event may be generic to type II supernovae.

The smoothness of the light curve is also indirect evidence of the need for mixing to occur (Arnett 1988; Woosley 1988; Shigeyama *et al.* 1988). Mixing of hydrogen into the center of the star allows for a time spread in the liberation of the recombination energy and thus helps to model the rather smooth and broad maximum observed in the light curve of SN 1987A (Shigeyama & Nomoto 1990). On the other side does mixing of heavies into the hydrogen

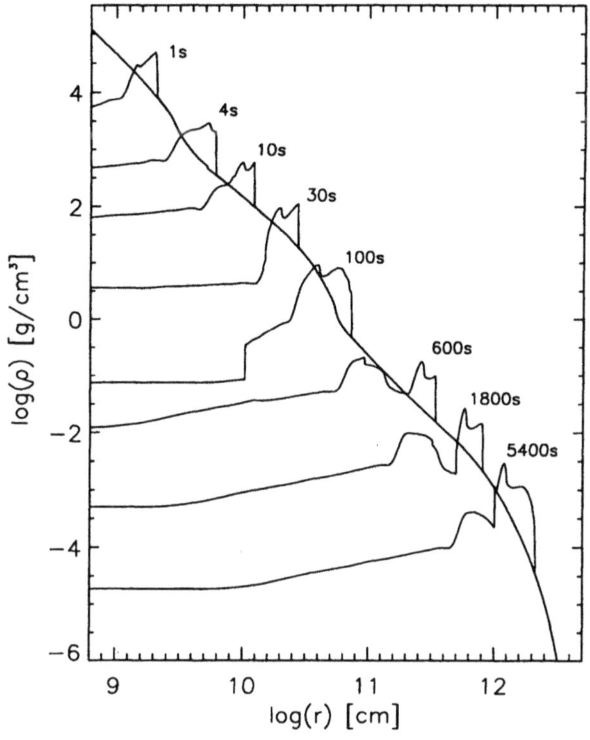

Fig. 3.10. Evolution of the density distribution of a $15M_{\odot}$ star caused by the propagation of the shock wave (launched during core collapse) through the stellar envelope. The rapid drop of the density near composition interfaces gives rise to a non-steady shock propagation which in turn causes density inversions

rich envelope lead to a homogenizing of the opacity, which again smoothes the light curve (*e.g.*, Arnett *et al.* 1989a; and references therein).

Finally, strong observational support for the generic nature of mixing processes in supernova envelopes (of SNe II and SNe Ib/c) comes from the bright SN IIb supernova SN 1993J in the nearby (3.6 Mpc) galaxy M81 (for a review on SN 1993J see, *e.g.*, Wheeler & Filippenko 1996). Mixing of the ejecta was inferred from observed asymmetric profiles of the [O I] and [Mg I] emission lines at late times (Spyromilio 1994; Wang & Hu 1994) and from light curve modeling (Shigeyama *et al.* 1994; Woosley *et al.* 1994b).

Theoretically, there exists no obvious reason that instabilities, mixing and clumping in supernova envelopes and ejecta are specific to SN 1987A. These processes will occur generically in core collapse supernovae, because the shock wave launched during core collapse will always encounter composition interfaces (*i.e.*, rapid drops of density) on its way out through the stellar envelope (Fig. 3.10). Hence, at these interfaces the acceleration of the shock, caused by its propagation down the density slope, reaches a local maximum. The relative deceleration of the shock wave past the composition interface causes a reverse shock and eventually an unstable density and pressure stratification. Consequently, instabilities and mixing are unavoidable. Only the strength and the number of instabilities (composition interfaces) will vary from star to star. Actually, observation of mixing, inferred from light curve and spectra, can provide interesting diagnostics of the internal structure of the progenitor (*e.g.*, Iwamoto *et al.* 1997).

3.4 Rayleigh–Taylor Instabilities in Supernova Envelopes

As stated by Chandrasekhar (1961): *The Rayleigh–Taylor instability derives from the character of the equilibrium of an incompressible heavy fluid of variable density. An important special case in this connexion is that of two fluids of different densities superposed one over the other (or accelerated towards each other).* In order to gain some insight in the Rayleigh–Taylor (henceforth RT) instability I will discuss two special cases in some detail.

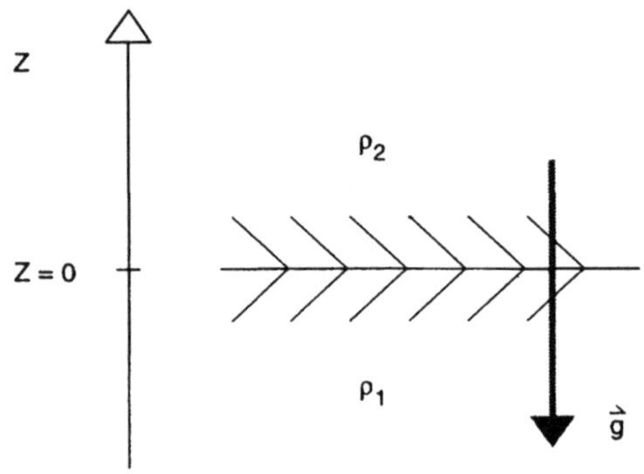

Fig. 3.11. Two uniform fluids of constant density ϱ_1 and ϱ_2 separated by a horizontal boundary at $z = 0$. The fluids are subject to an acceleration g (*e.g.*, due to gravity) acting in negative z-direction. If $\varrho_2 > \varrho_1$ the arrangement of fluids is Rayleigh–Taylor *unstable*

Following Chandrasekhar (1961) let us consider the static case of two uniform fluids of constant density ϱ_1 and ϱ_2 separated by a horizontal boundary at $z = 0$ (Fig. 3.11). The fluids are subject to an acceleration g (*e.g.*, due to gravity) acting in negative z-direction, and the interface between the fluids is subject to forces arising from surface tension T. Suppose further that this system is slightly perturbed. Then by a normal mode analysis of the disturbance, *i.e.*, seeking solutions of the form $\exp(ik_x x + ik_y y + nt)$, where k_x, k_y and n are constants, the following dispersion relation can be derived (Chandrasekhar 1961)

$$\tau^{-2} = gk \left\{ \frac{\varrho_2 - \varrho_1}{\varrho_2 + \varrho_1} - \frac{k^2 T}{g(\varrho_2 + \varrho_1)} \right\} . \tag{3.17}$$

Here the growth rate $\tau = n^{-1}$, and the absolute value of the wave vector of the perturbation $k = (k_x^2 + k_y^2)^{1/2}$. According to Eq. (3.17) the arrangement of fluids is *stable*, if $\varrho_2 < \varrho_1$, *i.e.*, when the lighter fluid is on top of the heavier fluid, because $n^2 < 0$ in that case. Otherwise, if $\varrho_2 > \varrho_1$, the arrangement of fluids is *unstable* for all wave numbers in the range $0 < k < k_c$, where $k_c = [(\varrho_2 - \varrho_1)\, g/T]^{1/2}$, but is stable for wavenumbers $k > k_c$, because surface tension succeeds in stabilizing short wavelength perturbations.

Let us consider a second simple case, namely that of a fluid confined between two rigid planes at $z = 0$ and $z = d$ which has an exponentially varying density $\varrho = \varrho_0 \exp(\beta z)$ with a typical (constant) density-scale height β^{-1}. The corresponding dispersion relation is (Chandrasekhar 1961)

$$\tau^{-2} = g\beta \left[1 + \frac{\frac{1}{4}\beta^2 d^2 + m^2\pi^2}{k^2 d^2}\right] , \tag{3.18}$$

where m is an integer ≥ 1. Obviously the stratification is ***stable if β is negative***, while it is ***unstable if β is positive.*** Note that the modes with shortest growth time have $m = 1$, and that if d, k and m are given τ is smallest, if $\frac{1}{4}\beta^2 d^2 = k^2 d^2 + m^2\pi^2$.

Since the energy given to the shock wave in a supernova explosion is much larger than the binding energy of the mass outside the collapsed core, gravity is dynamically unimportant for the propagation of the shock wave through the stellar envelope. Thus, as first pointed out by Chevalier, the only way to drive a RT instability is by ***pressure and density gradients of opposite signs***, the ***"effective acceleration"*** being the (negative) pressure gradient, *i.e.*,

$$g \Longrightarrow -\frac{1}{\varrho}\frac{\partial p}{\partial r} . \tag{3.19}$$

In order to be of any consequence for the supernova explosion the RT instability growth time, τ_{RT}, obviously must be shorter than the hydrodynamic time scale, $\tau_{\mathrm{hydro}} \equiv r_{\mathrm{sh}}/v_{\mathrm{sh}}$, where r_{sh} and v_{sh} are the radius and the velocity of the shock wave. Substituting g in Eq. (3.18) according to relation (3.19) Chevalier (1976) derived the following RT instability criterion for power-law density distributions:

$$\left[\left(\frac{H_1}{r_{\mathrm{sh}}}\right) \cdot \min\left\{\left(\frac{H_1}{r_{\mathrm{sh}}}\right), \left(\frac{H_2}{r_{\mathrm{sh}}}\right)\right\}\right]^{-1/2} \cdot \frac{c_{\mathrm{s}}}{v_{\mathrm{sh}}} > 1 , \tag{3.20}$$

where $H_1 \equiv \left|(\partial \ln P/\partial r)^{-1}\right|$ and $H_2 \equiv \left|(\partial \ln \varrho/\partial r)^{-1}\right|$ are the (absolute) pressure and density-scale heights, and where c_{s} is the local sound speed.

3.5 Simulations of RT Instabilities in Supernova Envelopes

Soon after the first observational evidence was obtained that instabilities and mixing had occurred in SN 1987A (see Sect. 3.3), several groups began to perform multi-dimensional simulations using polytropic and realistic stellar models (Arnett *et al.* 1989b; Müller *et al.* 1989; Benz & Thielemann 1990; Den *et al.* 1990; Hachisu *et al.* 1990; Yamada *et al.* 1990; Fryxell *et al.* 1991; Hachisu *et al.* 1992; Herant & Benz 1991; Müller *et al.* 1991a,b,c). In all these simulations the explosion was started artificially by liberating the typical type II supernova (kinetic) energy of about 10^{51} erg in the innermost zones of the stellar model. Thus, it was assumed (because of numerical reasons and because of our ignorance concerning the explosion mechanism) that the propagation of the shock wave through the stellar envelope can be considered without a detailed treatment of the explosion mechanism. We shall see below that this assumption has to be abandoned. In order to trigger the instability the initially spherically symmetric stellar models had to be perturbed, the amplitudes of the perturbations ranging from 0.1 to 10%.

The hydrodynamic simulations showed in particular that Rayleigh–Taylor (RT) instabilities occur in realistic stellar models near composition interfaces and that the instabilities reach the nonlinear stage characterized by narrow finger-like structures which are topped by "mushroom caps". About 5 hrs after the onset of the explosion the He-shell is shredded into clumpy fragments the density contrast exceeding a value of five (Fig. 3.12; for more details see, *e.g.*, Müller *et al.* 1991b,c). Three-dimensional calculations confirm the results obtained in two-dimensions (Müller *et al.* 1991a), and also show pronounced instabilities producing clumpy structures, which have a density up to a factor of ten higher than the surrounding matter.

Furthermore, Herant & Benz (1991, 1992) showed that the mixing induced by RT instabilities occurring during the first few hours is substantially modified at later times by the radioactive decay of ^{56}Ni and ^{56}Co, which increases the peak velocity of the nickel and of its decay product iron by approximately 30%. These "boosted" iron velocities are consistent with the velocities deduced from infra-red iron lines for the bulk of the observed Fe (Haas *et al.* 1990). However, they fail to explain the high velocity wings of the lines, which extend to velocities $\gtrsim 3000\,\mathrm{km/s}$, and which, according to Haas *et al.* (1990), represent between 8 and 30% of the iron. The 17.94 and 25.99 μm Fe II line profiles observed by Haas *et al.* (1990) and Colgan *et al.* (1994) also contain an unresolved $3-5\sigma$ emission feature, which can be interpreted as a high velocity clump of material containing $\approx 3\,\%$ of the total iron mass and moving with a velocity of about 3900 km/s.

The results of the hydrodynamic simulations strongly indicate that the mixing induced by the RT instabilities in the envelope is insufficient to explain the observations of SN 1987A and, in particular, the early discovery of γ-rays, the large velocities of iron, and the clumpy structure of the ejecta. This *quantitative* failure of the simulations may be taken as evidence that non-radial motion must have occurred in type-II supernovae already during the very early phases of the explosions, *e.g.*, induced by the convectively unstable hot bubble created by neutrino heating in the "delayed explosion" scenario (see Sects. 3.2.6 and 3.6). Concerning the simulations this also implies that the formation of the shock wave and its propagation cannot be studied independently, as it has been done in all investigations up to now.

Finally, multi-dimensional hydrodynamic simulations have shown that RT instabilities do not only occur in SNe II explosions of blue super-giants as SN 1987A, but do also occur in SNe IIP explosions of red super-giants (Herant & Woosley 1994; Shigeyama *et al.* 1996) and in helium star models of SNe Ib (Hachisu *et al.* 1991, 1994; Iwamoto *et al.* 1997).

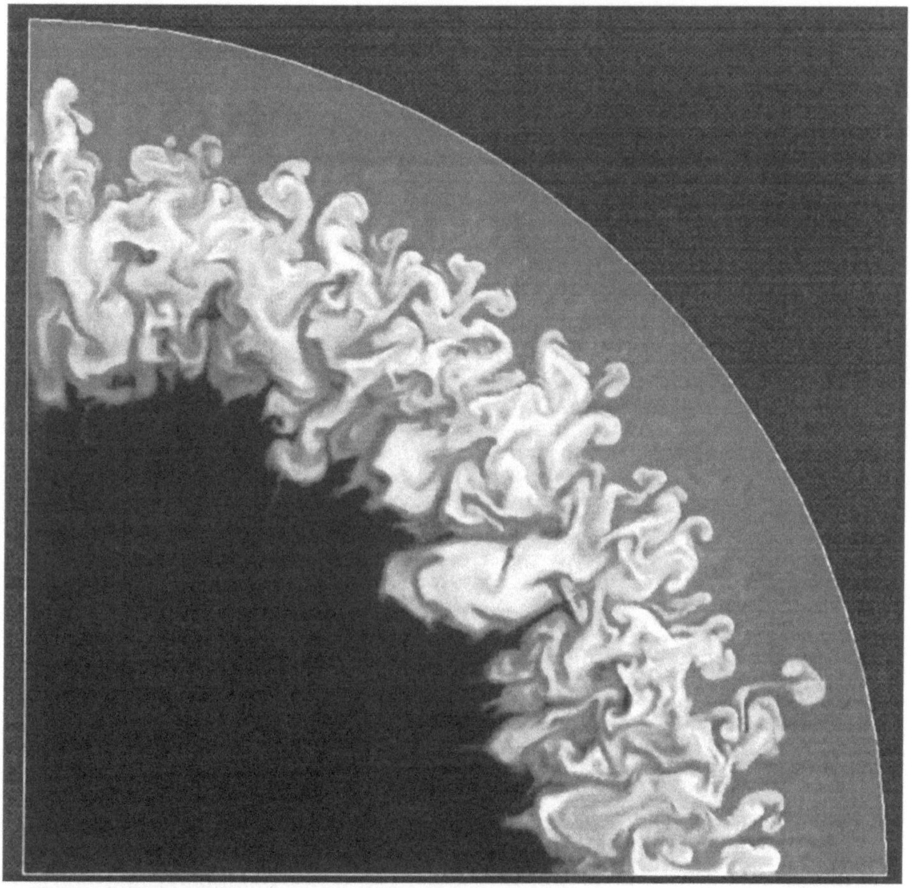

Fig. 3.12. Helium mass fraction computed on a moving grid of 800 × 400 zones in spherical coordinates (r, θ) for an explosion energy of 10^{51} erg and an initial random perturbation of 10 % amplitude about 4 hrs after the onset of the explosion. The plot shows only a part of the (moving) computational grid up to a radius of 2.7×10^{12} cm. The helium shell (bright grey) has been shredded by the RT instabilities. (From Müller *et al.* 1991b)

3.6 Neutrino Driven Convective Instabilities

The criterion for the occurrence of convection in matter with density ϱ, entropy per baryon S and electron number fraction Y_e is given by (Epstein 1979)

$$-\left.\frac{\partial \ln \varrho}{\partial \ln P}\right|_{S,Y_e} \left[\left.\frac{\partial \ln P}{\partial S}\right|_{\varrho,Y_e} \frac{\mathrm{d}S}{\mathrm{d}r} + \left.\frac{\partial \ln P}{\partial Y_e}\right|_{\varrho,S} \frac{\mathrm{d}Y_e}{\mathrm{d}r} \right] > 0 \,, \tag{3.21}$$

where r is the radial coordinate. Hence, a sufficient condition for convective motion is given, when both the gradients of entropy and electron number fraction are negative, *i.e.*, $\mathrm{d}S/\mathrm{d}r < 0$ and $\mathrm{d}Y_\mathrm{e}/\mathrm{d}r < 0$.

3.6.1 Convection Inside the Proto-Neutron Star. After the shock wave stalls ($\approx$ 10 ms after bounce) and before the delayed neutrino heating mechanism begins to become effective ($\approx$ 100 ms after bounce) the outer parts of the proto-neutron star are convectively unstable because of two reasons. Firstly, the deleptonization occurring in the shocked matter outside the neutrino-sphere produces a negative lepton gradient. Secondly, the weakening (and the eventual stalling) of the prompt shock wave gives rise to a negative entropy gradient in the same region. According to Eq. 3.21 this is a sufficient condition for convective motion to occur. This situation is commonly encountered in supernova simulations (Burrows 1987; Burrows & Lattimer 1988; Bruenn 1993).

The existence of convective instabilities inside the proto-neutron star has been demonstrated in two-dimensional (Burrows & Fryxell 1992, 1993; Janka & Müller 1993, 1996a; Mezzacappa *et al.* 1998a) and three-dimensional hydrodynamic simulations (Müller 1993; Müller & Janka 1994, 1997), which covered the onset and growth of the convective instability until $\approx$ 50 ms after core bounce. In all simulations an inner boundary condition at a fixed finite radius was used in order to save a significant amount of computer time. Hence, the flow was not simulated in the innermost part of the proto-neutron star.

The evolution of the convective instabilities in the mantle of proto-neutron stars is illustrated in Fig. 3.13, which is taken from Müller & Janka (1997) and which shows the time evolution of the electron number fraction Y_e in a region between 15 km (the radius of the fixed inner boundary) and 155 km. The shocked and neutronized material starts to develop finger-like structures on growth time scales of roughly 10 ms penetrating inward (right upper panel in Fig. 3.13). As the convective velocities reach or even exceed the local sound speed, significant undershooting and overshooting occurs at the inner and outer boundaries of the unstable region, respectively. The initially narrow fingers begin to merge into successively larger blobs. After another 5–10 ms the whole unstable region is involved in a convective overturn (left lower panel in Fig. 3.13). Finally, after about 20 ms the inner region is completely mixed and homogenized.

The extent of overshooting depends on whether the flow is simulated in two or three space dimensions. A mode analysis performed by Müller & Janka (1997) reveals that in 2D the convective structures are roughly a factor of two larger than in 3D, where the dominant modes have (angular) wavelengths of about $20^o - 30^o$ (*i.e.*, clump sizes of about $10^o - 15^o$). Consequently, the 2D structures have more inertia and give rise to a larger amount of overshooting (50% more in 2D than in 3D).

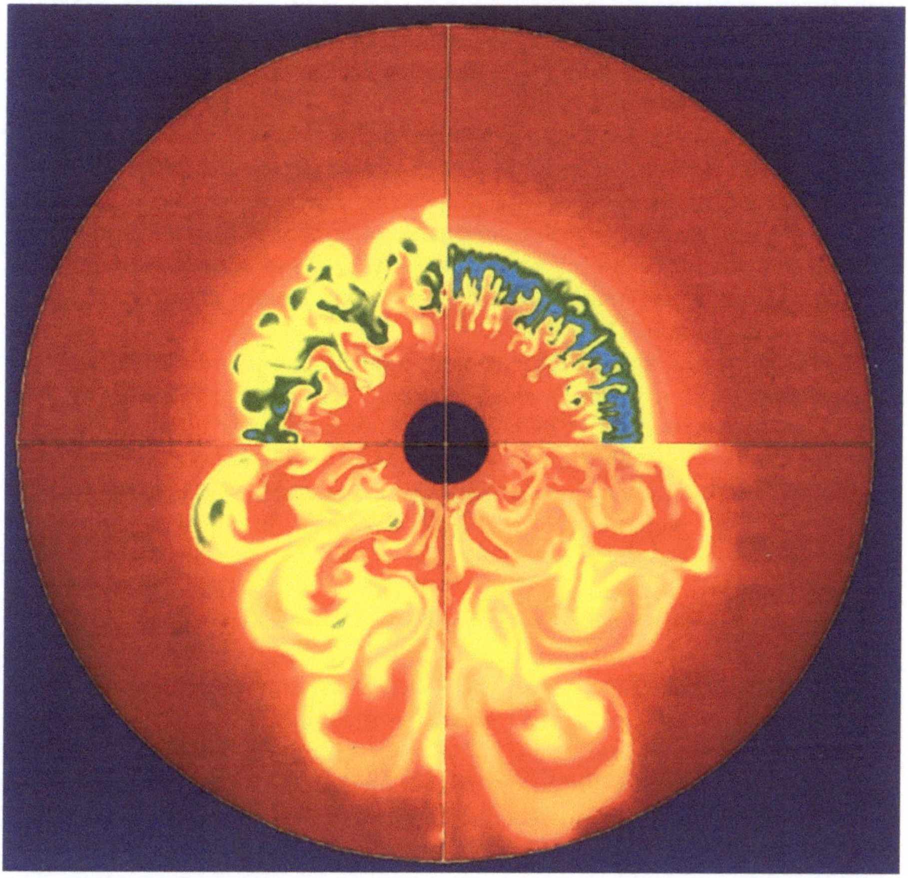

Fig. 3.13. Convection inside the nascent neutron star. The panels show the time evolution of the electron number fraction Y_e in a region between 15 km and 155 km. The panels are arranged in counter-clockwise order, starting from the right upper side, and show snapshots at 12 ms, 14 ms, 18 ms, and 21 ms after the start of the 2D simulation. The color levels correspond to Y_e-values in the range $[0.1, 0.5]$ increasing from blue, over green, yellow and orange to red. Blue regions have $Y_e \lesssim 0.12$, yellow regions correspond to $0.13 \lesssim Y_e \lesssim 0.16$, and orange regions to $0.16 \lesssim Y_e \lesssim 0.21$. (From Müller & Janka 1997)

As the convective velocities reach and even partially exceed the local sound speed ($v_{\rm conv} \approx 10^9$ cm/s) strong pressure waves and even weak shock waves are generated by the convective flow. This convective activity inside the proto-neutron star causes significant inhomogeneities in the temperature and density stratifications. The inhomogeneities vary in time and space implying non-radial mass motions and a time-dependent mass quadrupole moment associated with the outer layers of the proto-neutron star. Both effects generate gravitational radiation with typical frequencies in the range of several hundred Hz to about one kHz (see Sect. 3.8). An additional gravitational wave signal is produced by the anisotropic neutrino emission found in the multi-dimensional models (Epstein 1978; Burrows & Hayes 1996; Müller & Janka 1997).

The convective mixing found in the multi-dimensional simulations around and below the neutrino-spheric region near the proto-neutron star surface is accompanied by an increase of the neutrino luminosities during the early phase of the supernova explosion. Since the convection is dynamic and violent and the gas velocities are close to the local speed of sound, neutrinos are transported out of the dense interior of the newly-formed neutron star much faster than by diffusion.

Recently, the first two-dimensional hydrodynamic simulations of the deleptonization of a newly formed neutron star were performed by Keil (1997); see also Keil *et al.* (1996). These simulations cover about 1.2 s of the proto-neutron star's evolution. They were performed in spherical coordinates using a grid of 100 (moving) radial zones and 60 angular zones (1.5 degree angular resolution). Neutrino transport was computed in diffusion approximation in radial direction for every angular zone, *i.e.*, angular transport of neutrinos was neglected. Keil (1997) found that driven by negative lepton-fraction and entropy gradients, convection starts near the neutrinosphere about 20-30 ms after core bounce. This is in agreement with previous simulations (see above), which however ended at this stage. As Keil (1997) pushed his simulations further in time, he found that convection continues and moves deeper into the protoneutron star. After about one second the whole protoneutron star is convective. Keil's (1997) simulation further show that deleptonization of the proto-neutron star proceedes much faster, when convective transport is included. Compared to a 1D model, the lepton flux and the neutrino luminosities are increased by up to a factor of two in his 2D simulation. The convection below the neutrinosphere also raises the neutrinospheric temperatures and mean energies of the emitted neutrinos by 10-20%. These effects can have important implications for the supernova explosion mechanism and will change the detectable neutrino signal from the Kelvin–Helmholtz cooling of the proto-neutron star.

3.6.2 Convection in the Hot Bubble Region. Noting that the hot bubble is convectively unstable Colgate (Colgate 1989; Colgate *et al.* 1993) and Bethe (1990) pointed out that this situation may give rise to a dynamical overturn of hot, neutrino-heated, rising material and cold post-shock matter, and can lead to large-scale deviations from spherically symmetric supernova explosions. This instability was indeed found in numerical simulations (Burrows *et al.* 1995; Herant *et al.* 1992, 1994; Janka & Müller 1995a,b, 1996a,b; Mezzacappa *et al.* 1998b; Miller *et al.* 1993; Müller & Janka 1994; Shimizu *et al.* 1993; 1994; Yamada *et al.* 1993).

In order to describe the formation of the convective instability and to simulate the convective motion in the "hot bubble" region an adequate treatment of neutrino-matter interaction is necessary and the generation and transport of neutrinos must be modeled. This can only be done in some approximate manner, because a multi-dimensional, multi-flavor and energy-dependent treatment of neutrino transport is presently not feasible. Moreover, the neutrino luminosities from the collapsed stellar core are dependent on convective processes below and around the "neutrino sphere" (see previous subsection) and cannot be accurately determined without an elaborate treatment of neutrino transfer in this turbulent region. Possibly even methods to simulate neutrino transport in more than one-dimension have to be applied. Finally, the neutrino emission depends on the uncertain nuclear equation of state and on the neutrino interaction rates in the high-density material, which are only known with an accuracy of, may be, a factor of two.

Because of this lack of knowledge, Janka & Müller (1996b) have performed a sensitivity study instead of calculating a consistent but specific model. Certainly, consistency of the modeling is preferable. However, selecting a specific physical input, like EOS and neutrino opacities, as it was done by other workers in the field, provides only little insight into the physical conditions required for a successful explosion. Instead, Janka & Müller (1996b) took a published model of a stellar core shortly after core bounce (Bruenn 1993) and replaced the inner high-density region somewhat inside the neutrino-sphere by an inner boundary condition, which was allowed to contract in a controlled way and where the neutrino fluxes (and spectra) were set to chosen values. This allowed them to study the impact of the neutrino energy deposition and convection in the outer part of the collapsed core on the mechanism of the explosion.

Janka & Müller (1996b) obtained powerful explosions and cases of failure or very weak explosions in both one-dimensional and two-dimensional simulations. The crucial parameter turned out to be the neutrino luminosities of ν_e and $\bar{\nu}_e$ after bounce and their decrease with time. Strong explosions require sufficiently high neutrino fluxes for a sufficiently long period. The turbulent overturn of cold post-shock material and high-entropy, neutrino-heated gas provides an efficient mechanism to transport energy from the neutrino-heated region to the shock front and clearly aids the re-expansion of the stalled shock front. However, turbulence in the neutrino-heated layer is a crucial help for the explosion only in a narrow range of neutrino luminosities. For lower neutrino fluxes the neutrino-heating is too weak and powerful explosions cannot be obtained even with convection. For higher neutrino fluxes convection reduces the neutrino luminosity that is necessary to end up with a certain explosion energy (for more details see Janka & Müller 1996b).

The nonspherical stratification and mass flow resulting from the convective instabilities in the neutrino heated region are, like the ones occurring inside the proto-neutron star, a source of gravitational radiation. Because of the longer time scales (up to a few 100 ms) and the larger radii (up to several 1000 km) involved in the "hot bubble" convection, the typical frequencies of the emitted gravitational radiation are expected to be significantly lower and the corresponding amplitudes to be smaller (see Sect. 3.8).

The evolution of the convective instabilities in the neutrino heated region and the resulting supernova explosion are illustrated in Figs. 3.14, 3.15 and 3.16, which are taken from Janka & Müller (1996b). Behind the shock high-energy neutrinos from the hot interior transfer energy to their reaction targets and heat the stellar matter. Due to the heating an unstable entropy gradient builds up between the shock and the neutrinosphere. Depending on the imposed neutrino flux at the inner boundary, the neutrino heated region becomes convective after about 50–80 ms. This stage of the evolution is illustrated in Fig. 3.14, which shows two convective regions: (i) the lepton and entropy unstable layer inside the proto-neutron star just below the neutrinosphere, and (ii) the entropy-unstable neutrino heated region. Note that the black, central, spherical region visible in Fig. 3.14 corresponds to the inner $0.85 M_\odot$ of the proto-neutron star not treated in the simulation.

As a consequence of the transport of energy from the region of strongest neutrino-heating into the layers just behind the shock front, the pressure behind the shock rises, the shock speeds up, and the explosion of the star gains momentum. Fig. 3.15 shows the entropy distribution inside a radius of 1590 km at a time 170 ms after the start of the simulation. The supernova shock can be recognized as the deformed (outer) discontinuity. Post-shock material penetrates inward, while neutrino-heated, high-entropy gas rises in expanding bubbles. The further evolution (in a different simulation) is illustrated in Fig. 3.16, which shows the growth of the dynamically expanding region between the central neutron star and the supernova shock at 0.25 s, 0.45 s, and 0.65 s after shock formation (from top to bottom). The supernova shock is visible as the deformed discontinuity. Its (average) radial position grows from about 3×10^8 cm in the top figure to 6×10^8 cm in the middle and more than 9×10^8 cm in the lower figure. Continued heating by neutrinos increases the energy of the explosion during a period of several hundred milliseconds from 5×10^{49} erg at $t = 0.1$ s to 1.1×10^{51} erg at $t = 0.95$ s.

As the shock moves further out, the expansion of the post-shock region becomes faster, and the density between shock and neutron star drops rapidly. This is the moment ($t \approx 0.5$ s) when the explosion energy saturates and a very low-density, high-temperature bubble around the proto-neutron star appears (lower panel in Fig. 3.16). At the end of our calculation ($t = 0.95$ s) the shock wave has reached a radius of nearly 2×10^9 cm.

3.7 Rotational Core Collapse

It is well known from observations that massive stars (spectral types O–F) have surface equatorial rotational velocities on the main-sequence in the range $100\,\mathrm{km/s} \lesssim v \sin(i) \lesssim 300\,\mathrm{km/s}$, where i is the inclination angle between the axis of rotation and the light of sight (for a review, see *e.g.*, Tassoul 1978). However, the initial angular momentum distribution and its evolution

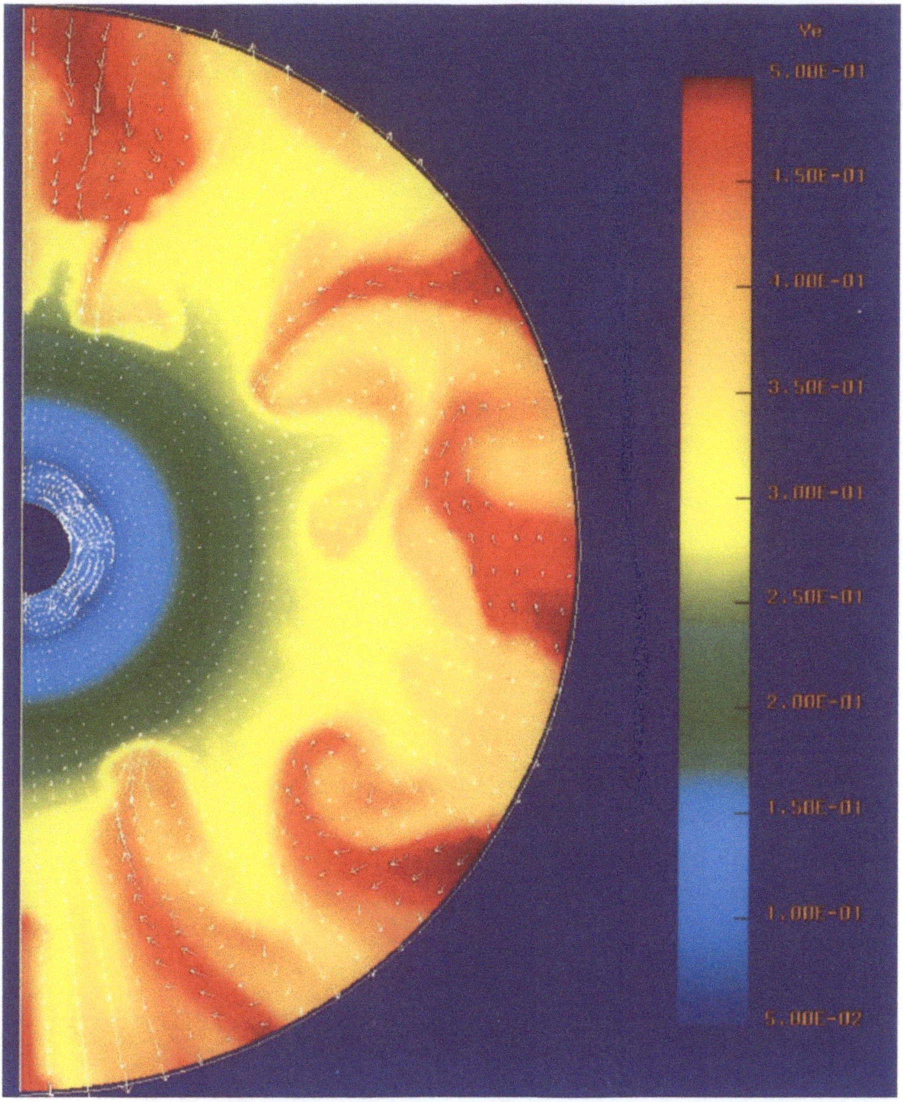

Fig. 3.14. The snapshot shows down-flows of cool, lepton-rich gas from the post-shock region (red to orange) and rising bubbles of neutron enriched, neutrino-heated gas in the inner part of the collapsed stellar core out to a radius of 333 km about 80 ms after the start of the simulation. The flow pattern is emphasized by velocity vectors. One can also recognize (blue region near the center) the convective activity below the neutrino-sphere. (From Janka & Müller 1996b)

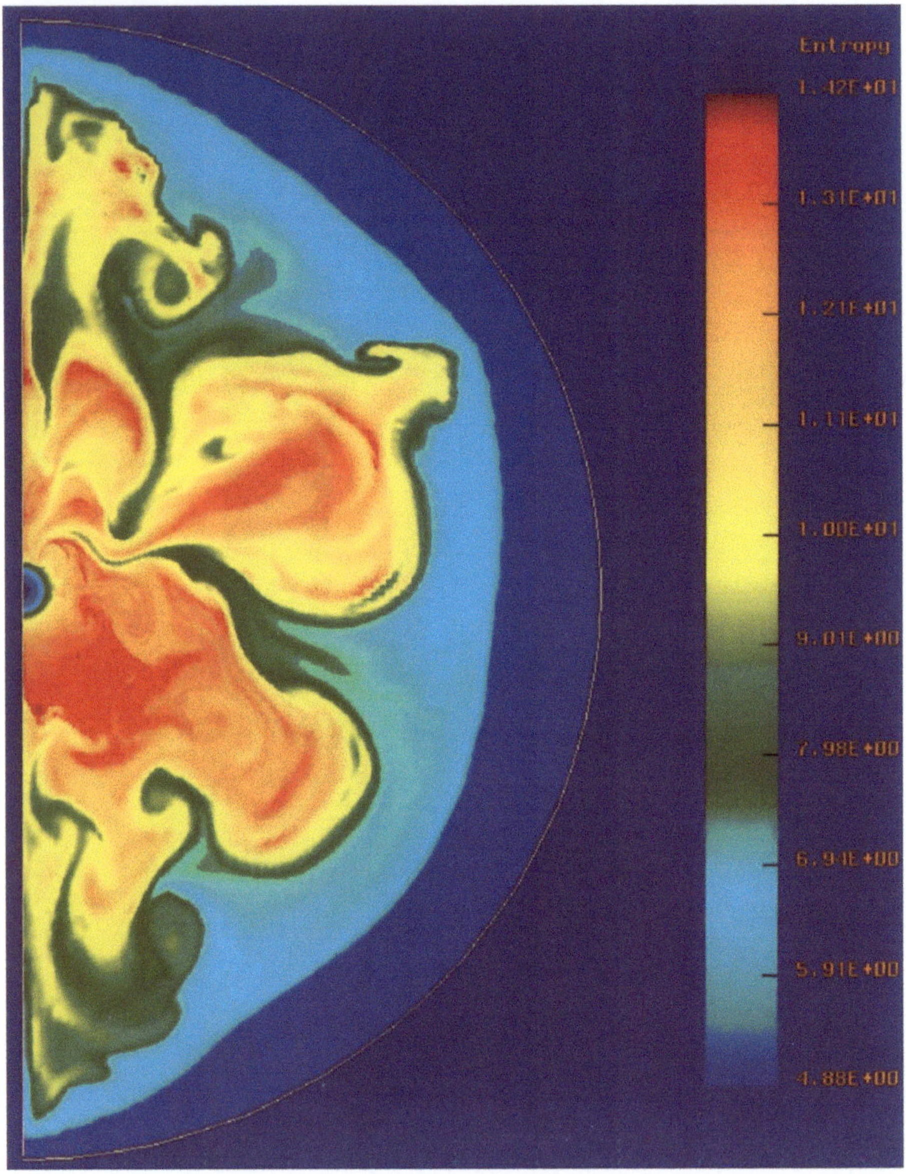

Fig. 3.15. Entropy (in k_B/nucleon) distribution in the central part of the star out to a radius of 1590 km at a time 170 ms after the start of the simulation. The supernova shock can be recognized as the deformed discontinuity between the dark blue and light blue regions. Post-shock material penetrates inward, while neutrino-heated, high-entropy gas rises in expanding bubbles. (From Janka & Müller 1996b)

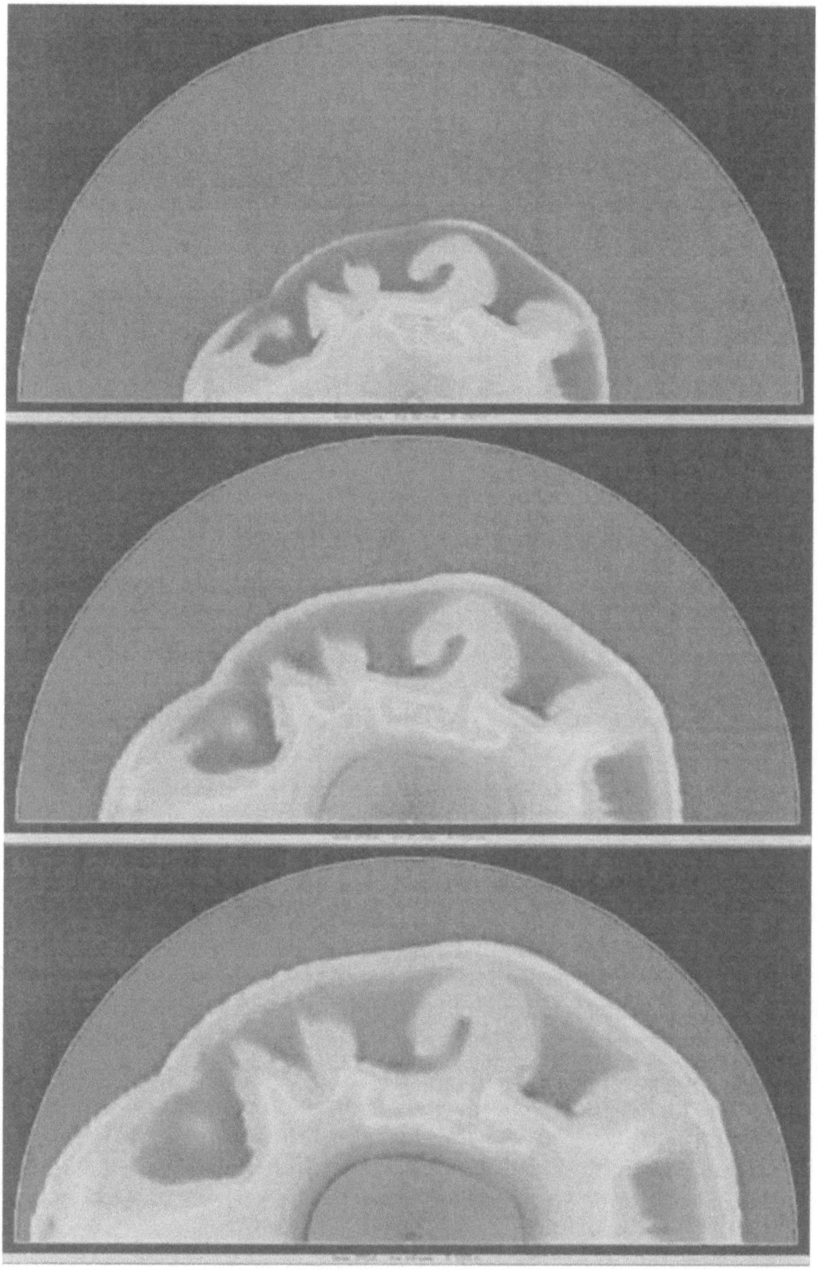

Fig. 3.16. Entropy (in k_B/nucleon) distribution in the central region of 12300 km radius at times $t = 433$ ms, 551 ms, and 649 ms after the start of the simulation (from top to bottom). The supernova shock is visible as the deformed discontinuity and moves from about 6000 km to 10 000 km within 200 ms. The region behind the shock shows anisotropy on large scales with contrasts in the entropy of about a factor of 3. The darker area visible at late times in the center of the star indicates the formation of a nearly spherical, low-density, high-entropy hot-bubble region around the proto-neutron star. (From Janka & Müller 1996b)

with time are unknown. The latter is determined by the amount of angular momentum transport occurring during the stars' lifetime. If transport of angular momentum is inefficient, the increasing compactification of the central regions of the evolving massive star would lead to relatively rapidly rotating iron cores, even if the star was rotating slowly on the main-sequence (Endal & Sofia 1977). If, on the other side, magnetic fields or shear viscosity are able to reduce the angular momentum in the interior of the star sufficiently, its iron core will rotate slowly. Unfortunately, 2D evolutionary calculations of rotating massive stars up to the stage of core collapse, which could help to narrow down the range of initial (*i.e.*, pre-collapse) angular momenta of iron cores, have not yet been performed (for first attempts in this direction, see *e.g.*, Langer *et al.* 1997).

In spite of the unknown rotational state of the iron core prior to collapse, several attempts have been made in the past two decades to study the influence of rotation on the dynamics of Type II supernova explosion by means of hydrodynamic simulations (LeBlanc & Wilson 1970; Müller *et al.* 1980; Tohline *et al.* 1980; Müller & Hillebrandt 1981; Symbalisty 1984; Mönchmeyer & Müller 1989; Finn & Evans 1990; Mönchmeyer *et al.* 1991; Marck & Bonazzola 1992; Mönchmeyer 1993; Yamada & Sato 1994; Zwerger 1995; Zwerger & Müller 1997). All of these simulations, except those recently performed by Zwerger & Müller (Zwerger 1995; Zwerger & Müller 1997), were restricted to a few specific core collapse models. They incorporated a more or less sophisticated treatment of the complicated microphysics and transport processes involved in the event (see Sects. 3.2 and 3.6).

Zwerger & Müller followed up a somewhat complementary approach. They restricted themselves to a simplified description of the microphysics approximating the iron cores by rotating polytropes and neglecting any transport effects. Thereby they were able to perform a truly comprehensive parameter study of the space spanned by possible pre-collapse iron core models. A similar approach has been pursued by Yamada & Sato (1994).

Concerning rotational core collapse calculations it is important to note two important facts. Firstly, all simulations up to now, which take into account the effects of rotation, have only followed the evolution up to the point when the prompt shock wave turns into an accretion shock. The influence of rotation on the delayed explosion mechanism, and in particular the interaction of rotation and convective instabilities, still has to be explored (for a first crude attempt see Shimizu *et al.* 1994). Secondly, all these simulations have been performed using Newtonian hydrodynamics, *i.e.*, relativistic effects have been neglected.

3.7.1 General Considerations. The most important difference between the collapse of rotating and non-rotating cores is that in rotating cores centrifugal forces may stop the collapse *before* nuclear matter density is reached. This possibility was pointed out by Shapiro & Lightman (1976) and discussed in more detail by Tohline (1984) and Eriguchi & Müller (1984). Numerical examples of such a low-density bounce have been given by several authors (Müller *et al.* 1980; Symbalisty 1984; Mönchmeyer & Müller 1989; Mönchmeyer 1993; Yamada & Sato 1994; Zwerger 1995; Zwerger & Müller 1997).

A theoretical argument for the stabilizing influence of rotation on (pseudo) radial modes of stars was derived by Ledoux (1945). To guarantee the stability of a configuration in rotational equilibrium the adiabatic index (at fixed S and Y_e; see Eq. (3.5) should fulfill the condition

$$\gamma > \gamma_{\rm crit} = \frac{2}{3}\frac{(2-5\beta)}{(1-2\beta)}\,, \tag{3.22}$$

the rotation parameter β being defined by

$$\beta = \frac{E_{\rm rot}}{|E_{\rm pot}|}\,. \tag{3.23}$$

Here $E_{\rm rot}$ and $E_{\rm pot}$ are the rotational and potential energy, respectively. Although Eq. (3.22) was derived for slow rigid rotators with γ independent of density, numerical calculations have shown that Eq. (3.22 holds under more general conditions, too (*e.g.*, Tassoul 1978). Moreover, Eq. (3.22) is only applicable if corrections due to general relativity are small (*i.e.*, if $GM/Rc^2 \ll 1$). Otherwise, the general relativistic analogue of Eq. (3.22) must be used (*e.g.*, Misner *et al.* 1973, Chap. 26). Compared to the Newtonian one the relativistic criterion gives a larger $\gamma_{\rm crit}$, because general relativity tends to destabilize configurations.

For a given average γ the critical value of β a rotating core in equilibrium must exceed to be stable against pseudo-radial modes is

$$\beta > \beta_{\rm crit} = \frac{1}{2}\frac{(4-3\gamma)}{(5-3\gamma)}\,. \tag{3.24}$$

This is also a necessary condition for a core bounce to occur at sub-nuclear densities (Tohline 1984; Mönchmeyer 1993). Actually, the stability and hydrodynamic evolution of a collapsing iron core is determined by the *effective* adiabatic index Γ (see Eq. (3.7)). Hence, γ must be substituted by Γ in Eq. (3.22) and (3.24).

Due to conservation of angular momentum the centrifugal forces increase significantly during collapse. Matter in the equatorial plane does not fall towards the center as fast as matter at the polar axis leading to a progressive flattening of the core. In addition, the collapse time scale is lengthened compared to non-rotating models.

Conservation of angular momentum may also lead to very rapidly rotating configurations, which are unstable against tri-axial deformations on secular or even dynamical time scales (Eriguchi & Müller 1984; Tohline 1984; Mönchmeyer *et al.* 1991; Zwerger 1995; Zwerger & Müller 1997). For MacLaurin spheroids, *i.e.*, for incompressible, rigidly rotating equilibrium configurations, secular and dynamical tri-axial instabilities grow, if $\beta \geq 13.75\,\%$ and

$\beta \geq 27.38\,\%$, respectively (*e.g.*, Tassoul 1978). Whether these instabilities do indeed occur (and if so, at which values of β) in compressible, differentially rotating stars is still an open question. If the equation of state at sub-nuclear densities is stiff, *i.e.*, if the adiabatic index is very close to 4/3, the core may be stabilized before its rotational energy exceeds the critical value. If, on the other side, the initial amount of rotation is small enough for the collapse to proceed to nuclear densities, $\beta > 0.14$ may not be reached before bounce.

After bounce a rotating core oscillates with a superposition of various radial and surface modes. The frequency of these modes is determined by the average density of the inner core. In contrast to a spherically symmetric core, which comes to rest soon after bounce, in a rotating core a certain fraction of its kinetic infall energy will be converted into oscillations, which are damped by nonspherical pressure waves. These oscillations strongly contribute to the gravitational wave signal of the bouncing core (see Sect. 3.8). Due to the asymmetric infall of matter the propagation of the shock wave at the pole differs from that near the equatorial plane.

Finally, it is very well known that rotation can drive meridional circulations in baroclinic regions (Tassoul 1978). For core collapse supernova models the interesting question arises, whether there exist large-scale circulation patterns behind the shock, that extend inside the neutrino-sphere and which could transport neutrinos towards the shock front much faster than diffusion (Müller & Hillebrandt 1981).

3.7.2 Simulations of Rotational Core Collapse I: Models with Detailed Microphysics. Up to now the most realistic iron core collapse models have been calculated by Mönchmeyer and co-workers (Mönchmeyer & Müller 1989; Mönchmeyer *et al.* 1991; Mönchmeyer 1993). They considered the axi-symmetric collapse of a $1.36 M_\odot$ iron core of a $20 M_\odot$ star. The spherically symmetric initial model was put into a state of rotation by adding a certain (small) amount of rotational energy to it, *i.e.*, the initial model was *not* in rotational equilibrium. Four different angular momentum distributions were examined. The initial rotation parameter was in the range $0.005 < \beta < 0.02$, and during collapse angular momentum conservation was assumed. A realistic equation of state (Hillebrandt & Wolff 1985) was used in the simulations, and electron captures on protons were taken into account. The neutrino transport was simplified by a leakage scheme. The main results obtained by Mönchmeyer *et al.* (1991) are:

- An initially fast rotating model ($\beta_\mathrm{i} = 0.02$) bounces well below nuclear matter density ($\varrho_\mathrm{b} = 3.6 \times 10^{13}\,\mathrm{g\,cm^{-3}}$) and evolves into a quasi-static rotating equilibrium configuration with a final rotation parameter $\beta_\mathrm{f} = 0.14$.
- If the initial rotation rate is moderate ($\beta_\mathrm{i} = 0.01$) the centrifugally caused bounce occurs just below nuclear matter density ($\varrho_\mathrm{b} \approx 10^{14}\,\mathrm{g\,cm^{-3}}$) and causes large-scale amplitude oscillations of the core, which strengthen the prompt shock wave.

- If the initial rotation rate is small ($\beta_i \lesssim 0.005$) the bounce occurs due to the stiffness of nuclear matter. In this case rotation "hurts" the prompt explosion mechanism. The collapse, which is fastest near the rotation axis, leads to the formation of an aspherical shock wave, which is strongest near the rotation axis. Hence, an entropy bubble forms at the polar axis, which is convectively unstable. The resulting convective flow weakens the prompt shock by transporting high entropy material from the polar region towards the equator.
- Measuring the gravitational wave signal allows one to discriminate between a bounce caused by the stiffness of nuclear matter and a bounce caused by centrifugal forces (see Sect. 3.8).

3.7.3 Simulations of Rotational Core Collapse II: Polytropic Models. The following discussion is restricted to the results of Zwerger & Müller (Zwerger 1995; Zwerger & Müller 1997), because their numerical study is the most comprehensive one (but see also Finn & Evans 1990; Yamada & Sato 1994). In their simulations the collapsing core was approximated by a rotating polytrope in order to examine a large set of initial conditions. They used a simplified analytic equation of state (Janka *et al.* 1993) neglecting all transport effects. The equation of state consisted of a "polytropic" part P_p and a "thermal" part P_th,

$$P = P_\mathrm{p} + P_\mathrm{th}. \tag{3.25}$$

The polytropic part P_p accounts for pressure contributions due to degenerate, relativistic electrons and (at high densities) due to the repulsive action of nuclear forces. P_th mimics the thermal pressure in the matter heated up by the prompt shock. Writing P as a sum of these two terms prevents matter elements from following the same P-ϱ-history during infall and expansion after bounce. Two different density regimes are considered, where

$$P_\mathrm{p} = \begin{cases} K_1 \cdot \varrho^{\Gamma_1}, & \text{if} \quad \varrho \leq \varrho_\mathrm{nuc} \\ K_2 \cdot \varrho^{\Gamma_2}, & \text{otherwise.} \end{cases} \tag{3.26}$$

The "nuclear density" $\varrho_\mathrm{nuc} = 2 \times 10^{14}\,\mathrm{g\,cm^{-3}}$, the supra-nuclear adiabatic index $\Gamma_2 = 2.5$, and the value of K_1 is fixed to

$$K_1 = \frac{3}{4}\left(\frac{\pi}{3}\right)^{2/3} \hbar c \left(\frac{Y_\mathrm{e}}{m_\mathrm{B}}\right)^{4/3} = 4.897 \times 10^{14}\,\mathrm{cgs}, \tag{3.27}$$

which holds for an ideal gas of ultra-relativistic, degenerate electrons ($\Gamma_1 = 4/3$). From the first law of thermodynamics the corresponding thermodynamically consistent energy density follows (for adiabatic changes) as

$$u_\mathrm{p} = \begin{cases} E_1 \cdot \varrho^{\Gamma_1}, & \text{if} \quad \varrho \leq \varrho_\mathrm{nuc} \\ E_2 \cdot \varrho^{\Gamma_2} + E_3 \cdot \varrho, & \text{otherwise}, \end{cases} \tag{3.28}$$

where E_1 and E_2 are related to K_1 and K_2 by

$$K_1 = (\Gamma_1 - 1) \cdot E_1 \quad \text{and} \quad \mathrm{K}_2 = (\Gamma_2 - 1) \cdot \mathrm{E}_2. \tag{3.29}$$

The constants E_2 (and hence K_2) and E_3 are fixed by requiring continuity of P_p and u_p at ϱ_{nuc}. This leads to

$$E_2 = \frac{\Gamma_1 - 1}{\Gamma_2 - 1} \cdot E_1 \cdot \varrho_{nuc}^{\Gamma_1 - \Gamma_2} \qquad \text{and} \qquad E_3 = \frac{\Gamma_2 - \Gamma_1}{\Gamma_2 - 1} \cdot E_1 \cdot \varrho_{nuc}^{\Gamma_1 - 1}. \tag{3.30}$$

The thermal pressure P_{th} is related to the thermal energy density u_{th} by an ideal gas relation

$$P_{th} = (\Gamma_{th} - 1) \cdot u_{th}, \tag{3.31}$$

with $\Gamma_{th} = 1.5$ corresponding to a mixture of relativistic and non-relativistic gases. The total energy density

$$u = u_p + u_{th}. \tag{3.32}$$

Initially u_{th} and P_{th} are zero and should remain so in the absence of shocks. Further details are given in Janka *et al.* (1993).

In their study Zwerger & Müller 1997 computed the evolution of models with $\Gamma_1 = 1.325$, 1.3225, 1.32, 1.315, 1.31, 1.30, 1.29 and 1.28, respectively. The pre-collapse models had rotation parameters $\beta_i = 0.0025$, 0.0045, 0.009, 0.018 and 0.04, respectively. Initially all models rotated according to the rotation law (Eriguchi & Müller 1984)

$$\Omega(\varpi) = \frac{\Omega_0}{1 + (\varpi/A)^2}, \tag{3.33}$$

where ϖ is the distance from the rotation axis. The constant Ω_0 is uniquely related to the parameter β_i, while A is a free parameter determining the initial angular momentum distribution. For large values of A (compared to the size of the initial model) one obtains almost rigidly rotating configurations, while small values of A correspond to strongly differentially rotating ones. Zwerger & Müller examined models with $A = 5 \times 10^9$ cm, 10^8 cm, 5×10^7 cm and 10^7 cm, respectively.

Contrary to all previous studies, except those of Bonazzola & Marck (1993, 1994), the initial models were in rotational equilibrium being calculated with the method of Eriguchi & Müller (1984). The pre-collapse models were $\Gamma = 4/3$ polytropes, whose collapse was induced by suddenly reducing Γ to one of the above specified values Γ_1. In the parameter space spanned by Γ_1, β_i and A, Zwerger & Müller computed a total of 81 initial models and followed their evolution well beyond core bounce.

A subset of models suffers a bounce caused by centrifugal forces at subnuclear densities. For a given β_i the bounce density decreases with increasing Γ_1 and decreasing A, *i.e.*, strongly differentially rotating models with Γ_1 close to 4/3 have the lowest bounce densities ϱ_b. Models with $A = 10^7$ cm and $\beta_i \geq 0.009$ bounce at $6.2 \times 10^{12}\,\mathrm{g\,cm^{-3}} \leq \varrho_b \leq 7.0 \times 10^{13}\,\mathrm{g\,cm^{-3}}$ the model with the lowest ϱ_b having $\beta_i = 0.04$ and $\Gamma_1 = 1.30$. But even for quite rigidly rotating configurations with $A = 10^8$ cm the collapse is stopped in some cases

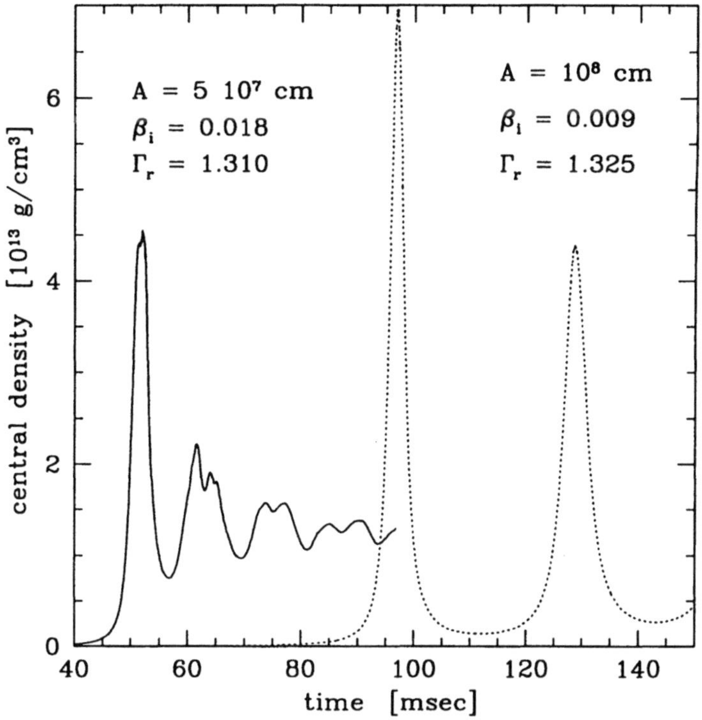

Fig. 3.17. Central density as a function of time for two models, which experience a bounce due to centrifugal forces at sub-nuclear densities (from Zwerger & Müller 1997)

well below nuclear matter density, e.g., $\varrho_{\rm b} = 3.6 \times 10^{13}\,{\rm g\,cm^{-3}}$ for $\beta_{\rm i} = 0.018$ and $\Gamma_1 = 1.32$, and $\varrho_{\rm b} = 7.0 \times 10^{13}\,{\rm g\,cm^{-3}}$ for $\beta_{\rm i} = 0.009$ and $\Gamma_1 = 1.325$.

Models suffering a bounce due to (or mainly due to) centrifugal forces show large amplitude oscillations of the inner core. The central density varies by more than a factor of ten in such models (Fig. 3.17). For example, in the model with $A = 5 \times 10^7$ cm, $\beta_{\rm i} = 0.018$ and $\Gamma_1 = 1.31$ (solid curve in Fig. 3.17) the bounce density is $\varrho_{\rm b} = 4.6 \times 10^{13}\,{\rm g\,cm^{-3}}$. Then the model re-expands, the central density dropping down to $\varrho = 7.5 \times 10^{12}\,{\rm g\,cm^{-3}}$. Eventually the model achieves a new equilibrium state with a central density of $\varrho = 1.3 \times 10^{13}\,{\rm g\,cm^{-3}}$. Even more extreme is the model with $A = 10^8$ cm, $\beta_{\rm i} = 0.009$ and $\Gamma_1 = 1.325$ (dotted curve in Fig. 3.17), where the central density drops by about a factor of 50 from $6.9 \times 10^{13}\,{\rm g\,cm^{-3}}$ (at bounce) to $1.4 \times 10^{12}\,{\rm g\,cm^{-3}}$ (at maximum re-expansion) and then rises again by a factor of about 30 to $4.4 \times 10^{13}\,{\rm g\,cm^{-3}}$ (during the second bounce Fig. 3.17).

Because of angular momentum conservation the effects of rotation become more important during collapse, which is reflected by an increase of the value of β. For several models β exceeds the critical value $\beta = 0.1375$, where MacLaurin spheroids become secularly unstable against tri-axial perturba-

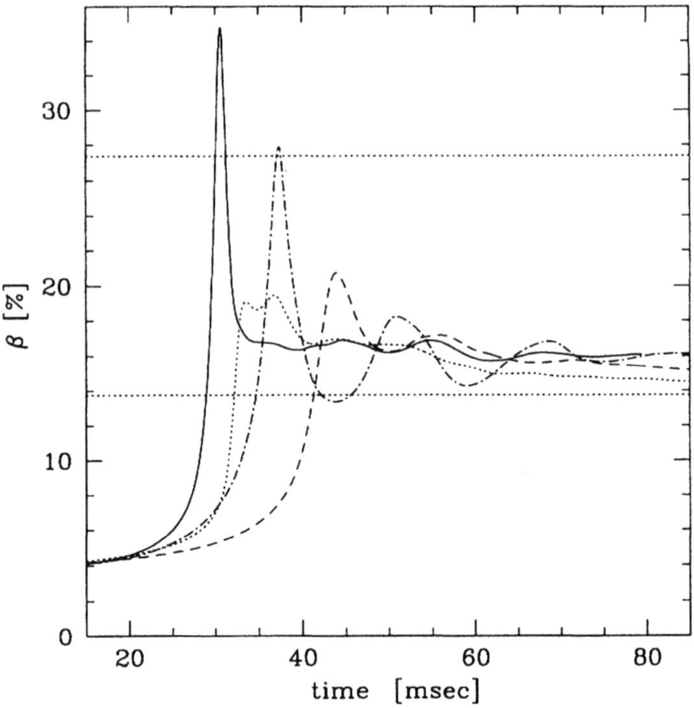

Fig. 3.18. Time evolution of the rotation parameter β for four models: $A = 10^7$ cm, $\beta_i = 0.04$ and $\Gamma_1 = 1.28$ (solid), $A = 10^7$ cm, $\beta_i = 0.04$ and $\Gamma_1 = 1.30$ (dash-dotted), $A = 5 \times 10^7$ cm, $\beta_i = 0.04$ and $\Gamma_1 = 1.28$ (dotted), $A = 5 \times 10^7$ cm, $\beta_i = 0.04$ and $\Gamma_1 = 1.30$ (dashed), The dotted horizontal lines mark the critical values β_{sec} (lower line) and β_{dyn} (upper line), where MacLaurin spheroids become secularly and dynamically unstable against non-axisymmetric perturbations. (From Zwerger & Müller 1997)

tions. Most of the secularly unstable models with $A \leq 10^8$ cm, $\beta_i \geq 0.009$ and $\Gamma_1 \leq 1.32$ remain in the unstable regime for several milliseconds, and for $\beta_i = 0.04$ even for several tens of milliseconds (Fig. 3.18). Whether this length of time is long enough to allow for a significant growth of the secular instability is, however, unclear. Two of the most differentially ($A = 10^7$ cm) and rapidly ($\beta_i = 0.04$) rotating models reach ($\Gamma_1 = 1.30$) and even exceed ($\Gamma_1 = 1.28$) the critical value for dynamical stability ($\beta = 0.2738$; Fig. 3.18). Although such large values of β only prevail for about one millisecond, this length of time is sufficiently long for a significant growth of the dynamical instability (see Sect. 3.8.6).

3.8 Gravitational Wave Signature of Core Collapse Supernovae

Until the year 2001 four long-baseline laser interferometric gravitational wave detectors will become operational (GEO600, LIGO, TAMA, VIRGO). These new instruments as well as several presently being built cryogenic resonant-mass detectors (including improved conventional bars cooled to temperatures of $\sim 50\,\mathrm{mK}$ and more recent large mass spherical detectors, like *e.g.*, TIGA (Merkowitz & Johnson 1995; see also Harry *et al.* 1996)) will achieve a (broad-band and/or narrow-band) sensitivity allowing astronomers for the first time in history to directly detect gravitational waves and to enter the area of gravitational wave astrophysics (for a review see, *e.g.*, Abramovici *et al.* 1995). Theoretical predictions of the gravitational wave signature (wave forms, spectra, duration) of potential astrophysical sources are thus urgently needed, *e.g.*, to improve the low signal-to-noise ratio. Among the potential sources core collapse supernovae (*i.e.*, SNe II and SNe Ib/Ic) involving the formation of a neutron star or a black hole are thought to emit a significant amount of gravitational radiation, if the collapse and/or the early phase of the explosion is very nonspherical (for a review see Thorne 1987; Piran 1990; Bonazzola & Marck 1994; Thorne 1995; Müller 1997). Non-sphericities can be caused by the effects of both rotation and convection leading to an overall coherent deviation from spherical symmetry or small-scale statistical fluctuations, respectively.

Measuring the gravitational wave signal of core collapse supernovae provides unique and direct information about the collapse dynamics. Note in this respect, that the electromagnetic "display" of a core collapse supernova only starts when the shock wave reaches the surface of the progenitor star hours after core bounce. Besides gravitational waves, the only other means to directly probe the collapse dynamics are neutrino observations. A measurement of both the neutrino and the gravitational wave signal would be most exciting, because it would provide extremely important and complementary information for the theoretical modeling of core collapse supernovae.

3.8.1 Quadrupole Formula. The gravitational wave signal is most commonly calculated in the post-Newtonian quadrupole approximation ($v \ll c$ and $\Phi \ll c^2$), in which the gravitational wave amplitude is proportional to the second time derivative of the mass-energy quadrupole moment. The numerically troublesome second-order time derivative of the quadrupole moment can be transformed into much better tractable spatial derivatives. In particular, the gravitational quadrupole radiation field, h^{TT}, can be calculated with an expression derived independently by Nakamura & Oohara (1989) and by Blanchet *et al.* (1990):

$$
\begin{aligned}
h_{ij}^{\mathrm{TT}}(\boldsymbol{X},t) &= \frac{2G}{c^4 R} P_{ijkl}(\boldsymbol{N}) \times \\
&\times \int \mathrm{d}^3x\ \varrho \left[2v^k v^l - x^k\ \partial_l \Phi - x^l\ \partial_k \Phi \right] ,
\end{aligned}
\tag{3.34}
$$

where $R = |\boldsymbol{X}|$ is the distance between the observer and the source, Φ is the Newtonian gravitational potential, ϱ is the mass-density and $\boldsymbol{v}$ is the velocity. The other quantities have their usual meanings except for $P_{ijkl}(\boldsymbol{N})$

(with $\boldsymbol{N} = \boldsymbol{X}/R$) which denotes the transverse-traceless (TT) projection operator onto the plane orthogonal to the outgoing wave direction $\boldsymbol{N}$, acting on symmetric Cartesian tensors according to

$$\begin{aligned} P_{ijkl}(\boldsymbol{N}) &= (\delta_{ik} - N_i N_k)\,(\delta_{jl} - N_j N_l) + \\ &- \frac{1}{2}(\delta_{ij} - N_i N_j)\,(\delta_{kl} - N_k N_l)\ . \end{aligned} \tag{3.35}$$

∂_i represents the partial derivative with respect to the x^i coordinate. The integrand in Eq. (3.34) is defined on a compact manifold and is known to the (2nd order) accuracy level of the numerical algorithm of the hydro-code. Eq. (3.34) can be shown to be equivalent to the standard representation

$$h_{ij}^{\mathrm{TT}}(\boldsymbol{X}, t) = \frac{2G}{c^4 R}\, P_{ijkl}(\boldsymbol{N})\, \frac{\partial^2}{\partial t^2} Q_{kl}\, (t - \frac{R}{c})\ , \tag{3.36}$$

where the trace-free part of the mass-quadrupole tensor of the matter distribution is given by

$$Q_{ij}(t) = \int \mathrm{d}^3 x\ \varrho(\boldsymbol{x}, t)\ \left(x_i x_j - \frac{1}{3}\delta_{ij}\boldsymbol{x}^2\right)\ . \tag{3.37}$$

It can be easily shown that evaluating the integral of Eq. (3.34) by an integration scheme (of at least 2nd order) is by one order of accuracy superior to twice applying numerical time-differentiation methods to quadrupole data given at discrete points of time (Finn & Evans 1990; Mönchmeyer *et al.* 1991).

3.8.2 Evaluation of the Quadrupole Formula I: Axisymmetric Source. The gravitational radiation field gives direct information about the second time derivative of the mass-quadrupole tensor Eq. (3.37). In case of axisymmetry the quadrupole moment, Q, is the only independent component of the quadrupole tensor. Its relation to the Cartesian component Q_{zz} of the radiative mass-quadrupole tensor is

$$Q \equiv \frac{3}{4}\sqrt{\frac{5}{\pi}}\, Q_{zz}\ . \tag{3.38}$$

For 2D (axisymmetric) simulations, which are most conveniently performed in spherical coordinates, it is natural to represent the (total) radiation field $\widetilde{h}^{\mathrm{TT}}$ in terms of the "pure-spin tensor harmonics" $T_{ij}^{\mathrm{E2,lm}}$ and $T_{ij}^{\mathrm{B2,lm}}$ with amplitudes A_{lm}^{E2} and A_{lm}^{B2} in the following way (Thorne 1980):

$$\begin{aligned} \widetilde{h}_{ij}^{\mathrm{TT}}(\boldsymbol{X}, t) = \frac{1}{R}\sum_{l=2}^{\infty}\sum_{m=-l}^{+l} \Big\{ & A_{lm}^{\mathrm{E2}}(t - \frac{R}{c})\ T_{ij}^{\mathrm{E2,lm}}(\theta, \phi) + \\ + \ & A_{lm}^{\mathrm{B2}}(t - \frac{R}{c})\ T_{ij}^{\mathrm{B2,lm}}(\theta, \phi) \Big\}. \end{aligned} \tag{3.39}$$

In spherical coordinates the coefficients A_{lm}^{E2} and A_{lm}^{B2} have especially simple integral representations over the source. By symmetry, there is only one non-vanishing quadrupole term in Eq. (3.39), namely A_{20}^{E2}. Higher-order terms are neglected in the quadrupole approximation $h^{\rm TT}$ of the gravitational radiation field $\widetilde{h}^{\rm TT}$. Transforming Eq. (3.34) to spherical coordinates and expressing v_i in terms of unit vectors in the r, θ and ϕ direction, one obtains by comparison of Eq. (3.34) with the lowest-order term of Eq. (3.39) for the quadrupole wave amplitude $A_{20}^{\rm E2}$ the expression

$$\begin{aligned} A_{20}^{\rm E2}(t) &= \frac{G}{c^4}\frac{16\pi^{3/2}}{\sqrt{15}}\int_{-1}^{1}\int_{0}^{\infty}\varrho(r,z,t)\Big[v_r v_r(3z^2-1) + \\ &\quad + v_\theta v_\theta(2-3z^2) - v_\phi v_\phi - 6 v_r v_\theta z\sqrt{1-z^2} + \\ &\quad - r\partial_r\Phi\,(3z^2-1) + 3\partial_\theta\Phi\, z\sqrt{1-z^2}\Big] r^2 {\rm d}r{\rm d}z\ , \end{aligned} \tag{3.40}$$

where $\partial_r = \partial/\partial r, \partial_\theta = \partial/\partial\theta$, and $z = \cos\theta$.

From the definition of $T_{ij}^{\rm E2,2m}$ (Thorne 1980, Eq. (2.39e)) one derives for the components of $h^{\rm TT}$ the formula

$$h_{\theta\theta}^{\rm TT} = \frac{1}{8}\sqrt{\frac{15}{\pi}}\,\sin^2\theta\,\frac{A_{20}^{\rm E2}(t)}{R}\ . \tag{3.41}$$

The only other nonzero component is $h_{\phi\phi}^{\rm TT} = -h_{\theta\theta}^{\rm TT}$. Note that due to the hypotheses of (i) axisymmetry and (ii) of post-Newtonian gravity $h_{\theta\theta}^{\rm TT} = h_+$ and $h_\times = 0$, where h_+ and $h_\times$ are the wave amplitudes corresponding to the two independent polarizations of the gravitational radiation field (see next subsection).

The total energy radiated in gravitational waves is given by the general expression

$$E = \frac{c^3}{G}\frac{1}{32\pi}\sum_{l,m}\int_{-\infty}^{+\infty}\left\{\left(\frac{{\rm d}A_{lm}^{\rm E2}}{{\rm d}t}\right)^2 + \left(\frac{{\rm d}A_{lm}^{\rm B2}}{{\rm d}t}\right)^2\right\}{\rm d}t\ . \tag{3.42}$$

The total energy radiated per unit frequency ${\rm d}E/{\rm d}\nu$ can be calculated from the amplitudes $A_{20}^{\rm E2}$ using the fast Fourier transform technique.

3.8.3 Evaluation of the Quadrupole Formula II: Genuine Three-Dimensional Source. If the source is of genuine three-dimensional nature, it is common to express the gravitational quadrupole radiation field, $h^{\rm TT}$, in the following tensorial form (Misner *et al.* 1973, Chap.35):

$$h_{ij}^{\rm TT}(\boldsymbol{X},t) = \frac{1}{R}\left(A_+\boldsymbol{e}_+ + A_\times\boldsymbol{e}_\times\right) \tag{3.43}$$

with the unit linear-polarization tensors

$$\boldsymbol{e}_+ = \boldsymbol{e}_\theta\otimes\boldsymbol{e}_\theta - \boldsymbol{e}_\phi\otimes\boldsymbol{e}_\phi\ , \tag{3.44}$$

$$\boldsymbol{e}_\times = \boldsymbol{e}_\theta\otimes\boldsymbol{e}_\phi + \boldsymbol{e}_\phi\otimes\boldsymbol{e}_\theta\ , \tag{3.45}$$

$\boldsymbol{e}_\theta$ and $\boldsymbol{e}_\phi$ being the unit polarization vectors in θ and ϕ-direction of a spherical coordinate system and $\otimes$ the tensor product. The amplitudes A_+ and $A_\times$ (or the dimensionless amplitudes $h_+ \equiv A_+/R$ and $h_\times \equiv A_\times/R$) represent the only two independent modes of polarization in the TT gauge, and are given by the following expressions for $\theta = 0, \phi = 0$:

$$A_+ = \ddot{I}_{xx} - \ddot{I}_{yy} , \tag{3.46}$$

$$A_\times = 2\ddot{I}_{xy} , \tag{3.47}$$

and for $\theta = \pi/2, \phi = 0$ by

$$A_+ = \ddot{I}_{zz} - \ddot{I}_{yy} , \tag{3.48}$$

$$A_\times = -2\ddot{I}_{yz} , \tag{3.49}$$

where

$$\ddot{I}_{ij} = \frac{G}{c^4} \int \mathrm{d}^3x \; \varrho \; (2v_i v_j - x_i \partial_j \Phi - x_j \partial_i \Phi) \; . \tag{3.50}$$

The total energy radiated in form of gravitational waves is given by

$$E = \frac{c^3}{5G} \int_{-\infty}^{+\infty} \left[\frac{\mathrm{d}}{\mathrm{d}t} \left(A_{ij} - \frac{1}{3} \delta_{ij} A_{ll} \right) \right]^2 \mathrm{d}t \tag{3.51}$$

$$= \frac{2c^3}{15G} \int_{-\infty}^{+\infty} \Big[\dot{A}_{xx}^2 + \dot{A}_{yy}^2 + \dot{A}_{zz}^2 - \dot{A}_{xx}\dot{A}_{yy} - \dot{A}_{xx}\dot{A}_{zz} + - \dot{A}_{yy}\dot{A}_{zz} + 3 \, (\dot{A}_{xy}^2 + \dot{A}_{xz}^2 + \dot{A}_{yz}^2) \Big] \; \mathrm{d}t \tag{3.52}$$

with

$$A_{ij} \equiv \ddot{I}_{ij} \; . \tag{3.53}$$

3.8.4 Gravitational Radiation from Convective Instabilities. The gravitational wave signature of convective instabilities inside the proto-neutron star and in the neutrino heated post-shock region has been investigated in detail by Müller & Janka (1997). They calculated the quadrupole waveforms, the energy spectra, and the total amount of the emitted gravitational wave energy due to the anisotropic mass distribution and mass motion. Moreover, they estimated the amplitude and energy of gravitational waves associated with the anisotropic neutrino emission that is caused by the convective transport of neutrinos and by aspherical perturbations of temperature and density in the neutrinospheric region. The investigation included the analysis of both a 2D and a 3D simulation of convection inside the proto-neutron star (Müller & Janka 1997), as well as three 2D simulations of the convective instability in the hot-bubble region covering the evolution for up to 1 sec past core bounce (Janka & Müller 1996b).

The only other related investigation up to now has been performed by Burrows & Hayes (1996). These authors conducted a single exploratory and experimental calculation of aspherical collapse, imposing a large amplitude (15%) density perturbation within 20° of the pole (*i.e.*, the symmetry axis) of their pre-collapse iron core model. Simulating the core's evolution in 2D until about 50 ms past core bounce, they determined the gravitational wave signature due to anisotropic neutrino radiation and mass motion. The 2D results of Müller & Janka (1997) are in rough qualitative agreement with the findings of Burrows & Hayes (1996), although there are quantitative differences.

Müller & Janka (1997) found that for a supernova located at a distance of 10 kpc the maximum dimensionless gravitational wave amplitudes $|h^{\mathrm{TT}}|$ associated with convective mass motions range from about 9×10^{-23} to 4×10^{-21}. The total amount of the emitted energy varies from $3 \times 10^{-14} M_{\odot} c^2$ to $5 \times 10^{-10} M_{\odot} c^2$. Convective motions inside the proto-neutron star involve more mass and are more violent and therefore produce a stronger gravitational wave signal with up to a factor of 10 larger wave amplitudes. Most of the gravitational radiation from convection inside the proto-neutron star is emitted in the frequency band 100–1000 Hz, while convective motions in the hot-bubble region generate waves from several 100 Hz down to a few Hz.

Comparing different 2D models Müller & Janka (1997) find that with increasing total neutrino luminosity and hence with increasing explosion energy the convective activity in the hot-bubble region changes from violent convective overturn associated with anisotropic accretion processes to rapid overall expansion and relatively slowly changing large-scale deformations of the expanding shells behind the outward propagating supernova shock. This change of the characteristics of non-radial motions in the hot-bubble region is directly reflected in the dominant frequencies of the gravitational wave signal. While turbulent overturn around the proto-neutron star produces gravitational waves with most power at frequencies of 100–200 Hz, the dominant frequencies are at only some 10 Hz when the period of convective activity is short and the non-sphericity of the model is determined by the explosive expansion. Thus, a measurement of the frequency of the wave signal would provide important insights into the explosion dynamics. Moreover, since the signal produced by the convection inside the proto-neutron star is typically of much higher frequency (about 1000 Hz), it would also be possible to discriminate the contributions from the two convection zones to the measured signal.

Interestingly, structures in the gravitational wave signal are well correlated with prominent features in the neutrino emission, if both gravitational wave and neutrino production are associated with dynamical processes in and around the nascent neutron star. Simultaneous information from both neutrino and gravitational wave measurements would therefore impose important constraints on core collapse models.

Anisotropic neutrino emission generates gravitational waves, too (Epstein 1978; Burrows & Hayes 1996). Müller & Janka (1997) estimated the degree of anisotropy from the density and temperature inhomogeneities associated with convective and turbulent processes in the neutrino-spheric region. They find that for typical post-bounce neutrino luminosities the gravitational wave amplitude can be larger than the wave amplitude due to mass motions by a factor of about 5, although the energy in the neutrino tidal field is only a minor contribution (few percent) to the total energy radiated in gravitational waves.

As emphasized by Müller & Janka (1997) the results for the gravitational wave signal associated with nonspherical neutrino emission depend sensitively on the duration of the phase of anisotropic neutrino loss and on the temporal evolution of the total luminosity in all kinds of neutrinos. These characteristics will vary with the properties of the exploding star and thus with the parameters of the forming neutron star and will also be sensitive to the details of the numerical scheme and physical input used for the simulations.

Müller & Janka (1997) further found that the 3D simulation of convection inside the proto-neutron star gives a strongly reduced gravitational wave signal compared to the corresponding 2D model. The main reason for this is that in three spatial dimensions the convective structures and elements are smaller ($l_{3D} \approx 0.5 l_{2D}$), move less fast ($v_{3D} \lesssim 5 \times 10^8$ cm/s $\approx 0.5 v_{2D}$), and, correspondingly, show less strong overshooting and undershooting (0.8 instead of 1.2 pressure-scale heights). The maximum quadrupole amplitudes due to mass motions are reduced by about a factor of 30, the gravitational wave amplitudes associated with the anisotropic neutrino emission by as much as a factor of 10. The total amount of energy radiated in form of gravitational waves is 2–3 orders of magnitude smaller in 3D. A similarly strong reduction of the signal strength is to be expected for the gravitational waves emitted by turbulent motions in the hot-bubble region when corresponding simulations will be performed in three spatial dimensions.

All neutrino driven supernova models that have been analyzed for their gravitational wave emission up to now do not include any rotational effects. Therefore, future supernova modeling will have to determine the strength and the form of the gravitational wave signal caused by the combined effects of rotation, convection inside the proto-neutron star, and neutrino-driven overturn in the hot-bubble region.

3.8.5 Gravitational Radiation from Axisymmetric Rotational Core Collapse. The gravitational wave signature of axisymmetric rotational core collapse has been computed by various authors (Müller 1982; Finn & Evans 1990; Mönchmeyer *et al.* 1991; Yamada & Sato 1995; Zwerger 1995; Zwerger & Müller 1997). In the following I mainly discuss the results of Zwerger & Müller (Zwerger 1995; Zwerger & Müller 1997), because they are the most comprehensive ones and because they more or less comprise the other results, too. However, two points from the other investigations are worth mentioning here.

Firstly, Mönchmeyer *et al.* (1991) found that there exist two structurally different types of gravitational wave signals emitted during rotational core collapse. "Standard" signals of type I exhibit a large wave amplitude at bounce, which is followed by a "ring-down" of the signal, *i.e.*, by oscillations with decreasing damped amplitudes. This "ring-down" is due to damped

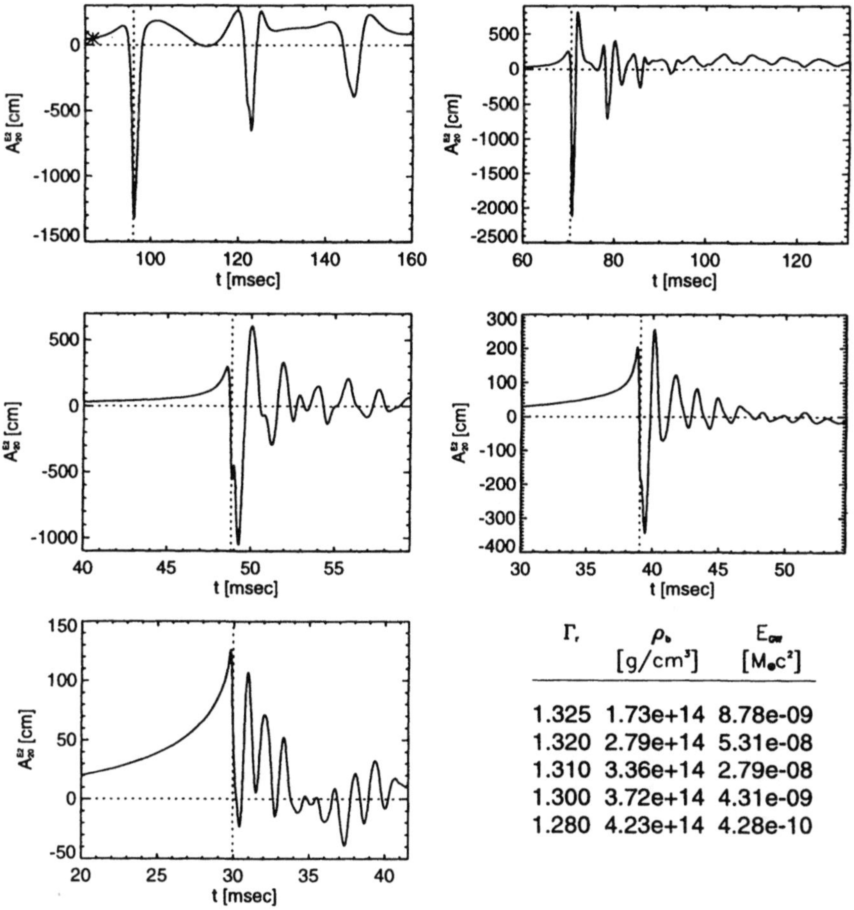

Γ_r	ρ_b [g/cm^3]	E_{GW} [M$_\odot$c^2]
1.325	1.73e+14	8.78e-09
1.320	2.79e+14	5.31e-08
1.310	3.36e+14	2.79e-08
1.300	3.72e+14	4.31e-09
1.280	4.23e+14	4.28e-10

Fig. 3.19. Quadrupole amplitudes A^{E2}_{20} produced by the collapse of almost rigidly rotating polytropes with $\beta_i = 0.009$. In each panel the vertical dotted line marks the time of bounce. The adiabatic index Γ_1, the central density at bounce and the total amount of energy radiated in form of gravitational waves are given for each model (by the line from top to bottom) in the table in the lower right part of the figure. (From Zwerger & Müller 1997)

volume and surface oscillations of the inner core after bounce. Signals of type I have been found in all other investigations, too (see top right panel of Fig. 3.19). They are produced by cores bouncing at nuclear density or at low central densities when the ratio of (radial) kinetic energy to rotational energy is small. Signals of type II are characterized by several pronounced peaks, which arise if the core bounces more than once (see top left panel of Fig. 3.19). Between the peaks the signal varies smoothly. Mönchmeyer *et al.* (1991) could not clarify which conditions are sufficient for the appearance of a single dominant coherent volume mode causing the type II signal. Hence,

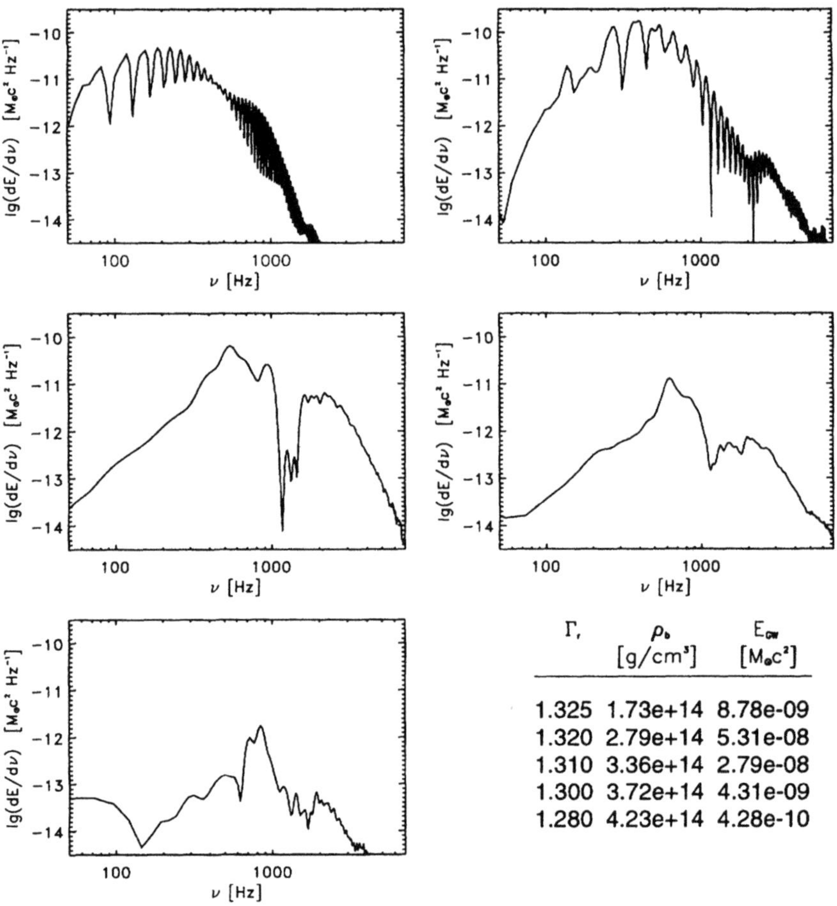

Γ_1	ρ_b [g/cm³]	E_{GW} [M⊙c²]
1.325	1.73e+14	8.78e-09
1.320	2.79e+14	5.31e-08
1.310	3.36e+14	2.79e-08
1.300	3.72e+14	4.31e-09
1.280	4.23e+14	4.28e-10

Fig. 3.20. Same as Fig. 3.19, but showing the spectral energy density dE/dν. (From Zwerger & Müller 1997)

they wondered whether such signals are only produced by models in a very narrow parameter range (see below).

Secondly, performing a set of nine collapse calculations of rotating polytropes, Yamada & Sato (1995) found that the peak amplitude of burst-like type I gravitational wave signals saturates for $q \gtrsim 0.5$, where $q = J/\frac{2GM}{c}$ is the dimensionless angular momentum of the core. They further found that the peak amplitude is sensitive to the stiffness of the equation of state just below nuclear matter density.

Zwerger & Müller (Zwerger 1995; Zwerger & Müller 1997) found both type I and type II gravitational wave signals. Which type occurs is solely determined by the adiabatic index, and in particular does not dependent on whether the central density of the core reaches nuclear matter density or not. Signals of type I are produced by models with a "soft" equation of state ($\Gamma_1 \lesssim$

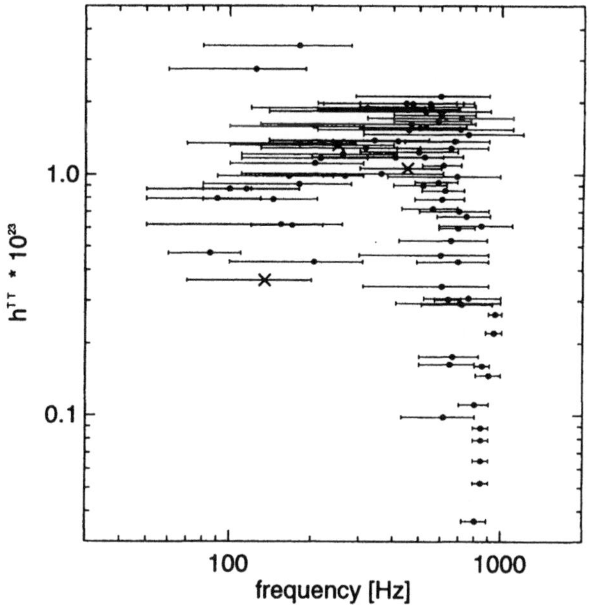

Fig. 3.21. The dimensionless gravitational wave amplitude of the models of Zwerger (1995). The error bar gives the frequency range, which contains most of the spectral power. It is assumed that the source is located at a distance of 10 Mpc. The amplitudes of the four models calculated by Mönchmeyer *et al.* (1991) are also shown and are marked by crosses. (From Zwerger & Müller 1997)

1.31), while signals of type II require a "stiff" equation of state ($\Gamma_1 \gtrsim 1.32$). The distinct amplitude spikes of the type II signals nicely correlate with the maxima of the central density. Moreover, decreasing the adiabatic index from 1.325 to 1.28 and keeping both other model parameters (*i.e.*, β_i and the initial rotation law) fixed, Zwerger & Müller observe a smooth transformation of the signal type. This is illustrated for a sequence of models with $\beta_i = 0.009$ and $A = 5 \times 10^9$ cm in Fig. 3.19. For $\Gamma_1 = 1.325$ the signal is of type II. It is of an intermediate type (*i.e.*, between type II and type I) for $\Gamma_1 = 1.32$, and turns into a "standard" type I signal for $\Gamma_1 = 1.31$ and 1.30. Finally, for $\Gamma_1 = 1.28$ a third type of signal is observed. Instead of showing a pronounced spike at bounce with a large negative wave amplitude, this so-called type III signal has a large positive and a somewhat smaller negative wave amplitude just before and after bounce, respectively. Signals of type III are not found for extremely differentially rotating initial models ($A = 10^7$ cm).

Zwerger & Müller also computed the amplitude and energy spectra of the gravitational radiation emitted by their models. The energy spectra cover a frequency range of 50 Hz $\lesssim \nu \lesssim$ 3 kHz, but most of the power is emitted between 500 Hz and 1 kHz. Models bouncing at sub-nuclear densities have spectra, which drop extremely rapidly above 1 kHz, and models with a type II wave signal have spectra, which show characteristic oscillations (Fig. 3.20). These oscillations vanish when the signal type changes to type I. Moreover, a detailed analysis reveals that the oscillatory spectral component appears immediately after the second bounce. Additional bounces produce no new spectral features. They only increase the overall spectral power. The results further show that the spectra are neither very sensitive to β_i nor to the degree of differential rotation.

The total amount of energy radiated in form of gravitational waves lies in the range $5 \times 10^{-11} M_\odot c^2 \lesssim E_{\rm GW} \lesssim 8 \times 10^{-8} M_\odot c^2$. The corresponding dimensionless wave amplitudes are in the range $4 \times 10^{-25} \lesssim h \lesssim 4 \times 10^{-23}$ for a source at a distance of 10 Mpc (Fig. 3.21). The largest signals are either produced by models which are initially slowly ($\beta_i \leq 0.009$) rotating and have an adiabatic index $\Gamma_1 \geq 1.32$, or which are initially rapidly ($\beta_i \geq 0.018$) and strongly differentially ($A = 10^7$ cm) rotating and have a relatively small adiabatic index ($\Gamma_1 \leq 1.30$). The first class of models experiences a bounce at densities above nuclear matter density with a fast deceleration of the collapsing core. This gives rise to a rapid change of the quadrupole moment of the core and consequently to a strong gravitational wave signal. In the other models the quadrupole moment is larger due to fast rotation, but changes less rapidly.

3.8.6 Gravitational Radiation from Non-axisymmetric Rotational Core Collapse. Concerning the collapse of initially rapidly rotating cores many authors have pointed out the possibility that if the core becomes secularly ($\beta > \beta_{\rm sec} \approx 0.14$) or dynamically ($\beta > \beta_{\rm dyn} \approx 0.27$) unstable to non-axisymmetric perturbations, a much stronger gravitational wave signal might be produced than by cores which remain axisymmetric during collapse (for a review, see *e.g.*, Piran 1990; Thorne 1995). According to this idea, the core will experience a centrifugal hang-up before nuclear matter density is reached and will be transformed into a bar-like configuration that spins end-over-end like an American football, if the rotation rate of the core exceeds the critical rotation rate(s). One has further speculated, whether the core might even break up into two or more massive pieces, if $\beta > \beta_{\rm dyn}$. In that case the resulting gravitational radiation *could* be almost as strong as that from coalescing neutron star binaries (Thorne 1995). The strength of the gravitational signal sensitively depends on (i) the radius at which the centrifugal hang-up occurs and (ii) what fraction of the angular momentum of the non-axisymmetric core goes into gravitational waves, and what fraction into hydrodynamic waves. These sound and shock waves are produced as the bar or lumps, acting like a twirling-stick, plow through the surrounding matter.

Nearly all investigations up to now did not consider the dynamics of non-axisymmetric collapse, but only studied the growth of secular and dynamical instabilities in already collapsed cores (*i.e.*, in nascent neutron stars) modeled as rapidly rotating, non-axisymmetric equilibrium polytropes or ellipsoids (Yoshida & Eriguchi 1995; Imamura *et al.* 1995; Lai & Shapiro 1995; Houser *et al.* 1994; Smith *et al.* 1996; see also Pickett *et al.* 1996). Note that, as the literature on this subject is extensive, I have only referenced the most recent studies here. Such an approach can be used to predict the gravitational radiation from rapidly rotating neutron stars, which might form as a

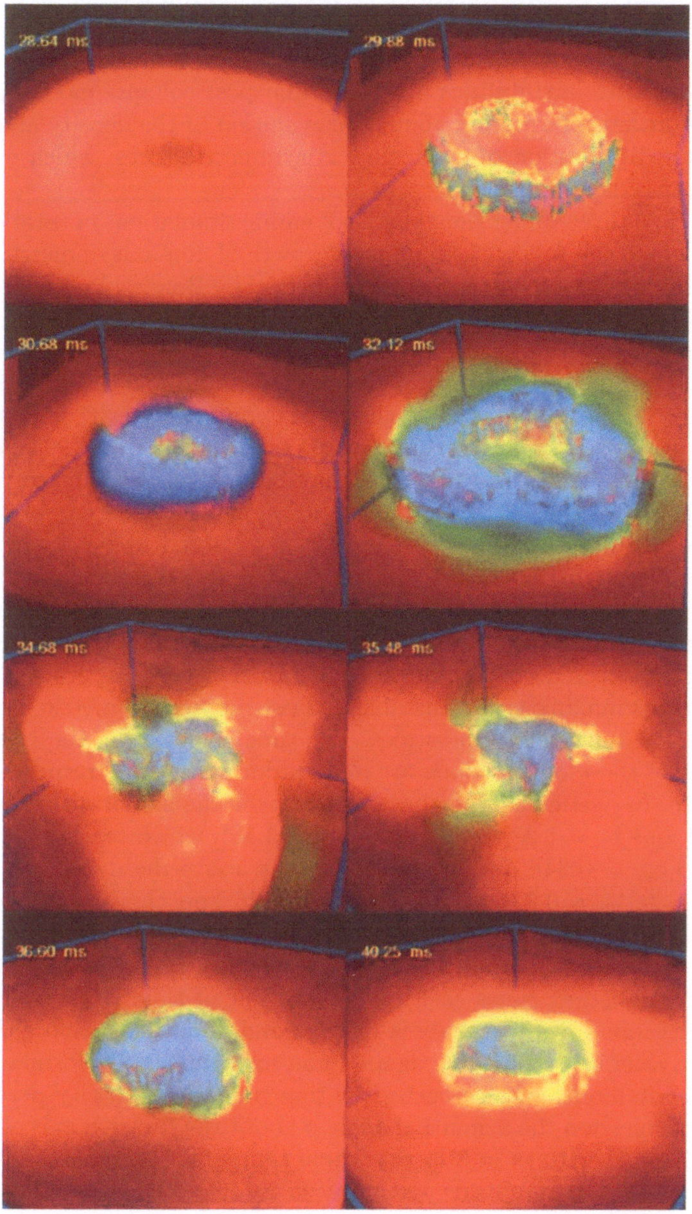

Fig. 3.22. Ray-cast images of the development of a non-axisymmetric instability in a collapsing, initially axisymmetric, rapidly rotating core. About 2 ms before the core becomes dynamically unstable against non-axisymmetric deformations the model has been perturbed by imposing a 5% density variation of the form $\sin(m\phi)$ with $m = 3$. The intensity is a measure of the density (brighter regions have larger densities) and the colors encode the thermal energy increasing from red, over yellow and green to blue. The size of the blue frame is $160 \times 160 \times 80$ km

consequence of core collapse. However, it is not appropriate for predicting the gravitational wave signature of the collapse of rapidly rotating cores. I stress this difference here, because it is often overlooked: Numerical simulations of non-axisymmetric instabilities in equilibrium post-collapse models are inadequate to obtain information about the evolution of the density and angular momentum distribution of collapsing axisymmetric cores, which eventually become non-axisymmetric.

Up to now three-dimensional hydrodynamic simulations of rotational core collapse, which can follow the nonlinear growth of non-axisymmetric instabilities *during* collapse, have only been performed by Bonazzola & Marck (1993, 1994), and by Marck & Bonazzola (1992). For their simulations they used a pseudo-spectral hydrodynamic code and a polytropic equation of state. They computed the evolution of several initial models and found that the gravitational wave amplitude is within a factor of two of that of 2D simulations for the same initial deformation of the core (Bonazzola & Marck 1994). However, their simulations were restricted to the pre-bounce phase of the collapse, and thus are less relevant.

Recently, Rampp *et al.* (1997) have also explored non-axisymmetric instabilities of collapsing, rapidly rotating cores and computed the resulting gravitational wave signal. Their starting point were the axisymmetric rotational core collapse models of Zwerger (Zwerger 1995; Zwerger & Müller 1997). They selected those models, which during collapse eventually reach rotation rates making them unstable against non-axisymmetric instabilities on secular or dynamical time scales. The time evolution of the rotation parameter β of the models, which are most promising in this respect, is shown in Fig. 3.18. In particular, they have investigated the model with $A = 10^7$ cm, $\beta_\mathrm{i} = 0.04$ and $\Gamma_1 = 1.28$ (solid curve in Fig. 3.18), which reaches a maximum value of $\beta_{\max} = 0.348$ during collapse and is above the dynamical stability limit ($\beta > \beta_{\mathrm{dyn}} = 0.27$) for 1.2 ms. At $t = 28.2$ ms they have mapped the axisymmetric model computed on a 2D spherical grid (360 radial and 90 angular zones) onto a 3D multiply-nested Cartesian grid (5 nested grids with $64^2 \times 32$ equidistant zones each). This moment of time is about 2.5 ms before the maximum rotation rate is reached and about 2 ms before the model becomes dynamically unstable against non-axisymmetric deformations. The axisymmetric model was perturbed by imposing a 5% density perturbation depending solely on the azimuthal angle ϕ about the rotation axis of the form $\delta\varrho/\varrho \propto \sin(m\phi)$ with $m = 3$ in the whole computational volume. Note that an additional small amplitude perturbation ($\lesssim 1\%$) is caused by the mapping procedure itself.

The development of the instability is shown in Fig. 3.22. Due to its rapid and strongly differential rotation the collapsing core develops a ring-like density distribution, which is clearly visible in the first snapshot. A few milliseconds after the model has been perturbed an $m = 3$ mode shows up in the density distribution. Note that although the $m = 3$ perturbation has been imposed on the whole computational grid, it grows only in the neighborhood of the ring-like density maximum. The further evolution is characterized by the development of three density maxima ("lumps") forming out of matter of the ring-like structure. The density contrast between the "lumps" and other matter at the same radius exceeds a factor of ten. Subsequently the three "lumps", which reflect the initially imposed $m = 3$ mode of perturbation, merge within about 5 ms into a bar-like (*i.e.*, $m = 2$) density structure, which according to linear stability analysis is the most unstable dynamical mode (*e.g.*, Tassoul 1978). The overall evolution is further characterized by intensive hydrodynamic activity produced by the twirling-stick action of first the "lumps" and later of the "bar". Most notably are trailing "spiral arms" causing mass and angular momentum to be transported away from the non-axisymmetric, high density regions (for further details see Rampp *et al.* 1997).

The gravitational wave signal obtained in the 3D simulation is comparable to that found in the corresponding axisymmetric model. Further simulations with different initial models and different types of perturbations as well as simulations covering later epochs of the core's evolution towards a rapidly(?) rotating, tri-axial(?) neutron star are definitely necessary. Studying the latter problem will, however, require other (implicit, relativistic) hydro-codes and a better treatment of the microphysics, particularly, of the equation of state and the neutronization of core matter.

4. Hydrodynamics and Thermonuclear Burning

The discussion in this section is an updated and modified version of that originally published in preprint form in an extensive review by Fryxell, Müller & Arnett (1989). The review was to be published as a contribution to a book on "Numerical Methods in Astrophysics" in 1989, but unfortunately the book has not appeared. Part of the review can also be found in Müller (1994) and is incorporated into chapters 4 and 11 of Arnett (1996).

Much of our understanding of astronomy and astrophysics is inseparably connected to our understanding of hydrodynamics and nuclear burning. Thermonuclear processes and gravitational binding are the two major sources of energy for astronomical phenomena. Where in astrophysics do we find nuclear burning? One of the most obvious examples is the Sun, representing the great majority of observed stars which calmly burn their nuclear fuel as objects in a state of quasi-hydrostatic equilibrium. Even the most placid of such stars can have its evolution changed by the action of slow mixing currents, acting over the long lifetime of the object. More spectacular are those objects exhibiting explosive phenomena and dynamic stages of evolution, such as novae, supernovae, thermonuclear shell flashes, X-ray bursts and γ-ray bursts. Here the flow is often supersonic.

Most of the observed mass in the universe is in the form of ionized plasma. The release of energy in such material gives rise to changing pressure gradients and so to hydrodynamic motion. Such fluid motion, in turn, will transport ashes and fuel to and from the flame zone. This gives rise to a subtle and difficult problem. The degree of difficulty of the problem is closely related to the nature of the coupling between the nuclear burning and the hydrodynamic flow. Special cases, in which this coupling is simpler, can be treated more reliably. First, nuclear reactions release energy in the form of radiation and kinetic energy of particles; at stellar densities this energy acts to heat the medium in which the reactions occur. The heating can modify the pressure, and thereby modify the flow. Second, nuclear reactions change the composition of the material, which in turn may modify its equation of state and its opacity, which in turn may modify the pressure and the heating, and again the flow. Conversely, the flow may remove ashes and bring fuel to the flame, countering the natural tendency of the flame to exhaust its fuel. The composition is specified by a set of abundances Y_i for i species of nuclei and particles, where it is convenient to define Y_i as the ratio of the number of particles of type i in some reference volume to the number of nucleons in the same volume. Often this coupling is dominated by a few nuclear species, and recognition of this gives rise to a set of approximations related to *minimum networks* and *post-processing*. The idea is to identify the key burning processes (*minimum network*), calculate them coupled to the hydrodynamic flow, then use those conditions to recalculate (*post-process*) the entire set of nuclear processes. Larger networks (*i.e.*, more processes) require more effort to calculate, but – if well chosen – give more accurate results. Doubling the size of a network without including a key process will simply give a more expensive error. For a discussion of minimum networks and their use in stellar evolutionary models see Arnett & Thielemann (1985), Thielemann & Arnett (1985) and Arnett (1996).

4.1 Time Scales

Analysis of the characteristic time scales is crucial for understanding the physics of the problem. To begin we consider the order of magnitude of some cross sections of interest.

For two particles of charge $Z_1 e$ and $Z_2 e$ we may define a Coulomb radius $r_{\rm Coul}$ from $Z_1 Z_2 e^2/2r_{\rm Coul} = E$, where E is an appropriate energy. For example, for an atom that energy would be the energy of the bound electron, so $E = p^2/2m_e$. Relating the momentum p to the radius, using the de Broglie wavelength, we have a cross section of order $\sigma_{\rm atom} = \pi a_0^2 Z_1 Z_2 = 0.88 \times 10^{-16} Z_1 Z_2\,{\rm cm}^2$, where a_0 is the radius of the first Bohr orbit. However, at temperatures at which nuclear processes can occur, atoms will be ionized. Taking $E = \frac{3}{2}kT$ gives $\sigma_{\rm Coul} = 1.01 \times 10^{-6} Z_1 Z_2/T^2\,{\rm cm}^2$, which for higher temperatures is smaller than the previous estimate. Similarly, if the nucleus has a radius $r = 1.4A^{\frac{1}{3}}$ Fermi, this gives a nuclear cross section of $\sigma_{\rm nuc} = \pi r^2 = 4.5 \times 10^{-26}(A_1^{1/3} + A_2^{1/3})^2\,{\rm cm}^2$.

These estimates are extremely crude. These quantum mechanical systems exhibit resonance behavior so that the cross sections vary enormously. Equating $\sigma_{\rm nuc}$ and $\sigma_{\rm Coul}$ for typical nuclei of astrophysical interest gives a temperature of about 10^{10} degrees Kelvin, higher than is generally of interest, which suggests that Coulomb scattering of ions is more common than nuclear interactions. A more accurate estimate generally gives stronger support for this idea. Consider an extreme example: $^{12}{\rm C} + ^{12}{\rm C}$ at a center of mass energy of 3.79 MeV has a measured reaction cross section of $\sigma = 3.1 \times 10^{-29}\,{\rm cm}^2$ (Mazarakis & Stephens 1973). This corresponds to a stellar temperature of $T = 4 \times 10^9$ K, and $\sigma_{\rm nuc}/\sigma_{\rm Coul} = 1.5 \times 10^{-3}$. Because this number is small, we expect the ions to establish an equilibrium velocity distribution, so that we have the great simplification of dealing with the reaction rates for a *thermal* gas.

At high temperatures ($T > 10^8$ K), radiative diffusion is dominated by Thomson opacity. Since $\sigma_{\rm Thom} = \frac{8}{3}\pi r_{\rm e}^2 = 0.665 \times 10^{-24}\,{\rm cm}^2$, this cross section equals $\sigma_{\rm Coul}$ at $T = (\frac{3}{2}Z_1 Z_2)^{1/2} \times 10^9$ K. For lower temperatures, typical of ignition conditions, heat diffuses faster than composition. The time scale for diffusion of heat through a distance δr is $\tau_{\rm diff} = (\delta r)^2/\lambda v$, where v is the velocity of diffusing objects and $\lambda = 1/N\sigma$ is the mean-free-path. For highly degenerate matter, electron conduction is more effective than radiative diffusion at transporting energy. The form of the time scale for diffusion of composition is identical to $\tau_{\rm diff}$ above, but because the velocities of the nuclei are smaller and their Coulomb cross sections higher, this time scale is longer. Thus we can concentrate on the ignition of new fuel by thermal heating rather than the action of superthermal particles, a significant simplification.

There are two separate time scales for burning which must be considered. The first is the *ignition time scale* of the fuel which is defined to be the temperature e-folding time

$$\tau_T = T/\dot{T} \approx C_V T/\dot{\varepsilon}_{\rm nuc}, \tag{4.1}$$

where $\dot{\varepsilon}_{\rm nuc}$ is the energy release rate of the nuclear processes, and C_V is the specific heat. Because charged-particle reactions are heavily modified by Coulomb barrier penetration, this time scale strongly decreases with increasing temperature. The second important time scale is the *burning time*, *i.e.*, the time to significantly reduce the abundance of fuel, which is defined as

$$\tau_i = X_i/\dot{X}_i = Y_i/\dot{Y}_i, \tag{4.2}$$

where X_i is the mass faction of species i, and Y_i is the mole number, obtained by dividing the mass fraction by the atomic weight of the species. In simple cases, this differs from τ_T by the ratio of the thermal energy content to the Q-value for the reaction (per unit mass burned). For example, for $^{12}{\rm C} + ^{12}{\rm C}$ this ratio is $0.25\,T_9$, where T_9 is temperature in units of 10^9 K; this is of order unity at explosive temperatures. For comparison, the ratio is about

0.005 for hydrostatic hydrogen burning by the CNO cycle. This ratio can also approach zero in explosive situations, if the matter is degenerate. In this case the specific heat approaches zero, so that consumption of a small amount of fuel gives a large change in temperature.

Finally, two time scales associated with hydrodynamic motion must be considered. The first is the time for a region to react to a pressure imbalance, which is taken to be the *sound travel time*

$$\tau_{\rm hyd} = \delta r / c_{\rm s}, \tag{4.3}$$

where $c_{\rm s}$ is the local sound speed. Finally, there is the time for a convective element to move through the region in which convection occurs (obviously an oversimplification of a complex process). This *convective time scale* is defined by

$$\tau_{\rm conv} = \delta r_{\rm conv} / v_{\rm conv}, \tag{4.4}$$

where $\delta r_{\rm conv}$ is the width of the convective zone and $v_{\rm conv}$ is the typical velocity of a convective blob.

Depending on the relative size of the various time scales, very different requirements for the numerical treatment arise. If the nuclear time scales τ_i are all large compared to $\tau_{\rm conv}$, one might approximate the convective zone as uniform in abundances, which slowly evolve on the nuclear time scales. If *some* of the τ_i are shorter than $\tau_{\rm conv}$, it is not correct to ignore the abundance gradients that this implies. In particular, these gradients may interact with the burning to modify the convective flow itself. Silicon burning is an interesting case of this: the time scales τ_i for α, n, and p, which maintain the quasi-equilibrium, are shorter than $\tau_{\rm conv}$, while the τ_i for ^{28}Si is longer. If the nuclear time scales are all small compared to $\tau_{\rm conv}$, the problem simplifies again, and each region is loosely coupled to its neighbors. For more dynamic problems, such as pulses or explosions, the convenient fiction of steady state convection is untenable, and the hydrodynamics must be treated as an equally important aspect of the problem.

4.2 Types of Burning

Many different types of burning occur in astrophysical problems. The type which occurs in a particular problem is determined by the ratios of the time scales described above. Burning can range from quiet hydrostatic burning, such as occurs in the center of most stars, to the explosive burning which can occur, for example, in models of thermonuclear supernova explosions (*i.e.*, SNe Ia). Burning can occur in laminar or turbulent flows, the fluid motion being produced either by external causes and/or by the burning itself. Hence, the interaction between burning and flow must be considered.

4.2.1 Hydrostatic Burning. This is the least violent form of burning and occurs in stars that are hydrostatic (in radiative burning zones). In this case, the reaction rate is slow enough that energy can be transported from the burning region by radiation or conduction as rapidly as it is generated. A similar case is a subsonically convective flame region, in which heat and ashes are removed by slow convection, and new fuel brought in. In other words, $\tau_i \gg \tau_{\rm diff}$ or $\tau_i \gg \tau_{\rm conv}$. These are the conditions which holds in stars such as the Sun over most of their lifetimes. The star remains in equilibrium, quietly burning its fuel, while the energy produced is transported to the star's surface, providing its luminosity. This process will continue until the fuel is exhausted, at which time the nuclear reactions can no longer maintain sufficient heat, and hence pressure support, to prevent the center of the star from contracting. The temperature and density at the center of the star will continue to increase until the conditions are reached at which the reaction products of the previous burning ignite. Depending on the situation, this burning can be either another hydrostatic phase or an explosive event.

4.2.2 Detonations. At the other extreme, detonations are the most violent form of burning one encounters. Astrophysical detonations generally occur only under degenerate conditions. During the initial stages of a thermonuclear runaway, before the temperature rises significantly, the reaction rate is still relatively small. In nondegenerate matter, the pressure increase produced by the reactions will cause the burning region to expand and cool, preventing the runaway from proceeding. In other words, $\tau_T > \tau_{\rm hyd}$. On the other hand, if the matter is degenerate, the temperature increase created by the burning will not create a significant increase in pressure. Thus the temperature will continue to increase until the matter becomes nondegenerate. At this point, the energy generation rate will be too large for hydrodynamic motion to stop it, and an explosion will result. If the resulting shock is sufficiently strong to raise the fuel above the ignition temperature, a detonation wave will propagate outwards from the point of ignition.

In its simplest form, detonation wave theory is easily understood (see *e.g.*, Courant & Friedrich 1948; Fickett & Davis 1979). As the shock propagates into the unburned fuel, it compresses and heats the material beyond the ignition point. Immediately behind the shock is the reaction zone, in which the fuel burns. In detonation theory the width of the reaction zone is frequently neglected, so that the detonation front is treated as a sharp discontinuity (Fig. 4.1). In this case jump conditions can be derived for the change in the hydrodynamic variables across the front in much the same way as is done for a simple shock. Note, however, that in order to obtain the post-detonation state one must first choose a detonation velocity, because unlike the case for simple shocks, the front velocity is not determined from the jump conditions. Moreover, one-dimensional steady detonation theory is oversimplified, as real detonations are observed to have inherent three-dimensional

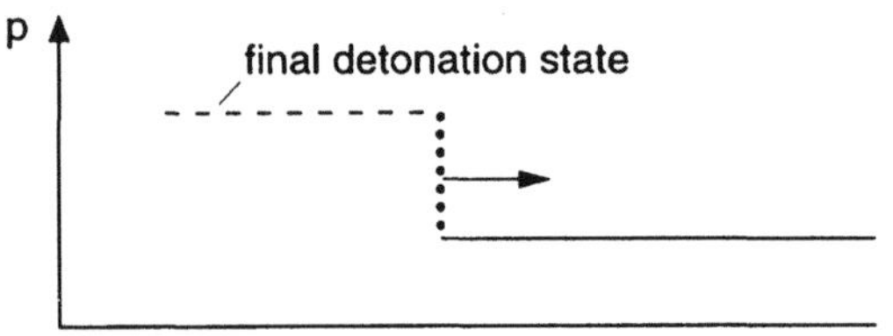

Fig. 4.1. Simplest model of a detonation wave: The width of the reaction zone is neglected, *i.e.*, the detonation front is treated as a sharp discontinuity

structure (cellular detonations) and can propagate in a non-steady manner (galloping detonations) (see, *e.g.*, Fickett & Davis 1979).

Depending on the value chosen for the detonation speed *no, one* or *two* solutions exist. In the latter case the two possible solutions correspond to *strong* and *weak* detonations. A strong detonation propagates at a speed slower than the post-detonation sound velocity with respect to the fluid behind the shock, *i.e.*, $D < u_2 + c_2$ (the subscript 2 denotes the post-detonation state), so that disturbances generated behind the front will eventually catch it. Thus this solution is unstable. Weak detonations, which propagate faster than the post-detonation sound velocity with respect to the fluid behind the shock, *i.e.*, $D > u_2 + c_2$, are generally considered to be unphysical except under certain special conditions (Fickett & Davis 1979).

The detonation which usually occurs in nature is the one corresponding to the propagation speed at which only one solution exists. This unique solution is called a *Chapman–Jouguet detonation.* It propagates with a speed equal to the sum of the post-detonation fluid velocity and sound speed, *i.e.*,

$$D_{\mathrm{CJ}} = u_2 + c_2 . \tag{4.5}$$

By assuming that the front propagates with this velocity, the post-detonation state is completely determined. Both the pressure and the density increase across the front, while in the frame in which the shock is stationary, the fluid velocity decreases.

Consider a detonation in a rigid tube followed by a piston with a fixed velocity. Two types of detonations which have somewhat different profiles can occur. If the velocity of the piston which generates the detonation is larger than the fluid velocity of the Chapman–Jouguet state, the result is an overdriven detonation. The structure in this case consists of the shock and reaction zone, followed by a constant state corresponding to the point at which all the fuel is depleted. This is a strong detonation, and therefore, the constant state behind the front does not correspond to the Chapman–Jouguet state. For smaller piston velocities, the detonation is unsupported. For this type of detonation, the reaction zone and the constant state behind the front are connected by a rarefaction, called the Taylor wave. Although this type of detonation propagates at the Chapman–Jouguet velocity, the Chapman–Jouguet state does not correspond to the constant state behind the front, but instead is the state at the top of the rarefaction.

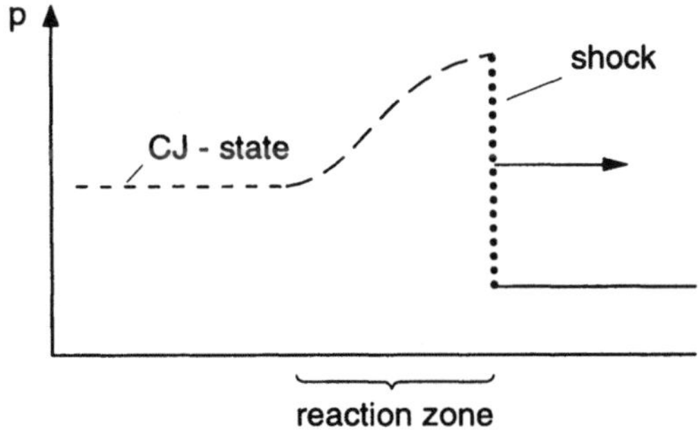

Fig. 4.2. ZND detonation wave: The shock, which is taken to be infinitely thin, is followed by a reaction zone of finite width

Although the above theory provides a satisfactory explanation for simple detonations, it has significant limitations because of the assumptions that the reaction rate is infinite and that the reaction zone has zero width. A slightly more complex treatment is used in the Zeldovich–von Neumann–Doering (ZND) model (Zeldovich 1940; von Neumann 1942; Doering 1943). This theory assumes that the shock, which is taken to be infinitely thin, is followed by a reaction zone of finite width (Fig. 4.2). The final state obtained after the fuel is completely burned is exactly the same as for the simpler theory described above. Additional complications in the theory arise due to multi-dimensional effects, such as cellular detonations and spinning detonations. However, these are beyond the scope of this discussion (*e.g.*, Fickett & Davis 1979).

4.2.3 Deflagrations. Deflagrations represent a much less violent form of burning than detonations, but in many ways are more complex (*e.g.*, Wilson 1985; Zeldovich *et al.* 1985). They result when the burning is unable to produce sufficient overpressure to create a shock which is strong enough to ignite the fuel. The motion of the front is usually very subsonic. Burning is initiated by the diffusive transfer of heat from the hot ashes behind the front into the cold fuel. Although, for the case of a thin front, deflagrations must obey the same jump conditions as detonations, the propagation velocity now depends on the rate of heat transfer.

Compared to detonations another major difference is that the pressure and density decrease behind the deflagration front, and in the reference frame in which the front is stationary, the velocity increases. In the case of a spherically symmetric deflagration, which begins at the origin, the velocity behind the front must eventually become zero to satisfy the boundary condition at the origin. The only way in which this can happen is if the deflagration is preceded by a compression wave which accelerates the material away from the front. The passage of the deflagration will then provide exactly the correct jump in velocity so that the material behind the front will come to rest. This can happen since the deflagration velocity is subsonic, and therefore, there is communication between the origin and the rest of the flow by sound waves.

In the simplest theory deflagrations are described as flow discontinuities, which propagate subsonically (*e.g.*, Courant & Friedrich 1948; Landau & Lifshitz 1959). If the flow is laminar, a physically more accurate description is that of a conductive or diffusive laminar flame, whose width and speed is determined by the rate of energy generation and the transport of energy and composition. In most situations, however, even this description is too simple, because deflagrations or subsonic flames are subject to a variety of instabilities, which can significantly influence their shape (wrinkling of the flame) and propagation speed. Moreover, because the propagation speed of the flame is subsonic, flame and flow do interact. In particular, if burning occurs in a convective or turbulent environment, depending on the relevant length scales and time scales, physically quite different combustion regimes must be considered (see next subsection).

Unlike the case of Chapman–Jouguet detonations, where it is possible to compute the exact propagation velocity, the propagation speed of deflagrations can only be estimated from dimensional reasoning (Landau & Lifshitz 1959; Buchler *et al.* 1979; Fryxell & Woosley 1982; Nozakura *et al.* 1984; Woosley 1986). For the simplest case of a laminar front which propagates as a result of radiative diffusion or conduction, it is fairly easy to obtain an order of magnitude estimate for the velocity of the wave. The width of the deflagration can be approximated by setting the diffusion time scale $\tau_{\rm diff}$ equal to the burning time scale τ_i. Thus the width of the front is given by

$$\delta \sim \sqrt{\lambda c \tau_i} \,, \tag{4.6}$$

where λ and c are the mean free path of photons or electrons and the speed of light, respectively. The (laminar) velocity of the deflagration can then be estimated as

$$D_{\rm lam} \sim \frac{\delta}{\tau_i} \sim \sqrt{\lambda c / \tau_i} \,. \tag{4.7}$$

Using an implicit radiation hydrodynamics code, Timmes & Woosley (1992) have determined the propagation velocity, the width, and the density structure of conductive deflagrations in degenerate carbon-oxygen and oxygen-neon-magnesium compositions under conditions typical of SN Ia progenitors ($\varrho \sim 10^{10}\,{\rm g\,cm^{-3}}$ and $T \sim 2 \times 10^9\,{\rm K}$). At these conditions, one expects from Eq. (4.6) that for both compositions the flame width δ is of the order of 10^{-5} cm. This implies a flame speed of a few $10\,{\rm km\,s^{-1}}$. In their numerical experiments, Timmes & Woosley (1992) constructed spheres with masses between 10 g and 100 g using extremely fine zoning ($\sim 10^{-6}$ cm). The spheres initially had constant temperature and density. The innermost 10 zones of a sphere were then artificially incinerated, giving (after a short initial transient) rise to a propagating isobaric, laminar, conductive flame, whose steady state physical properties were determined. Timmes & Woosley (1992) found that the laminar flame speed in a degenerate carbon-oxygen mixture is given with about 10% accuracy by the expression

$$D_{\text{lam}} = 92.0 \left(\frac{\varrho}{2 \times 10^9} \right)^{0.805} \left[\frac{X(^{12}C)}{0.5} \right]^{0.889} \text{km}\,\text{s}^{-1} \tag{4.8}$$

in the density range $10^7\,\text{g}\,\text{cm}^{-3} \leq \varrho \leq 10^{10}\,\text{g}\,\text{cm}^{-3}$, which is consistent with the earlier result of Woosley (1986).

As the laminar flame speed in a degenerate carbon-oxygen mixture is very subsonic ($D_{\text{lam}} \sim 10^{-2}\,c_s$), it is generally agreed that it is too slow to produce an explosion with the observed properties (see Sect. 3.1). However, in SNe Ia thermonuclear burning takes place in an environment in which gravity plays a role and drives Rayleigh–Taylor instabilities (Müller & Arnett 1982, 1986; see also Sect. 3.4). In addition, the flame is subject to the Landau–Darrieus instability (Landau 1944; Darrieus 1938) and to the Kelvin–Helmholtz instability (see, *e.g.*, Landau & Lifshitz 1959). Hence, as the Reynolds number of the flow Re $\sim 10^{14}$ in SNe Ia, turbulence develops and the flame front becomes wrinkled. The average rate of propagation is then no longer determined by the laminar speed, but by turbulence. As a result the velocity of the front is increased compared to that of a diffusively propagated deflagration ($D_{\text{turb}} > D_{\text{lam}}$) by a larger heat transfer rate in the distributed burning regime or by an increase in the surface area of the front in the flamelet burning regime (*e.g.*, Kerstein 1988; see next subsection for a definition of these burning regimes).

Whether the acceleration of the flame by turbulence and/or intrinsic instabilities is sufficient to explain the observations, is one of the, if not the single most important question in SNe Ia modeling. In order to answer this question, one has to consider turbulent combustion in a self-gravitating fluid (see next subsection).

4.2.4 Turbulent Combustion. When burning occurs in a turbulent flow (partially or not at all caused by the burning itself), the combustion process can no longer be modeled by a plane laminar conductive deflagration. Instead, a much more complicated type of burning must be considered, namely turbulent combustion, which is an extremely challenging and still unsolved problem (*e.g.*, Wilson 1985; Zeldovich *et al.* 1985; Pope 1987; Clavin 1994). A detailed and comprehensive discussion of this complicated flow phenomenon is far beyond the scope of my lecture and being no expert in the field would also surmount my scientific competence. Thus, I have restricted myself to describe some aspects of turbulent combustion, which are of relevance for thermonuclear supernovae (*i.e.*, SNe Ia). Recently, both Niemeyer & Woosley (1997) and Niemeyer & Kerstein (1997) have carefully addressed this problem. The subsequent discussion closely follows their analyses.

Turbulent combustion in Chandrasekhar mass models for Type Ia supernovae can occur in two distinct burning regimes (Niemeyer & Woosley 1997). If the laminar flame structure is unaffected by turbulence on small scales (comparable to the flame thickness), turbulent combustion proceeds in the *flamelet* regime (*e.g.*, Clavin 1994). Burning in the flamelet regime is characterized by a passively advected flame surface, which propagates through

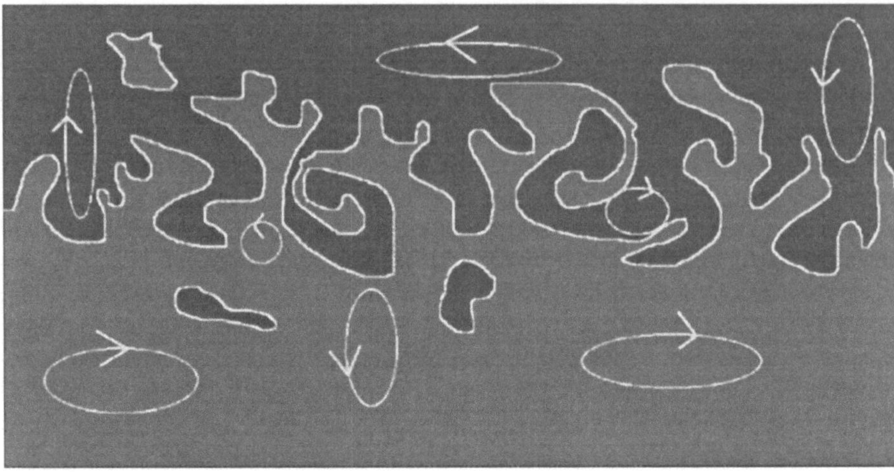

Fig. 4.3. Turbulent combustion in the flamelet regime. The thin white surface marks the wrinkled and non-contiguous laminar flame, which separates fuel (dark regions) from ashes (grey regions). The size of the convective eddies is larger than the width of the laminar flame and fuel consumption occurs in an extended turbulent flame brush behind the flame. (Plot by J. Niemeyer)

the medium at the laminar flame speed $D_{\rm lam}$, the large-scale structure of the flame surface being determined by turbulence, only (*i.e.*, independent of microphysics). Fuel consumption takes place in an extended region behind the fuel-ash boundary in the so-called turbulent flame-brush (*e.g.*, Pope 1987; Fig. 4.3). In Chandrasekhar mass models for SNe Ia burning occurs in the flamelet regime during the early phase of the explosion when the flame is thin and fast (Timmes & Woosley 1992), and turbulence is weak.

During a SN Ia explosion the laminar flame becomes thicker and slower as it proceeds to lower densities, while turbulence becomes stronger due to the nonlinear growth of Rayleigh–Taylor instabilities of the flame surface (see below). Eventually, the conditions for the flamelet regime are no longer satisfied, and burning enters the ***distributed*** or ***stirred*** burning regime (*e.g.*, Pope 1987). Turbulent eddies enter the flame and disrupt it. Turbulent transport dominates over conductive transport even on microscopic scales (Fig. 4.4).

The transition from flamelet to distributed burning is well known in the combustion literature, but has only recently caught the attention of astrophysicists (*e.g.*, Khokhlov *et al.* 1997; Niemeyer & Woosley 1997). Commonly, the Karlovitz number Ka is used to characterize the transition. This dimensionless number, which is of order unity at the transition, is either defined as a ratio of the reaction and flow length scales (Lewis & von Elbe 1961) or as a ratio of the reaction and flow time scales (Bradley 1993). Both definitions are equivalent for gaseous combustion provided the Prandtl number Pr$= \nu/\kappa$ is about unity, where ν is the kinematic viscosity and κ is a ther-

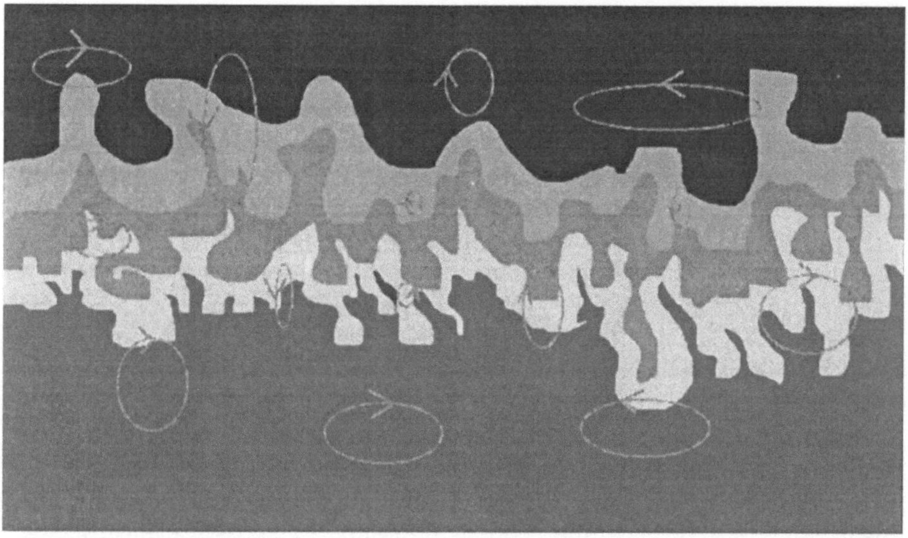

Fig. 4.4. Turbulent combustion in the distributed regime. Convective eddies penetrate the flame (white to grey layer), which is no longer well defined. Fuel (black) is consumed within turbulent eddies, which dominate over conductive transport. (Plot by J. Niemeyer)

mal diffusion coefficient. Recently, it has been pointed out by Niemeyer & Kerstein (1997) that in SNe Ia the Prandtl number is not of the order unity, but ranges from 10^{-5} to 10^{-1} (Nandkumar & Pethick 1984). Hence, a novel criterion for the transition to distributed turbulent combustion is required in degenerate white dwarfs. Niemeyer & Kerstein (1997) propose to use the eddy diffusivity, evaluated at the scale of the flame thickness as the governing parameter; if it exceeds the microscopic thermal conductivity, flamelets are considered to break up. In Chandrasekhar mass models of SNe Ia the density ϱ_{tr} at which the transition occurs is of the order $10^7\,\mathrm{g\,cm^{-3}}$. The actual value depends on the size of the turbulent velocity fluctuations u_{turb} and varies from $\varrho_{\mathrm{tr}} \approx 5 \times 10^6\,\mathrm{g\,cm^{-3}}$ for $u_{\mathrm{turb}} \approx 10^6\,\mathrm{cm/s}$ to $\varrho_{\mathrm{tr}} \approx 3 \times 10^7\,\mathrm{g\,cm^{-3}}$ for $u_{\mathrm{turb}} \approx 10^8\,\mathrm{cm/s}$. These transition densities are roughly a factor of 5 smaller than those determined previously by Niemeyer & Woosley (1997), which has important implications for the so-called delayed-detonation scenario of SNe Ia (Khokhlov 1991; Woosley & Weaver 1994).

Let us now consider the flame instabilities (see previous subsection), which generate the turbulence. In SNe Ia these instabilities act on vastly different length scales. They modify the flame structure on scales ranging from about $10^7\,\mathrm{cm}$, which is a significant fraction of the white dwarf's radius ($R_{\mathrm{WD}} \approx 10^8\,\mathrm{cm}$), down to microscopic scales of the order $10^{-3}\,\mathrm{cm}$, which are determined by the size of the laminar flame thickness and the viscous dissipation-scale.

The Rayleigh–Taylor (RT) instability dominates the dynamics of the flame on scales $10^6\,\mathrm{cm} \lesssim l_{\mathrm{RT}} \lesssim 10^7\,\mathrm{cm}$. It arises, because the density decreases and the temperature increases behind a subsonic flame (deflagration). When the flame propagates outwards from the center of a star against the force of gravity, one has a low density bubble of hot ashes trying to push outward through the denser and colder unburned fuel. This arrangement is unstable, similar to the case of a classical RT instability (see Sect. 3.4). The evolution of the RT instability is characterized by a short period of exponential growth during which the perturbation amplitudes become comparable to its wavelength, and then by the nonlinear stage where interaction among modes leads to a merging and fragmentation of structures (*e.g.*, Read 1984; Young 1984). The shear flow occurring at the surfaces of rising bubbles and sinking spikes produces Kelvin–Helmholtz (KH) unstable regions along these surfaces (Niemeyer & Hillebrandt 1995a). Eventually, a turbulent mixing layer forms (Snider & Andrews 1994). The growth of the layer can be estimated by the Sharp–Wheeler model (Sharp 1984). In this one-dimensional model the nonlinear stage of a RT unstable front is described by statistical means. Such an approach is justified, because the RT unstable front is characterized by a large number of structures with various scales. Numerical simulations based on the Sharp–Wheeler model (Glimm & Li 1988) show that the RT mixing region asymptotically grows at a rate of approximately

$$l_{\mathrm{RT}} \approx 0.05\, g_{\mathrm{eff}}\, t^2 \,, \tag{4.9}$$

where $g_{\mathrm{eff}} \equiv At\, g$ is the effective gravitational acceleration and $\mathrm{At} \equiv (\varrho_2 - \varrho_1)/(\varrho_2 + \varrho_1)$ is the Atwood number (see Sect. 3.4). This result agrees well with experiments (Read 1984) and full hydrodynamic simulations (Young 1984). According to Eq. (4.9), a RT unstable front propagates with a velocity

$$u_{\mathrm{RT}} \approx 0.1\, g_{\mathrm{eff}}\, t \propto \sqrt{g_{\mathrm{eff}}\, l_{\mathrm{RT}}} \,. \tag{4.10}$$

The RT structures contain most of the overall kinetic energy of the exploding star and evolve significantly during a SN Ia explosion, *i.e.*, they are very anisotropic and non-stationary. Hence, burning and hydrodynamic flow occurring on the RT unstable scales can probably not be described by a simple parameterized model, but must be simulated directly (Niemeyer & Hillebrandt 1995a).

The joint action of the large-scale RT and KH instabilities feeds a turbulent cascade (Richardson 1922), which extends all the way down to the Kolmogorov length η, at which scale dissipation due to microscopic viscosity becomes important. This scale is commonly called the ***dissipation scale.*** It can be estimated within the framework of Kolmogorov's (1941) turbulence model, which assumes homogeneous, isotropic and incompressible turbulence. One finds $\eta = l_{\mathrm{RT}}\,\mathrm{Re}^{-3/4}$ (Landau & Lifshitz 1959). For Reynolds numbers $\mathrm{Re} \approx 10^{14}$, which are typical of white dwarfs, η is of the order $10^{-3}\,\mathrm{cm}$, only.

In the intermediate range, *i.e.*, for scales with $\eta \ll l \ll l_{\mathrm{RT}}$, the hydrodynamic behaviour is dominated by the inertia of the fluid. In this so-called *inertial range* the famous Kolmogorov scaling law for incompressible turbulent velocity fluctuations holds (Kolmogorov 1941; Landau & Lifshitz 1959) provided two assumptions are fulfilled. Firstly, the properties of the fluid in this range must be independent of both the microscopic viscosity and the large-scale geometry. Secondly, the transport of kinetic energy from large to small scales is completely determined by its constant mean energy dissipation value ε. The scaling law, obtained under these two assumptions from dimensional analysis, is

$$\varepsilon \propto u(l)^3/l \,, \tag{4.11}$$

and hence the velocity fluctuations $u(l)$ must scale consistently as

$$u(l) \propto l^{1/3} \,. \tag{4.12}$$

As pointed out by Niemeyer & Kerstein (1997) turbulence generated by the Rayleigh–Taylor instability creates parallel cascades of kinetic and potential energy. The former cascade resembles that of Kolmogorov turbulence, where large eddies fragment into smaller ones. The potential energy cascade, which is only present in RT generated turbulence, is produced by the disruption of buoyant regions of burned, hot matter leading to an additional flux of density fluctuations from large to small scales. This density cascade may dominate over the eddy cascade (L'vov & Falkovitch 1992) and gives rise to scaling laws for the density and velocity fluctuations given by (Bolgiano 1959; Obukhov 1959)

$$\varrho(l) \propto l^{1/5} \tag{4.13}$$

and

$$u(l) \propto l^{3/5}, \tag{4.14}$$

respectively. Experimental and analytical evidence in support of and against the above Bolgiano-Obukhov scaling can be found in Borue & Orzag (1997).

Another length scale of importance in turbulent combustion is the *Gibson* scale l_{Gibs} introduced in the context of turbulent flame modeling by Peters (1988). At scales below the Gibson scale hydrodynamic effects on the flame are negligible, because the slow, small turbulent eddies present at those scales are burned before they can alter the flame structure significantly. The Gibson scale is defined by the relation

$$u(l_{\mathrm{Gibs}}) = D_{\mathrm{lam}} \,. \tag{4.15}$$

On microscopic scales the flame is intrinsically unstable due to the Landau–Darrieus (LD) instability. For the case of SNe Ia, this instability has been explored in detail both analytically and numerically by Blinnikov & Sasarov (1996) and Bychkov & Liberman (1995), and its onset was simulated (in 2D) by Niemeyer & Hillebrandt (1995b). These studies show that on scales larger than the Markstein, or critical length $l_{\rm crit} \sim 100\delta$ (Markstein 1951), flames propagating towards the surface of the white dwarf are unstable to the LD instability on all wavelengths. In the nonlinear regime the LD instability leads to the formation of cusps at the points where the flame front self-intersects. This gives rise to a quadratic damping term (Zeldovich *et al.* 1985), which stabilizes the flame surface once cells with a stationary, scale-independent amplitude form. The speed of the cellular surface is determined by the increased surface area (*e.g.*, Niemeyer & Woosley 1997). In the absence of gravity, there is no known upper limit for the largest scale of cell formation for a cellular flame front driven purely by the LD instability. This is no longer true, when the effect of gravity is taken into account. Then the cell structure becomes scale dependent and the nonlinear stabilization of the flame becomes ineffective (Khokhlov 1995). In a turbulent environment cell formation ceases on the scale l where the turbulent velocity fluctuations $u(l)$ become comparable to the speed of the cellular front $u_{\rm cell}(l)$. The eddy turnover time is then comparable to its interaction time with the front, or in other words, turbulent perturbations of the front are comparable to the cell size. Hence, the transition between cellular and turbulent burning occurs at the respective Gibson scale defined as $u(l_{\rm Gibs}) = u_{\rm cell}(l_{\rm Gibs})$.

This concludes the discussion of the instabilities operational in SNe Ia. However, in order to describe the evolution of thermonuclear burning in SNe Ia, we need to address a final point. How is the effective speed of the burning front in the turbulent regime ($D_{\rm turb}$) related to the turbulent velocity (u; *e.g.*, Eq. 4.14). This is a very important and non-trivial question, as the two velocities refer to *a priori* different physical processes. If the turbulent velocities are large (*i.e.*, $u \gg D_{\rm lam}$), one commonly assumes $D_{\rm turb} = u$. Hence, one assumes that the effective burning velocity scales with thc turbulent velocity. This scaling assumption is confirmed, *e.g.*, by Niemeyer & Kerstein (1997).

From the discussion above the following picture for the propagation of thermonuclear burning in Chandrasekhar mass models of SNe Ia is suggested. (Niemeyer & Woosley 1997). Initially, burning propagates in form of a laminar conductive flame of thickness δ with speed $D_{\rm front} = D_{\rm lam}$ (see Eqs. (4.7) and (4.8)), until the smallest cells appear due to the Landau–Darrieus instability at $l_{\rm crit} \approx 100\delta$. In this so-called cellular regime the propagation speed of the front is no longer scale independent, but approximately scales as $D_{\rm front}(l) \sim l^{0.1}$ (Blinnikov & Sasarov 1996; Niemeyer & Woosley 1997). At the Gibson length $l_{\rm Gibs}$ (Eq. (4.15)) the turbulent velocities generated by the (macroscopic) gravity driven RT instability become comparable to the cellular flame speed. Nonlinear cellular stabilizations fails, and burning changes from the cellular regime to the (inertial) Kolmogorov regime. Under the scaling assumption $D_{\rm turb} = u$, this implies that for scales $l > l_{\rm Gibs}$ the turbulent flame speed is determined by the eddy turnover time on every scale. Hence, $D_{\rm front}(l) \propto l^{1/3}$ (Eq. (4.12)) in case of Kolmogorov scaling, or $D_{\mathrm{front}}(l) \propto l^{3/5}$ (Eq. (4.14)) if Bolgiano-Obukhov scaling applies. Finally, above $l_{\mathrm{turb/RT}} \approx 10^6\,\mathrm{cm}$, which marks the boundary between the turbulent burning regime and the buoyancy-driven RT regime (Niemeyer & Woosley 1997), burning propagates according to $D_{\mathrm{front}}(l) \propto l^{1/2}$ (Eq. (4.10)). If turbulence obeys Bolgiano-Obukhov scaling, the division into inertial and RT range becomes obsolete, because buoyancy effects are already taken into account in that scaling. The actual values of the length scales at which the various transitions occur depend on density. Finally note that the above description of the evolution changes, if burning enters the distributed burning regime at some stage.

4.3 Nuclear Reaction Networks

Consider an element of matter at sufficiently high temperature and density so that thermonuclear reactions occur. In astrophysical situations a number of different types of reactions can be important. For example, consider reactions of the sort $X(p,\gamma)Y$, or equivalently $X + p \rightarrow Y + \gamma$, where X is the target nucleus, Y the product nucleus, p the bombarding particle (a proton) and γ an exit channel particle (a gamma ray). The reactions which one must deal with most often are

$$\begin{array}{cc} (p,\gamma) & (p,n) \\ (p,\alpha) & (n,\gamma) \\ (n,\alpha) & (\alpha,\gamma) \end{array}$$

and their inverses

$$\begin{array}{cc} (\gamma,p) & (n,p) \\ (\alpha,p) & (\gamma,n) \\ (\alpha,n) & (\gamma,\alpha) \end{array}$$

although several other types of reactions may be important as well. During hydrogen burning, reactions involving ^{2}H and ^{3}He must be considered. The "heavy ion" reactions $^{12}\mathrm{C} + {}^{12}\mathrm{C}$, $^{12}\mathrm{C} + {}^{16}\mathrm{O}$ and $^{16}\mathrm{O} + {}^{16}\mathrm{O}$ dominate carbon and oxygen burning stages; the more important exit channels involve emission of (1) a proton, (2) an alpha, (3) a neutron and (4) a deuteron (for oxygen burning). The most important three body reaction is $3(^4\mathrm{He}) \rightarrow {}^{12}\mathrm{C} + \gamma$. The slower weak interactions are of fundamental importance because they alone change the ratio of neutrons to protons in the system.

Consider the rate of change in the number density n_i of species i. The nonlinear differential equation which describes the behavior of n_i may be written (e.g., Arnett 1996)

$$\begin{aligned} \dot{n}_i &= \sum_j c_i(j)\lambda_j n_j + \sum_{j,k} c_i(j,k) < jk > n_j n_k \\ &\quad + \sum_{j,k,l} c_i(j,k,l) < jkl > n_j n_k n_l \,. \end{aligned} \tag{4.16}$$

The first term in Eq. (4.16) describes changes of n_i due to β-decays, electron captures (EC) and photo-disintegrations, the second term describes changes due to two-body reactions, and finally, the third term describes changes due to three-body reactions. Accordingly λ_j is the weak interaction rate (β-decay, EC) or the photo-disintegration rate, and $< jk >$ and $< jkl >$ denote the thermally averaged products of cross section and relative velocity in the center of mass system for the two-body and three-body reactions, respectively (e.g., Fowler et al. 1975). The coefficients c_i are short notations for

$$\begin{aligned} c_i(j) &= \pm N_i \, , \\ c_i(j,k) &= \pm \frac{N_i}{N_j! N_k!} \, , \\ c_i(j,k,l) &= \pm \frac{N_i}{N_j! N_k! N_l!} \, , \end{aligned} \tag{4.17}$$

where N_j, N_k, N_l are absolute numbers indicating how many particles of type i, j, k, l are involved in the reaction and the $N_j!$ *etc.*, prevent double counting of reactions with identical particles. The sign depends upon whether the nucleus i is produced (+) or destroyed (−) in the reaction.

If an expansion or contraction occurs, as is generally the case in astrophysical applications, Eq. (4.16) must be modified. In order to avoid inconvenient and perpetual bookkeeping, it is useful to consider the number of species i per unit of some conserved quantity, instead of per unit volume. Nucleon number is such a quantity, being conserved even in the relativistic limit.

Avogadro's number N_{A} is defined as the number of atoms of some species i which makes W_i grams, where W_i is the atomic weight of species i. For helium, W_{He} is 4.0026 for example. If m_i is the mass of a nucleus of species i as measured in the lab, then for a mixture consisting only of species i, the mass density is $\varrho_0 = n_i \, m_i$ and $N_{\mathrm{A}} \, m_i = W_i$. For a mixture of species, $\varrho_0 = \Sigma_i n_i \, m_i$.

This quantity changes with composition and is not relativistically invariant. However the related quantity

$$\varrho = \frac{\Sigma_i n_i \, A_i}{N_{\mathrm{A}}} \tag{4.18}$$

avoids these difficulties. We have replaced the mass m_i by A_i (the atomic number) times the atomic mass unit. Under laboratory conditions ϱ almost equals ϱ_0. We now define the nucleon number fraction for species i to be

$$X_i = \frac{n_i \, A_i}{\varrho N_{\mathrm{A}}} \, , \tag{4.19}$$

which is just the fraction of the nucleons tied up in the form of particles of species i. In the astrophysical literature this and a similar quantity involving ϱ_0, not ϱ, are both called the *mass fraction*. From Eqs. (4.18) and (4.19),

$$\Sigma_i X_i = 1 \, . \tag{4.20}$$

A related quantity is

$$Y_i = \frac{X_i}{A_i} = \frac{n_i}{\varrho N_{\mathrm{A}}}\ , \tag{4.21}$$

which is a ratio of the number of nuclei of species i to the total number of nucleons in the system. Now, the total number of nucleons is

$$n = \Sigma_i n_i\, A_i\ , \tag{4.22}$$

so

$$\Sigma_i Y_i = \frac{\Sigma_i n_i}{n}\ , \tag{4.23}$$

which is not equal to unity, in general. In actual calculations the distinction between ϱ and ϱ_0 is usually ignored. Using Eq. (4.21), Eq. (4.16) becomes

$$\begin{aligned}\dot{Y}_i \quad = \quad & \sum_j c_i(j)\lambda_j Y_j + \sum_{j,k} c_i(j,k)\varrho N_{\mathrm{A}} < jk > Y_j Y_k \\ & + \sum_{j,k,l} c_i(j,k,l)\varrho^2 N_{\mathrm{A}}^2 < jkl > Y_j Y_k Y_l\ .\end{aligned} \tag{4.24}$$

Unlike the number density n_i, the specific abundance Y_i does not change when only an expansion or contraction occurs, but requires the action of nuclear processes or mixing. Thus we have a set of coupled, nonlinear ordinary differential equations. Such a system of equations is said to define a *nuclear reaction network.*

4.3.1 Explicit and Implicit Solution Methods. The reaction network equations often behave as *stiff* differential equations (e.g., Gear 1971). In general they have rate terms which contain second and higher powers of the abundances being sought. Thus they are nonlinear. Only numerical techniques can provide solutions to a general set of reaction network equations. Consider the system of equations

$$\frac{\mathrm{d}Y_i}{\mathrm{d}t} = f_i(\varrho, T, Y_1, \ldots, Y_N)\ ,\ i = 1, \ldots, N\ , \tag{4.25}$$

where $f_i(\varrho, T, Y_1, Y_2, Y_3 \ldots)$ is the right hand side of Eq. (4.24) and N is the number of species considered in the reaction network. Suppose one approximates the time derivatives by finite differences

$$\frac{\mathrm{d}Y_i}{\mathrm{d}t} \approx \frac{Y_i(t+\delta t) - Y_i(t)}{\delta t}\ . \tag{4.26}$$

This implicitly assumes the change in variables to be small during the time interval δt. Two simple choices for evaluating f_i are apparent:

$$\frac{Y_i(t+\delta t) - Y_i(t)}{\delta t} \quad = \quad f_i(t) \tag{4.27}$$

or

$$= \quad f_i(t+\delta t)\ , \tag{4.28}$$

i.e., , one can evaluate the reaction rate terms at time t, or at time $t + \delta t$. Both are accurate to first order in δt; Eq. (4.27) results from a Taylor expansion from time t, while Eq. (4.28) results from a similar expansion about time $t + \delta t$. One might also use some average of the two, or estimate $f_i(t + \delta t/2)$, and arrive at an expression accurate to second order in δt. One must distinguish between *accuracy* and *stability*; both are needed for a correct solution. Because one approximates a derivative using a truncated Taylor expansion, the resulting difference equation may have extra solutions in addition to those of the differential equation (*e.g.*, Potter 1973). One could say that the resulting numerical garbage is due to *numerical instability* and consider it to be a mathematical problem, or regard the difference equations to be incorrectly formulated to represent the physics involved. The latter point of view is rarer, but often of more practical value.

With the choice denoted by Eq. (4.27), called *forward* or *explicit* differencing, one has N uncoupled equations for N unknowns. All Y_i, density and temperature at $t + \delta t$ may be found individually in terms of the Y_i, density and temperature at time t. The latter quantities are just those one starts with. Although the solution of these equations is truly elementary, difficulties are encountered near steady state and equilibrium (*e.g.*, Arnett 1996).

The approximation implied by Eq. (4.28) is called the *backwards differencing*, *implicit differencing* or *implicit* difference form. In this case one has to solve a coupled, nonlinear system of equations by matrix manipulation in order to obtain the abundances at the new time step. A further complication arises when the network has to be solved for time-dependent density and temperature (see Sect. 4.4.2).

4.3.2 Linearization and Matrix Inversion. Let us consider the solution of the network equations at a constant density and temperature . If one uses implicit differencing, the $f_i(t+\delta t)$'s contain terms which are at least quadratic in the unknowns. Products of the sort $Y_j(t+\delta t)\, Y_k(t+\delta t)$ give a nonlinearity, which means that no simple general solution can be obtained. However, in introducing finite differences to replace time derivatives in Eq. (4.26), one had to assume that

$$\left|\frac{\mathrm{d}Y_i}{\mathrm{d}t}\right| \delta t \ll Y_i(t) \qquad \text{or} \qquad Y_i(t + \delta t)\ . \tag{4.29}$$

Denote

$$\Delta_j = Y_j(t + \delta t) - Y_j(t)\ . \tag{4.30}$$

Then,

$$\begin{aligned} Y_j(t + \delta t)\, Y_k(t + \delta t) &= [Y_j(t) + \Delta_j]\,[Y_k(t) + \Delta_k] \\ &\approx Y_j(t)\Delta_k + Y_k(t)\Delta_j + Y_j(t)Y_k(t) \\ &\approx Y_j(t + \delta t)\, Y_k(t) + Y_j(t)\, Y_k(t + \delta t) \\ &\quad - Y_j(t)\, Y_k(t) \end{aligned} \tag{4.31}$$

if one neglects terms above first order in Δ_j and Δ_k. Because this approximation is only that which would be required for reasonable accuracy of numerical integration, the process of linearizing the reaction network equations does not really introduce any significant new restrictions. Essentially, one moves forward in time by analytic continuation; at each new time interval one updates the nonlinear coefficients, so that one obtains the nonlinear solutions. For three-body reactions, this linearization can be more restrictive, but predictor-corrector schemes or multistep iterations can solve these problems. Now upon linearization, one obtains

$$\begin{aligned}\Delta_i = \delta t \Big\{ \quad & \textstyle\sum_j \quad c_i(j)\,\lambda_j(\Delta_j + Y_j) + \\ & \textstyle\sum_{j,k} \quad c_i(j,k)\,\varrho N_{\mathrm{A}} < jk > (Y_j\Delta_k + Y_k\Delta_j + Y_jY_k) + \\ & \textstyle\sum_{j,k,l} \quad c_i(j,k,l)\,\varrho^2 N_{\mathrm{A}}^2 < jkl > \\ & \qquad (Y_jY_k\Delta_l + Y_jY_l\Delta_k + Y_kY_l\Delta_j + Y_jY_kY_l)\Big\}\,,\end{aligned} \tag{4.32}$$

where all the abundances are evaluated at time t.

Such an equation, linear in the unknowns (here the Δ_i's), may be written for all N species. The equation for a species i contains a term in Δ_k for each species k which is connected to i by a reaction. Obviously the equations for "mediating" particles such as p, n and alphas have many terms. One may write all these equations succinctly in matrix form:

$$\boldsymbol{\Delta A} = \boldsymbol{B}\,, \tag{4.33}$$

where $\boldsymbol{\Delta}$ is a column matrix in the Δ's, $\boldsymbol{B}$ a column matrix in constants, and $\boldsymbol{A}$ a square matrix (of dimension N). To better understand the structure of $\boldsymbol{A}$, consider a system in which there is only an *alpha particle chain*; that is

$$C(\alpha,\gamma)\,D(\alpha,\gamma)\,E(\alpha,\gamma)\,F(\alpha,\gamma)\,G(\alpha,\gamma)H\,. \tag{4.34}$$

If the order is alphas first, followed by six nuclei, the structure is illustrated visually as follows:

$$\boldsymbol{A} = \begin{pmatrix} 1 & 1 & 1 & 1 & 1 & 1 & 1 \\ 1 & 1 & 1 & 0 & 0 & 0 & 0 \\ 1 & 1 & 1 & 1 & 0 & 0 & 0 \\ 1 & 0 & 1 & 1 & 1 & 0 & 0 \\ 1 & 0 & 0 & 1 & 1 & 1 & 0 \\ 1 & 0 & 0 & 0 & 1 & 1 & 1 \\ 1 & 0 & 0 & 0 & 0 & 1 & 1 \end{pmatrix}\,. \tag{4.35}$$

Here the 1's denote that the respective matrix element is nonzero. Because of the coupling of the alphas to all the nuclei, the matrix is not *tridiagonal* (*i.e.*, all elements zero except those on the diagonal and their nearest neighbors). The fast solution schemes for tridiagonal matrices are not directly applicable (although the structure of the matrix can be used to assist solution). Note that as the dimensions increase for a matrix with this or many similar structures, the fraction of zero elements increases also, resulting in a sparse matrix. Because the time needed to solve a general matrix equation such as Eq. (4.33) increases rapidly with the dimension, it is best to use as few reactions in a network as physically plausible. This method was first used in astrophysics for silicon burning (Truran *et al.* 1967); and presented in detail by Arnett & Truran (1969). Wagoner (1969) used a second-order version for Big Bang nucleosynthesis. Using standard methods for solving linear equations, one calculates

$$\boldsymbol{\Delta} = \boldsymbol{B}\boldsymbol{A}^{-1} . \tag{4.36}$$

As the new abundances are $Y_i(t + \delta t) = Y_i(t) + \Delta_i$ for each species i, one has advanced one time step δt. Examining the fractional change Δ_i/Y_i and other criteria such as change in density, temperature or the size of the last time step, a new time step is chosen. New temperatures and densities are calculated (if they were not included in the matrix solution), and new reaction rates obtained. The new matrices $\boldsymbol{A}$ and $\boldsymbol{B}$ are constructed, and solved, and so on around the cycle until the evolution of the abundances has gone on as far as desired.

4.3.3 Iterative Solution. Instead of linearizing Eq. (4.24) as explained in the previous section, one can also solve the network by Newton's method. This of course requires the calculation of the Jacobian of the nonlinear system. For constant density and temperature, Eq. (4.24) can be rewritten in vector form as

$$\boldsymbol{g}\left[\boldsymbol{Y}(t^{n+1})\right] \equiv \frac{\boldsymbol{Y}(t^{n+1}) - \boldsymbol{Y}(t^n)}{\delta t} - \boldsymbol{f}\left[\boldsymbol{Y}(t^{n+1})\right] = 0 , \tag{4.37}$$

where $\boldsymbol{g}, \boldsymbol{f}$ and $\boldsymbol{Y}$ are column vectors (of dimension N), and where the superscript on t denotes the time step number. According to Newton's method, Eq. (4.37) leads to

$$\begin{aligned} G^l_{ij}\Delta^{l+1}_j &= \left(\frac{\partial g^l_i}{\partial Y^l_j(t^{n+1})}\right)\left(Y^{l+1}_j(t^{n+1}) - Y^l_j(t^{n+1})\right) \\ &= -g_i\left[\boldsymbol{Y}^l(t^{n+1})\right] , \end{aligned} \tag{4.38}$$

which is a system of linear equations for the unknown (column) vector

$$\boldsymbol{\Delta}^{l+1} = \boldsymbol{Y}^{l+1}(t^{n+1}) - \boldsymbol{Y}^l(t^{n+1}) . \tag{4.39}$$

G^l_{ij} is the Jacobian of the l-th iteration step and is given by

$$G^l_{ij} = \frac{\delta_{ij}}{\delta t} - \frac{\partial f_i\left[\boldsymbol{Y}^l(t^{n+1})\right]}{\partial Y^l_j(t^{n+1})} . \tag{4.40}$$

Using an initial guess $\boldsymbol{Y}^0(t^{n+1}) = \boldsymbol{Y}(t^n)$ and an appropriate δt, Eq. (4.38) will typically converge to a relative accuracy of 10^{-6} after about $l = 3$ iterations. The linearization of the network discussed in the previous section is the equivalent of performing just one Newton iteration. However, the size of the time step must then be limited such that the omitted higher order terms are negligible (Arnett & Truran 1969).

4.4 Coupling Reaction Networks and Hydrodynamics

If a nuclear reaction network is used within a hydrodynamic code, various interdependencies of the rate equations and the hydrodynamic equations exist.

4.4.1 Local Coupling. First, there is a (spatially *local*) coupling of both subsets of equations via the energy release (or consumption) due to nuclear burning. The reactions lead to an increase (or decrease) of the temperature and consequently of the pressure described by the source term in the energy or entropy equation, which in turn leads to an increase (or decrease) of the reaction rates. Since the reaction rates are extremely sensitive to temperature variations, a very strong nonlinear coupling results. The change of the internal energy, ε, due to this coupling is given by (*e.g.*, Müller 1986)

$$\frac{\partial \varepsilon}{\partial t} = -9.644 \times 10^{17}\, \varrho \sum_i \Delta m_i c^2 \frac{\partial Y_i}{\partial t} \quad [\mathrm{erg\,cm^{-3}\,s^{-1}}] , \tag{4.41}$$

where $\Delta m_i = (m_i - A_i m_\mathrm{u})c^2$ and m_u are the mass excess in MeV and the atomic mass unit, respectively.

In stellar evolution calculations the standard procedure is to solve the network separately from the other evolution equations, *i.e.*, Eq. (4.24) is solved for given density, temperature and abundances obtained previously in the time step. The resulting energy release and abundance changes are taken into account in the next time step (*e.g.*, Weaver *et al.* 1978).

This type of coupling (or better decoupling) of network and hydrodynamic equations can lead to instabilities when calculating nuclear transmutations while entering or leaving the regime of nuclear statistical equilibrium (NSE; for an explanation of NSE, see *e.g.*, Clayton 1968). Therefore, Müller (1986) proposed the following method, which avoids these instabilities. To illustrate this method, let us write the network equations and the source term (for a fixed density) as

$$\frac{\partial Y_i}{\partial t} - f(Y_1, \ldots, Y_N, T) = 0 , \tag{4.42}$$

$$\frac{\partial \varepsilon}{\partial t} - h(Y_1, \ldots, Y_N, T) = 0 , \tag{4.43}$$

where $i = 1 \ldots N$, $j = 1 \ldots N$, and N is the number of species. f and h are nonlinear functions of the arguments. If Eqs. (4.42) and (4.43) are discretized implicitly (see Eq. (4.28)), the variables from time level t^n are advanced to time level $t^{n+1} = t^n + \delta t$ according to

$$Y_i^{n+1} - Y_i^n - \delta t\, f(Y_1^{n+1}, \ldots, Y_N^{n+1}, T^{n+1}) = 0 , \tag{4.44}$$

$$\varepsilon^{n+1} - \varepsilon^n - \delta t\, h(Y_1^{n+1},\ldots,Y_N^{n+1},T^{n+1}) = 0\,. \tag{4.45}$$

As the energy itself is some known function of Y_i and T (via the equation of state), Eqs. (4.44) and (4.45) are a nonlinear system of equations for the $N+1$ unknown variables $Y_1^{n+1},\ldots,Y_N^{n+1}$ and T^{n+1}. Contrary to previous methods, in Müller's (1986) approach the whole system is solved with the Newton technique, *i.e.*, abundances *and* temperature are updated *together*. The resulting method is very robust and allows one to integrate the network equations into and out of the regime of nuclear statistical equilibrium without numerical problems.

4.4.2 Spatial Coupling. A second type of (spatially *non-local*) coupling arises, because the composition as well as the density and the temperature are modified by the hydrodynamics and in particular by diffusion, convective mixing and by advection in the case of an Eulerian hydrodynamics code. In the following, the latter two cases are considered in more detail

The conservation form for the equations of Eulerian hydrodynamics in two-dimensional Cartesian geometry with nuclear burning can be expressed by the vector equation (see also lecture by Le Veque)

$$\frac{\partial \boldsymbol{U}}{\partial t} + \frac{\partial \boldsymbol{F}(\boldsymbol{U})}{\partial x} + \frac{\partial \boldsymbol{G}(\boldsymbol{U})}{\partial y} = \boldsymbol{H}(\boldsymbol{U})\,. \tag{4.46}$$

The vectors $\boldsymbol{U}$, $\boldsymbol{F}$, $\boldsymbol{G}$, and $\boldsymbol{H}$ are defined by

$$\boldsymbol{U} = \begin{pmatrix} \varrho \\ \varrho u \\ \varrho v \\ \varrho E \\ \varrho X_i \end{pmatrix}\,, \qquad \boldsymbol{F}(\boldsymbol{U}) = \begin{pmatrix} \varrho u \\ \varrho u^2 + P \\ \varrho v u \\ (\varrho E + P)u \\ \varrho X_i u \end{pmatrix}\,, \tag{4.47}$$

$$\boldsymbol{G}(\boldsymbol{U}) = \begin{pmatrix} \varrho v \\ \varrho u v \\ \varrho v^2 + P \\ (\varrho E + P)u \\ \varrho X_i v \end{pmatrix}\,, \qquad \boldsymbol{H}(\boldsymbol{U}) = \begin{pmatrix} 0 \\ 0 \\ 0 \\ \varrho \dot{e} \\ \varrho \dot{X}_i \end{pmatrix}\,, \tag{4.48}$$

where u and v are the x- and y-component of the fluid velocity $\boldsymbol{v}$, E is the total energy per unit mass, X_i is the mass fraction of species i, and P is the pressure. The quantities $\dot{e}$ and $\dot{X}_i$ represent the time rate of change of the specific internal energy $e = \varepsilon/\varrho$ and mass fractions resulting from nuclear reactions.

Eqs. (4.46), (4.47) and (4.48) show that the change of the abundances (source term $\varrho\dot{X}_i$) is coupled spatially via the flow. Formulated differently, the reaction network now involves a comoving derivative

$$\dot{Y}_i = \frac{dY_i}{dt} = \frac{\partial Y_i}{\partial t} + \boldsymbol{v}\cdot\boldsymbol{\nabla} Y_i\,, \tag{4.49}$$

where the LHS is given by Eq. (4.24).

The $\boldsymbol{v} \cdot \boldsymbol{\nabla} Y_i$ operator couples the flow to the fuel abundance. For a region of linear size δr, the flow time scale is $\tau_{\text{flow}} = v/\delta r$. If this is short compared to all of the nuclear burning time scales, then $\boldsymbol{v} \cdot \boldsymbol{\nabla} Y_i \approx 0$, and the flow region approaches homogeneity. If we further assume that the flow is turbulent, cascading from large eddies down to a dissipation scale (*e.g.*, Landau & Lifshitz 1959), then diffusion at these small scales will remove the composition irregularities. This is the limit of complete microscopic mixing.

For the first thermonuclear stages of stellar evolution, hydrogen and helium burning, cooling occurs by radiative diffusion. This sets the dissipation scale for small eddies in a convection zone. From our previous discussion (Sect. 4.1), this implies that the diffusion length for nuclei (*i.e.*, composition) is longer than the scale for small eddies, but not enormously so. Complete mixing may be the appropriate approximation, at least for non-explosive phenomena.

Later burning stages (C, Ne, O and Si burning) are cooled by $\nu_e \bar{\nu}_e$ emission. This speeds the burning and decouples the dissipation from the diffusion process. Instead of a homogeneous, mixed region, a more realistic picture of the convective zone might be an ensemble of independent "blobs" which move in and out of the flame, with little microscopic mixing. Most stellar evolutionary calculations to date still assume homogeneous (microscopic) mixing.

Oxygen shell burning is crucial for the late evolution of massive stars. Its nature affects, *e.g.*, the size of the core which will collapse (Arnett 1994). In the oxygen-burning shell the sound travel time τ_{hyd}, the convective turnover time τ_{conv}, and the nuclear burning time τ_i (see Sect. 4.1) are all of the same order as the evolutionary time, and nuclear energy sources and neutrino cooling are ongoing in convective flows. Meanwhile, it has become possible to directly simulate the complex thermonuclear and hydrodynamic behaviour of oxygen-burning shells in two space dimensions (Arnett 1994; Bazan & Arnett 1994; 1997). These simulations show the nature of convective burning being dramatically different from that assumed in 1D models. One finds mixing beyond convective boundaries determined by mixing-length theory (*i.e.*, overshooting), hot spots of burning fuel in the oxygen shell and a convective velocity structure being dominated by plumes.

Silicon burning, the last burning stage before core collapse, is even more difficult to simulate, because further complications arise due to quasi-equilibrium and electron capture (*e.g.*, Arnett 1996). For silicon burning there are two classes of time scales: those relating to the slow transformation of Si to Fe, and those faster ones relating to the maintenance of quasi-equilibrium among the Si to Ca nuclei and free neutrons, protons and alpha particles (Clayton 1968). In a typical Si-burning convective zone, these two time scales straddle the convective turnover time, so the homogeneous approximation is incorrect. As a first step toward correcting for this effect, most evolutionary calculations have assumed local quasi-equilibrium, and transported only quasi-equilibrium global parameters (*e.g.*, the Si/Fe ratio, or Si-Ca and Fe-peak abundances).

In astrophysics, the most complicated interplay of nuclear burning and hydrodynamics is encountered when trying to simulate turbulent combustion (see Sect. 4.2.4). In thermonuclear supernovae this combustion processes has become a major research activity in astrophysics during the past decade. Various groups have performed two and three-dimensional simulations of detonations, deflagrations, and deflagration-to-detonation transitions (Müller & Arnett 1982, 1986; Khokhlov 1991; Livne 1993; Arnett & Livne 1994a, b; Khokhlov 1994; Bravo & García-Saenz 1995; Khokhlov 1995; Niemeyer & Hillebrandt 1995a, b; Khokhlov *et al.* 1997; Niemeyer & Woosley 1997; Ruiz-Lapuente *et al.* 1997 and references therein). Another major activity of multi-dimensional turbulent combustion modeling in astrophysics concerns novae (*e.g.*, Shankar *et al.* 1992; Shankar & Arnett 1994; Glasner & Livne 1995; Glasner *et al.* 1997).

Simulating these and other multi-dimensional combustion flows one always has to grapple with the dilemma, whether one chooses a Lagrangian or an Eulerian description of the flow.

Lagrangian methods avoid artificial mixing of nuclear species and diffusion of heat, and thus eliminate one of the major errors present in Eulerian codes (see also Sect. 2.2.1). However, severe numerical difficulties arise when using Lagrangian codes for problems where flows in more than one spatial dimension have to be simulated. The formation of vortices in the flow will cause large distortions in the Lagrangian grid and perhaps even grid tangling. Although this can be alleviated by periodic rezoning, the major advantage of the Lagrangian approach is then lost. The sudden mixing of species which occurs during the rezoning may provide even worse effects than the gradual mixing which occurs during the entire Eulerian calculation.

Thus, for two or three-dimensional calculations, Eulerian methods are probably to be preferred, since the grid remains regular. However, the errors associated with the mixing of species must then be considered. The amount of mixing can be controlled to some extent by using an adaptive grid which moves with the species discontinuities. In addition, an implicit method may be required to move the grid correctly, which enormously increases the complexity of the code, as well as the amount of computer time required. Local adaptive mesh refinement can help to keep the size of the mixed region small, but can not eliminate the error entirely. If the fluid in the mixed region explodes due to the numerical error, the results will still be qualitatively wrong. Multi-fluid calculations, which track the location of the interfaces between the various fluids, can perhaps solve the problem in simple cases, but for flows containing a large number of species, in which every zone may have a different composition due to burning, such methods seemed to be impractical. However, recently a new efficient capturing-tracking hybrid scheme for deflagration discontinuities based on the level-set approach (Markstein 1951) has been proposed, which could make such multi-fluid flow problems tractable (Smiljanovski *et al.* 1997).

4.4.3 Conservation of Species. When simulating hydrodynamic problems involving nuclear burning it is crucial to maintain conservation of the partial density of each species, both globally and locally. If one uses a conservative formulation of the species equations (Eq. (4.46)), global conservation is guaranteed. However, changes to some of the difference methods are required to ensure that the sum of the partial densities equals the total density in each zone (Fryxell *et al.* 1989). This problem has also been considered by Larrouturou (1991), who has proposed respective modifications for numerical schemes based on the multi-component *approximate* Riemann solvers of Roe (1981) and Osher (see Abgrall & Montagné 1989).

To illustrate the problem, let us consider, *e.g.*, the hydrodynamics code PROMETHEUS, which is based on a direct Eulerian implementation of the piecewise parabolic method (PPM) of Colella & Woodward (1984) and which was developed by Fryxell *et al.* (1989). In PROMETHEUS the values of the mass fraction of each species at the zone interfaces are obtained by interpolation including contact discontinuity detection. The parabolic distribution of each mass fraction is then constructed and monotonized. Since the monotonicity constraint in PPM is nonlinear, *i.e.*, the constraint operates on each abundance in a different way, the sum of the resulting interface values of the mass fractions will not, in general, be equal to unity. As a result, the sum of the zone averaged mass fractions at the new time will also not be unity. This local non-conservation error can have serious effects on the results of the calculation, at times even leading to unphysical features in the flow.

One way to guarantee that the sum of the mass fractions remains unity is to omit one of the species equations. The value of the partial density for that species would then be determined by subtracting the sum of the partial densities of the other species from the total density. This has the undesirable consequence of concentrating the total error in one abundance. If the errors which occur become large (the error can frequently exceed 100%) the value of this last abundance will be meaningless. Fortunately, there is a better solution to this problem, as was shown by Fryxell *et al.* (1989). When the error in the sum of the mass fractions at the zone interface after applying monotonicity is in error by more than some specified amount, the zone interface values are set equal to the zone average value. This reduces the species advection to first-order in that zone and causes more artificial mixing of species, but these errors are usually less severe than those obtained without the correction. In choosing the maximum amount of error allowed before flattening the zone structure, there is a tradeoff between the error in conservation and the amount of mixing. The maximum error allowed in the calculations of Fryxell *et al.* (1989) was 10^{-7}. The time averaged value of the mass fraction at the zone interface is obtained by advecting the average value in the domain of dependence along the streamline.

4.4.4 Time Step Restrictions. Making an intelligent choice for the value of the time step δt is one of the most critical aspects in efficiently solving problems which couple hydrodynamics and nuclear burning. Using too small a value will require the computation of an unnecessarily large number of time steps, whereas using too large a value will lead either to unacceptably large errors or, if an iterative method is used to solve the reaction network, an unconverged solution.

When a reaction network is incorporated into an explicit hydro-code, several restrictions (besides the well-known CFL-condition; see Sect. 2.2.2) must be imposed on the time step. First, the temperature and the mass fraction of each nuclear species must not be allowed to change by too much during the time step. Since the non-iterative solution of the network requires linearization of the equations (see Sect. 4.3.2), if too large a change in any variable is permitted, the linear approximation will no longer be valid, and large errors will result. In addition, if an iterative procedure is used (see Sect. 4.3.3), the solution will not converge. These restrictions take the form

$$\delta t_T^{n+1} = \alpha_T \, \delta t^n \, \frac{T^{n+1}}{|T^{n+1} - T^n|} \tag{4.50}$$

and

$$\delta t_X^{n+1} = \alpha_X \, \delta t^n \, \frac{X_i^{n+1}}{|X_i^{n+1} - X_i^n|}, \tag{4.51}$$

where the values of the dimensionless constants α_T and α_X depend on the type of network used for the calculation and how violent the burning is. Typically, one allows changes in the temperature of 3% or less, while abundance changes of 5% to 10% can usually be tolerated. When an iterative method of solution is used for the network, another time step restriction is achieved by requiring that the solution converges in a small number of iterations. This restriction can be written as

$$\delta t_I^{n+1} = \delta t^n \, \mathrm{Min} \left\{ 1, \left(\frac{3}{I} \right)^2 \right\}, \tag{4.52}$$

where I is the number of iterations required for convergence on the previous solution of the network. This restriction has the effect of reducing δt whenever the number of iterations required exceeds 3.

The value of δt is chosen to be the minimum obtained from each of the above equations in the most restrictive zone. There will be times when even this value of δt will be too large for the network to converge. When this happens, there is an additional procedure which one can use to continue the calculation. The values of temperature, energy, and mass fraction are stored before the network is solved. If the network does not converge, the value of δt can be reduced and the network recomputed starting from the original values. Since a smaller value of δt was used for the network than for the hydrodynamics, the network must be solved several times until the variables have been advanced to the correct time level. Equivalently, one could recalculate the entire hydrodynamics and burning step with the same value of δt, but this would require storing old values of all of the hydrodynamic variables.

4.4.5 A Computational Aspect. As we have seen above solving a reaction network in an (1D) stellar evolution code requires the solution of a nonlinear system of equations independently for each grid point, if no mixing occurs. This parallelism can be utilized on vector and parallel processors. The reaction rates, the elements of the Jacobian, and the solution of the linear systems arising during each iteration step should be calculated for all, or at least a large number, of grid points in parallel.

Whereas the vectorization or parallelization of the linear systems poses a more challenging problem, the calculation of the reaction rates and of the elements of the Jacobian can be easily vectorized or parallelized by straightforward programming techniques. It should be noted here that this task is extremely simplified if all reaction rates used in the network are given by a common fit-formula with rate-dependent coefficients.

According to the idea of Gustavson *et al.* (1970), the sparse linear system arising within the Newton iteration of the network equations can be solved very efficiently. A computer program which uses symbolic processing generates another program which contains the FORTRAN statements for the optimal Gaussian elimination routine in the sense that it explicitly takes into account the sparsity pattern of the matrix by omitting all operations involving zeroes. It should be stressed here, however, that this approach assumes that no pivoting is required, which indeed is the case for the network equations.

Besides eliminating unnecessary operations, this approach allows one to vectorize or parallelize the problem, because at all grid points the matrices have the same structure, differing only by the values of the elements. If the matrix elements for a whole set of grid points are stored in a three-dimensional array, where the last two indices give the row and column index of the element and the first index specifies the grid point, the sparse Gaussian elimination routine (and the program generator) can be designed in such a way that the innermost loop runs over the number of grid points, thereby fully exploiting the parallelism of the problem. Such a program generator, called GIFT, is available from the author upon request.

4.5 Some Instructive Numerical Experiments

Fryxell *et al.* (1989) have studied the coupled processes of hydrodynamics and nuclear burning in a general geometry. They critically examined some of the numerical tools available to the theorist performing an extensive set of test calculations of one-dimensional shock tubes with and without burning. In their study, Fryxell *et al.* (1989) considered several different Eulerian schemes including a second-order flux-vector splitting scheme (SADIE; Arnold 1985) and the PROMETHEUS code based on the PPM scheme of Colella & Woodward (1985). They further tested a Lagrangian PPM scheme so that the effects of artificial mixing could easily be seen. Although all tests problems considered by Fryxell *et al.* (1989) were one-dimensional, they clearly showed the types of errors one will encounter in performing multi-dimensional calculations. Some of their results will be discussed in this section (see also Müller 1994).

The equation of state used in the calculations of Fryxell *et al.* (1989) was chosen to provide the correct qualitative behavior for a realistic astrophysical gas, but yet be simple enough to be used for easily reproducible test calculations. Three contributions to the pressure are included – ideal gas, radiation, and a term which simulates a degenerate electron gas as a simple gamma-law. The effects of radiation pressure will strongly limit the temperature which the combustion products reach, and the effects of electron degeneracy can determine whether the fuel will burn explosively, producing a detonation or deflagration, or quietly, such as in the center of the Sun.

All codes considered by Fryxell *et al.* (1989) were written in strict conservation form (see Sect. 3 of the lecture by Le Veque). In most circumstances, this is the best approach, since it guarantees that shock jumps will be correct. However, using a conservative energy equation can lead to large errors if the kinetic energy is much larger than the thermal energy. In this case, the resulting pressure and temperature may be very inaccurate. When nuclear burning is included in the calculation, such large temperature errors could be disastrous. Using a nonconservative equation for only the internal energy is probably better under these circumstances. However, when such a situation does not occur, conservative methods should provide more accurate results.

Two separate reaction networks were considered by Fryxell *et al.* (1989). The first was a very simple network consisting of two species, ^{12}C and ^{56}Ni, and a single exothermic reaction. In reality, carbon does not burn directly to nickel. However, the approximation was made for this network that after the ^{12}C(^{12}C, α)^{20}Ne reaction is completed, the reaction products burn instantly to ^{56}Ni. Thus they used the rate for the initial reaction only but took the energy release obtained from burning the carbon completely to nickel.

In some of their test problems they used a second more complex reaction network, which contained the 13 species of the complete α-chain up to ^{56}Ni. These nuclei (^{4}He, ^{12}C, ^{16}O, ^{20}Ne, ^{24}Mg, ^{28}Si, ^{32}S, ^{36}Ar, ^{40}Ca, ^{44}Ti, ^{48}Cr, ^{52}Fe and ^{56}Ni) are linked by 27 reactions, which include the eleven (α,γ) reactions from ^{12}C$(\alpha,\gamma)^{16}$O to ^{52}Fe$(\alpha,\gamma)^{56}$Ni, the corresponding 11 endothermic photo-disintegration reactions, the three heavy-ion reactions ^{12}C(^{12}C,α)^{20}Ne, ^{12}C(^{16}O,α)^{24}Mg, and ^{16}O(^{16}O,α)^{28}Si, and the triple-α reaction and its inverse.

4.5.1 Detonations Caused by Numerical Errors. In one set of test calculations Fryxell *et al.* (1989) showed that due to numerical errors artificial detonations can be obtained. The initial conditions for this set of degenerate shock tube calculations with burning were given by

$$\begin{aligned} \varrho_L &= 2 \times 10^9\,\mathrm{g\,cm^{-3}} & \varrho_R &= 2 \times 10^9\,\mathrm{g\,cm^{-3}} \\ u_L &= 0 & u_R &= 0 \\ T_L &= 5 \times 10^9\,\mathrm{K} & T_R &= 5 \times 10^8\,\mathrm{K} \end{aligned}$$

The composition of the material was ^{56}Ni on the left side of the grid and ^{12}C on the right. The simple two species nuclear reaction network for burning carbon to nickel was included, and the radiation pressure term was discarded in the equation of state (see Sect. 4.5.1). The calculations were performed on a grid of length 0.1 cm with the initial jump located at 0.05 cm.

The ^{12}C to the right of the contact discontinuity is initially at too low a temperature to burn, and the shock is far too weak to raise the temperature to the ignition point. There is no fuel to the left of the contact discontinuity where the temperature is high, so there is no burning in that region either. In fact, if the contact discontinuity remained perfectly sharp, there would be no energy generation from burning during the entire calculation. However, since the discontinuity is spread slightly by Eulerian difference schemes, a small amount of the fuel is diffused into the region where the temperature is high, and some of the heat is conducted from the hot region into the fuel. Thus there is a narrow region where nuclear burning occurs. The amount of diffusion is, however, small enough that the energy generation is not dynamically important, and the results are qualitatively correct. The flow never reaches a configuration where the temperature and the abundance of ^{12}C are large in the same zone. In fact, at the end of the calculation, not even one zone has burned completely.

When using a flux-vector differencing scheme (*i.e.*, SADIE) instead of PPM (*i.e.*, PROMETHEUS) a qualitatively different result was obtained. Unlike the other schemes in the comparison, SADIE has the property of rapidly spreading slowly moving (and even stationary) contact discontinuities. Thus the amount of mixing between the cold fuel and the hot ashes across the contact is much larger than for the other codes. Under these conditions, the amount of mixing was sufficient to cause the fuel to detonate. The temperature of the fuel is raised to more than 4×10^{11} K producing an energy generation rate of $\approx 10^{34}\,\mathrm{erg\,g^{-1}\,s^{-1}}$. The resulting detonation appears physically correct, traveling at the correct speed and producing the correct post-detonation conditions. However, the origin of the detonation was due completely to a numerical error.

To complicate matters further Fryxell *et al.* (1989) performed a calculation using Eulerian PPM with slightly modified initial conditions. The fluid in the entire grid was given an initial velocity $u_L = u_R = 10^8\ \mathrm{cm\,s^{-1}}$. The results should be exactly the same as in the previous experiment except shifted to the right by a distance $u_L\,t$. However, the results looked almost identical to those obtained with SADIE. The only reason SADIE gives different results from the other codes was that the contact discontinuity was nearly stationary. When the discontinuity is moved across the grid, every Eulerian code will spread it to some extent. Even with PPM, the code which spreads discontinuities the least of those tested, the spreading was sufficient to cause the ^{12}C to detonate, producing a qualitatively incorrect solution. This shows the amount of caution which is needed in calculating reactive flows with an Eulerian hydrodynamic code.

4.5.2 Detonations with Single Exothermic Reaction. Another set of test problems Fryxell *et al.* (1989) considered was one containing a detonation, which, unlike the previous example, is not of numerical origin. The fluid to the left of the interface was given a finite velocity which produces a sufficiently strong shock to ignite the fuel to the right of the interface. The equation of state for this example included all three terms (ideal gas, radiation, and simplified electron degeneracy). Nuclear energy generation was provided by the simple reaction network containing only ^{12}C and ^{56}Ni. This is the simplest possible situation to analyze since there is only a single exothermic reaction to deal with. The initial conditions used for this set of calculations were

$$\begin{aligned} \varrho_L &= 1.25 \times 10^9\,\mathrm{g\,cm^{-3}} & \varrho_R &= 1 \times 10^9\,\mathrm{g\,cm^{-3}} \\ u_L &= 5 \times 10^8\,\mathrm{cm\,s^{-1}} & u_R &= 0 \\ T_L &= 2 \times 10^{10}\,\mathrm{K} & T_R &= 8 \times 10^7\,\mathrm{K} \end{aligned}$$

The initial composition of the material to the left of the jump was pure ^{56}Ni with pure ^{12}C to the right. The calculations were performed on grids of length $2, 4, 8$ and 16×10^{-4} cm, with the initial jump always being in the middle of the grid.

In this set of calculations, the number of zones in the grid was fixed at 200, while the length of the grid was increased. This provides an indication of how coarse a grid can be used while still obtaining acceptable results. In doubling the grid length from 2×10^{-4} cm to 4×10^{-4} cm, the only noticeable difference in the results is in the resolution of the peak behind the shock. The speed of the front and post-detonation state remain unchanged. However, when the grid length is doubled again, the results change qualitatively. The shock now propagates to the right with a velocity considerably higher than the Chapman–Jouguet velocity, and an additional plateau appears in the profile. The values at the second plateau are still the same as the post-detonation values obtained in the more finely zoned calculations, as they must be in a conservative scheme. However, because of the additional unphysical wave which forms, the results are totally unacceptable. If the grid size is doubled again, the situation becomes worse. The speed of the shock front increases further and the discrepancy between the post-shock values of the variables and the correct post-detonation values is even larger. Similar behavior was noticed by Colella *et al.* (1986).

This example clearly shows the fallacy of the argument that, since the jump conditions in a detonation, as in a shock, are based only on conservation laws, detonations can be calculated on any mesh with a conservative difference scheme. In ideal hydrodynamics, the shock front is infinitely thin (see Sect. 4.2.2; Fig. 4.1). Thus, the fuel spends no time within the shock and does not start to burn until it reaches the post-shock state. The problem arises when the time which the fuel spends within the shock front, which is given by the width of the shock produced by the difference scheme divided by the velocity of the shock, becomes comparable to or longer than the burning time of the fuel. When this happens, nearly all of the fuel is burned within the shock front. Thus, when the numerical errors become large enough that the physics of the problem is violated, it is not surprising that an unphysical solution is obtained. If the energy production within the lower portion of the shock front is large enough, the pressure in these zones will be raised sufficiently to create a bulge in the shock profile. This bulge then propagates to the right ahead of the real front, producing the extra wave.

The grid size at which each code in the test made the transition to the unphysical solution was different and depended on the shock width and amount of mixing of the fuel and ashes.

SADIE and Eulerian PPM worked properly until the grid length reached 8×10^{-4} cm. The Lagrangian PPM method produced unacceptable results only when a grid size of 16×10^{-4} cm or larger was used.

It was possible, however, to obtain acceptable results on large grids by making a small modification to the codes. The incorrect solution was obtained because the burning was occurring within the shock instead of behind it (see Sect. 4.2.2; Fig. 4.2). This error can be avoided by not allowing the fuel to burn in any zone which contains a shock. Such zones are easily detected in a hydrodynamic code. Any zone which contains a sufficiently large pressure gradient and a negative velocity divergence (to ensure that the zone is compressing) can be considered to be within a shock. The detonation wave obtained using this algorithm with the Eulerian PPM code moved at the correct velocity and had the proper post-detonation state for grid lengths, at which the unmodified algorithm failed before.

4.5.3 Detonation with α-Network. Fryxell *et al.* (1989) simulated another shock tube containing a detonation. However, the energy generation was now produced by the 13 species, 27 reactions α-network described above. This calculation provides an example of the type of results which can be obtained when using a moderately complex network with an Eulerian hydrodynamic code.

For a detonation with an α-network it is impossible to obtain a solution which is even close to being converged using the techniques examined in the survey of Fryxell *et al.* (1989). Since the burning times for the various reactions differ by several orders of magnitude, there is no way to resolve the time and length scales for each of the burning processes. In this sense, the detonation wave is no longer thin. If one chooses to resolve the structure of the front, as in the previous example, then the calculation cannot proceed long enough to follow the completion of the burning. If, on the other hand, the calculation is performed to see what the final products of the burning will be, the structure of the front cannot be resolved. This also makes calculating the correct qualitative behavior easier than with the simple network. Since the front is wider for the α-network, qualitatively correct detonations can be computed on grids large enough that meaningful calculations of stellar size problems can be performed.

The initial conditions for the detonation with an α-network computed by Fryxell *et al.* (1989) were

$$\begin{aligned} \varrho_L &= 2.5 \times 10^9\,\mathrm{g\,cm^{-3}} & \varrho_R &= 1 \times 10^9\,\mathrm{g\,cm^{-3}} \\ u_L &= 5 \times 10^8\,\mathrm{cm\,s^{-1}} & u_R &= 0 \\ T_L &= 8 \times 10^9\,\mathrm{K} & T_R &= 8 \times 10^7\,\mathrm{K} \end{aligned}$$

The initial composition of the material to the left of the jump was pure ^{56}Ni with pure ^{12}C to the right. All other mass fractions were set to an initial value of 10^{-10} in each zone, and all three terms in the equation of state were included. Results were obtained on grids of two different lengths, 0.01 cm (jump at $x = 0.005$ cm) and 1 cm (jump at $x = 0.5$ cm), so that the structure of the front could be seen on two different time scales ($t = 6 \times 10^{-12}$ s and $t = 6 \times 10^{-10}$ s, respectively).

The structure of the detonation obtained using Eulerian PPM on a 1 cm grid of 800 zones at $t = 6 \times 10^{-10}$ s is characterized by a small poorly resolved peak immediately behind the shock front. Behind the peak is a plateau which is not quite flat, since there is still nuclear burning occurring in this region. At the left of the grid is a second plateau. The density and pressure in this region have been reduced slightly by a rarefaction wave which has propagated to the left from the initial discontinuity. The two plateaus are joined by a contact discontinuity. This discontinuity is uncharacteristically wide for PPM because of the nonlinear interaction of the diffusion of species across the discontinuity with the burning processes. The enhanced rate of species diffusion is a result of the flattening of the zone structure required to maintain local conservation of the abundances. The structure of the detonation obtained on the 10^{-2} cm grid is qualitatively similar.

The time evolution of the abundances shows three distinct regions in the post-detonation flow. The region between the shock and the contact discontinuity contains the early time products of carbon burning. To the left of the contact discontinuity is a region in which the initial composition of almost pure ^{56}Ni is beginning to photo-disintegrate to form an equilibrium distribution of abundances. At the contact discontinuity, there is a third narrow region which contains a mixture of species from the other two regions. This mixing is unphysically large and results completely from the diffusion present in the Eulerian hydrodynamic method.

The evolution of the abundances proceeds as follows. Since the material ahead of the shock is pure ^{12}C, the only reaction in the network which can operate is ^{12}C(^{12}C,^{4}He)^{20}Ne. Thus the dominant abundances immediately behind the shock should be ^{4}He and ^{20}Ne. This composition layer is just barely detectable on the smaller grid. Further behind the shock, ^{20}Ne photo-disintegrates into ^{16}O and ^{4}He. With the formation of a significant abundance of ^{16}O, the three heavy-ion reactions become dominant, producing ^{20}Ne, ^{24}Mg, and ^{28}Si. The abundance of ^{32}S, which is produced by the $^{28}\mathrm{Si}(\alpha,\gamma)^{32}\mathrm{S}$ reaction, is also becoming significant near the contact discontinuity on the smaller grid. The remaining (α,γ) reactions are just beginning at this time ($t \approx 10^{-10}$ s), producing trace abundances of ^{36}Ar, ^{40}Ca, ^{44}Ti, and ^{48}Cr. At a somewhat later time ($t = 6 \times 10^{-10}$ s) ^{12}C and ^{16}O are depleted and the products of the (α,γ) reactions are increased. The mass fraction of the heaviest element in the network, ^{56}Ni, has increased to about 10^{-4} near the contact discontinuity at this time.

The abundances obtained on a coarser grid of 200 zones are qualitatively similar to those discussed above. The primary difference is that the region where mixing has taken place is much wider. Fortunately, the mixing in this case does not seem to have a significant effect on the qualitative results of the calculation. Although the burning was far from being complete at the end of the test calculation of Fryxell *et al.* (1989), the propagation velocity of the detonation front was approximately correct, and the variables behind the detonation appeared to be approaching the correct values. However, Fryxell *et al.* (1989) pointed out that in similar calculations performed with more diffusive schemes the mixing of species will dominate the abundance profiles over more than half the region between the shock and contact discontinuity, making it difficult, if not impossible, to follow the abundance changes caused by the various burning processes in this part of the flow.

The result obtained by Fryxell *et al.* (1989) using the Lagrangian PPM code were much cleaner, since there was no mixing region in the center to corrupt the abundance profiles. The boundary between the two burning regions was very sharply defined. Even so, except for this central region, the results obtained by the Eulerian code agreed very well with the Lagrangian results. Thus, it appears that even with a network containing many species, a good Eulerian code can produce reasonable results for many problems. For more diffusive Eulerian methods the mixing in the central region could be so severe that it would corrupt the abundance profiles over the entire grid, making it virtually impossible to observe the abundance changes due to the various burning processes.

5. Simulation of Astrophysical Jets

Jets, *i.e.*, highly collimated supersonic outflows, are a common astrophysical phenomenon. They are found associated with young stellar objects (YSOs; Mundt 1988; Eislöffel 1996), planetary nebulae (Mellema 1996), short-period cataclysmic variables (Shabaz *et al.* 1997), compact stars (Mirabel & Rodriguez 1994; Vermeulen 1996; Bell 1997) and active galactic nuclei (AGNs;

e.g., Bridle & Perley 1984; Muxlow & Garrington 1991). Despite their ubiquitity and decades of intense observational and theoretical efforts, the formation, collimation and acceleration of jets is still not understood (see the articles in the books edited by Hughes (1991), Bugarella *et al.* (1993), Hardee *et al.* (1996) and Kundt (1996).

Many of the ideas and models, which have been proposed to explain the formation of astrophysical jets, involve accretion onto a central object, such as a young star surrounded by the remainder of the molecular cloud in which it formed, or a rotating super massive black hole in an active galactic nucleus being fed by interstellar gas and gas from tidally disrupted stars (*e.g.*, Begelman *et al.* 1984). However, accretion seems to be no necessary prerequisite for the formation of jets, because there exists observational evidence that isolated pulsars and neutron stars can produce jets (Bell 1997). Concerning the (initial) collimation and acceleration of astrophysical jets most models involve MHD processes at or near the inner edge of a rotating magnetized accretion disk with corona. The combined action of thermal and magnetic pressure gradients, of centrifugal force and of gravity drives an outflow, which does collimate itself (Blandford & Payne 1982). General relativistic effects seem to be crucial for a successful launch of the jet (for a review see, *e.g.*, Camenzind 1997).

Numerical studies of extragalactic jets can be divided into several distinct categories (*e.g.*, Lovelace & Romanova 1996). Ordered with increasing distance to the source these categories are the formation and collimation of jets at sub-parsec-scales, and the propagation of jets at parsec-scales and at kiloparsec-scales, respectively. Up to now most simulations have dealt with the propagation of jets in the framework of Newtonian hydrodynamics. A Newtonian treatment is justified, however, only at distances larger than several kiloparsec. At smaller scales, and in particular at parsec-scales, jets propagate at relativistic speeds (see Sect. 5.1). Hence, special relativistic hydrodynamic simulations must be used in order to describe their propagation correctly (see Sect. 5.4). Simulating the formation and collimation of jets requires both more complicated physics and numerics, because hydromagnetic and general relativistic effects must not be neglected. Up to now, only a few such simulations have been performed (*e.g.*, Shibata & Uchida 1986; Bell 1996; Stone & Norman 1994; Koldoba *et al.* 1995; Meier *et al.* 1996; Ouyed *et al.* 1997; Romanova *et al.* 1997).

As a general discussion of astrophysical jets is far beyond the scope of my lecture, I will focus on hydrodynamic simulations of propagating (non-magnetized) extragalactic jets, *i.e.*, of jets entering the computational domain with supersonic speed. Firstly, however, I will briefly review some observational findings concerning extragalactic jets.

5.1 Observations of Extragalactic Jets

In 1918 H.D. Curtis photographed the giant elliptical galaxy M87 and noticed a "curious straight ray", which was "apparently connected with the nucleus of the galaxy by a thin line of matter". This was the first discovery of a so-called extragalactic jet. Nowadays several hundreds of these extragalactic jets are known, which are detectable by the synchrotron and inverse Compton radiation they emit at radio frequencies. Very high resolution radio observations (VLBI imaging) provide direct evidence that the jets transport enormous amounts of energy from the nuclei of radio galaxies and quasars to intergalactic space thereby powering the observed extended regions of radio emission (called radio lobes; Fig. 5.1) of these sources (Blandford & Rees 1974; Scheuer 1974; Bridle & Perley 1984).

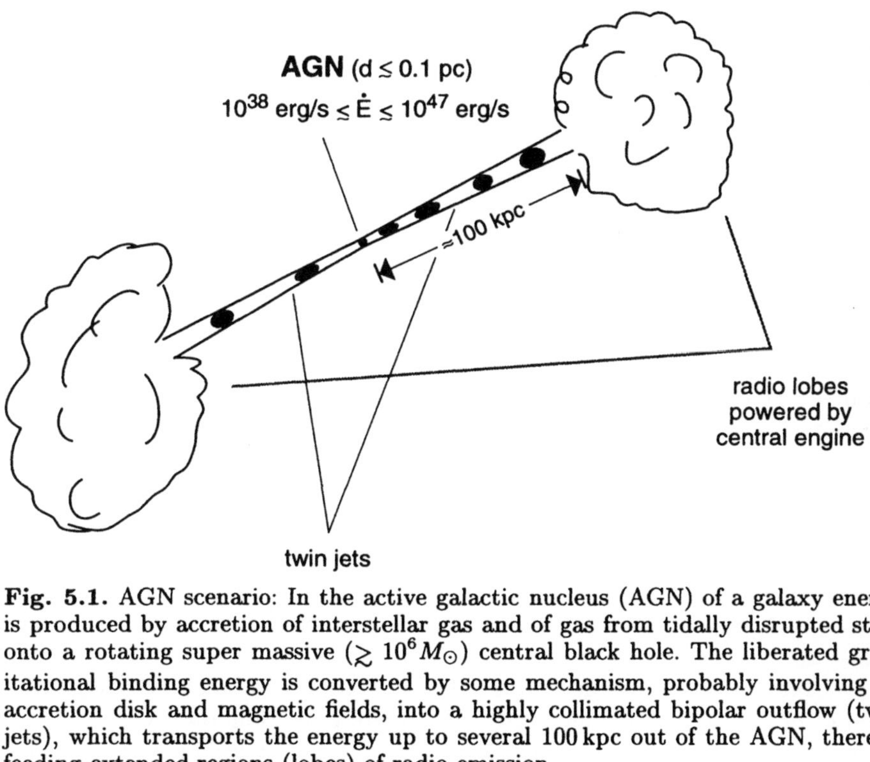

Fig. 5.1. AGN scenario: In the active galactic nucleus (AGN) of a galaxy energy is produced by accretion of interstellar gas and of gas from tidally disrupted stars onto a rotating super massive ($\gtrsim 10^6 M_\odot$) central black hole. The liberated gravitational binding energy is converted by some mechanism, probably involving an accretion disk and magnetic fields, into a highly collimated bipolar outflow (twin jets), which transports the energy up to several 100 kpc out of the AGN, thereby feeding extended regions (lobes) of radio emission

High resolution radio interferometry has shown that the jets are extremely well collimated with opening angles of a few degrees only, extending from the very center of the galaxy (*i.e.*, distances of less than a parsec) all the way out to distances of up to several hundreds of kiloparsecs (Bridle & Perley 1984; Muxlow & Garrington 1991). Whereas motion is observed on parsec-scales, except for M87 (*e.g.*, Biretta 1996), none of the sources in which jets have been detected shows any direct evidence for motion on kiloparsec-scales. Nevertheless, the observed continuity of the jets from parsec to kiloparsec-scales, supports the commonly accepted view that jets are collimated, continuous outflows. Whether the jet "fluid" consists of a baryonic plasma (*i.e.*, protons and electrons), an electron-positron pair plasma, a mixture of both or predominantly of electromagnetic Poynting flux is still debated controversially (*e.g.*, the books edited by Harry *et al.* 1996 and Kundt 1996).

At kpc scales, the morphology of the observed jets varies from weak sources (Fanaroff–Riley Class I sources; Fanaroff & Riley 1974) to powerful sources (Fanaroff–Riley Class II radio galaxies and lobe-dominated radio-loud quasars). The most remarkable difference is that FR I sources tend to have (at kpc scales) prominent smooth continuous two-sided jets running into the lobe structures, whereas powerful sources have usually one-sided (with a jet/counterjet flux density ratio $> 4{:}1$), knotty jets with bright outer hot spots (*e.g.*, Muxlow & Garrington 1991). In core dominated sources, where the luminous core is usually associated with a bright one-sided radio jet, interferometric imaging of very long baseline (VLBI) has revealed a continuity between small and large-scale structures with inner pc scale one-sided jets extending to outer kpc scale jets.

Features in the bright one-sided jets of FR II sources are observed to be (apparently) moving away from the core at superluminal speeds in more than 40 sources (Ghisellini *et al.* 1993) to distances beyond 100 pc (as, *e.g.*, in the quasar 3C273; Davis *et al.* 1991) indicating relativistic flow. Additional, independent observational support of highly relativistic speeds can be derived from the intra-day radio variability occurring in more than a quarter of all compact extragalactic radio sources (Krichbaum *et al.* 1992). According to Begelman *et al.* 1994, if the observed intra-day radio variability is intrinsic and results from incoherent synchrotron radiation, the associated jets must have bulk Lorentz factors in the range $\sim 30 - 100$. Relativistic outflow with apparent superluminal motion has also been observed in the Galaxy, namely in the two galactic "micro-quasars" GRS 1915+105 (Mirabel & Rodriguez 1994) and GRO J1655-40 (Tingay *et al.* 1995). These galactic X-ray sources consist of a binary, where one member of the binary is most likely a black hole.

Within the nowadays accepted model of compact radio sources (*e.g.*, Blandford *et al.* 1977) the apparent superluminal motion observed in these sources, as well as their one-sidedness, are explained by assuming that one of the twin jets propagates with relativistic speed at a small angle to the line of sight towards the observer (Fig. 5.2). In that case the apparent time duration (between two events) seen by a distant observer is compressed, because the source "chases behind" its own radiation. Hence, the apparent velocity of any transverse motion can appear to exceed the speed of light. In particular, this holds for the radio condensations ("radio blobs") ejected from the

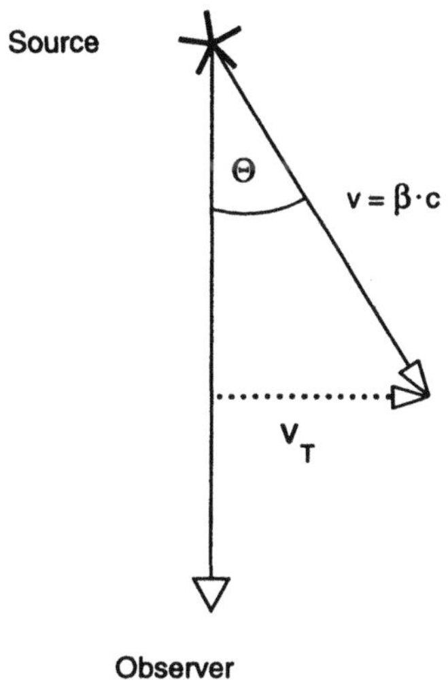

Fig. 5.2. When a source of radiation propagates with relativistic speed $v = \beta c$ at a small angle Θ to the line of sight towards the observer, its transverse velocity v_T can apparently exceed the speed of light c when measured by the observer. This purely kinematic effect is a consequence of the relativistic *transverse* Doppler effect

source with apparent superluminal speeds. This purely kinematic effect is a consequence of the relativistic *transverse* Doppler effect. A distant observer measures an apparent transverse velocity $v_{obs} = \beta_{obs} c$ for any feature moving with a space velocity $v = \beta c$ at an angle Θ to the line of sight towards the observer, which is given by (Blandford & Königl 1979)

$$\beta_{obs} = \frac{\beta \sin \Theta}{1 - \beta \cos \Theta} . \tag{5.1}$$

This relation implies that the apparent transverse velocity has a maximum for $\cos \Theta = \beta$ with

$$\beta_{obs}^{max} = W\beta , \tag{5.2}$$

where

$$W \equiv \frac{1}{\sqrt{1 - \beta^2}} \tag{5.3}$$

is the Lorentz factor of the feature. According to Eq. (5.2) $\beta_{obs}^{max} > 1$ for $\beta > 1/\sqrt{2}$. Hence, the apparent transverse velocity can become superluminal for large jet velocities (*i.e.*, β close to 1) and sufficiently small angles Θ.

The one-sidedness of jets observed in FR II sources can be explained by relativistic motion at a small angle to the line of sight, too. Due to relativistic aberration radiation emitted at some angle to the flow direction in the source frame, is seen at a smaller angle in the observer's frame. In addition, the emission from the jet propagating towards the observer is blue shifted while that of the counterjet is red shifted. Both effects give rise to the so-called *Doppler beaming* of the emission in the direction of motion. The resulting asymmetry in the luminosity of the twin jets can be quantified through a Doppler factor given by (Blandford & Königl 1979)

$$\mathcal{D} = W^{-1}\,(1 - \beta\cos\Theta)^{-1}\,. \tag{5.4}$$

For an optically thin source, the observed surface-brightness $\Sigma_{\rm obs}$ is related to the surface-brightness Σ that would be measured by a comoving observer by

$$\Sigma_{\rm obs}(\nu) \propto \Sigma(\nu)\,\mathcal{D}^{2+\alpha}\,, \tag{5.5}$$

where α is the spectral index of the emitted radiation. Thus, the relative Doppler boosting of jet and counter-jet is proportional to

$$\left[\frac{1+\beta\cos\Theta}{1-\beta\cos\Theta}\right]^{2+\alpha}\,. \tag{5.6}$$

For $\beta = 0.99$ (*i.e.*, a Lorentz factor $W \approx 7$) and $\Theta \leq 5$ degrees the fraction in Eq. (5.6) is larger than 144 (and less than 199). The resulting jet–counter-jet surface-brightness ratio is larger than 2×10^5, since $0.5 \lesssim \alpha \lesssim 1$.

Doppler beaming also provides an explanation for the Laing–Garrington correlation. According to Laing, Garrington and colleagues (Garrington *et al.* 1988; Laing 1988) the near side lobe of a powerful radio source (*i.e.*, FR II) containing a jet is almost always less depolarized than its counterpart on the far side of the nucleus. This asymmetry can be explained by assuming that the jets are intrinsically two-sided and relativistic. Hence, the nearer jet appears brighter and radiation from its lobe passes through less of the depolarizing medium around the source.

Whether the relativistic flows inferred from radio jets at sub-parsec and parsec-scales extend to kiloparsec distances is still an unsolved issue. The common belief is, however, that jets in FR II radio galaxies and quasars which show significant flux asymmetries between jet and counterjet are at least mildly relativistic on large scales. In fact, bulk flow velocities of the order of $0.7\,c$ have been found for several sources (1928+738: Hummel *et al.* 1992; 3C179: Akujor 1992; 1055+201, 1830+285, 2209+080: Hooimeyer *et al.* 1992). The observed asymmetries indicate that relativistic motion extends out to kpc scales, although with smaller values of the overall bulk speeds (Bridle *et al.* 1994). This slow-down of the jets makes it necessary to look for adequate models of flow deceleration between parsec and kiloparsec-scales.

In the case of FR I sources, direct measurements of the proper motion of knots in the M87 jet performed with the VLA (Biretta & Owen 1990; Biretta *et al.* 1995) have proven the existence of apparent superluminal motions out to a distance of about one kiloparsec. According to these authors, the measurements are compatible with a kinematic model for the jet in M87 consisting of an inner jet (from the nucleus to knot A located at ≈ 1 kpc) containing a high Mach number flow with a Lorentz factor $\gtrsim 3$ and a slower outer jet (Lorentz factor $\lesssim 2$) beyond knot A. Such a relativistic inner jet could be responsible for the one-sidedness of the jet base seen in many FR I sources (Muxlow & Garrington 1991). Recent observations support the idea that the jets of all FR I radio sources are initially relativistic (on pc-scales) and are slowed down to sub-relativistic speeds at distances of $\sim 100\,\mathrm{pc}$ (*e.g.*, Komissarov 1990a,b; Bicknell 1994, 1995; Urry & Padovani 1995; Laing 1996). Applying such a decelerating relativistic jet model to the archetypical FR I source 3C31 leads to an almost constant beam deceleration from $0.8\,c$ at distances less than two kpc from the nucleus, to $0.1\,c$ at distances $\approx 7\,\mathrm{kpc}$ (Laing 1996). Furthermore, data from HST optical and MERLIN radio observations of the quasar 3C264 are consistent with a model where an initially highly relativistic jet ($v \sim 0.98\,c$) decelerates dramatically to a velocity $\sim 0.4\,c$ at a projected distance of $\sim 300\,\mathrm{pc}$ (Baum *et al.* 1997), which coincides with the outer boundary of the galactic disk.

5.2 Newtonian Hydrodynamic Simulations of Extragalactic Jets

Although the collisional mean free path is very large in the jet plasma, a hydrodynamic description of the jet flow is justified, because of the presence of micro-Gauss magnetic fields which provide the collisional coupling of the plasma. In extragalactic jets the Larmor radii and Debye lengths of positrons (or protons) and electrons are several orders of magnitude smaller than the jet widths (Begelman *et al.* 1984). For this reason, hydrodynamic simulations have become an important tool in our understanding of the morphology and dynamics of extragalactic jets ever since Norman *et al.* (1982) verified the jet model of Blandford & Rees (1974) in their pioneering investigation (for a review see, *e.g.*, Norman 1993, 1996a).

5.2.1 Numerical Set-up and Parameter Space. Simulating the propagation of non-relativistic or Newtonian jets consists of solving the following initial-boundary value problem: A continuous collimated beam of gas density ϱ_b, velocity v_b and pressure P_b is injected through an aperture of radius R_b into a computational domain filled with an ambient medium gas of density ϱ_m and pressure P_m moving with a velocity v_m. In the simplest numerical setup one assumes that the beam is introduced with zero opening angle (*i.e.*, perfect collimation), the ambient medium is homogeneous, and that both the jet material and the ambient medium are an ideal gas with a specific heat ratio $\gamma = \mathrm{const}$. Accordingly, the initial-boundary value problem, and hence the flow, is specified by six quantities $\varrho_\mathrm{b}, v_\mathrm{b}, P_\mathrm{b}, \varrho_\mathrm{m}, v_\mathrm{m}, P_\mathrm{m}$ and the choice of a length or time scale, or a reference frame.

In astrophysical jet simulations, the hydrodynamic equations are commonly solved in normalized form, *i.e.*, the three conservation equations for mass, momentum and energy are scaled to make them dimensionless (*e.g.*, Norman *et al.* 1982). The units of length, velocity and density are chosen to be the jet (beam) radius R_b, the sound velocity c_m and the density ϱ_m of the ambient medium. Then, for example, the unit of time is $t_0 = R_\mathrm{b}/c_\mathrm{m}$. Fixing the frame of reference by assuming the ambient medium to be at rest (*i.e.*, $v_\mathrm{m} = 0$) the flow is completely characterized by three dimensionless parameters: the density ratio

$$\eta \equiv \frac{\varrho_\mathrm{b}}{\varrho_\mathrm{m}} \tag{5.7}$$

and the pressure ratio

$$K \equiv \frac{P_\mathrm{b}}{P_\mathrm{m}} \tag{5.8}$$

of the incoming beam gas to the ambient medium, and the Mach number of the jet

$$\mathrm{Ma_b} \equiv \frac{v_\mathrm{b}}{c_\mathrm{b}} = \sqrt{\frac{\eta}{K}}\,\frac{v_\mathrm{b}}{c_\mathrm{m}}\,, \tag{5.9}$$

where c_b is the sound velocity in the beam gas. Note that the Mach number is chosen as the third parameter, because $v_\mathrm{m} = 0$. In most simulations of astrophysical jets the parameter K is assumed to be unity, *i.e.*, only pressure-matched jets are considered.

5.2.2 Simulations. The majority of jet simulations has been (and is still being) performed in two spatial dimensions. The computational domain in these simulations is either of rectangular (Cartesian geometry) or cylindrical shape (axial symmetry). The latter geometry is a natural one for studying the propagation and morphology of axially symmetric jets. When 3D simulations cannot be afforded to investigate the behaviour of jets against non-axisymmetric instabilities, 2D simulations in Cartesian geometry are an alternative using planar or slab jets. The size of the computational domain in jet simulations typically encompasses $\sim 100R_\mathrm{b}$ in the direction of the jet propagation and $\sim 10 - 20R_\mathrm{b}$ in the perpendicular direction. The spatial resolution varies from $5 - 10$ zones per beam radius in simulations with poor resolution to ~ 100 zones per beam radius in the largest 2D simulations. The first three-dimensional simulations (in Cartesian coordinates) used relatively coarse computational grids of $\sim 10^5$ zones (Williams & Gull 1984) and $\sim 10^6$ zones (Arnold 1985; Arnold & Arnett 1986), respectively. Meanwhile 3D simulations have been performed with $\sim 10^7$ zones (Norman 1996b; Müller & Brinkmann 1997).

It has been found that 10 zones per beam radius are adequate to capture the essentials of the jet flow with a modern higher order-accurate shock-capturing finite volume scheme such as MUSCL, PPM or TVD (*e.g.*, Norman 1996a). Twenty zones are required to resolve additional internal structure in the jet and to correctly describe the dominant modes of the Kelvin–Helmholtz instability responsible for bending and filamenting of the jet (Hardee & Clarke 1992). In order to begin to get an accurate estimate of the mass entrainment caused by the onset of turbulence in an extragalactic jet a resolution of at least 40 zones per beam radius is necessary (Boris *et al.* 1992; see also Loken *et al.* 1996).

Since the pioneering work of Norman *et al.* (1982) hydrodynamic simulations have been used successfully to investigate the morphology and dynamics of propagating Newtonian jets (see next subsection). Magnetohydrodynamic simulations have shown the importance of toroidal magnetic fields for the confinement of jets (*e.g.*, Clarke *et al.* 1986, 1989; Lind *et al.* 1989; Kössl *et al.* 1990a, b), and for the appearance of genuine new MHD features, like *e.g.*, the formation of a "nose cone" (Clarke *et al.* 1989). If magnetic fields are dynamically important, they inhibit the formation of large cocoons, which suggests that magnetic fields in the lobes of classical double radio sources are probably dynamically unimportant (Clarke *et al.* 1989).

On kiloparsec-scales, the assumptions of axisymmetry and of a homogeneous environment become doubtful. Consequently, 2D (planar) and 3D hydrodynamic as well as MHD jet simulations have been performed to study the stability of jets (*e.g.*, Hardee & Clarke 1992; Bodo *et al.* 1994; Hardee & Clarke 1995; Hardee *et al.* 1995; Hardee 1996). Recently these simulations have been extended to consider the growth of the Kelvin–Helmholtz instability in (optically thin) radiatively cooling jets (Stone *et al.* 1997). Models for narrow-angle-tail sources whose complex morphologies are the result of the interaction of a supersonic jet with a cross wind (Begelman *et al.* 1979) have been tested by means of 3D simulations (*e.g.*, Williams & Gull 1984; Balsara & Norman 1992). The propagation of a jet into an oblique magnetic field (Koide *et al.* 1996), across an oblique interstellar medium/intra-cluster medium interface (*e.g.*, Hooda & Wiita 1996), and the deflection of a jet by an off-center collision with dense clouds of gas (de Young 1991) are a few other "environmental" issues investigated by hydrodynamic simulations.

Jet simulations have to be continued long enough in time to account properly both for the long-term evolution of powerful extragalactic sources with typical life-times $\sim 10^8$ yr (corresponding to a distance of a few 10^2 jet radii) and the observed deceleration of (initially relativistic) jets at distances beyond several 10^2 pc (see Sect. 5.1). This requirement poses a severe numerical problem, because a very large computational domain and hence a large number of grid zones are necessary for such studies. Long-term 2D Newtonian hydrodynamic simulations have been performed by Hooda *et al.* (1994) which cover a period corresponding to $\sim 10^8$ yr. They considered conical low-density jets propagating several hundred times their original radii with jet parameters believed to be suitable for powerful radio galaxies. However, besides being Newtonian, these simulations were concerned with what happens to jets when they cross pressure-matched interfaces between the interstellar and the intra-cluster medium, and not with the deceleration of the jets (see also Sect. 5.6).

5.3 Morphology and Dynamics

In a series of papers Norman and collaborators (Norman *et al.* 1982, 1983, 1984, 1985) presented a comprehensive numerical study of pressure-matched ($K = 1$; Eq. (5.8)), axisymmetric, supersonic ($M_{\rm b} > 1$) classical jets propagating into a homogeneous medium. Initial jet models were constructed by varying the beam Mach number, $M_{\rm b}$, and the density ratio η (Eq. 5.7). The study covered one decade in M_b (from 1.5 to 12) and three decades in η (from 0.01 to 10; Fig. 1 of Norman *et al.* 1983) including light or diffuse ($\eta < 1$) and heavy or dense ($\eta > 1$) jets. A detailed understanding of the morphology and dynamics of supersonic axisymmetric jets was derived from the study, which is summarized in the following.

5.3.1 Overall Morphology. Supersonic pressure-matched light ($\eta < 1$) jets consist of a supersonic *beam* of nearly constant diameter, which stays well collimated due to the presence of internal oblique shock waves and of plane and centered rarefaction waves (Figs. 5.3 and 5.4; see also Sect. 5.3.5). The beam ends in a contact discontinuity called the *working surface* (Blandford & Rees 1974), which separates ambient gas shocked by a leading *bow shock* from beam gas decelerated to the speed of the contact discontinuity by a trailing terminal shock configuration called a *Mach disk.* The Mach disk, the contact discontinuity and the intermediate beam cap gas form the head of the jet.

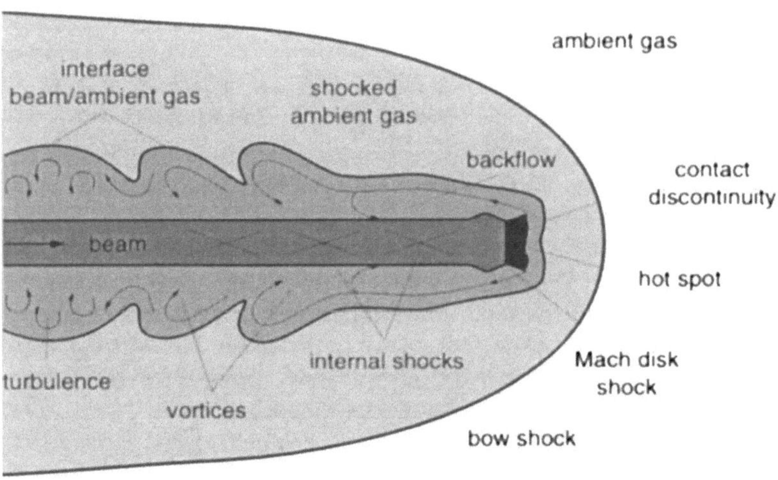

Fig. 5.3. Morphology of an axisymmetric supersonic light jet propagating into a homogeneous ambient medium (see text for details)

The propagation speed of the working surface determines the velocity of the jet, *i.e.*, its advance through the ambient medium. The jet velocity should not be confused with the actual velocity of the material in the beam which is referred to as the beam flow velocity.

Part of the shocked, high pressure beam cap gas is deflected sideways and flows backward along the beam, feeding the cocoon. The deflection process involves the repeated formation of a strong vortex at the jet head, which is then shed of the beam cap and becomes part of the backflow in the cocoon. Simultaneously, a new vortex is generated. The backflow in the cocoon is

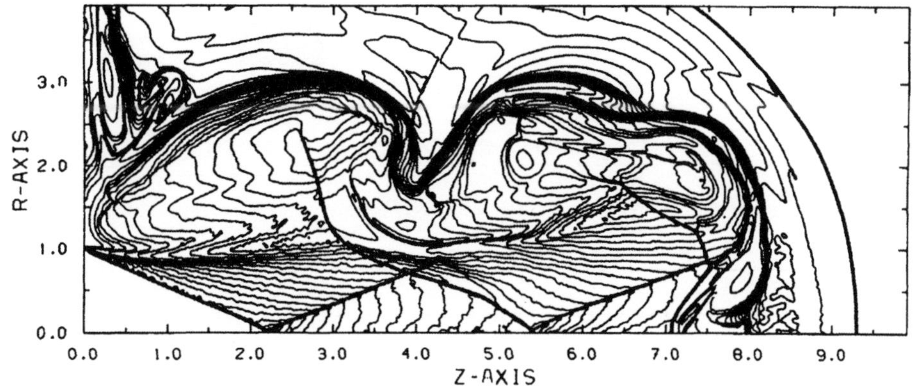

Fig. 5.4. Snapshot of the density distribution (logarithmic scale) of an axisymmetric, supersonic ($\mathrm{Ma}_b = 6$), light ($\eta = 0.1$), pressure-matched jet propagating into a homogeneous ambient medium (from Kössl & Müller, 1988). The simulation was performed with a very high grid resolution of 100 zones per beam radius. The z-axis corresponds to the symmetry axis of the jet

supersonic except for the regions between the shedded vortices. At the cocoon boundary the subsonic shear flow gives rise to Kelvin–Helmholtz instabilities, which eventually lead to a turbulent cocoon (see below).

5.3.2 One Dimensional "Jets". Except for the cocoon and the oblique shocks and rarefactions in the beam, the main structural features of a supersonic jet (bow shock, working surface and Mach disk) can be understood from a simple one-dimensional jet analogue. Consider the situation defined in the upper part of Fig. 5.5, where two pressure-matched states are separated by a diaphragm. The gas on the left side of the diaphragm has a supersonic velocity towards the right (*i.e.*, in positive x-direction), while the gas on the opposite side is at rest. The moving gas has a lower density than the gas at rest. This collision of a supersonic flow with a gas at rest exactly resembles the conditions on the symmetry axis of a light axisymmetric pressure-matched jet propagating into a quiet ambient medium.

When the diaphragm is suddenly removed three waves are produced (see lower part of Fig. 5.5): a shock and contact discontinuity propagating in positive x-direction, and a reflected shock propagating towards the left with respect to the contact discontinuity (but still in positive x-direction!). The ambient gas passing through the shock is compressed, heated and accelerated to a finite velocity. The beam gas passing through the reflected shock is decelerated to the post-shock velocity of the ambient gas and compressed to the corresponding post-shock pressure. The contact discontinuity separates the very dense shocked ambient gas from the decelerated, less dense compressed beam gas. Obviously, the shock of the "1D jet" corresponds to the bow shock in an axisymmetric jet, while the contact discontinuity and the reflected shock resemble the working surface and the Mach disk, respectively.

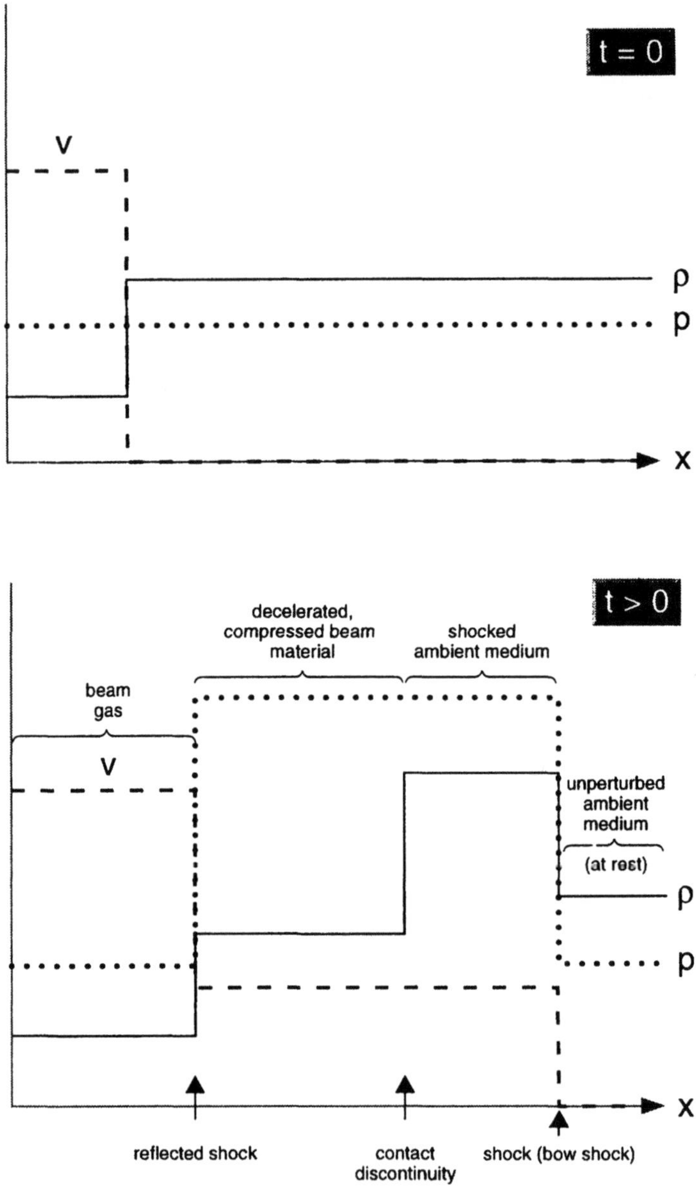

Fig. 5.5. One-dimensional pressure-matched light "jet": A supersonic flow collides with a homogeneous gas at rest. The gas in the flow is initially in pressure equilibrium with the gas at rest and its density is lower than that of the ambient gas. The panel at the top shows the initial setup, while the lower panel illustrates the qualitative nature of the flow at some later time

5.3.3 Propagation Speed. The propagation velocity of the jet through the ambient medium can be estimated using a one-dimensional argument, too. One equates the (one-dimensional) momentum flux of the beam gas and the ambient medium in a reference frame moving with the working surface. For pressure-matched jets an estimate of the Newtonian jet velocity, $v_{\mathrm{j}}^{\mathrm{N}}$, is then given by (Norman *et al.* 1983)

$$v_{\mathrm{j}}^{\mathrm{N}} = \frac{\sqrt{\eta}}{1+\sqrt{\eta}} v_{\mathrm{b}} \,. \tag{5.10}$$

According to this formula, light jets ($\eta < 1$) propagate at a small fraction of the beam flow velocity and only very dense jets ($\eta \gg 1$) have propagation speeds close to the beam velocity. Note that Eq. (5.10) has been obtained on the basis of the head-on momentum transfer between beam and ambient gas disregarding any multi-dimensional effects as, *e.g.*, the sideways expansion of the head of the jet. Hence, the estimate has to be considered as an upper bound of the actual jet propagation velocity. In the large sample of models studied by Norman *et al.* (1983), dense jets are seen to propagate most efficiently. Their speed, however, never exceeds 80 % of the one-dimensional estimate. Low Mach number light jets are even less efficient with a computed jet velocity of $\approx 40\,\%$ of the 1D estimate. Only hypersonic ($M_{\mathrm{b}} \geq 30$), sufficiently dense ($\eta \geq 0.03$) jets, as considered by Massaglia *et al.* 1996, seem to have propagation efficiencies equal one.

5.3.4 Cocoon Prominence. The mass flux across the terminal shock (which when integrated over time gives to a good approximation the mass within the cocoon) is proportional to the beam fluid velocity relative to the Mach shock. Taking into account that the Mach shock propagates almost at the advance speed of the jet, Eq. (5.10) allows one to determine the relative prominence of the cocoon in classical jet models.

Diffuse or light jets ($\eta < 1$) propagating at a small fraction of the beam flow velocity create a pile-up of shocked beam material which feeds an extended and turbulent cocoon. The prominence of the cocoon diminishes with increasing η, *i.e.*, when the jet becomes more and more dense, and when reducing the beam Mach number at a fixed beam density, *i.e.*, when the jet becomes less and less supersonic. Dense or heavy jets ($\eta > 1$), on the contrary, have a small mass flux into the cocoon leading to morphologies in which the beam is the most prominent element ("naked beams").

5.3.5 Jet Collimation and Stability. The shear flow at the beam cap generates an intense vortex, which drives a strong shock into the beam. This shock is reflected back and forth between the axis and the beam boundary leading to a regular shock structure in the beam with alternating incident and reflected shocks (Fig. 5.4), where incident and reflected refers to the symmetry axis. These internal shocks are responsible for the collimation of the beam, even if there is a substantial overpressure in the beam, compared to the cocoon (Falle 1987). The reason for this is that in an oblique shock only the velocity component normal to the shock is reduced, while the tangential component is steady (Fig. 5.6 upper panel). This simple picture is complicated by the fact, that the shocks are boundary lines of plane rarefaction waves towards the axis and of centered rarefaction waves towards the beam boundary (Fig. 5.4). The centers of the latter are at the reflection points of the internal shocks at the beam boundary. In these waves the velocity vector is steadily deflected away from the axis (Fig. 5.6 lower panel; *e.g.*, Landau & Lifshitz 1959). The net effect of shock collimation and centered rarefaction decollimation is an almost constant beam diameter. In the plane rarefaction waves inside the shock structure the magnitude of density and of pressure decreases, while the absolute value of the velocity increases. Thus the density and pressure have local minima just upstream and local maxima just downstream of the reflection points at the axis and vice versa for the velocity.

During the propagation of light jets direct interactions between Kelvin–Helmholtz instabilities in the cocoon and the beam can be observed. These interactions considerably disturb the internal structure of the beam, which becomes highly time dependent. The initially stationary internal shocks start propagating downstream and cause reflected shocks when they interact. Thus, shock waves propagating upstream can be observed all over the beam. In addition, due to the interaction of the non-stationary internal shocks and the Mach disk, the latter is temporarily substituted by a weaker oblique internal shock, which decelerates the beam flow less than the Mach disk does. Consequently, the beam cap is slowed down and reaccelerated almost periodically. The resulting non-steady propagation of the jet leads to Rayleigh–Taylor unstable conditions at the working surface (similar to the non-steady propagation of the shock in a supernova envelope; see Sect. 3.4), which give rise to the growth of vortices at the beam cap.

Whenever the beam cap is accelerated a smaller secondary bow shock is formed, which outruns the leading bow shock. This interaction produces a disturbance in the bow shock, which propagates away from the axis. Because of the non-steady propagation of the jet higher and higher modes of the Kelvin–Helmholtz instability are excited in the cocoon. These Kelvin–Helmholtz instabilities give rise to a feed-back loop operating on a time scale determined by the sound crossing time of the cocoon, or respectively, by the shock propagation time along the beam. The feed-back loop is triggering and maintaining the observed beam "pumping" (*i.e.*, the non-steady propagation of the jet). It is probably also the cause for the continued decrease of the wavelength of the Kelvin–Helmholtz instability. The final outcome of all this is that at least the cocoon (*i.e.*, the jet boundary layer) will eventually become turbulent.

The situation is quite different for heavy jets. Their beams almost completely lack internal structure, because of pressure equilibrium between the beam and its surroundings (*i.e.*, the ambient medium, as there exists no cocoon). Therefore, the beam/cocoon interface of heavy jets is very sta-

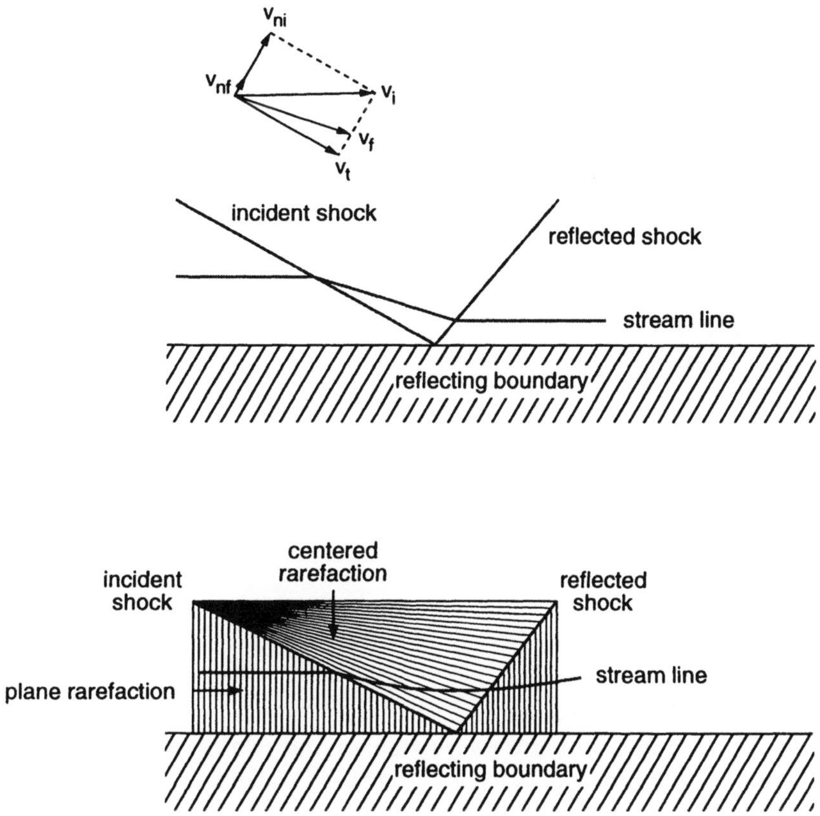

Fig. 5.6. Collimation of a supersonic jet: The top panel shows that oblique shocks alone tend to focus the flow (tangential and normal components of velocity are indicated). However, the combined action of oblique shocks and of plane and centered rarefaction waves (lower panel) gives rise to an almost constant beam diameter. (From Kössl & Müller 1988)

ble against the growth of pinch instabilities that would evolve into internal shocks.

5.4 Relativistic Simulations

5.4.1 Basic Equations. The equations of special relativistic hydrodynamics are a set of conservation equations (units are chosen such that the speed of light $c = 1$; see also Sect. 8 of the lecture by Le Veque)

$$\frac{\partial D}{\partial t} + \nabla(D\boldsymbol{v}) = 0\,, \tag{5.11}$$

$$\frac{\partial S^i}{\partial t} + \nabla(S^i \boldsymbol{v}) + (\nabla p)^i = 0 , \qquad \mathrm{i} = 1,2,3 , \tag{5.12}$$

$$\frac{\partial \tau}{\partial t} + \nabla[(\tau + p)\boldsymbol{v}] = 0 , \tag{5.13}$$

for the conserved rest-mass density D, momentum density $\boldsymbol{S} = (S^1, S^2, S^3)^{\mathrm{T}}$ and energy density τ, respectively. These variables, which are defined in a fixed frame, are related to quantities in the local rest frame of the fluid through

$$D = \varrho W , \tag{5.14}$$

$$\boldsymbol{S} = \varrho h W^2 \boldsymbol{v} , \tag{5.15}$$

$$\tau = \varrho h W^2 - p - D , \tag{5.16}$$

where ϱ, p, $\boldsymbol{v}$, W and h are the proper rest-mass density, the pressure, the flow velocity, the flow Lorentz factor and the specific enthalpy, respectively. The latter two quantities are given by

$$W = \frac{1}{\sqrt{1 - \boldsymbol{v} \cdot \boldsymbol{v}}} \tag{5.17}$$

and

$$h = 1 + \varepsilon + \frac{p}{\varrho} , \tag{5.18}$$

where ε is the specific internal energy.

An important difference between Newtonian and relativistic dynamics is the presence of a maximum velocity, *i.e.*, the speed of light in vacuum in the latter case. This implies that relativistic flows can no longer be scaled in space and time separately, but instead both scales are related by the speed of light. Hence, beyond K, η and M_{b} (see Sect. 5.2.1), an additional fourth parameter (for a given adiabatic index) is required to completely specify a relativistic jet propagating into a homogeneous medium. Commonly, one chooses the beam flow velocity, v_{b} (measured in units of c) as the additional fourth flow parameter.

The existence of an additional flow parameter reflects the fact that a jet can be relativistic in two ways: in a kinematically relativistic jet the beam gas moves with relativistic velocity ($W \gg 1$), while in a thermodynamically relativistic jet the specific internal energy of the beam gas is large compared to its rest mass energy ($h \gg 1$). Hence, the latter jets are called "hot" jets, and the former ones "cold" or "highly supersonic" jets (Martí *et al.* 1996). Note that the effective inertia of the flow increases in both cases, which is important for a qualitative understanding of the simulation results (see below).

For large specific internal energies the local sound speed, v_{s}, approaches a maximum value of

$$v_{\mathrm{s}}^{\mathrm{max}} = \sqrt{\gamma - 1}\, c = \frac{c}{\sqrt{3}} , \tag{5.19}$$

where the second equality in Eq. (5.19) holds for an ultra-relativistic gas with $\gamma = 4/3$. The corresponding maximum *proper* sound speed is given by

$$c_s^{\max} \equiv W_s^{\max} v_s^{\max} = \frac{c}{\sqrt{2}} , \tag{5.20}$$

where ($W_s^{\max}$) W_s is the (maximum) Lorentz factor corresponding to the (maximum) local sound speed ($v_s^{\max}$) v_s (Königl 1980).

These considerations show that relativistic effects arising from the presence of large flow velocities are directly controlled by choosing an appropriate beam flow velocity v_b. Once v_b is specified, the value of the beam Mach number, $M_b \equiv v_b/v_s$, must be larger than

$$M_b^{\min} = \frac{v_b}{c\sqrt{\gamma - 1}} = \sqrt{3}\,\frac{v_b}{c} , \tag{5.21}$$

which is the minimum Mach number for a flow with adiabatic index γ moving at speed v_b. Values of M_b near $M_b^{\min}$ are obtained for "hot" jets. The minimum *proper* Mach number is given by

$$\mathcal{M}_b^{\min} \equiv \frac{W_b}{W_s^{\max}} M_b^{\min} = \sqrt{2} W_b \frac{v_b}{c} , \tag{5.22}$$

which plays the role of the classical Mach number in two-dimensional, steady relativistic hydrodynamics, and which only depends on the beam flow velocity. Note, that the second equalities in Eqs. (5.21) and (5.22) again hold for an ultra-relativistic gas with $\gamma = 4/3$.

5.4.2 Numerical Methods and Simulations. Until recently, severe numerical difficulties arising in the integration of the special relativistic hydrodynamic equations (Eqs. (5.11) – (5.13)) had restricted the research of relativistic jets to the stationary regime (Wilson 1987; Daly & Marscher 1988; Dubal & Pantano 1993; Bowman 1994). The major difficulty is caused by the strong coupling of the equations via the Lorentz factor, which contains all velocity components and appears in all equations (Norman & Winkler 1986). Older numerical schemes were also less accurate and relied on artificial viscosity to handle shock waves (e.g., Potter 1973). However, an accurate treatment of shocks is crucial in relativistic hydrodynamics, because there exists no limit to the compression ratio across a shock as the upstream Lorentz factor tends to infinity. Thus, relativistic shocks resemble isothermal shocks in Newtonian hydrodynamics producing very thin dense regions, which are difficult to resolve adequately (Norman & Winkler 1986).

The development of high-resolution shock-capturing finite volume schemes (see Sect. 3.7 of the lecture by Le Veque) and their application to special relativistic flows has made it possible to overcome the numerical difficulties arising in the older methods (e.g., Martí & Müller 1994, 1996; Martí 1998). There are two main reasons for this success. Firstly, modern schemes do not require any artificial viscosity. Secondly, and even more important, the new methods explicitly take into account the non–linear character of the set of hyperbolic partial differential equations using exact or approximate (relativistic) Riemann solvers. Modern schemes are able to simulate multi-dimensional relativistic flows with Lorentz factors $W \sim 10^2$ (*i.e.*, $v \sim 0.9999\,c$) without numerical problems (Martí 1998).

During the past five years a series of 2D and 3D hydrodynamic and magnetohydrodynamic simulations of relativistic jets have been performed (van Putten 1993; Eulderink & Mellema 1994; Duncan & Hughes 1994; Martí *et al.* 1994, 1995; Duncan *et al.* 1996; Hughes *et al.* 1996; Koide *et al.* 1996; Martí *et al.* 1996; Komissarov & Falle 1996a, b; van Putten 1996; Martí *et al.* 1997a, b; Nishikawa *et al.* 1997). The collimation of relativistic spherical outflows by thick accretion disks was proven by Eulderink & Mellema (1994). Van Putten (1993) simulated a relativistic slab jet with a beam Lorentz factor $W_b = 3.25$. Although he computed the propagation of the jet for a distance of seven beam radii only, he was able to verify the formation of a Mach disk at the end of the relativistic beam. Martí *et al.* (1994) and Duncan & Hughes (1994) performed longer simulations in cylindrical and planar symmetry of low beam Mach number models (*i.e.*, hot jets). Both groups derived similar conclusions concerning the morphology of these relativistic models (see below). A highly relativistic ($W = 22$) and highly supersonic axisymmetric jet was simulated by Martí *et al.* (1995).

Martí *et al.* (1997a) performed the most comprehensive parameter study of the morphology and dynamics of hot and cold relativistic axisymmetric jets up to now. They used a conservative shock-capturing finite volume scheme in cylindrical coordinates (r, z). Numerical fluxes at cell interfaces are calculated by means of an approximate Riemann solver that uses the complete characteristic information contained in the Riemann problems between adjacent cell (Donat & Marquina 1996). The solver is based on the spectral decomposition of the Jacobian matrices of the relativistic system of equations (Font *et al.* 1994). The spatial accuracy of the algorithm is improved up to second-order accuracy by means of a conservative monotonic parabolic reconstruction of the pressure, proper rest-mass density and flow velocity following the work of Colella & Woodward (1984; see Martí & Müller 1996 for the explicit formulae). Integration in time is done simultaneously in both spatial directions using a total-variation-diminishing (TVD) Runge–Kutta scheme of high-order accuracy (Shu & Osher 1988). A one-dimensional Newton–Raphson iteration is used to compute in each time step the primitive variables $\{\varrho, \varepsilon, p, v_r, v_z\}$ from the conserved ones (see Martí & Müller 1996 for details). The simulations covered the evolution in a region extending up to $50\,R_b$ in z-direction and $7\,R_b$ in r-direction. The computational domain was covered by a uniform numerical grid consisting of 1000×140 zones, corresponding to a spatial resolution of 20 zones per beam radius. In order to delay and weaken the effect of an erroneous, numerically caused partial reflection of the bow shock at the top (outflow) boundary, the equidistant grid was extended in radial direction by 28 geometrically spaced zones covering a region from $r = 7\,R_\text{b}$ to $r = 10.5\,R_\text{b}$.

Most simulations of relativistic jets have been performed using an ideal gas equation of state with a constant adiabatic index of $\gamma = 5/3$ or $\gamma = 4/3$, respectively. The effects of the inclusion of a variable adiabatic index on the internal structure of relativistic jets have been explored by Duncan *et al.* (1996), who used a pair-plasma equation of state, and by Komissarov & Falle (1997), who considered a relativistic electron-proton gas.

The superluminal knots of VLBI sources are widely believed to be shocks traveling along relativistic jets. Hughes *et al.* (1996) have modulated the Lorentz factor of the inflowing material to model the propagation of these knots responsible for variability. Their radiative transfer calculations show that nested bow shocks resulting from the varying inflow speed appear as a series of axial knots for observers close to the line of sight, closely resembling VLBI maps. Gòmez *et al.* (1997) and Komissarov & Falle (1996a) constructed steady state jets which were perturbed by superimposing variations of the inflow velocity, and then computed the synchrotron emission from the jets assuming an ad hoc magnetic field with an energy density proportional to the gas pressure. This procedure allows one to study the combined effect of standing and traveling shocks on the appearance of a relativistic jet. The results show that the complex behaviour observed in many sources can be accounted for by this model (Gómez *et al.* 1997; see also Sect. 5.7).

Meanwhile, one has also begun to consider the effects of a non-homogeneous environment on the propagation of relativistic jets. Plewa *et al.* (1997) have simulated a relativistic jet propagating into an "atmosphere" whose density gradient is inclined to the jet direction. This mimics an extragalactic jet, whose direction of propagation does not coincide with one of the symmetry axes of the galaxy from which it is ejected. In order to save computational resources Plewa *et al.* (1997) have applied the technique of adaptive mesh refinement (AMR; see Sect. 5.3 of the lecture by Le Veque) in a relativistic jet simulation.

First simulations of relativistic MHD jets have been performed by Koide *et al.* (1996) using a simplified total variation diminishing (TVD) method and by van Putten (1996) using his own so-called weak smoothing method. Koide *et al.* (1996) considered 2D slab jets which are injected into a magnetic field oriented parallel to the jet direction. Their results, obtained on a grid of $20\,R_\text{b} \times 20\,R_\text{b}$ with a resolution of 4 zones per beam radius, show that both relativistic effects and the parallel magnetic field help to collimate the jet. The axisymmetric simulations of van Putten (1996) performed on a grid of 512×128 zones revealed a stagnation point–nozzle–Mach disk morphology. It is produced by an oscillation of the Mach disk at the "hot spot", which may explain the correlated optical and radio emission in the quasar 3C 273.

While writing this article a 3D variant of the 2D MHD simulation of Koide *et al.* (1996) was published by Nishikawa *et al.* (1997). The relativistic jet ($W_b = 4.56$) is again injected into a magnetic field oriented parallel to the jet direction. The cubic computational domain has an edge size of $20\,R_b$ and is covered by 101^3 zones equivalent to a spatial resolution of five zones per beam radius. According to Nishikawa *et al.* (1997) the resulting structure and kinematics of the 3D jet is noticeably different from that of the 2D planar one: the beam decelerates more efficiently, develops no internal structures, and the head speed is smaller.

5.5 Morphology and Dynamics of Relativistic Jets

Generally speaking, the gross morphological features found in Newtonian calculations (bow shock, beam, working surface, cocoon) are present in relativistic jets, too. Important, but not surprising differences were revealed between hot ($h \gg 1$) and cold ($W \gg 1$) relativistic jets (Martí *et al.* 1997a; Figs. 5.7 and 5.8).

Hot relativistic jets have "naked" beams surrounded by very thin cocoons and more extended lobes near the head of the jet (Fig. 5.7). The Mach shock at the jet head is permanently present during the whole simulation giving rise to a very steady propagation of the jet. In hot jets the backflow appearing at the working surface is minimized or even non-existing. Moreover, hot jets almost completely lack internal structure, as was first pointed out by Martí *et al.* (1994) and Duncan & Hughes (1994). The absence of shocks in hot beams is a consequence of the pressure equilibrium between the beam and the surrounding cocoon/shocked ambient medium. Hot jets are also very stable. Neither the reflecting mode nor the fundamental, or ordinary, pinch mode (typical of low beam Mach number flows; *e.g.*, Norman *et al.* 1984) seem to develop in hot jets (at least for the time scales covered by the simulations). This result is in accordance with a study of Kelvin–Helmholtz instabilities in the linear regime by Ferrari *et al.* (1978), who found relativistic, hot jets to be unconditionally stable.

Cold or highly supersonic jets display a very rich and complex structure within the beam and the cocoon (Fig. 5.8). They have beams with internal oblique shocks separating compressed and rarefied regions and collimating the flow. The angle formed by the oblique shocks with the jet axis decreases with the proper beam Mach number. Cold jets are further dominated by a thick cocoon which tends to become less prominent as the beam Lorentz factor increases. Vortex shedding at the head of the jet feeds the cocoon with vortices, which grow as they are advected upstream and eventually give rise to a turbulent thick cocoon (Fig. 5.8). The cocoon is overpressured with respect to the beam, the overpressure being roughly proportional to the corresponding beam flow Lorentz factor. The existence of such overpressured cocoons was already pointed out by Begelman & Cioffi (1989) and confirmed by Loken *et al.* (1992) in the case of Newtonian hypersonic jets (see also Massaglia *et al.* 1996). The overpressured cocoon helps to confine the jet during its early evolution and is probably of relevance for the formation of knots observed

Fig. 5.7. Color-coded contour plots of the logarithm of the proper rest-mass density (top) and pressure (bottom) of a hot, relativistic jet ($\eta = 0.01, v_b = 0.99, \mathcal{M}_b = 1.72$, and $\gamma = 4/3$). The maximum values are coded in white ($\log \varrho = 0.62; \log p = 0.50$). Decreasingly smaller values are coded in green, bright blue, dark blue, red and the minimum values are coded in black ($\log \varrho = -3.04; \log p = -1.77$). The jet has a "naked" beam and almost completely lacks internal structure. The beam is surrounded by lobes (near the jet head) instead of a cocoon and is in pressure equilibrium with the shocked ambient medium. (From Martí *et al.* 1997a)

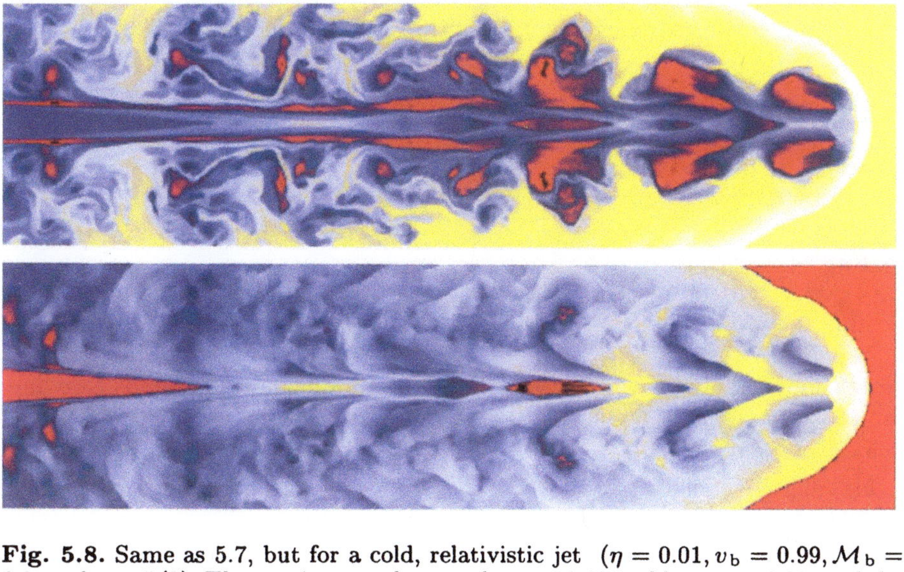

Fig. 5.8. Same as 5.7, but for a cold, relativistic jet ($\eta = 0.01, v_b = 0.99, \mathcal{M}_b = 6.0$, and $\gamma = 5/3$). The maximum values are $\log \varrho = 0.63$ and $\log p = -0.82$, and the minimum ones are $\log \varrho = 2.93$ and $\log p = -4.17$. The jet has a beam with internal oblique shocks and an extended turbulent cocoon. (From Martí *et al.* 1997a)

in large-scale jets of many powerful radio sources. In models with $\gamma = 4/3$ the contact discontinuity is positioned very close to the bow shock (Martí *et al.* 1997a). This gives rise to bow shock perturbations causing changes in its overall shape, which are reminiscent of the "nose cones" observed in classical MHD jets (Clarke *et al.* 1989; Lind *et al.* 1989).

5.5.1 Propagation Efficiency. Similar to the classical case, an estimate of the jet velocity can be obtained by equating the momentum flux of the beam and the ambient gas in the frame of the working surface (Eq. (5.10)). Because both, large specific internal energies (compared to c^2) and relativistic velocities are encountered, the momentum flux balance yields (Martí *et al.* 1997a)

$$\varrho_b h_b W_b'^2 v_b'^2 = \varrho_m h_m W_m'^2 v_m'^2, \tag{5.23}$$

for a beam in pressure equilibrium with the ambient medium. The symbol ($'$) indicates that the corresponding velocities are measured in the reference frame of the working surface. The estimated (relativistic) velocity of the jet in the rest frame of the ambient medium, $V_j^R = -v_m'$, can be introduced in Eq. (5.23) taking into account that the following relation between primed and unprimed velocities holds

$$W_b' v_b' = W_j W_b (v_b - V_j^R) \,. \tag{5.24}$$

Then one has

$$\varrho_b h_b W_b^2 (v_b - V_j^R)^2 = \varrho_m h_m (V_j^R)^2 \,, \tag{5.25}$$

which after some rearrangement gives (Martí *et al.* 1997a)

$$V_j^R = \frac{\sqrt{\eta_R^*}}{1 + \sqrt{\eta_R^*}} v_b \,, \tag{5.26}$$

where η_R^* is defined as

$$\eta_R^* = \eta_R W_b^2 \tag{5.27}$$

and η_R is given by

$$\eta_R = \frac{\varrho_b h_b}{\varrho_m h_m} . \tag{5.28}$$

Note that for fluids with non-relativistic internal energies η_R tends to η and for those having both non-relativistic internal energies and non-relativistic velocities η_R^* tends to η.

In the classical case only dense (*i.e.*, $\eta \gg 1$) jets have propagation speed estimates close to the beam flow velocity (Eq. 5.10)). However, for a relativistic jet the condition $V_j^R \approx v_b$ is fulfilled whenever $\sqrt{\eta_R^*} \gg 1$, *i.e.*, for ultra-relativistic and/or extremely hot jets. This result is easily understood, because in relativistic dynamics both the internal energy and the Lorentz factor contribute to the increase of the *inertial mass density*.

The study of Martí *et al.* (1997a) yielded estimates of the jet propagation speed in the range $0.17 \leq V_{\rm j}^{\rm R} \leq 0.94$. The corresponding propagation efficiency $\delta \equiv V_{\rm j}^{\rm R,num}/V_{\rm j}^{\rm R}$ (*i.e.*, the ratio of the numerically computed and the analytically estimated velocity of a relativistic jet) was found to lie in the range $0.76 \leq \delta \leq 1.24$. Efficiencies very close to one (*i.e.*, $\delta = 1.00 \pm 0.02$) are obtained for hot jets. The lack of internal structure within the beam and the insignificant or non-existing cocoon are responsible for the almost one-dimensional behaviour of these jets. Models which are highly supersonic and have a small adiabatic index ($\gamma = 4/3$) have efficiencies larger than 100%. This remarkable result is caused by an acceleration of those jets during an early stage of their evolution after the first oblique internal shock has formed within the beam.

The propagation efficiencies of relativistic jets are significantly larger than those of Newtonian jets. Norman *et al.* (1983), for example, obtained efficiencies in the range $0.49 \leq \delta \leq 0.90$ for their comprehensive set of Newtonian jet models including light and heavy jets as well as highly supersonic ones ($M_{\rm b} \leq 12$). The high efficiencies found by Martí *et al.* (1997a) are only comparable with those of the hypersonic Newtonian models ($M_{\rm b} \approx 30 - 300$) computed by Massaglia *et al.* (1996).

5.5.2 Estimate of the Cocoon Mass. Martí *et al.* (1997a) also derived an estimate of the total mass within the cocoon determined by the time integral of the mass flux across the terminal Mach shock. For a cylindrical flow of cross section $\pi R_{\rm b}^2$ and for a constant beam flow characterized by the parameters, η, $v_{\rm b}$, the estimate is

$$m_{\rm cocoon} = \pi R_{\rm b}^2 \varrho_{\rm b} v_{\rm b}' W_{\rm b}' t' \,, \tag{5.29}$$

where t' is the time measured in the reference frame of the Mach shock. In terms of quantities measured in the rest frame of the ambient medium the previous equation reads

$$m_{\rm cocoon} = \pi R_{\rm b}^2 \varrho_{\rm b} W_{\rm b} (v_{\rm b} - V_{\rm M}) t \,, \tag{5.30}$$

where $V_{\rm M}$ is the velocity of the Mach shock measured in the rest frame of the ambient medium.

If one wants to compare different models at some epoch, say when the jet has propagated along a distance L, then for each model the appropriate evolutionary time to be considered is $t = L/V_{\rm j}^{\rm R}$. Under these conditions the total mass within the cocoon scales as

$$m_{\rm cocoon} \propto \eta W_{\rm b} \frac{v_{\rm b} - V_{\rm M}}{V_{\rm j}^{\rm R}} \,. \tag{5.31}$$

Taking into account that $V_{\rm M} \approx V_{\rm j}^{\rm R}$ and substituting the propagation speed of the jet by the estimate of Eq. (5.26), an extremely simple expression for the cocoon mass in terms of initial beam parameters can be derived:

$$m_{\rm cocoon} \propto \frac{\eta}{\sqrt{\eta_{\rm R}}} \,. \tag{5.32}$$

Consistent with this expression is the result that hotter jets ($\eta_R \gg \eta$) will have increasingly less prominent cocoons. Moreover, the fact that the cocoon mass does not depend on the beam flow velocity, and the fact that the density of the shocked beam gas increases with increasing beam Lorentz factors, explain the thinner cocoons of faster jets.

5.5.3 Comparison with Newtonian Jets. Qualitatively speaking, the properties of relativistic jets resemble those of dense Newtonian jets, because relativistic effects enhance the effective inertia of the jet in a twofold way (see Sect. 5.5.1). As the enhanced inertia makes it more difficult to perturb relativistic jets, they are more stable than Newtonian jets. That is also the reason why the one-dimensional propagation speed estimates agree so well with the numerically obtained jet speeds, *i.e.*, why relativistic jets propagate so efficiently compared to their Newtonian counterparts.

When comparing Newtonian and relativistic jets quantitatively an ambiguity arises. Relativistic jets are completely characterized by four parameters (three in the case of pressure-matched jets), while Newtonian jets require one parameter less (see Sect. 5.4.1). Hence, it is not *a priori* clear which parameter one allows to vary in the comparison.

Martí *et al.* (1997a) compared models which had the same rest mass density ratio η and the same (classical) beam Mach number M_b, *i.e.*, the beam flow velocity or the beam Lorentz factor was allowed to vary. This is a perfectly reasonable choice, if one only wants to study the generic properties of relativistic jets. Komissarov & Falle (1996b, 1997) argue that this choice is not appropriate for a comparison between relativistic and Newtonian jets. If one holds all flow parameters fixed except for the Lorentz factor, as in the work of Martí *et al.* (1997a), one compares jets with different power and thrust. Instead, Komissarov & Falle (1996b, 1997) suggest that one should compare Newtonian and relativistic jets with the same beam velocity and power, the latter quantity being the most relevant to extragalactic jets. Komissarov & Falle (1997) found that for their choice of parameters the behaviour of relativistic and classical jets is very similar, except that the relativistic jets generate lighter cocoons because of their smaller mass flux. This difference is not very dramatic, however, and probably indistinguishable by observations.

5.6 Long Term Evolution of Relativistic Jets

Jets in powerful extragalactic radio sources (FR II radio galaxies and quasars) seem to have been feeding the lobes in these sources for periods as long as 10^7 yr with kinetic powers in the range 10^{44}–$10^{47}\,\mathrm{erg\,s^{-1}}$ (Rawlings & Saunders 1991; Daly 1995). These powerful jets are supposed to be relativistic on parsec-scales (see Sect. 5.1). On larger scales, flux asymmetries between jets and counter-jets indicate that relativistic motion extends out to kpc-scales, although with smaller values of the overall bulk speeds (Bridle *et al.* 1994), making it necessary to look for adequate models of flow deceleration between both scales. Finally, an effective deceleration should occur at the terminal hot spots for which advance speeds in the range $0.01\,c - 0.1\,c$ are inferred (Liu *et al.* 1992; Daly 1995).

Numerical simulations of jets typically cover the propagation of the jet for a few tens of initial beam radii, corresponding to time scales much smaller than those in real sources. Recently however, a first step towards a numerical study of the deceleration mechanism has been made by Martí *et al.* (1997b), who simulated the evolution of a relativistic jet modeling that of the powerful radio source Cygnus A. The initial conditions were chosen to be $v_{\rm b} = 0.986\,c$ (*i.e.*, $W_{\rm b} = 6.0$), $\eta = 3.4 \times 10^{-3}$ and $M_{\rm b} = 9.0$. An ideal gas equation of state with $\gamma = 5/3$ was used. The jet was assumed to propagate through a uniform, static atmosphere in pressure equilibrium with the beam at the injection point. The model corresponds to a powerful jet (beam kinetic power $10^{47}\,{\rm erg\,s^{-1}}$) when scaled according to a beam radius of $R_{\rm b} = 0.35\,{\rm kpc}$ and an ambient medium density $\varrho_{\rm a} = 10^{-26}\,{\rm gcm^{-3}}$. From Eq. (5.26) the estimated hot spot propagation speed is $v_{\rm hs}^{\rm e} = 0.26\,c$. This is significantly larger than the hot spot advance speeds inferred from powerful radio sources ($0.01\,c$ to $0.1\,c$; Liu *et al.* 1992). The evolution of the jet was simulated for a period of $\approx 3 \times 10^6\,{\rm yr}$, which is about one tenth of the estimated spectral age of many powerful radio sources.

The overall morphology found by Martí *et al.* (1997b) was typical of a light, highly-supersonic jet characterized by an extended overpressured cocoon and a beam with internal conical shocks. After a short one-dimensional phase the evolution is characterized by a strong deceleration followed by a long-term phase of steady evolution. During the deceleration phase the jet speed decreases from about $0.26\,c$ to less than $0.10\,c$, because the beam flow is slowed down through internal shocks and broadened near the hot spot. The numerical resolution (4 zones per beam radius) was insufficient for capturing the process of mass entrainment (*e.g.*, Norman 1996a). Because of the deceleration of the jet, beam gas starts to inflate large lobes around the jet's head, similar to those observed in many powerful radio sources (*e.g.*, in Cygnus A; Carilli *et al.* 1996).

The supersonic propagation of the jet through the external medium drives a bow shock that pushes the ambient gas, causing the jet to propagate in a region of low density (the cavity). The evolution of the cavity pressure can be inferred from the evolution of the thermal energy within the cavity, $E_{\rm cav}^{\rm th}$, and from the cavity's volume (Begelman & Cioffi 1989)

$$p_{\rm cav} \approx E_{\rm cav}^{\rm th}\, l_{\rm j}^{-1}\, r_{\rm cav}^{-2}\,, \tag{5.33}$$

where $l_{\rm j}$ and $r_{\rm cav}$ are the length of the jet and the width of the cavity, respectively.

With $E_{\rm cav}^{\rm th} \propto t$, $l_{\rm j} \propto t$ (assuming $v_{\rm hs} = v_{\rm hs}^{\rm e}$) and $r_{\rm cav} \propto t^{0.5}$ (approximating the bow shock, which defines the cavity's volume, as a strong shock) one finds from Eq. (5.33) that the pressure inside the cavity scales as $p_{\rm cav} \propto t^{-1}$. From their simulation, Martí *et al.* (1997b) obtain $p_{\rm cav} \propto t^{-0.6}$, the difference being mainly due to the slower increase of the cavity volume as a consequence of the deceleration of the hot spot ($l_j \propto t^{0.6}$ instead of $\propto t$; see also Cioffi & Blondin 1992).

5.7 Simulation of Parsec-Scale Jets

Until recently, the simulation of parsec-scale jets and superluminal radio components has been restricted to simplified stationary models (Marscher & Gear 1985; Hughes *et al.* 1989, 1991; Gómez *et al.* 1994a, b). The presence of radiation emitting flows at almost the speed of light enhances the importance of relativistic effects in the appearance of these sources (relativistic Doppler boosting, light aberration, time delays). Hence, models require the use of a combination of hydrodynamics and synchrotron radiation transfer when trying to explain the observations. Additional assumptions must be made to determine the radio emission (*i.e.*, synchrotron emissivity and opacity) associated with a given flow.

Modeling the radiation signature of superluminal components in parsec-scale relativistic jets Gómez *et al.* (1997) performed hydrodynamic calculations using the relativistic time-dependent code of Martí *et al.* (1997a). In order to compute the radio emission from their hydrodynamic jet models, Gómez *et al.* (1997) distribute the internal energy density among the electrons following a power law and assume that the magnetic energy remains a fixed fraction of the particle energy density. The emission is then calculated by integrating the synchrotron transfer equations while accounting for relativistic effects such as Doppler boosting and light aberration (for more details see Gómez *et al.* 1995). Komissarov & Falle (1996b, 1997) solve the transfer equations for the total intensity in the optically thin limit. In order to account for light-travel time delays within the jet, retarded emission and absorption coefficients are considered in the transfer equations.

In the hydrodynamic models discussed by Gómez *et al.* (1997) and by Komissarov & Falle (1996b, 1997) moving radio components are obtained from perturbations of steady relativistic jets. The jets are assumed to propagate through an atmosphere of decreasing pressure. This environment causes an expansion of the jets in accordance with the observed jet opening angles. Where pressure mismatches exist between the jet and the surrounding atmosphere reconfinement shocks are produced. These shocks are either isolated (Komissarov & Falle 1996b,1997) or form a sequence (Gómez *et al.* 1995, 1997). The energy density enhancement produced downstream from the shocks gives rise to stationary radio knots similar to those observed in many VLBI sources. Superluminal components are produced by triggering small perturbations in the steady jets which propagate almost at the beam flow velocity. Typical perturbation setups involve sinusoidal variations or flow discontinuities. Perturbations of the first kind were imposed by Mioduszewski *et al.* 1997 (see also Hughes *et al.* 1996), who considered sinusoidal variations of the jet Lorentz factor between 1 and 10. However, because of their large amplitude, the perturbations propagate only at barely relativistic speeds.

In the simulated radio maps the perturbations appear as regions of enhanced emission moving downstream at superluminal apparent velocities. Interactions of the perturbations with the underlying steady jet result in changes in the internal brightness distribution of the superluminal components (Gómez *et al.* 1997). This shows that the dynamic interaction between the induced traveling shocks and the underlying steady jet can account for the complex behavior observed in many extragalactic radio sources. A list of these observations and their possible interpretation within the hydrodynamic model of compact extragalactic radio sources can be found in Gómez *et al.* (1997).

6. Smoothed Particle Hydrodynamics

It is commonly known that it is very efficient to compute multi-dimensional integrals using the Monte Carlo method. Based on this fact Lucy (1977) and independently Gingold & Monaghan (1977) developed a Monte Carlo approach to solve the multi-dimensional hydrodynamic equations. The approach is called Smoothed Particle Hydrodynamics (SPH). The main advantage of the method is that is does not require a computational grid and that it is a Lagrangian method. Thus, in multi-dimensional calculations no computational power is wasted in simulating large voids which is often necessary when grid-based finite volume methods are used. The SPH algorithm is not too complex, a simple implementation typically requiring about 1000 lines of FORTRAN code and, moreover, the scheme is very robust. This robustness, however, implies some potential risk: If the code contains a severe bug, most finite volume schemes tend to crash after a few time steps, while SPH keeps on running and produces unphysical results (*e.g.*, superluminal speeds). Furthermore, for some problems SPH seems to be able to get very accurate results with a surprisingly small number of particles ($\sim 10^3$ or even less). This does not imply that SPH is able to solve supercomputer problems on workstations without a loss of accuracy, although some authors seem to argue in that direction (*e.g.*, the articles about SPH and the following roundtable discussion in Trease *et al.* 1991). On the other side, if indeed more than 10^4 particles have to be used for certain problems, and if self-gravity has to be included, such a simulation would exceed even the capabilities of modern supercomputers, because like in any N-body algorithm the computational power required grows as the square of the particle number N.

Three modifications of the original algorithm have increased the potential of SPH remarkably: (i) the introduction of a variable smoothing length has considerably enhanced (for a given number of particles) the resolution capabilities (*e.g.*, Hernquist & Katz 1989; Benz 1990); (ii) the use of individual time steps for each particle allows one to overcome global time step limitations due to stability criteria like the Courant condition (Porter 1985; Jernigan & Porter 1989; Hernquist & Katz 1989); and (iii) the use of the hierarchical tree method to calculate the gravitational potential has reduced the original N^2-algorithm to a $N \log N$-algorithm without sacrificing the free Lagrange nature of SPH (Porter 1985; Barnes & Hut 1986).

Concerning the computational performance of SPH, several authors have shown that it is possible to vectorize both the hierarchical tree method and the SPH algorithm (Hernquist 1990; Makino 1990a; Barnes 1990; Haddad *et al.* 1991). Makino & Hut (1988) and Theuns & Rathsack (1993) have implemented the hierarchical tree method on a massively parallel computer. Thus, SPH can efficiently be used on different kinds of supercomputers, but the original simplicity of the algorithm has partially been lost. However, compared with multi-grid or adaptive-grid finite volume methods SPH is still remarkably simple.

In the following the SPH algorithm is briefly reviewed (for a more thorough discussion see, *e.g.*, Hernquist & Katz 1989; Benz 1990; Monaghan 1985, 1992 and the references therein). A detailed description of a specific implementation of SPH which uses the hierarchical tree method to handle self-gravity can also be found in Steinmetz & Müller (1993). The latter work also contains a set of demanding test problems (see Sect. 6.6) and astrophysical applications (see also, *e.g.*, Steinmetz 1996b), which exhibit both the capabilities and limits of SPH. A related investigation has been performed by Davies *et al.* 1993, who compared the performance and accuracy of the SPH implementation of Benz (1990) with that of the PPM multi-grid implementation of Ruffert (1992) when simulating stellar collisions.

Recent developments concerning the SPH algorithm focus on the incorporation of Riemann based solutions of the equations of hydrodynamics (Monaghan 1997), on the extension of SPH to ultra relativistic hydrodynamics (Chow & Monaghan 1997), on an improved prescription for the artificial viscosity in SPH (Balsara 1995; Watkins *et al.* 1996; Selhammar 1997a), on the use of nonspherical kernels in SPH (Fulbright *et al.* 1996; Shapiro *et al.* 1996; Owen *et al.* 1997; Selhammar 1997b), and on the stability of SPH (Swegle *et al.* 1995; Moris 1996).

6.1 The SPH formalism

Consider an ensemble of N particles with position $\boldsymbol{r}_i$ and mass m distributed in space according to the density $\varrho(\boldsymbol{r})$. Following Gingold & Monaghan (1977, 1982) a smoothed estimate of the density ϱ is given by

$$\begin{aligned} \langle \varrho(\boldsymbol{r}) \rangle &= \int \mathrm{d}^3 r' \varrho(\boldsymbol{r}') W(\boldsymbol{r} - \boldsymbol{r}', h) \\ &\approx \sum m W(\boldsymbol{r} - \boldsymbol{r}_i, h) , \end{aligned} \tag{6.1}$$

where $W(\boldsymbol{r}, h)$ is the so-called kernel, a function strongly peaked around $|\boldsymbol{r} - \boldsymbol{r}'| = 0$, and h is the so-called smoothing length. This approach can be generalized to an arbitrary physical quantity $A(\boldsymbol{r})$:

$$\begin{aligned}\langle A(\boldsymbol{r})\rangle &= \int \mathrm{d}^3 r' A(\boldsymbol{r}') W(\boldsymbol{r}-\boldsymbol{r}',h) \\ &\approx \sum m \frac{A(\boldsymbol{r}_i)}{\varrho(\boldsymbol{r}_i)} W(\boldsymbol{r}-\boldsymbol{r}_i,h)\,. \end{aligned} \tag{6.2}$$

By a Taylor series expansion of the integrand in Eq. (6.2), one can show that A and its smoothed analogue differ by terms proportional to h^2, *i.e.*, the scheme is second-order accurate in space. Because the volume integrals extend over all space, one easily obtains the smoothed estimate of the time derivative of A as

$$\frac{\mathrm{d}}{\mathrm{d}t}\langle A(\boldsymbol{r})\rangle = \langle \frac{\mathrm{d}A(\boldsymbol{r})}{\mathrm{d}t}\rangle\,. \tag{6.3}$$

Using integration by parts spatial derivatives of a physical quantity can be transformed into spatial derivatives of the kernel, *i.e.*,

$$\begin{aligned}\nabla\langle A(\boldsymbol{r})\rangle &= \langle \nabla A(\boldsymbol{r})\rangle \\ &\approx \sum m \frac{A(\boldsymbol{r}_i)}{\varrho(\boldsymbol{r}_i)} \nabla W(\boldsymbol{r}-\boldsymbol{r}',h)\,. \end{aligned} \tag{6.4}$$

Note that in the case of a variable smoothing length the operator $\langle\rangle$ does not commutate with $\mathrm{d}/\mathrm{d}t$ and ∇, *i.e.*, the time derivative and the gradient of the smoothing length occur in the equations above. Furthermore, it is not clear, if the scheme is then still second-order accurate in space.

After these considerations it still remains to discuss the particular shape of the function $W(\boldsymbol{r}-\boldsymbol{r}',h)$. Various types of spherically symmetric kernels have been suggested over the years (Monaghan 1985; Benz 1990). Among those the spline kernel (Monaghan & Lattanzio 1985), mostly used in current SPH-codes, yields the best results:

$$W(\boldsymbol{r}-\boldsymbol{r}',h) = \frac{1}{\pi h^3}\begin{cases} 1-\frac{3}{2}u^2+\frac{3}{4}u^3\,, & 0\le u\le 1 \\ \frac{1}{4}(2-u)^3\,, & 1<u\le 2 \\ 0\,, & u>2 \end{cases} \tag{6.5}$$

with $u = |\boldsymbol{r}-\boldsymbol{r}'|/h$. It reproduces constant densities exactly in 1D, if the particles are placed one a regular grid of spacing h, and has compact support.

Anisotropic kernels have been first applied by Bicknell & Gingold (1985). Since a few years they are investigated more thoroughly (Fulbright *et al.* 1996; Shapiro *et al.* 1996; Owen *et al.* 1997; Selhammar 1997b). The advantage of such anisotropic kernels is that they can more fully adapt to flow problems with a large degree of anisotropy, as *e.g.*, in the case of large-scale cosmological structure formation, and provide an optimal spatial resolution. The improved spatial resolution is achieved, however, at the price of having to deal with tensor smoothing. Consequently, the particle forces are no longer spherically symmetric (Eq. (6.6)), and thus angular momentum conservation cannot be guaranteed.

Given the smoothed estimates of physical quantities and their derivatives above there is no problem to find the SPH formulation of the hydrodynamic equations (*e.g.*, Monaghan 1985; Benz 1990). However, if one applies the smoothing procedure in a straightforward way, the forces are not anti-symmetric, *i.e.*, the force exerted by a particle i on another particle j is not equal to the negative of the force exerted by particle j on particle i. This violates the conservation of linear and angular momentum. Using adequate operator identities one can derive several SPH formulations of the hydrodynamic equations with anti-symmetric forces (Monaghan 1985, 1992). Starting, for example, from the Lagrangian of hydrodynamics (*e.g.*, Soper 1976), one substitutes the matter density by its SPH approximation (Eq. (6.1)), and minimizes the resulting SPH Lagrangian with respect to the streamlines. This procedure yields the SPH equation of motion

$$\frac{\mathrm{d}\boldsymbol{v}_i}{\mathrm{d}t} = -\sum_j m \left(\frac{P_i}{\varrho_i^2} + \frac{P_j}{\varrho_j^2} + Q_{ij} \right) \nabla_i W(|\boldsymbol{r}_i - \boldsymbol{r}_j|, h) + \nabla_i \Phi(\boldsymbol{r}_i) , \qquad (6.6)$$

where ∇_i, d/dt and Q_{ij} are the gradient operator with respect to the coordinates of the i-th particle, the Lagrangian time derivative and the artificial viscosity tensor, respectively. The artificial viscosity is required in SPH to handle shock waves. It poses a major obstacle in extending SPH to relativistic flows, because it is not clear what form it should take (Israel 1987; Chow & Monaghan 1997).

Because in Eq. (6.6) the (spherically symmetric) kernel W only depends on the absolute value of $\boldsymbol{r}_i - \boldsymbol{r}_j$ (Note that this does not hold for anisotropic kernels!), both the pressure term and the gravity term consist of a superposition of central forces. Consequently, both linear and angular momentum are (globally) exactly conserved. The gradient of the gravitational potential is given by

$$\nabla_i \Phi(\boldsymbol{r}_i) = G \sum_{j=1}^{N} m_i \int_0^{r_{ij}} \mathrm{d}r\, r^2 W(r, h) \, \frac{\boldsymbol{r}_i - \boldsymbol{r}_j}{r_{ij}^3} , \qquad (6.7)$$

where $r_{ij} \equiv |\boldsymbol{r}_i - \boldsymbol{r}_j|$.

Using the first law of thermodynamics and applying the SPH formalism the energy equation can be written (in terms of the specific internal energy ε) as

$$\begin{aligned} \frac{\mathrm{d}\varepsilon}{\mathrm{d}t} &= \frac{P_i}{\varrho_i^2} \sum_j m(\boldsymbol{v}_i - \boldsymbol{v}_j) \cdot \nabla_i W(|\boldsymbol{r}_i - \boldsymbol{r}_j|, h) \\ &+ \frac{1}{2} \sum_j m_j Q_{ij} (\boldsymbol{v}_i - \boldsymbol{v}_j) \cdot \nabla_i W(|\boldsymbol{r}_i - \boldsymbol{r}_j|, h) , \qquad (6.8) \end{aligned}$$

where the second term describes the heat generation due to the artificial viscosity.

Monaghan & Gingold (1983) proposed the following form for the artificial viscosity tensor:

$$Q_{ij} = \begin{cases} \frac{1}{\varrho_{ij}}(-\alpha c_{ij}\mu_{ij} + \beta\mu_{ij}^2) \,, & (\boldsymbol{r}_i - \boldsymbol{r}_j)\cdot(\boldsymbol{v}_i - \boldsymbol{v}_j) \leq 0 \\ \\ 0 \,, & \text{otherwise} \end{cases} \tag{6.9}$$

with

$$\mu_{ij} = \frac{h_{ij}(\boldsymbol{v}_i - \boldsymbol{v}_j)(\boldsymbol{r}_i - \boldsymbol{r}_j)}{(r_{ij}^2 + \eta^2)} \,, \tag{6.10}$$

where c_{ij}, h_{ij} and ϱ_{ij} are the arithmetic means of the sound velocity c, the smoothing length h, and the density ϱ, respectively. The quantity $\eta \approx 0.1h$ is a smoothing parameter to prevent Q_{ij} from becoming infinite, if $|\boldsymbol{v}_i - \boldsymbol{v}_j| \neq 0$ and $r_{ij} \to 0$. The parameters α and β mimic a first and a second Navier–Stokes viscosity coefficient. Usually one uses $\alpha = 0.5$ and $\beta = 1$, but for problems involving strong shocks a choice of $\alpha = 1$ and $\beta = 2$ is more appropriate to avoid post-shock oscillations. Because of the specific form of μ_{ij} the Monaghan & Gingold (1983) viscosity tensor is non-zero for shear flows, and hence does not guarantee local angular momentum conservation. An improved form of the artificial viscosity tensor has been proposed by Balsara (1995). In case of a compressive, shear-free flow (*i.e.*, $\nabla \cdot \boldsymbol{v} \neq 0$ and $\nabla \times \boldsymbol{v} = 0$) it is identical to the artificial viscosity of Monaghan & Gingold (1983), but it is greatly reduced in shear flows. Consequently, unphysical transport of angular momentum can be reduced significantly.

Even though the energy equation is not explicitly symmetric in i and j, one can show that nevertheless energy is conserved (Benz 1990). However, in one aspect the conservation of energy and momentum differ. If one integrates the equations in time by means of finite differences momentum is automatically conserved due to the anti-symmetry of the forces. In contrast, energy is only conserved if one uses time-centered velocities in Eq. (6.8), *i.e.*, replacing the time-centered velocities by the respective forward extrapolated velocities leads to energy non-conservation. The conservation laws are also violated, if the hierarchical tree method is used to determine the gravitational potential, because the forces are no longer anti-symmetric for distant particles. However, the resulting errors are not large. They are of the order of a few percent after several thousand time steps in a simulation with a few hundred particles, and they become negligible if the number of particles exceeds a few thousand (Steinmetz & Müller 1993).

6.2 Self-gravity

Generally, self-gravity plays an important role in astrophysical problems and seldomly can be ignored. In principle, it is quite simple to include self-gravity in SPH. In order to obtain the gravitational force acting on a particle i one has to sum up the gravitational forces exerted by all other particles j on that particle. However, this process needs $\propto N^2$ force evaluations. Two different approaches have been proposed which approximate the gravitational forces and which allow to handle large particle numbers efficiently. The probably most attractive approach is the hierarchical tree method (Appel 1981; Porter 1985, Barnes & Hut 1986; Hernquist 1987), because it preserves one of the main advantages of SPH, namely being grid-less. In this method neighbouring particles are systematically grouped to clusters, neighbouring clusters to clusters of clusters and so on, until only one big cluster, the so-called root-cluster containing all particles is left. This hierarchy of clusters is organized in a tree. The center of mass (COM) and the monopole and quadrupole of each cluster is calculated. The force on a particle i is then obtained as follows. Beginning with the root-cluster one checks, whether the COM of the cluster is well separated from the particle i, *i.e.*,

$$\frac{S_k}{r_{ik}} \leq \theta \ , \tag{6.11}$$

where S_k and r_{ik} are the size of the cluster k and its COM-distance to the particle i, respectively. Typically, the opening angle $\theta = 0.8$. It determines the accuracy of the scheme: Small values of θ yield more accurate results, but also require more computational time, whereas a vanishing θ corresponds to the original direct summation scheme. If the cluster is well separated, the force on particle i is calculated by using the monopole and quadrupole moment of the cluster k. Otherwise, the cluster k is divided into its sub-clusters and the previous procedure is repeated. The whole process requires a computational time proportional to $N \log N$.

To construct the tree, two different methods are commonly used. The Barnes & Hut (1986) algorithm (henceforth BH-tree) initially assumes a cube containing all particles of the system. This cube is divided into 8 sub-cubes, each sub-cube into another 8 sub-sub-cubes, and so on. This procedure is continued until each sub-cube contains only one or no particle (empty cubes are not stored). All particles inside a sub-cube of higher level are combined to a cluster. As noted by Hernquist (1990) both the tree construction and the tree walking as well as the force summation can be vectorized. The other approach is a binary tree based on the relationship of mutual nearest neighbours (Appel 1981; Porter 1985; Benz *et al.* 1990). In a first step the nearest neighbour of each particle i is determined. Mutual nearest neighbours are replaced by a cluster, whose COM, size, monopole and quadrupole moment are calculated. In the next step the system of the still unpaired particles and the already constructed clusters is considered. Again mutual nearest neighbours are combined to clusters. The procedure is repeated until only the root-cluster remains.

A comparison of these two tree construction methods is subject of a paper by Makino (1990b). The results can roughly be summarized as follows. Tree walking and force calculation can be vectorized in both methods in a similar way. Therefore, the force calculation needs approximately the same time in both methods. The BH-tree seems to be a bit faster, while the accuracy of the binary tree is somewhat higher. However, whereas the tree construction can also be vectorized for the BH-tree, this is not possible for the binary tree, mainly because the time consuming part of getting the mutual nearest neighbour cannot be vectorized if the method of Benz *et al.* (1990) is used. Therefore, the tree construction time is comparable for both methods on workstations, but about ten times larger for binary trees on vector machines. However, in contrast to the BH-tree it is not necessary to construct the binary tree at every time step, because one can use the same tree for several time steps. This property of the binary tree becomes especially advantageous when a multiple time step scheme is used (see above).

The fact that the tree is newly constructed at each time step leads to a discontinuous change of the error in the force determination. Therefore, hierarchical tree methods are inadequate when used together with higher-order time integration schemes. Moreover, the forces of distant particles are only approximated by a multipole expansion of the force of the cluster to which the particles belong. Upper bounds on the largest possible errors that are introduced into a tree-code calculation by use of various multipole acceptability criteria have been derived by Salmon & Warren (1994). Because of the finite multipole expansion, the forces between two particles i and j in a tree-code calculation are no longer anti-symmetric in i and j, and thus violate both the conservation of linear and angular momentum and the conservation of energy as well. However, these errors decrease with increasing particle numbers (for a detailed discussion see Hernquist 1987; Barnes & Hut 1988). Because tree construction and tree walking require much computational time on its own, sophisticated direct summation methods should be used for particle numbers less than a few thousand.

6.2.1 The GRAPE Project. A completely different approach to perform large N-body simulations was chosen by Sugimoto and collaborators at the University of Tokyo (for an overview see Sugimoto 1993a,b): The calculation of one force interaction is a combination of a very few specific arithmetic operations: three coordinate differences, three squares, one sum, one square root, one inverse and three multiplications. Finally, the forces of all particles have to be added together. Furthermore, many of these operations do not depend on each other and can be done in a pipeline. Since the number of operations for the total force on one particle is the same for every particle, it is possible to parallelize it with a very good load balancing.

Sugimoto *et al.* have designed a series of special purpose hardware boards GRAPE (GRAvity PipE) to calculate Eq. (6.7). Furthermore, a list of particles within a sphere of a given radius h_i of particle i is returned, too, which is especially attractive if one wants to run a SPH code on such a board. The board is connected with a standard workstation via a VME interface. With a library it is possible to use GRAPE via subroutine calls in normal FORTRAN or C code. The prototype machine GRAPE-1 reached 240 Mflops in 1989. Meanwhile there exist three series of GRAPE boards: GRAPE-1 (240 Mflops), GRAPE-1A (240 Mflops), GRAPE-3 (15 Gflops) and GRAPE 3A (5 Gflops/board) are low precision boards ($\approx 1\,\%$ in the force), the precision, however, being sufficient for most astrophysical applications. GRAPE-2 (40 Mflops), HARP-1 (180 Mflops), HARP-2 (2 Gflops) and GRAPE-4 (1.1 Tflops) are working with 32 and 64 bit arithmetic and are designed to calculate molecular systems and specific stellar dynamical problems. Finally, GRAPE-2A (180 Mflops) and MD-GRAPE (4 Gflops) are versions which include a user-loadable force look-up table that can be used for arbitrary central force laws (for more details see, *e.g.*, Hut 1997). Recently, an initial "point design study" of the feasibility of constructing hardwired Petaflops-class computational engines (GRAPE-6) has been completed (Hut *et al.* 1997).

In a series of publications, the Tokyo group and others have shown that it is indeed possible to perform large N-body simulations on such a board with a speed close to its peak performance (for a review see, *e.g.*, Hut 1997). Furthermore, simulations involving SPH and GRAPE have been performed, too (*e.g.*, Steinmetz & Müller 1995; Steinmetz 1996a, b). In these GRAPESPH simulations, the gravitational force and the neighbour list is obtained with GRAPE, while the evaluation of the hydrodynamic force and the solution of the equation of state is done on a workstation. Similarly, tree codes can be implemented on GRAPE, too. Again, the force evaluation is done on the board, but the tree construction and the determination of the interaction list has to be done on the front end. Thus, a powerful workstation is necessary in order to use a combination of a SPH and/or a tree code with GRAPE efficiently.

6.3 Variable Smoothing Length

In SPH the smoothing length h limits the resolution, *i.e.*, all structures on scales less than h are strongly smoothed out. In the early implementations of SPH the smoothing length was assumed to be fixed in space and time, and its size was determined by the mean density of the system in consideration (see also Gingold & Monaghan 1977, 1982). The upper limit of the density is then given when the center of all particles coincide, whereas the lower limit is obtained when all particles have a separation larger than $2h$, *i.e.*,

$$
\begin{aligned}
\langle \varrho_i \rangle_{\mathrm{min}} &= mW(0,h)\,, \\
\langle \varrho_i \rangle_{\mathrm{max}} &= \sum_{i=1}^{N} mW(0,h)\,.
\end{aligned}
\tag{6.12}
$$

Thus, the maximum dynamic (density) range is N. However, this estimate is too optimistic. Firstly, one needs at least 10 particles to avoid very large fluctuations in velocity, density and energy, and secondly the positions of the particles must not be degenerated. The latter point means that the particle positions have to differ, because otherwise only a density profile can result which is proportional to the kernel W. Hence, the dynamic range is more realistically approximated by about $N/50 \dots N/100$.

However, the resolution can be increased dramatically, if one assumes a spatially and temporally varying smoothing length $h(\boldsymbol{r}, t)$, which is determined by the local instead of the mean density. Whereas the introduction of a time-dependent h is straightforward, there exist two possibilities to introduce a spatially variable h, the so-called gather and scatter interpretation (*e.g.*, Hernquist & Katz 1989): To guarantee the conservation laws of energy, momentum and angular momentum, a mixed interpretation is commonly used. Benz (1990) has symmetrized the smoothing length by using $W_{ij} = W(\boldsymbol{r}_i - \boldsymbol{r}_j, (h_i + h_j)/2)$, while Hernquist & Katz (1989) symmetrized the kernel itself $W_{ij} = 0.5(W(\boldsymbol{r}_i - \boldsymbol{r}_j, h_i) + W(\boldsymbol{r}_i - \boldsymbol{r}_j, h_j))$. Note that if the smoothing lengths of two particles are very different, the corresponding symmetrized kernel of Benz (1990) has an effective width only half as large as that proposed by Hernquist & Katz (1989). Steinmetz & Müller (1993) have tested both methods and got similar results in most cases. However, the symmetrization of Hernquist & Katz (1989) shows a somewhat better damping of post-shock oscillations in the case of strong shock waves.

Using a variable smoothing length can lead to a violation of the fluid conservation laws (Hernquist 1993; Nelson & Papaloizou 1994), because terms proportional to ∇h and $\partial h/\partial t$ occur in the equation of motion (*e.g.*, Hernquist & Katz 1989; Benz 1990). In extreme situations the violation of the conservation laws can reach the $\approx 10\%$ level (Hernquist 1993). Hence, SPH simulations done with a variable smoothing length should be interpreted with caution. Test calculations, designed to compare the effects of including and ignoring the ∇h terms (Nelson & Papaloizou 1994) show that their inclusion has no detrimental effects on the ability of SPH to model known problems (shock tubes, adiabatic collapse of cold gas spheres) with reasonable accuracy, but leads to a dramatic improvement on problems where SPH has shown to perform badly because of poor energy conservation (*e.g.*, head-on collision between polytropes). The SPH algorithm is also no longer a priori second-order accurate when a variable smoothing length is used (Benz 1990).

One problem still remains, namely how does one determine h? Because h has the dimension of a length, it is quite natural to couple h to $\varrho(\boldsymbol{r})$ via

$$h(\boldsymbol{r}) = \zeta \left(\frac{m}{\varrho(\boldsymbol{r})} \right)^{\frac{1}{3}}, \tag{6.13}$$

where ζ is a parameter of order unity. However, then the problem arises that h must be known in order to compute ϱ and vice versa. Miyama *et al.* (1984) have solved this problem using the density of the previous time step. Benz (1990) and Evrard (1988) propose another approach, where Eq. (6.13) is differentiated with respect to time and the continuity equation is used to rewrite the resulting right hand side:

$$\frac{\mathrm{d}h}{\mathrm{d}t} = \frac{1}{3} h \nabla \boldsymbol{v} \ . \tag{6.14}$$

The smoothing length h is then evolved in time similar to any other hydrodynamic quantity. Alternatively, one can couple h to the number of neighbouring particles $N_{\text{neighbour}}$ within a sphere of size $2h$ and try to keep this number fixed (Hernquist & Katz 1989). However, this method is quite time consuming, because a complete tree-walk is necessary to determine $N_{\text{neighbour}}$. This is partially circumvented by allowing $N_{\text{neighbour}}$ to vary within a given tolerance.

In their test calculations Steinmetz & Müller (1993) have found that the simple procedure of Miyama *et al.* (1984) to update the smoothing length leads to instabilities in many cases. The approaches of Benz (1990) and of Hernquist & Katz (1989) give quite good results for one-dimensional problems, or when the particle positions almost resemble a regular grid. But in the case of a randomly sampled three-dimensional problem, the smoothness of h is not guaranteed and the following effects are observed: Small density fluctuations lead to fluctuations in h, which in turn amplify the density fluctuations and so on. This effect is also well known in finite-difference methods when adaptive grids are used. In order to avoid it, one should determine h from global trends in the motion of the fluid, and not from local features.

In the method of Hernquist & Katz (1989), the number of neighbours $N_{\text{neighbour}}$ is a free parameter. Thus it is an open question, whether the instabilities observed by Steinmetz & Müller (1993) occur for all choices of $N_{\text{neighbour}}$ or whether there exists a critical value of $N_{\text{neighbour}}$ above which the method is stable. In the latter case it is unclear whether this value of N_{neigbour} is problem-dependent or not. Note in this context that Balsara (1995) has argued that the number of neighbours commonly used in SPH simulations ($N_{\text{neighbour}} \approx 40$) probably is too small for a correct description of the propagation of sound waves. Considering the above problems Steinmetz & Müller (1993) have proposed a more satisfactory algorithm to update the smoothing length, which gives remarkably good results.

6.4 Time Integration, Initial Model

In most SPH implementations, time integration is performed with a second-order accurate scheme, which consists of a multiple time step leap-frog integrator. Velocities can be predicted as described in Hernquist & Katz (1989). As noted by Anzer *et al.* (1987) and by Monaghan & Varnas (1988) this scheme has a very slowly growing instability, if the viscosity parameter $\alpha \lesssim 0.1$ (Eq. (6.9)). However, this instability is not very troublesome, because in many astrophysical applications shock waves occur, *i.e.*, artificial viscosity has to be switched on. The use of multiple time steps is very effective the computational times being reduced at least by a factor of two. In the case that a highly condensed subregion evolves, the reduction in computational time can exceed a factor of ten or even more!

To get an initial model one can use the method proposed by Lucy (1977). The N particles are sampled randomly according to the density distribution $\varrho(\boldsymbol{r})$ one wants to model. The particles are assigned a mass m and a specific energy according to the energy profile $E(\boldsymbol{r})$. However, the number of particles which currently can be used in a simulation are much too small and do not allow for a satisfactory approximation of ϱ and E, *i.e.*, the presence of large pressure gradients in an initial model gives rise to random motion with large kinetic energies. Following Lucy (1977), one should then allow the random motion to relax into equilibrium by solving the equation of motion with an additional damping term. However, the relaxation method has the severe disadvantage that the star needs several sound crossing times to reach the equilibrium state. Especially in simulations with a large number of particles the relaxation process can thus become as time consuming as the whole simulation itself! On the other hand, the same initial model is often used for a whole sequence of simulations, so this disadvantage is not very severe. For non-equilibrium initial models, one can skip the relaxation process provided the motion rapidly becomes highly dynamic, *i.e.*, $E_{\text{kin}} \gg |E_{\text{pot}}|, E_{\text{int}}$. For moderately dynamic systems, the stretched grid method is to be preferred (*e.g.*, Hernquist & Katz 1989).

It is important to note, that whatever method is used, one has to check that all physical quantities (in particular the energy and the velocity) of the initial model are smooth. Note that the relaxation process of Lucy (1977) guarantees such a smooth behaviour, because during the relaxation process the SPH algorithm is already applied to the different physical quantities. Otherwise, the simplest way to achieve this smoothness is to apply the SPH formalism to the initial model (see also Hernquist & Katz 1989; Benz 1990). Disregard of this point gives rise to unphysical oscillations close to steep density gradients and to an unstable behaviour (*e.g.*, small-scale oscillations in energy and density) of almost hydrostatic models.

6.5 Computational Aspects

For SPH codes being based on the hierarchical tree method two strategies exist to vectorize the algorithm. Hernquist (1990) proposes to vectorize over all nodes of a level in the tree during tree walking and over all force terms during the force calculation. The resulting typical vector length is short (between some tens and some hundreds). In Makino's (1990a) method, all particles walk through the tree in parallel and one force term is added to all particles in parallel, too. This results in a much larger vector lengths (about N) but requires more bookkeeping and thus complicates the structure of the program. Test calculations by Steinmetz & Müller (1993) show that the computational time of both methods is practically identical on CRAY vector computers, as these reach a large percentage of their maximum theoretical speed even for short vector lengths. Thus, because of its simplicity and greater flexibility, Hernquist's (1990) method should be used to vectorize the force calculations. Barnes (1990) proposed an alternative vectorization method, which is based on the fact that nearby particles have similar interaction lists. Therefore, one can use the same interaction list for a group of particles. However, as a price for the increased degree of vectorization one has to compute a larger number of force terms, which makes the method less efficient than the methods of Hernquist (1990) or Makino (1990a).

The typical CPU-time required for a binary tree based SPH simulation of a few times 10^4 particles and 10^3 time steps ranges from a few hours to up to 20 hours on one processor of a CRAY-YMP (333 Mflops peak performance). The CPU-time scales almost linearly with the number of particles.

Finally, concerning cosmological studies it should be noted that a collisionless dark-matter component can be included into a SPH simulation in an easy and consistent way as an additional kind of particles which interact purely gravitationally.

6.6 How dissipative is SPH?

One of the main advantages of one-dimensional Lagrangian finite difference methods is their small numerical diffusion, which results from the fact that in Lagrangian coordinates no advection terms are present in the hydrodynamic equations (see Sect. 2.2.1). This attractive property makes Lagrangian methods optimally suited to handle problems involving chemical or nuclear kinetics (see Sect. 4). Unfortunately, this property is lost in multi-dimensional flow problems, because the appearance of shear flows and/or vortices repeatedly requires remapping the tangled and twisted grid. This remapping introduces some amount of numerical diffusion into most multi-dimensional Lagrangian schemes, which exceeds that of "state of the art" Eulerian schemes.

Concerning the diffusivity of SPH the situation is more complicated. On the one hand according to the Monte-Carlo nature of SPH each physical quantity is determined by a weighted superposition of neighbouring particles with both neighbours and weights changing from time step to time step. Moreover, the SPH particles can penetrate each other. These two effects can be looked upon as a kind of intrinsic remap of a free Lagrangian grid, which would imply a sizeable numerical diffusivity. On the other hand SPH should be less diffusive than Eulerian methods, because due to its Lagrangian nature mass, energy and velocity are particle properties which need not to be advected. This conception is supported by the large amount of artificial viscosity required in the shock tube calculations to damp-out post-shock oscillations. However, shock tube calculations are usually performed with a one-dimensional version of SPH where the particles are initially on a regular grid, *i.e.,* particles hardly penetrate each other and the neighbours of a

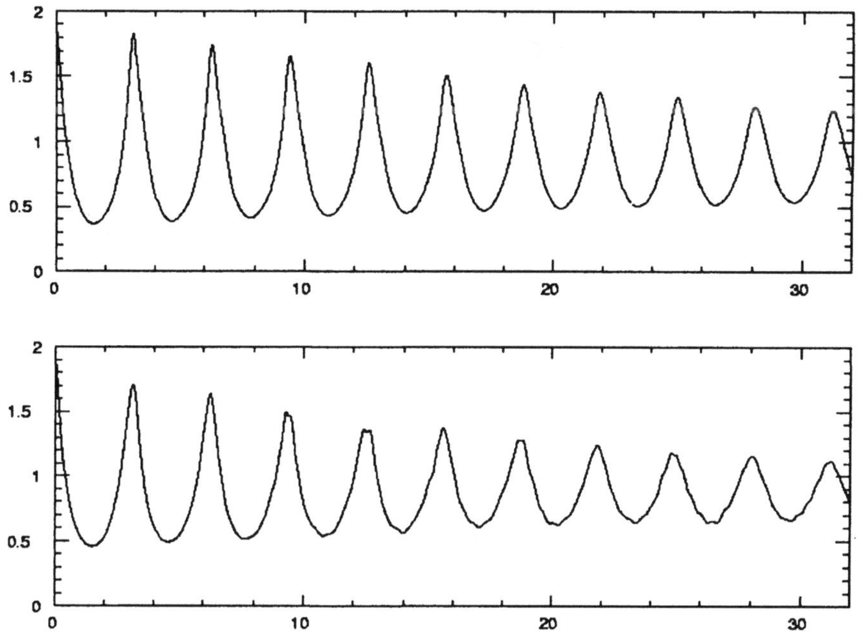

Fig. 6.1. Central density of a pulsating $n = 1.5$ polytrope as a function of time normalized to the central density of the equilibrium model. The pulsation is initiated by a spherical homologous contraction of the equilibrium configuration by a factor of 0.8 in radius. In the upper calculation h is constant in space, while a spatially variable smoothing length was used in the lower simulation. Obviously, the increase in resolution due to a spatially variable smoothing length is accompanied by a larger amount of numerical diffusion. (From Steinmetz & Müller 1993)

specified particle remain almost unchanged. Thus, the size of the numerical diffusivity of SPH remains unclear.

In order to clarify the diffusivity issue Steinmetz & Müller (1993) have examined the development of strong pulsations of a $n = 1.5$ polytrope. The pulsation is initiated by imposing a homologous contraction on an equilibrium model. The resulting evolution of the central density is shown in Fig. 6.1 for a simulation with a spatially variable (top) and a spatially constant (bottom) time-dependent smoothing length. Obviously, the pulsation amplitude decreases in both simulations. Approximately one half of the observed decrease is due to the artificial viscosity required to stabilize the model, while the other half is due to the SPH inherent numerical diffusivity. Note that the radial pulsation does not decay into non-radial modes, but that kinetic energy is rather converted into thermal energy. Furthermore, at first glance it seems surprising that the model with the spatially constant smoothing length is less diffusive than that with the variable h. However, this can be explained. If h is spatially constant, the density contrast within the star can only be approximated by varying the number of particles as a function of radius, while for a spatially variable smoothing length both the number and the size can be adjusted. Thus, for a spatially constant h contributions of many more particles are required to approximate the central density than for a spatially variable h, *i.e.*, a change in the number of neighbours is statistically more significant in the latter case. The resulting energy diffusion is mainly determined by the diffusion in the innermost parts of the star, because the thermal energy grows more than linear with density. This interpretation is proven by the fact, that when Steinmetz & Müller (1993) used an explicit adiabatic model where the pressure is directly computed from the density (*i.e.*, no energy equation is used), the numerical diffusion of the scheme was reduced drastically.

6.7 How Well Does SPH Treat Shocks?

Astrophysical flows often involve shock waves. Thus, it is important that shocks are accurately described by the numerical code used for the simulation of the flow. The respective requirements on the code fall into two different categories. In the first category of problems the structure of the shock front itself is important, *i.e.*, a high spatial resolution of the discontinuity is crucial. Typical examples for this category of problems are hydrodynamic flows with nuclear burning, where an insufficient spatial resolution can lead to quantitatively very inaccurate and in some cases to even qualitatively incorrect solutions (see Sect. 4). In the second category of problems the time scales of processes triggered by the shock wave are comparable or larger than the hydrodynamic time scale. Then mainly an accurate description of the two states on both sides of the shock is important, while the structure of the discontinuity matters less.

In early publications about SPH (Gingold & Monaghan 1982; Monaghan 1985; Loewenstein & Mathews 1986) it was shown, that SPH is able to solve Riemann or shock tube test problems in a satisfactory way. However, in publications dealing with finite volume methods, Riemann problems are only looked upon as a kind of trivial test problems, which are not demanding at all. For this reason Steinmetz & Müller (1993) have tested their SPH code on a more severe hydrodynamic test problem, namely the interaction of two strong shock or blast waves (Woodward 1982), and have compared their SPH results with those obtained with the Piecewise Parabolic Method (PPM) of Colella & Woodward (1984).

The initial state of the interacting blast wave problem consist of three constant states of a $\gamma = 1.4$ gas in the interval $0 \le x \le 1$ with reflecting boundaries on both sides. The density is unity everywhere and the pressure is assumed to be

$$P(x) = \begin{cases} 1000, & 0 \le x \le 0.1\,, \\ 0.01, & 0.1 < x \le 0.9\,, \\ 100, & 0.9 < x \le 1\,. \end{cases} \tag{6.15}$$

The initial conditions give rise to two shock waves propagating into the cold (*i.e.*, low pressure) gas and two rarefaction waves moving into the hot (*i.e.*, high pressure) gas. The Mach numbers M of the two shocks are very large in the left and right shock the Mach numbers being $M \approx 170$ and $M \approx 51$, respectively. The reflection of rarefaction waves at the boundaries and at contact discontinuities, the weakening of the shocks caught by the reflected rarefactions, and finally the shock collision lead to a quite complex behaviour (for more details see Woodward 1982; Woodward & Colella 1984). Up to the time when the two shocks collide the density and velocity profiles can be described by a set of straight lines.

In the SPH calculations of Steinmetz & Müller (1993) the initial state was obtained putting the particles on a uniform grid with spacing h. The boundaries are simulated using "ghost-particles", *i.e.*, for each particle extending across a boundary an analogous particle is created on the other side of the boundary having the same distance to the boundary, but the opposite velocity. The smoothing length is time-dependent and spatially variable according to the method of Benz (1990; see also Eq. (6.14)). The evolution of the density and the velocity field are shown in Fig. 6.2 (for $N = 400$) and in Fig. 6.3 (for $N = 2000$). In both figures the times of the snapshots are chosen from Woodward & Colella (1984) and hence correspond to (dimensionless) times equal to 0.010, 0.016, 0.026, 0.028, 0.030, 0.032, 0.034 and 0.038, respectively.

In the low resolution run the shock is quite accurately resolved at the beginning of the calculation (Fig. 6.2; snapshots 1-3). However, the transitions between constant states and rarefaction waves are rather smooth and not very sharp. Moreover, some oscillations in the constant state between the shock and the contact discontinuity as well as in the rarefaction waves are visible (Fig. 6.2; snapshots 2-3). But the states themselves are well represented and are even quantitatively in agreement with the results of Woodward & Colella (1984). All shortcomings of the low resolution run are removed in the high resolution run (Fig. 6.2; snapshots 1-3). Except for a very narrow region at the edges of the different states the profiles seem to be represented by a set of straight lines.

At $t = 0.28$ (Fig. 6.2 and 6.3; snapshot 4) the two blast waves collide near $x = 0.7$ giving rise to a very narrow density maximum with $\varrho \approx 30$. In the low resolution run the maximum density is overestimated by about 15%, and in the high resolution run it is still too large by more than 5%. According to Woodward & Colella (1984) the density maximum should decrease and broaden with time. This behaviour is reproduced by the SPH results (Figs. 6.2 and 6.3; snapshots 4-8), but they also show an additional erroneous density spike immediately to the left of the density maximum which is most clearly seen in the snapshots at late times (Figs. 6.2 and 6.3; snapshots 7-8). Also at late times one recognizes another incorrect density feature slightly to the right of the density minimum at $x = 0.75$ where the density undershoots. Both defects of the solution are caused by the usage of a variable smoothing

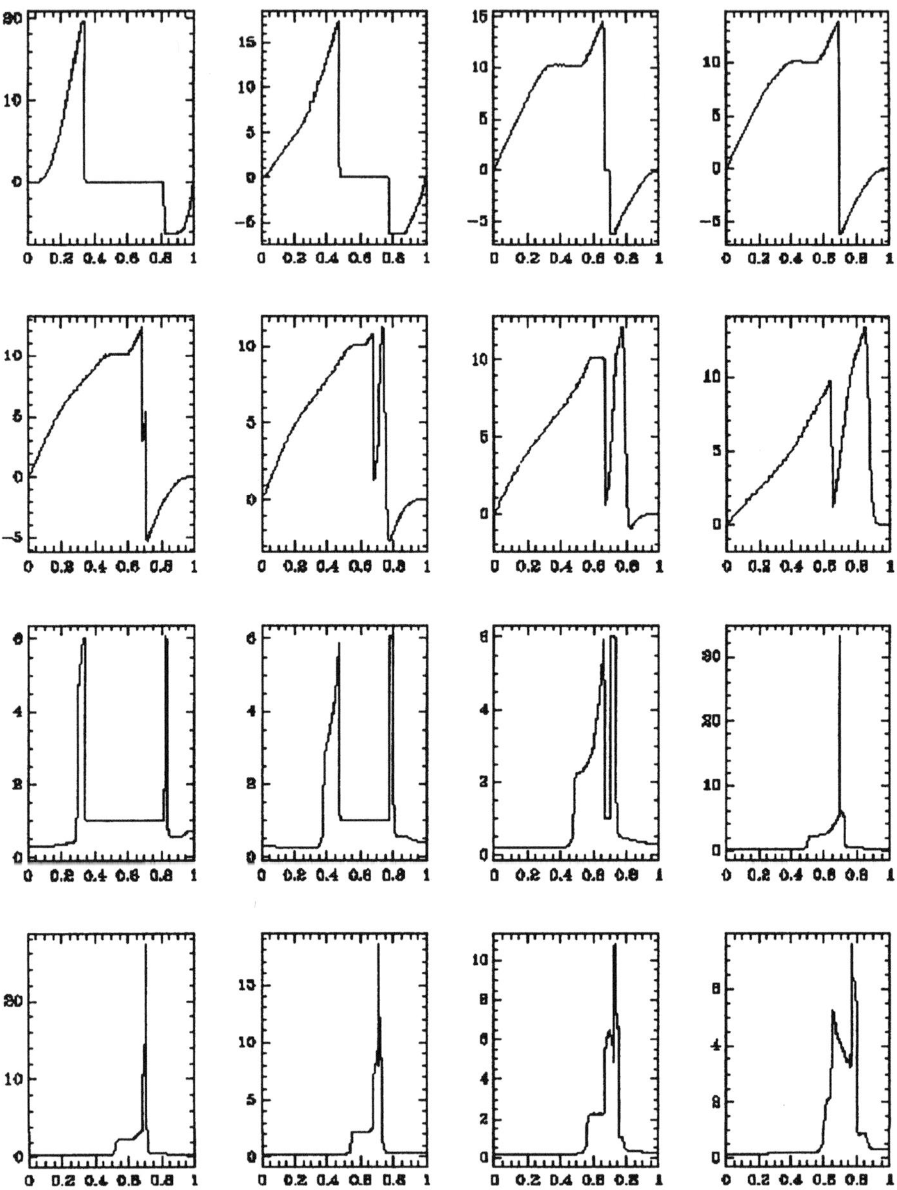

Fig. 6.2. Snapshots of velocity (top two rows) and density (lower two rows) for the Woodward double shock tube problem computed with SPH using 400 particles. The states to the left and to the right of shocks and contact discontinuities are accurately reproduced. However, the discontinuities themselves are strongly smeared out. Furthermore, some small post-shock oscillations are visible. (From Steinmetz & Müller 1993)

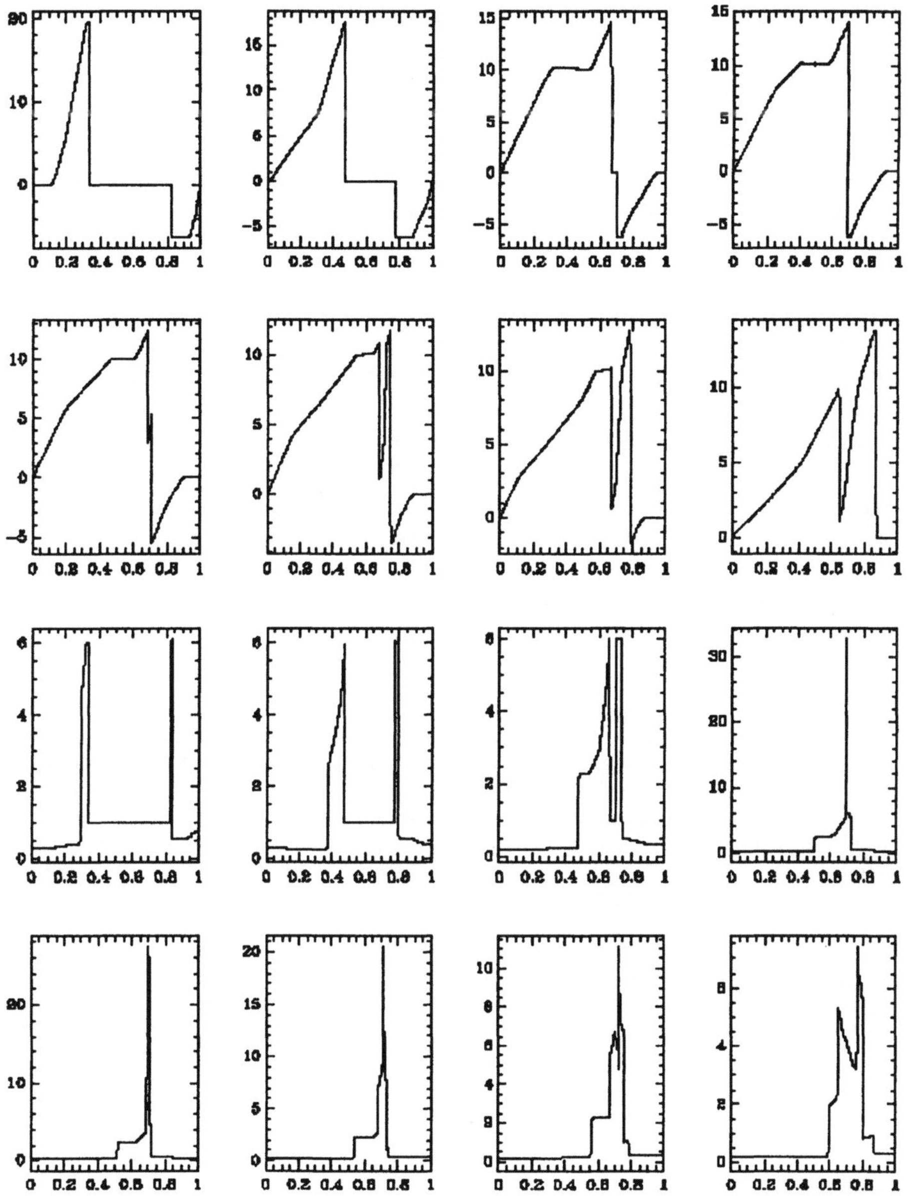

Fig. 6.3. Same as Fig. 6.2, but using 2000 particles. No post-shock oscillations are present and the discontinuities are much more crisp. Only a small overshoot in density is still visible where the two shock waves collided. (From Steinmetz & Müller 1993)

length. The large density jump occurring during the shock collision yields a very small smoothing length for particles in the collision region, *i.e.*, the resulting intrinsic density of a particle $\varrho_{\rm intr} = 2m/3\,\pi\,h$ is extremely large (for a detailed discussion of this point see Loewenstein & Mathews 1986; Evrard 1988).

In the low resolution run some small additional oscillations occur near the density minimum between the two maxima (Fig. 6.2; snapshot 8). All features of the solution of Woodward & Colella (1984) are present in the $N = 400$ calculation, although some of them are strongly smeared out, as *e.g.*, the discontinuity at $x \approx 0.85$ (Fig. 6.2; snapshot 8). In the calculation using 2000 particles the oscillations are not present. All features are much more crisp, and the results are comparable with those of quite sophisticated finite volume methods like for example FCT (Flux-Corrected Transport). However, the quality of PPM cannot be reached.

Thus, in conclusion, it is possible to handle even demanding hydrodynamic test problems involving strong shocks and their interaction with SPH provided a sufficiently large number of particles is used in the simulation. In my opinion, SPH and finite volume methods should be looked upon not as competing but as complementary methods to solve the hydrodynamic equations each of which has its own merits and defects.

Acknowledgement. This work has greatly benefited from many years of fruitful and enjoyable collaborations with Dave Arnett, Bruce Fryxell, Wolfgang Hillebrandt, Thomas Janka, Chema Martí, Max Ruffert, Gerhard Schäfer and Matthias Steinmetz. I would like to thank Jens Niemeyer and Wolfgang Hillebrandt for sharing their insight into turbulent combustion with me. Special thanks go to Wolfgang Hillebrandt and Chema Martí for reading parts of my lecture notes and for giving me their feed-back on how to improve them. I further like to express my gratitude to Alfred Gautschy and Oskar Steiner for inviting me to Les Diablerets and for organizing this expedient 27th Saas-Fee Advanced Course. The final version of these lecture notes was written, while the author was visiting the Institute for Theoretical Physics (ITP) of the University of California at Santa Barbara participating in the 1997 ITP workshop on "Supernovae". I would like to thank the director and the staff of the ITP for their hospitality. This research was supported in part by the National Science Foundation under Grant No. PHY94-07194.

References

Abgrall, R. & Montagné, J.L., 1989, *Rech. Aèrosp.* **4**, 1

Abramovici, A. *et al.*, 1995, in *Particle and Nuclear Astrophysics and Cosmology in the next Millenium* (Proc. of the 1994, Snowmass Summer Study), eds. E.W. Kolb and R. Peccei, World Scientific, Singapore, 398

Akujor, C.E., 1992, *Astron. Astrophys.* **259**, L61

Anzer, U., Börner, G. & Monaghan, J.J., 1987, *Astron. Astrophys.* **176**, 235

Appel, A.W., 1981, Undergraduate thesis, Princeton Univ., Preprint No.156

Arnett, D., 1969, *Astrophys. Spac. Sci.* **5**, 180

Arnett, D., 1988, *Astrophys. J.* **331**, 337

Arnett, D., 1994, *Astrophys. J.* **427**, 932

Arnett, D., 1996, *Supernovae & Nucleosynthesis*, Princeton Univ. Press, Princeton

Arnett, D. & Livne, E., 1994a, *Astrophys. J.* **427**, 315

Arnett, D. & Livne, E., 1994b, *Astrophys. J.* **427**, 330

Arnett, D. & Thielemann, F.-K., 1985, *Astrophys. J.* **295**, 589

Arnett, D. & Truran, J.W., 1969, *Astrophys. J.* **157**, 339

Arnett, W.D., Bahcall, J.N., Kirshner, R.P. & Woosley, S.E., 1989a, *Ann. Rev. Astron. Astrophys.* **27**, 629.

Arnett, W.D., Fryxell, B.A. & Müller, E., 1989b, *Astrophys. J. Lett.* **341**, L63

Arnold, C.N., 1985, *Ph. D. thesis*, Univ. Microfilms, Ann Arbor, Michigan

Arnold, C.N. & Arnett, W.D., 1986, *Astrophys. J. Lett.* **305**, L57

Axelrod, T.S., 1980, Ph.D. Thesis, Univ. California, Santa Cruz; and in *Type I Supernovae*, ed. J.C. Wheeler, University of Texas, Austin, 80

Axelrod, T.S., 1988, in *IAU Coll. 108 on Atmospheric Diagnostics of Stellar Evolution: Chemical Peculiarity, Mass Loss, and Explosion*, ed. K. Nomoto, Springer-Verlag, Berlin, 375

Baade, W. & Zwicky, F., 1934, *Phys. Rev.* **45**, 138

Balsara, D.S., 1995, *J. Comp. Phys.* **121**, 357

Balsara, D.S. & Norman, M.L., 1992, *Astrophys. J.* **393**, 631

Barbon, R., Ciatti, F. & Rosino, L., 1979, *Astron. Astrophys.* **72**, 287

Barnes, J.E., 1990, *J. Comp. Phys.* **87**, 161

Barnes, J.E. & Hut, P., 1986, *Nature* **324**, 446

Barnes, J.E. & Hut, P., 1988, *Astrophys. J. Suppl.* **70**, 389

Baron, E. & Cooperstein, J., 1990, *Astrophys. J.* **353**, 597

Baron, E., Cooperstein, J. & Kahana, S., 1985a, *Phys. Rev. Lett.* **155**, 126

Baron, E., Cooperstein, J. & Kahana, S., 1985b, *Nucl. Phys.* **A440**, 744

Barthelmy, S., Gehrels, N., Leventhal, M., MacCullum, C.J., Teegarden & B.J., Tueller, J., 1989, *IAU Circular* **4764**

Baum, S.A., O'Dea, C.P., Giovannini, G., Biretta, J., Cotton, W.B., de Koff, S., Feretti, L., Golombek, D., Lara, L., Macchetto, F.D., Miley, G.K., Sparks, W.B., Venturi, T. & Komissarov, S.S., 1997, *Astrophys. J.* **483**, 178

Bazan, G. & Arnett, D., 1994, *Astrophys. J. Lett.* **433**, L41

Bazan, G. & Arnett, D., 1997, *Astrophys. J.* , in press

Begelman, M.C. & Cioffi, D.F., 1989, *Astrophys. J. Lett.* **345**, L21

Begelman, M.C. & Sarazin, C.L., 1986, *Astrophys. J. Lett.* **302**, L59

Begelman, M.C., Rees, M.J. & Blandford, R.G., 1979, *Nature* **279**, 770

Begelman, M.C., Blandford, R.D. & Rees, M.J., 1984, *Rev. Mod. Phys.* **56**, 255

Begelman, M.C., Rees, M.J. & Sikora, M., 1994, *Astrophys. J.* **429**, L57

Bell, A.R., 1996, *Phys. Plasmas* **1**, 1643

Bell, J., 1997, *Vistas in Astron.* **41**, 87

Benz, W., 1990, in *The Numerical Modelling of Nonlinear Stellar Pulsations, Problems and Prospects*, ed. J.R. Buchler, Kluwer, Dordrecht, 269

Benz, W. & Thielemann, F.-K., 1990, *Astrophys. J. Lett.* **348**, L17

Benz, W., Bowers, R.L., Cameron, A.G.V. & Press, W.H., 1990, *Astrophys. J.* **3**48, 647

Bertola, F., 1964, *Ann. d'Astrophys.* **27**, 319

Bethe, H.A., 1990, *Rev. Mod. Phys.* **62**, 801

Bethe, H.A. & Wilson, J.R., 1985, *Astrophys. J.* **295**, 14

Bethe, H.A., Brown, G.E., Applegate, J.H. & Lattimer, J., 1979, *Nucl. Phys.* **A324**, 487

Bicknell, G.V., 1994, *Astrophys. J.* **422**, 542

Bicknell, G.V., 1995, *Astrophys. J. Suppl.* **101**, 29
Bicknell, G.V. & Gingold, R.A., 1985, *Astrophys. J.* **273**, 749
Biretta, J.A., 1996, in *Energy Transport in Radio Galaxies and Quasars*, eds. P.E. Hardee, A.H. Bridle and J.A. Zensus, ASP Conference Series, Vol. 100, 187
Biretta, J.A. & Owen, F.N., 1990, in *Parsec-scale Radio Jets*, eds. J.A. Zensus and T.J. Pearson, Cambridge Univ. Press, Cambridge, 125
Biretta, J.A., Zhou, F. & Owen, F.N., 1995, *Astrophys. J.* **447**, 582
Blanchet, L., Damour, T. & Schäfer, G., 1990, *Mon. Not. Roy. Astronom. Soc.* **242**, 289
Blandford, R.D. & Königl, A., 1979, *Astrophys. J.* **232**, 34
Blandford, R.D. & Payne, D.G., 1982, *Mon. Not. Roy. Astronom. Soc.* **199**, 883
Blandford, R.D. & Rees, M.J., 1974, *Mon. Not. Roy. Astronom. Soc.* **169**, 395
Blandford, R.D., McKee, C.F. & Rees M.J., 1977, *Nature* **267**, 211
Blinnikov, S.I. & Sasarov, P.V., 1996, *Phys. Rev.* **E53**, 4827
Bludman, S.A., Mochkovitch, R. & Zinn-Justin, J. (eds.), 1994, *Supernovae* (Les Houches Summer School, Session LIV, 1990), Elsevier, Amsterdam
Bludman, S.A., Feng, D.H., Gaisser, T. & Pittel, S. (eds.), 1995, *The Physics of Supernovae* (Symposium held at the Franklin Institute in Philadelphia, USA, April 7-8, 1994), *Phys. Rep.* **256**
Bodo, G., Massaglia, S., Ferrari, A. & Trussoni, E., 1994, *Astron. Astrophys.* **283**, 655
Bolgiano, R.,Jr., 1959, *J. Geophys. Res.* **64**, 2226
Bonazzola, S. & Marck, J.-A., 1993, *Astron. Astrophys.* **267**, 623
Bonazzola, S. & Marck, J.-A., 1994, *Ann. Rev. Nucl. Part. Sci.* **45**, 655
Boris, J.P., Grinstein, F.F., Oran, E.S. & Kolb, R.L., 1992, *Fluid Dyn. Res.* **10**, 199
Borue, V. & Orzag, S.A., 1997, preprint
Bowman, M., 1994, *Mon. Not. Roy. Astronom. Soc.* **269**, 137
Bradley, D., 1993, in *24th Int. Symp. Combust.*, Combustion Institute, Pittsburgh, 247
Branch, D., Nomoto, K. & Filippenko, A.V., 1991, *Comm. Astrophys.* **15**, 221
Bravo, E. & García-Saenz, D, 1995, *Astrophys. J. Lett.* **450**, L17
Bridle, A.H. & Perley, R.A., 1984, *Ann. Rev. Astron. Astrophys.* **22**, 319
Bridle, A.H., Hough, D.H., Lonsdale, C.J., Burns, J.O. & Laing, R.A., 1994, *Astronom. J.* **108**, 766
Brown, G.E., Bethe, H.A. & Baym, G., 1982, *Nucl. Phys.* **A375**, 481
Bruenn, S.W., 1985 *Astrophys. J. Suppl.* **58**, 771
Bruenn, S.W., 1993, in *Nuclear Physics in the Universe*, eds. M.W. Guidry and M.R. Strayer, IOP, Bristol, 31
Buchler, J.R., Colgate, S.A. & Mazurek, T. J., 1979, *J. Phys.* **C2**, 159
Bugarella, D, Livio, M. & O'Dea, C.P. (eds.), 1993, *Astrophysical Jets* (Space Telescope Science Institute Symposium Series 6, Baltimore, May 1992), Cambridge Univ. Press, Cambridge
Burrows, A., 1987, *Astrophys. J. Lett.* **318**, L57
Burrows, A. & Fryxell, B.A., 1992, *Science* **258**, 430
Burrows, A. & Fryxell, B.A., 1993, *Astrophys. J.* **418**, L33
Burrows, A. & Hayes, J., 1996, *Phys. Rev. Lett.* **76**, 352
Burrows, A. & Lattimer, J.M., 1988, *Phys. Rev.* **163**, 51
Burrows A., Hayes J. & Fryxell B.A., 1995, *Astrophys. J.* **450**, 830
Bychkov, V.V. & Liberman, M.A., 1995, *Astron. Astrophys.* **302**, 727
Camenzind, M., 1997, in *Relativistic Astrophysics* (Proc. 162. Heraeus Seminar, Bad Honnef, Germany, 1996), eds. H. Riffert, H. Ruder, H.-P. Nollert and F.W. Hehl, Vieweg-Verlag, Weinheim, in press

Cappellaro, E., Turatto, M., Benetti, S., Tsvetkov, D.Yu., Bartunov, O.S. & Makarova, I.N., 1993, *Astron. Astrophys.* **273**, 383

Cappellaro, E., Turatto, M., Tsvetkov, D.Yu., Bartunov, O.S., Pollas, C., Evans, R. & Hamuy, M., 1997, *Astron. Astrophys.* **322**, 431

Carilli, C.L., Perley, R.A., Barthel, N. & Dreher, J.W., 1996, in *Cygnus A, Study of a Radio Galaxy*, eds. C.L. Carilli and D.E. Harris, Cambridge University Press, Cambridge, 76

Chalabaev, A.A. & Cristiani, S., 1991, in *SN1987A and Other Supernovae*, eds. I.J. Danziger and K. Kjär, ESO Workshop and Conference Proceedings No.37, ESO, Garching, FRG, 655

Chandrasekhar, S., 1939, *An Introduction to the Study of Stellar Structure*, Dover Publisher, New York, Chapter XI.

Chandrasekhar, S., 1961, *Hydrodynamic and Hydromagnetic Stability*, Clarendon Press, Oxford, Chap. X

Chevalier, R., 1976, *Astrophys. J.* **207**, 872

Chevalier, R., 1984, *Ann. N. Y. Acad. Sci.* **422**, 215

Chow, E. & Monaghan, J.J., 1997, *J. Comp. Phys.* **134**, 296

Chugai, N.N., 1991, *Sov. Astron.* **35**, 171

Cioffi, D.F. & Blondin, J.M., 1992, *Astrophys. J.* **392**, 458

Clark, D.H., Andrews, P.J. & Smith, R.C., 1981, *Observatory* **101**, 203

Clarke. D.A., Norman, M.L. & Burns, J.O., 1986, *Astrophys. J. Lett.* **311**, L63

Clarke. D.A., Norman, M.L. & Burns, J.O., 1989, *Astrophys. J.* **342**, 700

Clavin, P., 1994, *Ann. Rev. Fluid Mech.* **26**, 321

Clayton, D.D., 1968, *Principles of Stellar Evolution and Nucleosynthesis*, McCraw-Hill, New York.

Clocchiatti, A. & Wheeler, J.C., 1997, *Thermonuclear Supernovae*, eds. P. Ruiz-Lapuente, R. Canal and I. Isern, Kluwer, Dordrecht, 863

Clocchiatti, A., Wheeler, J.C., Brotherton, M.S., Cochran, A.L., Wills, D. & Barker, J.W., 1996, *Astrophys. J.* **462**, 462

Colella, P. & Woodward, P.R., 1984, *J. Comp. Phys.* **54**, 174

Colella, P., Majda, A. & Roytburd, V., 1986, *SIAM J. Sci. Stat. Comput.* **7**, 1059

Colgan, S.W.J., Haas, M.R., Erickson, E.F., Lord, S.D. & Hollenbach, D.J., 1994, *Astrophys. J.* **427**, 874

Colgate, S.A., 1989, *Nature* **341**, 489

Colgate, S.A., Herant, M. & Benz, W., 1993, *Phys. Rep.* **227**, 157

Cook, W.R. *et al.*, 1988, *Astrophys. J. Lett.* **334**, L87

Cooperstein, J. & Baron, E., 1990, in *Supernovae*, ed. A. Petschek, Springer-Verlag, New York, 213

Courant, R. & Friedrich, K. O., 1948, *Supersonic Flow and Shock Waves*, Interscience, New York

Daly, R.A., 1995, *Astrophys. J.* **454**, 580

Daly, R.A. & Marscher, A.P., 1988, *Astrophys. J.* **334**, 539

Darrieus, G., 1938, in *La Technique Moderne*, unpublished

Davies, M.B., Ruffert, M., Benz, W. & Müller, E., 1993, *Astron. Astrophys.* **272**, 430

Davis, R.J., Unwin, S.C. & Muxlow, T.W.B., 1991, *Nature* **354**, 374

Den, M., Yoshida, T. & Yamada, Y., 1990, *Prog. Theor. Phys.* **83**, 723

de Young, D.S., *Astrophys. J.* **371**, 69

Doering, W., 1943, *Ann. Phys.* **43**, 421

Doggett, J.B. & Branch, D., 1985, *Astronom. J.* **90**, 2303

Donat, R. & Marquina, A., 1996, *J. Comp. Phys.* **125**, 42

Dotani, T. *et al.*, 1987, *Nature* **330**, 230

Dubal, M.R. & Pantano, O., 1993, *Mon. Not. Roy. Astronom. Soc.* **261**, 203

Duncan, G.C. & Hughes, P.A., 1994, *Astrophys. J. Lett.* **436**, L119
Duncan, C., Hughes, P. & Opperman, J., 1996, in *Energy Transport in Radio Galaxies and Quasars*, eds. P.E. Hardee, A.H. Bridle and J.A. Zensus, ASP Conference Series, Vol. 100, 143
Eislöffel, J., 1996, in *Jets from Stars and Galactic Nuclei*, ed. W. Kundt, Springer-Verlag, Berlin, 104
El Eid, M.F. & Hillebrandt, W., 1980, *Astron. Astrophys. Suppl.* **42**, 215
Elias, J.H., Mathews, K., Neugebauer, G. & Persson, S.E., 1985, *Astrophys. J.* **296**, 379
Endal, A.S. & Sofia, S., 1977, *Phys. Rev. Lett.* **39**, 1429
Ensman, L.M. & Woosley, S.E., 1988, *Astrophys. J.* **333**, 754
Epstein, R.I., 1978, *Astrophys. J.* **223**, 1037
Epstein, R.I., 1979,*Mon. Not. Roy. Astronom. Soc.* **188**, 305
Erickson, E.F., Haas, M.R., Colgan, S.W.J., Lord, S.D., Burton, M.G., Wolf, J., Hollenbach, D.J. & Werner, M., 1988, *Astrophys. J.* **330**, L39
Eriguchi, Y. & Müller, E., 1984, *Astron. Astrophys.* **147**, 161
Eulderink, F. & Mellema, G., 1994, *Astron. Astrophys.* **284**, 654
Evrard, A.E., 1988, *Mon. Not. Roy. Astronom. Soc.* **235**, 911
Falle, S.A.E.G., 1987, in *Astrophysical Jets and Their Engines*, ed. W. Kundt, Reidel, Dordrecht, 163
Fanaroff, B.L. & Riley, J.M., 1974, *Mon. Not. Roy. Astronom. Soc.* **167**, 31
Ferrari, A., Trussoni, E. & Zaninetti, L., 1978, *Astron. Astrophys.* **64**, 43
Fickett, W. & Davis, W.C., 1979, *Detonation*, Univ. California Press, Berkeley
Filippenko, A.V., 1988, *Astronom. J.* **96**, 1941
Filippenko, A.V., 1991, in *Supernovae and Stellar Evolution*, eds. A. Ray and T. Velusamy, World Scientific, Singapore, 34
Filippenko, A.V., 1997, in *Thermonuclear Supernovae*, eds. P. Ruiz-Lapuente, R. Canal and J. Isern, Kluwer, Dordrecht, 1
Filippenko, A.V. & Sargent, W.L.W., 1986, *Astronom. J.* **91**, 691
Filippenko, A.V. & Sargent, W.L.W., 1989, *Astrophys. J. Lett.* **345**, L43
Filippenko, A.V. *et al.*, 1995, *Astrophys. J.* **450**, L11
Finn, L.S. & Evans, C.R., 1990, *Astrophys. J.* **351**, 588
Flowers, E. & Itoh, N., 1976, *Astrophys. J.* **206**, 218
Font, J. A., Ibáñez, J.Mª, Marquina, A. & Martí, J.Mª, 1994, *Astron. Astrophys.* **282**, 304
Fowler, W.A., Caughlin, G.R. & Zimmermann, B.A., 1975, *Ann. Rev. Astron. Astrophys.* **13**, 69
Fransson, C., 1988, in *IAU Coll. 108 on Atmospheric Diagnostics of Stellar Evolution: Chemical Peculiarity, Mass Loss, and Explosion*, ed. K. Nomoto, Springer-Verlag, Berlin, 383
Freedman, D.Z., 1974, *Phys. Rev.* **D9**, 1389
Fryxell, B.A. & Woosley, S.E., 1982, *Astrophys. J.* **261**, 332
Fryxell, B.A., Müller, E. & Arnett, W.D., 1989, *Max-Planck-Institut für Astrophysik, Preprint* **449**, Garching
Fryxell, B.A., Müller, E. & Arnett, W.D., 1991, *Astrophys. J.* **367**, 619
Fulbright, M.S., Benz, W. & Davies, M.B., 1996, *Astrophys. J.* **440**, 254
Fuller, G.M., 1982, *Astrophys. J.* **252**, 741
Garrington, S.T., Leahy, J.P., Conway, R.G. & Laing, R.A., 1988, *Nature* **331**, 147
Gaskell, C.M., Cappellaro, E., Dinerstein, H.L., Garnett, D., Harkness, R.P. & Wheeler, J.C., 1986, *Astrophys. J. Lett.* **306**, L77
Gear, J.W., 1971, *Numerical Initial Value Problems in Ordinary Differential Equations*, Prentice-Hall, Englewood Cliffs, New Jersey

Gehrels, N., Leventhal, M. & MacCallum, C.J., 1988, in *Nuclear Spectroscopy of Astrophysical Sources*, eds. N. Gehrels and G. Share, AIP, New York, 87

Ghisellini, G., Padovani, P., Celotti, A. & Maraschi, L., 1993, *Astrophys. J.* **407**, 65

Gingold, R.A. & Monaghan, J.J., 1977, *Mon. Not. Roy. Astronom. Soc.* **181**, 375

Gingold, R.A. & Monaghan, J.J., 1982, *J. Comp. Phys.* **46**, 429

Glasner, S.A. & Livne, E., 1995, *Astrophys. J. Lett.* **445**, L149

Glasner, S.A., Livne, E. & Truran, J.W., 1997, *Astrophys. J.* **475**, 754

Glimm, J. & Li, X.L., 1988, *Phys. Fluids* **31**, 2077

Goldreich, P. & Weber, S.V., 1980, *Astrophys. J.* **238**, 991

Gómez, J.L., Alberdi, A. & Marcaide, J.M., 1994a, *Astron. Astrophys.* **284**, 51

Gómez, J.L., Alberdi, A., Marcaide, J.M., Marscher, A.P. & Travis J.P, 1994b, *Astron. Astrophys.* **292**, 33

Gómez, J.L., Martí, J.Mª, Marscher, A.P., Ibáñez, J.Mª & Marcaide, J.M., 1995, *Astrophys. J. Lett.* **449**, L19

Gómez, J.L., Martí, J.Mª, Marscher, A.P., Ibáñez, J.Mª & Alberdi, A., 1997, *Astrophys. J. Lett.* **482**, L33

Gustavson, F.G., Liniger, W., Willoughby, R., 1970, *J. Assoc. Comput. Mach.* **17**, 87

Haas, M.R., Colgan, S.W.J., Erickson, E.F., Lord, S.D., Burton, M.G. & Hollenbach, D.J., 1990, *Astrophys. J.* **360**, 257

Hachisu, I., Matsuda, T., Nomoto, K. & Shigeyama, T., 1990, *Astrophys. J. Lett.* **358**, L57

Hachisu, I., Matsuda, T., Nomoto, K. & Shigeyama, T., 1991, *Astrophys. J. Lett.* **368**, L27

Hachisu, I., Matsuda, T., Nomoto, K. & Shigeyama, T., 1992, *Astrophys. J.* **390**, 230

Hachisu, I., Matsuda, T., Nomoto, K. & Shigeyama, T., 1994, *Astron. Astrophys. Suppl.* **104**, 341

Haddad, B., Clausset, F. & Combes, F., 1991, *J. Comp. Phys.* **97**, 107

Hanuschik, R.W. & Thimm, G.J., 1990, *Astron. Astrophys.* **231**, 77

Hanuschik, R.W., Spyromilio, J., Stathakis, R., Kimeswenger, S., Gochermann, J., Seidensticker, K.J. & Meurer, G., 1993, *Mon. Not. Roy. Astronom. Soc.* **261**, 909

Hardee, P.E., 1996, in *Energy Transport in Radio Galaxies and Quasars*, eds. P.E. Hardee, A.H. Bridle and J.A. Zensus, ASP Conference Series, Vol. 100, 273

Hardee, P.E. & Clarke, D.A., 1992, *Astrophys. J. Lett.* **400**, L9

Hardee, P.E. & Clarke, D.A., 1995, *Astrophys. J.* **449**, 119

Hardee, P.E., Clarke, D.A. & Howell, D.A., 1995, *Astrophys. J.* **441**, 644

Hardee, P.E., Bridle, A.H. & Zensus, J.A. (eds.), 1996, *Energy Transport in Radio Galaxies and Quasars* (Proc. Workshop, Tuscaloosa, Alabama, September 1995), ASP Conference Series 100

Harkness, R.P. & Wheeler, J.C., 1990, *Supernovae*, ed. A.G. Petschek, Springer-Verlag, New York, 1

Harry, G.M., Stevenson, T.R. & Paik, H.J., 1996, *Phys. Rev.* **D 54**, 2409

Herant, M. & Benz, W., 1991, *Astrophys. J. Lett.* **345**, L412

Herant, M. & Benz, W., 1992, *Astrophys. J.* **387**, 294

Herant, M. & Woosley, S.E., 1994, *Astrophys. J.* **425**, 814

Herant, M., Benz, W. & Colgate, S.A., 1992, *Astrophys. J.* **395**, 642

Herant, M., Benz, W., Hix, W.R., Fryer, C.L. & Colgate, S.A., 1994, *Astrophys. J.* **435**, 339

Hernquist, L., 1987, *Astrophys. J. Suppl.* **64**, 715

Hernquist, L., 1990, *J. Comp. Phys.* **87**, 137

Hernquist, L., 1993, *Astrophys. J.* **404**, 717
Hernquist, L. & Katz, N., 1989, *Astrophys. J. Suppl.* **70**, 419
Hillebrandt, W., 1987, in *High Energy Phenomena Around Collapsed Stars*, ed. F. Pacini, Reidel, Dordrecht, 73.
Hillebrandt, W., 1990, in *High-Presssure Equation of State: Theory and Applications*, eds. S. Eliezer and R.A. Ricci, North Holland, Amsterdam, 314.
Hillebrandt, W., 1994, in *Supernovae*, eds. S.A. Bludman, R. Mochkovitch and J. Zinn-Justin, Elsevier, Amsterdam, 251
Hillebrandt, W. & Höflich, P., 1989, *Rep. Progress Phys.* **52**, 1421
Hillebrandt, W. & Müller, E., 1981, *Astron. Astrophys.* **103**, 147
Hillebrandt, W. & Wolff, R.G., 1985, in *Nucleosynthesis: Challenges and New Developments*, eds. W.D. Arnett and J.W. Truran, Univ. Chicago Press, Chicago, 131.
Hooda, J.S. & Wiita, P.J., 1996, *Astrophys. J.* **470**, 211
Hooda, J.S., Mangalam, A.V. & Wiita, P.J., 1994, *Astrophys. J.* **423**, 116
Hooimeyer, J.R.A., Schilizzi, R.T., Miley, G.K. & Barthel, P.D., 1992, *Astron. Astrophys.* **261**, 25
Houser, J.L., Centrella, J.M. & Smith, S.C., 1994, *Phys. Rev. Lett.* **72**, 1314
Hoyle, F. & Fowler, W.A., 1960, *Astrophys. J.* **132**, 565
Hughes, P.A. (ed.), 1991, *Beams and Jets in Astrophysics*, Cambridge Univ. Press, Cambridge
Hughes, P., Aller, H.D. & Aller, M.F., 1989, *Astrophys. J.* **341**, 54
Hughes, P., Aller, H.D. & Aller, M.F., 1991, *Astrophys. J.* **374**, 57
Hughes, P., Duncan, C. & Mioduszewski, A., 1996, in *Energy Transport in Radio Galaxies and Quasars*, eds. P.E. Hardee, A.H. Bridle and J.A. Zensus, ASP Conference Series, Vol. 100, 137
Hummel, C.A., Schalinski, C.J., Krichbaum, T.P., Rioja, M.J., Quirrenbach, A., Witzel, A., Muxlow, T.W.B., Johnston, K.J., Matveyenko, L.I. & Shevchenko, A., 1992, *Astron. Astrophys.* **257**, 489
Hut, P., 1997, in *Computational Astrophysics* (Proc. 12th 'Kingston meeting' on Theoretical Astrophysics), eds. D.A. Clarke and M.J. West, ASP Conference Series, Vol. XXX, San Francisco, in press; also astro-ph/9704277
Hut, P., Arnold, J.M., Makino, J., McMillan, S.L.W. & Sterlin, T.L., 1997, in *Proc. 1997 Petaflops Algorithms Workshop (PAL'97)* (April 13-18, 1997, Williamsburg, Virginia), in press; also astro-ph/9704183
Imamura, J.N., Toman, J., Durisen, R.I., Pickett, B.K. & Yang, S., 1995, *Astrophys. J.* **444**, 363
Israel, W., 1987, in *Relativistic Fluid Dynamics*, eds. A. Anile & Y. Choquet-Bruhat, Springer-Verlag, Berlin, 152
Iwamoto, K., Young, T.R., Nakasato, N., Shigeyama, T., Nomoto, K., Hachisu, I. & Sato, H., 1997, *Astrophys. J.* **477**, 865
Janka, H.-Th., 1993, in *Frontier Objects in Astrophysics and Particle Physics*, eds. F. Giovannelli and G. Mannocchi, Società Italiana di Fisica, Bologna, 345
Janka, H.-Th. & Hillebrandt, W., 1989a, *Astron. Astrophys.* **224**, 49
Janka, H.-Th. & Hillebrandt, W., 1989b, *Astron. Astrophys. Suppl.* **78**, 375
Janka, H.-Th. & Müller, E., 1993, in *Frontiers of Neutrino Astrophysics*, eds. Y. Suzuki and K. Nakamura, Universal Academy Press, Tokyo, 203.
Janka, H.-Th. & Müller, E., 1995a, *Phys. Rep.* **256**, 135
Janka, H.-Th. & Müller, E., 1995b, *Astrophys. J. Lett.* **448**, L109.
Janka, H.-Th. & Müller, E., 1996a, in *Supernovae and Supernovae Remnants* (IAU Colloquium 145, Xian, China, May 25-29, 1993), eds. R. McCray and Z. Wang, Cambridge Univ. Press, Cambridge, 109
Janka, H.-Th. & Müller, E., 1996b, *Astron. Astrophys.* **306**, 167.

Janka, H.-Th., Zwerger, T. & Mönchmeyer, R., 1993, *Astron. Astrophys.* **268**, 360
Jernigan, J.G. & Porter, D.H., 1989, *Astrophys. J. Suppl.* **71**, 871
Keil, W., 1997, *Ph.D. Thesis*, Technische Universität München, 1997, unpublished
Keil, W., Janka, H.-Th. & Müller, E., 1996, *Astrophys. J. Lett.* **473**, L111
Kerstein, A.R., 1988, *Combust. Sci. Tech.* **50**, 441
Koide, S., Nishikawa, K. & Mutel, R.L., 1996, *Astrophys. J. Lett.* **463**, L71
Khokhlov, A.M., 1991, *Astron. Astrophys.* **245**, 114
Khokhlov, A.M., 1994, *Astrophys. J. Lett.* **424**, L115
Khokhlov, A.M., 1995, *Astrophys. J.* **449**, 695
Khokhlov, A.M., Oran, E.S. & Wheeler, J.C., 1997, *Astrophys. J.* **478**, 678
Koldoba, A.V., Ustyugova, G.V., Romanova, M.M., Chechetkin, V.M. & Lovelace, R.V.E., 1995, *Astrophys. J. Suppl.* **232**, 241
Kolmogorov, A.N., 1941, *Dokl. Akad. Nauk. SSSR* **30**, 299
Komissarov, S.S., 1990a, *Astrophys. Spac. Sci.* **165**, 325
Komissarov, S.S., 1990b, *Astrophys. Spac. Sci.* **171**, 105
Komissarov, S.S. & Falle, S.A.E.G., 1996a, in *Energy Transport in Radio Galaxies and Quasars*, eds. P.E. Hardee, A.H. Bridle and J.A. Zensus, ASP Conference Series, Vol. 100, 165
Komissarov, S.S. & Falle, S.A.E.G., 1996b, in *Energy Transport in Radio Galaxies and Quasars*, eds. P.E. Hardee, A.H. Bridle and J.A. Zensus, ASP Conference Series, Vol. 100, 173
Komissarov, S.S. & Falle, S.A.E.G., 1997, *Mon. Not. Roy. Astronom. Soc.* , in press
Königl, A., 1980, *Phys. Fluids* **23**, 1083
Kössl, D. & Müller, E., 1988, *Astron. Astrophys.* **206**, 204
Kössl, D., Müller, E. & Hillebrandt, W., 1990a, *Astron. Astrophys.* **229**, 378
Kössl, D., Müller, E. & Hillebrandt, W., 1990b, *Astron. Astrophys.* **229**, 401
Krichbaum, T.P., Quirrenbach, A. & Witzel, A., 1992, in *Variability of Blazars*, ed. E. Valtaoja and M. Valtonen, Cambridge Univ. Press, Cambridge, 331
Kundt, W. (ed.), 1996, *Jets from Stars and Galactic Nuclei* (Proc. Workshop, Bad Honnef, Germany, July 3-7), Springer-Verlag, Berlin
Lai, D. & Shapiro, S.L., 1995, *Astrophys. J.* **442**, 259
Laing, R.A., 1988, *Nature* **331**, 149
Laing, R.A., 1996, in *Energy Transport in Radio Galaxies and Quasars*, eds. P.E. Hardee, A.H. Bridle and J.A. Zensus, ASP Conference Series, Vol. 100, 241
Landau, L.D., 1944, *Acta Physiochim. URSS* **19**, 77
Landau, L.D. & Lifshitz, E.M., 1959, *Fluid Mechanics*, Pergamon, Oxford
Langer, N., Fliegner, J., Heger, A. & Woosley. S.E., 1997, in *Nuclei in the Cosmos IV*, ed. M. Wiescher, *Nucl. Phys.* **A**, in press
Larrouturou, B., 1991, *J. Comp. Phys.* **95** 59
Lattimer, J.M., Pethick, C.J., Ravenhall, D.G. & Lamb, D.Q., 1985, *Nucl. Phys.* **A432**, 646
LeBlanc, J.M. & Wilson, J.R., 1970, *Astrophys. J.* **161**, 541
Ledoux, P., 1945, *Astrophys. J.* **102**, 143
Leibundgut, B., 1995, *The Lives of Neutron Stars*, eds. M.A. Alpar, Ü. Kiziloğlu and L. van Paradijs, Kluwer, Dordrecht, 3
Leibundgut, B., 1996, *Compact Stars in Binaries*, eds. J. van Paradijs, E.P.J. van den Heuvel and E. Kuulkers, Kluwer, Dordrecht, 107
Lewis, B. & von Elbe, G., 1961, *Combustion, Flames and Explosions of Gases*, Academic Press, London
Li, H., McCray, R. & Sunyaev, R., 1993, *Astrophys. J.* **419**, 824
Lind, K.R., Payne, D.G., Meier, D.L. & Blandford, R.D., 1989, *Astrophys. J.* **344**, 89
Litvinova, I.Yu. & Nadyozhin, D.K., 1983, *Astrophys. Spac. Sci.* **89**, 89

Liu, R., Pooley, G. & Riley, J.M., 1992, *Mon. Not. Roy. Astronom. Soc.* **257**, 545

Livne, E., 1993, *Astrophys. J. Lett.* **406**, L17

Loewenstein, M. & Mathews, W.G., 1986, *J. Comp. Phys.* **62**, 414

Loken, C., Burns, J.O., Clarke, D.A. & Norman, M.L., 1992, *Astrophys. J.* **392**, 54

Loken, C., Burns, J., Bryan, G. & Norman, M.L., 1996, in *Energy Transport in Radio Galaxies and Quasars*, eds. P.E. Hardee, A.H. Bridle and J.A. Zensus, ASP Conference Series, Vol. 100, 267

Lovelace, R.V.E. & Romanova, M.M., 1992, in *Energy Transport in Radio Galaxies and Quasars*, eds. P.E. Hardee, A.H. Bridle and J.A. Zensus, ASP Conference Series, Vol. 100, 25

Lucy, L.B., 1977, *Astronom. J.* **82**, 1013

L'vov, V.S. & Falkovitch, G.E., 1992, *Physica D* **57**, 361

Mahoney, W.A. *et al.*, 1988, *Astrophys. J. Lett.* **334**, L81

Makino, J., 1990a, *J. Comp. Phys.* **87**, 148

Makino, J., 1990b, *J. Comp. Phys.* **88**, 393

Makino, J. & Hut, P., 1988, *Comput. Phys. Rep.* **9**, 199

Marck, J.-A. & Bonazzola, S., 1992, in *Approaches to Numerical relativity*, eds. R. D'Inverno, Cambridge Univ. Press, Cambridge, 247

Marscher, A.P. & Gear, W.K., 1985, *Astrophys. J.* **298**, 114

Markstein, G.H., 1951, *Journal of the Aeronautical Siences* March, 199

Martí, J.Mª, 1998, in *Numerical Methods for Wave Propagation Phenomena* (Proc. Workshop, Manchester, 1996), eds. E.F. Toro and J.F. Clarke, Kluwer, New York, in press

Martí, J.Mª & Müller, E., 1994, *J. Fluid Mech.* **258**, 317

Martí, J.Mª & Müller, E., 1996, *J. Comp. Phys.* **123**, 1

Martí, J.Mª, Müller, E. & Ibáñez, J.Mª, 1994, *Astron. Astrophys.* **281**, L9

Martí, J.Mª, Müller, E., Font, J.A., & Ibáñez, J.Mª, 1995, *Astrophys. J. Lett.* **448**, L105

Martí, J.Mª, Font, J.A., Ibáñez, J.Mª & Müller, E., 1996, in *Energy Transport in Radio Galaxies and Quasars*, eds. P.E. Hardee, A.H. Bridle and J.A. Zensus, ASP Conference Series, Vol. 100, 149

Martí, J.Mª, Müller, E., Font, J.A., Ibáñez, J.Mª & Marquina, A., 1997a, *Astrophys. J.* **479**, 151

Martí, J.Mª, Müller, E. & Ibáñez, J.Mª, 1997b, in *Open Problems on Astrophysical Jets* (Proc. Workshop, Turin, Nov. 1996), eds. S. Massaglia and G. Bodo, Gordon & Breach, in press

Massaglia, S., Bodo, G. & Ferrari, A., 1996, *Astron. Astrophys.* **307**, 997

Matz, S.M., Share, G.H., Leising, M.D., Chupp, E.L., Vestrand, W.T., Purcell, W.R., Strickman, M.S. & Reppin, C., 1988, *Nature* **331**, 416

Maza, J. & van den Bergh, S., 1976, *Astrophys. J.* **204**, 519

Mazarakis, M. & Stephens, W., 1973, *Phys. Rev.* **C7**, 1280

Mazurek, T.J., 1974, *Nature* **252**, 287

Mazurek, T.J., 1982, in *Supernovae: A Survey of Current Research*, eds. M.J. Rees and R.J. Stoneham, Reidel, Dordrecht, 371

McCray, R. & Wang, Z. (eds), 1996, *Supernova and Supernovae Remnants* (IAU Colloquium 145, Xian, China, May 25-29, 1993), Cambridge Univ. Press, Cambridge

Meier, D.L., Payne, D.G. & Lind, K.R., 1996, in *IAU Symp. 175: Extragalactic Radio Sources*, eds. R. Ekers, C. Fanti and L. Padrielli, Kluwer, Dordrecht, 433

Mellema, G., 1996, in *Jets from Stars and Galactic Nuclei*, ed. W. Kundt, Springer-Verlag, Berlin, 149

Merkowitz, S.M. & Johnsons, W.W., 1995, *Phys. Rev.* **D 51**, 2546

Mezzacappa, A., Calder, A.C., Bruenn, S.W., Blondin, J.M., Guidry, M.W., Strayer, M.R. & Umar, A.S., 1998a, *Astrophys. J.* , in press (January 1st issue); also astro-ph/9709184
Mezzacappa, A., Calder, A.C., Bruenn, S.W., Blondin, J.M., Guidry, M.W., Strayer, M.R. & Umar, A.S., 1998b, *Astrophys. J.* , in press; also astro-ph/9709188
Miller, D.S., Wilson, J.R. & Mayle, R.W., 1993, *Astrophys. J.* **415**, 278
Minkowski, A.M., 1941, *Publ. Astron. Soc. Pac.* **53**, 224
Mioduszewski, A.J., Hughes, P.A. & Duncan, G.C., 1997, *Astrophys. J.* **476**, 649
Mirabel, I.F. & Rodriguez, L.F., 1994, *Nature* **371**, 46
Misner, C.W., Thorne, K.S. & Wheeler, J.A., 1973, *Gravitation*, Freeman, San Francisco
Miyama, S.M., Hayashi, C. & Narita, S., 1984, *Astrophys. J.* **279**, 621
Mönchmeyer, R., 1993, *Ph.D. thesis*, Technical University München, unpublished
Mönchmeyer, R. & Müller, E., 1989, in *Timing Neutron Stars*, eds. H. Ögelman and E. van den Heuvel, Kluwer, Dordrecht, 549
Mönchmeyer, R., Schäfer, G., Müller, E. & Kates, R., 1991, *Astron. Astrophys.* **246**, 417
Monaghan, J.J., 1982, *SIAM J. Sci. Stat. Comput.* **3**, 422
Monaghan, J.J., 1985, *Comput. Phys. Rep.* **3**, 71
Monaghan, J.J., 1992, *Ann. Rev. Astron. Astrophys.* **30**, 543
Monaghan, J.J., 1997, *J. Comp. Phys.* , in press
Monaghan, J.J. & Gingold, R.A., 1983, *J. Comp. Phys.* **52**, 374
Monaghan, J.J. & Lattanzio, J.C., 1985, *Astron. Astrophys.* **149**, 135
Monaghan, J.J. & Varnas, S.R., 1988, *Mon. Not. Roy. Astronom. Soc.* **231**, 515
Morris, J.P., 1996, *Publ. Astron. Soc. Austr.* **13**, 97
Müller, E., 1982, *Astron. Astrophys.* **114**, 53
Müller, E., 1986, *Astron. Astrophys.* **162**, 103
Müller, E., 1990, *J. Phys. G* **16**, 1571
Müller E., 1993, in *Proc. of the 7th Workshop on Nuclear Astrophysics* (Ringberg Castle, March 22-27, 1993), eds. W. Hillebrandt and E. Müller, *Max-Planck-Institut für Astrophysik, Preprint* **P7**, Garching, 27
Müller E., 1994, in *Supernovae*, eds. S.A. Bludman, R. Mochkovitch and J. Zinn-Justin, Elsevier, Amsterdam, 395.
Müller E., 1997, in *Gravitational Radiation* (Proc. Les Houches Summer School, 1995), eds. S. Bonazzola and J.-A. Marck, Cambridge Univ. Press, Cambridge, in press
Müller, E. & Arnett, W.D., 1982, *Astrophys. J. Lett.* **261**, L109
Müller, E. & Arnett, W.D., 1986, *Astrophys. J.* **307**, 619
Müller, E. & Brinkmann, W., 1997, *Astron. Astrophys.* , submitted
Müller, E. & Hillebrandt, W., 1981, *Astron. Astrophys.* **103**, 358
Müller, E. & Janka, H.-Th., 1994, in *Reviews in Modern Astronomy* **7**, ed. G.Klare, Astronomische Gesellschaft, Hamburg, 103
Müller, E. & Janka, H.-Th., 1997, *Astron. Astrophys.* **317**, 140
Müller, E., Różyczka, M. & Hillebrandt, W., 1980, *Astron. Astrophys.* **81**, 288
Müller, E., Hillebrandt, W., Orio, M., Höflich, P. Mönchmeyer & R., Fryxell, B.A., 1989, *Astron. Astrophys.* **220**, 167
Müller, E., Fryxell, B.A. & Arnett, W.D., 1991a, in *Chemical and Dynamical Evolution of Galaxies*, eds. F. Ferrini, J. Franco and F. Matteucci, ETS Editrice, Pisa, 394
Müller, E., Fryxell, B.A. & Arnett, W.D., 1991b, in *SN1987A and Other Supernovae*, eds. I.J. Danziger and K. Kjär, ESO Workshop and Conference Proceedings No.37, ESO, Garching, 99
Müller, E., Fryxell, B.A. & Arnett, W.D., 1991c, *Astron. Astrophys.* **251**, 505

Mundt, R., 1988, in *Formation and Evolution of Low Mass Stars*, eds. A.K. Dupree and M.T.V.T. Lago, Springer-Verlag, Berlin, 252

Muxlow, T.W.B. & Garrington, S.T., 1991, in *Beams and Jets in Astrophysics*, ed. P.A. Hughes, Cambridge Univ. Press, Cambridge, 52

Nakamura, T. & Oohara, K., 1989, in *Frontiers in Numerical Relativity*, eds. C.R. Evans, L.S. Finn and D.W. Hobill, Cambridge Univ. Press, 254.

Nandkumar, R. & Pethick, C.J., 1984, *Mon. Not. Roy. Astronom. Soc.* **209**, 511

Nelson, R.P. & Papaloizou, J.C.B., 1994, *Mon. Not. Roy. Astronom. Soc.* **270**, 1

Niemeyer, J.C. & Hillebrandt, W., 1995a, *Astrophys. J.* **452**, 769

Niemeyer, J.C. & Hillebrandt, W., 1995b, *Astrophys. J.* **452**, 779

Niemeyer, J.C. & Kerstein, A.R., 1997, *New Astronomy* **2**, 239

Niemeyer, J.C. & Woosley, S.E., 1997, *Astrophys. J.* **475**, 740

Nishikawa, K., Koide, S., Sakai, J., Christodoulou, D.M., Sol, H. & Mutel, R.L., 1997, *Astrophys. J. Lett.* **483**, L45

Nomoto, K. & Sugimoto, D., 1977, *Publ. Astron. Soc. Japan* **29**, 165

Nomoto, K. & Hashimoto, M., 1988, *Phys. Rep.* **163**, 13

Nomoto, K., Filippenko, A.V., Shigeyama, T., 1990, *Astron. Astrophys.* **240**, L1

Nomoto, K., Yamaoka, H., Pols, O.R., van den Heuvel, E.P.J., Iwamoto, K., Kumagai, S. & Shigeyama, T., 1994, *Nature* **371**, 227

Nomoto, K., Iwamoto, K. & Suzuki, T., 1995, *Phys. Rep.* **256**, 173

Nomoto, K., Iwamoto, K., Young, T.R., Nakasato, N. & Suzuki, T., 1997, in *Thermonuclear Supernovae*, eds. P. Ruiz-Lapuente, R. Canal and J. Isern, Kluwer, Dordrecht, 839

Norman, M.L., 1993, in *Astrophysical Jets*, eds. D. Burgarella, M. Livio and C. O'Dea, Cambridge Univ. Press, Cambridge, 211

Norman, M.L., 1996a, in *Energy Transport in Radio Galaxies and Quasars*, eds. P.E. Hardee, A.H. Bridle and J.A. Zensus, ASP Conference Series, Vol. 100, 405

Norman, M.L., 1996b, in *Energy Transport in Radio Galaxies and Quasars*, eds. P.E. Hardee, A.H. Bridle and J.A. Zensus, ASP Conference Series, Vol. 100, 319

Norman, M.L. & Winkler, K.-H.A., 1986, in *Astrophysical Radiation Hydrodynamics*, eds. M.L. Norman and K.-H.A. Winkler, Reidel, Dordrecht, 449

Norman, M.L., Smarr, L., Winkler, K.-H.A. & Smith, M.D., 1982, *Astron. Astrophys.* **113**, 285

Norman, M.L., Winkler, K.-H.A. & Smarr, L., 1983, in *Astrophysical Jets*, eds. A. Ferrari and A.G. Pacholczyk, Reidel, Dordrecht, 227

Norman, M.L., Winkler, K.-H.A. & Smarr, L., 1984, in *Physics of Energy Transport in Extragalactic Radio Sources* (NRAO Workshop No.9), eds. A.H. Bridle and J.A., Eilek, NRAO, Green Bank, West Virginia, USA, 150

Norman, M.L., Smarr, L. & Winkler, K.-H.A., 1985, in *Numerical Astrophysics*, eds. J. Centrella, J.M. LeBlanc, R.L. Bowers and J.R. Wilson (hon.), Jones & Bartlett, Boston, 88

Nozakura, T., Ikeuchi, S. & Fujimoto, M. Y., 1984, *Astrophys. J.* **286**, 221

Obukhov, A.M., 1959, *Dokl. Akad. Nauk SSSR* **125**, 1246 [Sov. Phys. Dokl. 4, 61 (1959)]

Ouyed, R., Pudritz, R.E. & Stone, J.M., 1997, *Nature* **385**, 409.

Owen, J.M., Villumsen, J.V., Shapiro, P.R. & Martel, H., 1997, *Mon. Not. Roy. Astronom. Soc.* , in press

Pendleton, G.N., Paciesas, W.S., Fishman, G.J., Meegan, C.A. & Wilson, R.B., 1995, *Astrophys. J.* **439**, 963

Peters, N., 1988, in *21st Int. Symp. Combust.*, Combustion Institute, Pittsburgh, 1232

Petschek, A.G. (ed.), 1990, *Supernovae*, Springer-Verlag, New York

Pickett, B.K., Durisen, R.H. & Davis, G.A., 1996, *Astrophys. J.* **458**, 714

Piran, T., 1990, in *Supernovae*, eds. J.C. Wheeler, T. Piran and S. Weinberg, World Scientific, Singapore, 303

Plewa, T., Martí, J.M$^{\underline{a}}$, Müller, E., Różyczka, M. & Sikora, M, 1997, in *Relativistic Jets in AGNs* (Proc. Workshop Cracow, May 1997), eds. M. Ostrowski, M. Sikora, G. Madejski and M. Begelman, in press; also astro-ph/9706124

Pope, S.B., 1987, *Ann. Rev. Fluid Mech.* **19**, 237

Porter, D.H., 1985, Ph.D. thesis, Univ. California, Berkeley

Porter, A. & Filippenko, A., 1987, *Astronom. J.* **93**, 1372

Potter, D., 1973, *Computational Physics*, Wiley, New York

Rampp, M., Müller, E. & Ruffert, M., 1997, *Astron. Astrophys.* , submitted

Rank, D.M. *et al.*, 1988, *Nature* **331**, 505

Rawlings, S. & Saunders, R., 1991, *Nature* **349**, 138

Ray, A. & Velusamy, T. (eds.), 1991, *Supernovae and Stellar Evolution* (School and Workshop, Goa, India, March 10-17, 1989), World Scientific, Singapore

Read, K.I., 1984, *Physica D* **12**, 45

Richardson, L., 1922, *Weather Predictions by Numerical Process*, Cambridge Univ. Press, Cambridge

Roe, P.L., 1981, *J. Comp. Phys.* **43**, 357

Romanova, M.M., Ustyugova, G.V., Koldoba, A.V., Chechetkin, V.M. & Lovelace, R.V.E., 1997, *Astrophys. J.* **482**, 708

Ruffert, M., 1992, *Astron. Astrophys.* **265**, 82

Ruiz-Lapuente, P., Canal, R. & Isern, J. (eds), 1997,*Thermonuclear Supernovae* (NATO ASI C486, Parador dÁiguablava, Spain, June 20-30, 1995), Kluwer, Dordrecht

Salmon, J.K. & Warren, M.S., 1994, *J. Comp. Phys.* **111**, 136

Sandie, W.G. *et al.*, 1988, *Astrophys. J. Lett.* **334**, L91

Sato, K., 1975a, *Prog. Theor. Phys.* **53**, 595

Sato, K., 1975b, *Prog. Theor. Phys.* **54**, 1325

Schaeffer, R., Casse, M. & Cahen, S., 1987, *Astrophys. J. Lett.* **316**, L31

Scheuer, P.A.G., 1974, *Mon. Not. Roy. Astronom. Soc.* **166**, 513

Selhammar, M., 1997a, *Astron. Astrophys.* **325**, 857

Selhammar, M., 1997b, astro-ph/9706071

Shabaz, T., Livio, M., Southwell, K.A. & Charles, P.A., 1997, *Astrophys. J. Lett.* **484**, L59

Shankar, A. & Arnett, D., 1994, *Astrophys. J.* **433**, 216

Shankar, A., Arnett, D. & Fryxell, B.A., 1992, *Astrophys. J. Lett.* **394**, L13

Shapiro, P.R., Martel, H., Villumsen, J.V. & Owen, M.J., 1996, *Astrophys. J. Suppl.* **103**, 269

Shapiro, S.L. & Lightman, A.P., 1976, *Astrophys. J.* **207**, 263

Sharp, D.H., 1984, *Physica D* **12**, 3

Shibata, K. & Uchida, K., 1986, *Publ. Astron. Soc. Japan* **38**, 631

Shigeyama, T. & Nomoto, K., 1990, *Astrophys. J.* **360**, 242

Shigeyama, T., Nomoto, K. & Hashimoto, M., 1988, *Astron. Astrophys.* **196**, 141

Shigeyama, T., Nomoto, K., Yamaoka, H., Tsujimoto, T. & Hashimoto, M., 1990, *Astrophys. J.* **361**, L23

Shigeyama, T., Suzuki, T., Kumagai, S., Nomoto, K., Saio, H & Yamaoka, H., 1994, *Astrophys. J.* **420**, 341

Shigeyama, T., Iwamoto, K., Hachisu, I. & Nomoto, K., 1996, in *Supernovae and Supernovae Remnants* (IAU Colloquium 145, Xian, China, May 25-29, 1993), eds. R. McCray and Z. Wang, Cambridge Univ. Press, Cambridge, 129

Shimizu, T., Yamada, S. & Sato, K., 1993, *Publ. Astron. Soc. Japan* **45**, L53

Shimizu, T., Yamada, S. & Sato, K., 1994, *Astrophys. J. Lett.* **432**, L119
Shu, C. & Osher, S., 1988, *J. Comp. Phys.* **77**, 439
Smiljanovski, V., Moser, V. & Klein, R., 1997, *Combust. Theor. and Modelling.* **1**, 183
Smith, S.C., Houser, J.L. & Centrella, J.M., 1996, *Astrophys. J.* **458**, 236
Snider, D.M. & Andrews, M.J., 1994, *Phys. Fluids* **6**, 3324
Soper, D.E., 1976, *Classical Field Theory*, Wiley, New York
Spyromilio, J., 1991, *Mon. Not. Roy. Astronom. Soc.* **253**, 25
Spyromilio, J., 1994, *Mon. Not. Roy. Astronom. Soc.* **266**, L61
Spyromilio, J. & Meikle, W.P.S., 1990 *Mon. Not. Roy. Astronom. Soc.* **242**, 669
Spyromilio, J. & Pinto, P.A., 1991, in *SN1987A and Other Supernovae*, eds. I.J. Danziger and K. Kjär, ESO Workshop and Conference Proceedings No.37, ESO, Garching, 423
Spyromilio, J., Stathakis, R.A. & Meurer, G.R., 1993 *Mon. Not. Roy. Astronom. Soc.* **263**, 530
Stathakis, R.A., Dopita, M.A., Cannon, R.D. & Sadler, E.M., 1991, in *Supernovae*, ed. S.E. Woosley, Springer-Verlag, New York, 95
Steinmetz, M., 1996a, *Mon. Not. Roy. Astronom. Soc.* **278**, 1005
Steinmetz, M., 1996b, in *Dark Matter in the Universe* (Proc. Internatl. School of Physics "Enrico fermi" Course CXXXII), eds. S. Bonometto, J.R. Primack and A. Provenzale, IOS Press, Amsterdam, 479
Steinmetz, M. & Müller, E., 1993, *Astron. Astrophys.* **268**, 391
Steinmetz, M. & Müller, E., 1995, *Mon. Not. Roy. Astronom. Soc.* **276**, 549
Stone, J.M. & Norman, M.L., 1994, *Astrophys. J.* **433**, 746
Stone, J.M., Xu, J. & Hardee, P.E., 1997, *Astrophys. J.* **483**, 136
Sugimoto, D., 1993a, *Publ. Astron. Soc. Japan* **45**, 269
Sugimoto, D., 1993b, *Physics World* **11**, 32
Sunyaev, R. *et al.*, 1987, *Nature* **330**, 227
Swegle, J.W., Hicks, D.L. & Attaway, S.W., 1995, *J. Comp. Phys.* **116**, 123
Symbalisty, E., 1984, *Astrophys. J.* **285**, 729
Tammann, G.A., 1974, in *Supernovae and Supernova Remnants*, ed. C.B. Cosmovici, Reidel, Dordrecht, 155
Tammann, G.A., 1982, in *Supernovae: A Survey of Current Research*, eds. M.J. Rees and R.J. Stoneham, Reidel, Dordrecht, 371
Tammann, G.A., 1994, in *Supernovae*, eds. S.A. Bludman, R. Mochkovitch and J. Zinn-Justin, Elsevier, Amsterdam, 1
Tammann, G.A., Löffler, W. & Schröder, A., 1994, *Astrophys. J. Suppl.* **92**, 487
Tassoul, J.-L., 1978, *Theory of Rotating Stars*, Princeton Univ. Press, New Jersey
Teegarden, B.J., Barthelmy, S.D., Gehrels, N., Tueller, J., Leventhal, M. & MacCallum, C.J., 1989, *Nature* **339**, 122
Theuns, T. & Rathsack, M.E., 1993, *Comput. Phys. Comm.* **76**, 141
Thielemann, F.-K. & Arnett, W.D., 1985, *Astrophys. J.* **295**, 604
Thorne, K.S., 1980, *Rev. Mod. Phys.* **52**, 299
Thorne, K.S., 1987, in *Three Hundred Years of Gravitation*, eds. S.W. Hawking and W. Israel, Cambridge Univ. Press, Cambridge, 330.
Thorne, K.S., 1995, in *Particle and Nuclear Astrophysics and Cosmology in the next Millenium* (Proc. of the 1994, Snowmass Summer Study), eds. E.W. Kolb and R. Peccei, World Scientific, Singapore, 160
Timmes, F.X. & Woosley, S.E., 1992, *Astrophys. J.* **396**, 649
Tingay, S.J. *et al.*, 1995, *Nature* **374**, 141
Tohline, J.E., 1984, *Astrophys. J.* **285**, 721
Tohline, J.E., Schombert, J.M. & Boss, A.P., 1980, *Space Sci. Rev.* **27**, 555

Trease, H.E., Fritts, M.J. & Crowley, W.P. (eds.), 1991, *Advances in the Free-Langrange Method, Lecture Notes in Physics* **395**, Springer-Verlag, Berlin
Truran, J.W., Arnett, W.D. & Cameron, A.G.W., 1967, *Canad. J. Phys.* **45**, 2315
Turatto, M., Cappellaro, E. & Benetti, S., 1994, *Astronom. J.* **202**, 108
Uomoto, A. & Kirshner, R.P., 1985, *Astron. Astrophys.* **149**, L7
Urry, C.M. & Padovani, P., 1995, *Publ. Astron. Soc. Pac.* **107**, 803
van den Bergh, S. & Tammann, G.A., 1991, *Ann. Rev. Astron. Astrophys.* **29**, 363
van Putten, M.H.P.M., 1993, *Astrophys. J. Lett.* **408**, L21
van Putten, M.H.P.M., 1996, *Astrophys. J. Lett.* **467**, L57
van Riper, K.A., 1978, *Astrophys. J.* **221**, 304
van Riper, K.A. & Lattimer, J.M., 1982, *Astrophys. J.* **249**, 270
Vautherin, D., 1994, *Supernovae*, eds. S.A. Bludman, R. Mochkovitch and J. Zinn-Justin, Elsevier, Amsterdam, 63.
Vermeulen, R.C., 1996, in *Jets from Stars and Galactic Nuclei*, ed. W. Kundt, Springer-Verlag, Berlin, 122
von Neumann, J., 1942, in *John von Neumann, collected works*, ed. A.J. Taub, Vol.6, 203
Wagoner, R.V., 1969, *Astrophys. J. Suppl.* **18**, 247
Wang, L. & Hu, J.Y., 1994, *Nature* **369**, 380
Watkins, S.J., Bhattal, A.S., Francis, N., Turner, J.A. & Whitworth, A.P, 1996, *Astron. Astrophys. Suppl.* **119**, 177
Weaver, T.A., Zimmerman, G.B. & Woosley, S.E., 1978, *Astrophys. J.* **225**, 1021
Weiler, K.W. & Sramek, R.A., 1988, *Ann. Rev. Astron. Astrophys.* **26**, 295
Wheeler, J.C. & Filippenko, A.V., 1996, in *Supernova and Supernovae Remnants*, eds., R. McCray and Z. Wang, Cambridge Univ. Press, Cambridge, 241
Wheeler, J.C. & Harkness, R.P., 1990, *Rep. Progress Phys.* **53**, 1467
Wheeler, J.C. & Levreault, R.M., 1985, *Astrophys. J. Lett.* **294**, L17
Wheeler, J.C., Harkness, R.P. & Cappellaro, E., 1987, in *Proc. 13th Texas Symposium on Relativistic Astrophysics*, ed. M.P. Ulmer, World Scientific, Singapore, 402
Williams, A.G. & Gull, S.F., 1984, *Nature* **310**, 33
Williams, F.A, 1985, *Combustion Theory*, Benjamin/Cummings, Menlo Park, California
Wilson, J.R., 1985, in *Numerical Astrophysics*, eds. J.M. Centrella, J.M. LeBlanc and J.R. Wilson, Jones and Bartlett, Boston, 422
Wilson, J.R. & Mayle, R., 1988, *Phys. Rep.* **163**, 63
Wilson, J.R., Mayle, R., Woosley, S.E. & Weaver, T.A., 1986, *Ann. N. Y. Acad. Sci.* **479**, 267
Wilson, M.J., 1987, *Mon. Not. Roy. Astronom. Soc.* **226**, 447
Wilson, M.J. & Falle, S.A.E.G., 1985, *Mon. Not. Roy. Astronom. Soc.* **216**, 971
Wilson, R.B. *et al.*, 1988, in *Nuclear Spectroscopy of Astrophysical Sources*, eds. N. Gehrels and G. Share, AIP, New York, 66
Witteborn, F., Rank, D.M., Bregman, J.D., Pinto, P.A., Wooden, D. & Axelrod, T.S., 1989, *Astrophys. J. Lett.* **338**, L9
Woodward, P.R., 1982, in *Parallel Computations*, ed. G. Rodrigue, Academic Press, New York
Woodward, P.R. & Colella, P., 1984, *J. Comp. Phys.* **54**, 115
Woosley, S.E., 1986, in *Nucleosynthesis and Chemical Evolution*, eds. B. Hauck, A. Maeder and G. Meynet, Swiss Soc. Astrophys. Astronom., Geneva Observatory, 1
Woosley, S.E., 1988, *Astrophys. J.* **330**, 218
Woosley, S.E., 1990, in *Supernovae*, ed. A.G. Petschek, Springer, Berlin, 182

Woosley, S.E. & Eastman, R.G., 1997, *Thermonuclear Supernovae*, eds. P. Ruiz-Lapuente, R. Canal and I. Isern, Kluwer, Dordrecht, 821

Woosley, S.E. & Weaver, T.A., 1986, *Ann. Rev. Astron. Astrophys.* **24**, 205

Woosley, S.E. & Weaver, T.A., 1988, *Phys. Rep.* **163**, 79

Woosley, S.E. & Weaver, T.A., 1994, *Supernovae*, eds. S.A. Bludman, R. Mochkovitch and J. Zinn-Justin, Elsevier, Amsterdam, 63.

Woosley, S.E., Pinto, P.A., Martin, P.G & Weaver, T.A., 1987, *Astrophys. J.* **318**, 664

Woosley, S.E., Langer, N. & Weaver, T.A., 1993, *Astrophys. J.* **411**, 823

Woosley, S.E., Langer, N. & Weaver, T.A., 1994a, *Astrophys. J.* **448**, 315

Woosley, S.E., Eastman, R.G., Weaver, T.A. & Pinto, P.A., 1994b, *Astrophys. J.* **429**, 300

Yahil, A., 1983, *Astrophys. J.* **265**, 1047

Yahil, A. & Lattimer, J.M., 1982, in *Supernovae: A Survey of Current Research*, eds. M.J. Rees and R.J. Stoneham, Reidel, Dordrecht, 53

Yamada, S. & Sato, K., 1994, *Astrophys. J.* **434**, 268

Yamada, S. & Sato, K., 1995, *Astrophys. J.* **450**, 245

Yamada, S., Shimizu, T. & Sato, K., 1993, *Prog. Theor. Phys.* **89**, 1175

Yamada, Y., Nakamura, T. & Oohara, K., 1990, *Prog. Theor. Phys.* **84**, 436

Yoshida, S. & Eriguchi, Y., 1995, *Astrophys. J.* **438**, 830

Young, D.L., 1984, *Physica D* **12**, 32

Zeldovich, Ya. B, 1940, *Zh. Eksp. Teor. Fiz.* **10**, 542 (English translation: NACA TM 1261 (1960))

Zeldovich, Ya.B., Barenblatt, G. I., Librovich, V. B. & Makhviladze, G. M., 1985, *The Mathematical Theory of Combustion and Explosions*, Plenum, New York

Zwerger, T., 1995, *Ph.D. Thesis*, Technische Universität München, 1995, unpublished

Zwerger, T. & Müller, E., 1997, *Astron. Astrophys.* **320**, 209

Zwicki, F., 1938, *Astrophys. J.* **88**, 529

Zwicki, F., 1965, in *Stellar Structure*, ed. L.H. Aller and D.B. McLaughlin, Univ. Chicago Press, Chicago, 367

Index

MIX
Papier aus verantwortungsvollen Quellen
Paper from responsible sources
FSC® C105338

If you have any concerns about our products,
you can contact us on
ProductSafety@springernature.com

In case Publisher is established outside the EU,
the EU authorized representative is:
Springer Nature Customer Service Center GmbH
Europaplatz 3, 69115 Heidelberg, Germany

Printed by Libri Plureos GmbH
in Hamburg, Germany